Beltsville Symposia in Agricultural Research

A Series of Annual Symposia sponsored by
THE BELTSVILLE AGRICULTURAL RESEARCH CENTER
Agricultural Research Service
United States Department of Agriculture
with the cooperation of
Friends of Agricultural Research
Beltsville, Inc.

[14] The Rhizosphere and Plant Growth

SYMPOSIA IN THIS SERIES

[1] Virology in Agriculture
May 10–12, 1976

[2] Biosystematics in Agriculture
May 8–11, 1977

[3] Animal Reproduction
May 14–17, 1978

[4] Human Nutrition Research
May 8–9, 1979

[5] Biological Control in Crop Production
May 18–21, 1980

[6] Strategies of Plant Reproduction
May, 17–20, 1981

[7] Genetic Engineering Applications to Agriculture
May 16–19, 1982

[8] Agricultural Chemicals of the Future
May 16–19, 1983

[9] Frontiers of Membrane Research
May 20–24, 1984

[10] Biotechnology for Solving Agricultural Problems
May 5–9, 1985

[11] Instrumentation in Agricultural Research
May 4–9, 1986

[12] Biomechanisms Regulating Growth and Development
May 3–7, 1987

[13] Biotic Diversity and Germplasm Preservation
May 9–11, 1988

[14] The Rhizosphere and Plant Growth
May 8–11, 1989

[15] Remote Sensing in Agriculture
May 16–18, 1990

[16] Photomorphogenesis in Plants: Emerging Strategies for Crop Improvement
Sept. 23–26, 1991

Beltsville Symposia in Agricultural Research

[14] The Rhizosphere and Plant Growth

Donald L. Keister and Perry B. Cregan
Editors

USDA, ARS, PSI, NFSGL
Bldg.011, HH19, BARC-West
Beltsville, MD 20705, USA

Papers presented at a Symposium held
May 8–11, 1989, at the Beltsville Agricultural
Research Center (BARC), Beltsville, Maryland

Donald L. Keister, Chairman

Sponsored by
THE BELTSVILLE AGRICULTURAL RESEARCH CENTER
Agricultural Research Service
United States Department of Agriculture
with the cooperation of
Friends of Agricultural Research
Beltsville. Inc.

Chapters indicated with an asterisk in the table of contents were first published in *Plant and Soil*, Volume 129 (1990)

Kluwer Academic Publishers
DORDRECHT / BOSTON / LONDON

Library of Congress Cataloging-in-Publication Data

The Rhizosphere and plant growth : papers presented at a symposium
held May 8-11, 1989, at the Beltsville Agricultural Research Center
(BARC), Beltsville, Maryland, U.S.A. / [edited] by Donald L.
Keister, Perry B. Cregan.
p. cm. -- (Beltsville symposia in agricultural research ; v.
14)
Includes index.
ISBN 0-7923-1032-2 (HB : acid free paper)
1. Rhizosphere--Congresses. 2. Plant-soil relationships-
-Congresses. I. Keister, Donald L. II. Cregan, Perry B.
III. Beltsville Agricultural Research Center. IV. Series:
Beltsville symposia in agricultural research ; 14.
QK644.R45 1991
581.1'0428--dc20 90-24248

ISBN 0-7923-1032-2

Published by Kluwer Academic Publishers,
P.O. Box 17, 3300 AA Dordrecht, The Netherlands.

Kluwer Academic Publishers incorporates
the publishing programmes of Martinus Nijhoff,
Dr W. Junk, D. Reidel, and MTP Press.

Sold and distributed in the U.S.A. and Canada
by Kluwer Academic Publishers,
101 Philip Drive, Norwell, MA 02061, U.S.A.

In all other countries, sold and distributed
by Kluwer Academic Publishers Group,
P.O. Box 322, 3300 AH Dordrecht, The Netherlands.

Printed on acid-free paper

Printed in the Netherlands

CONTENTS

Short Papers

Short Papers

Short Papers

Foreword

The Beltsville Symposium was established in 1976 to provide a forum for discussion among scientists involved in research that has the potential for advancing agriculture and agricultural sciences. This symposium, the fourteenth in this series, focuses on the rhizosphere and plant growth.

Plant growth and development are controlled largely by the soil environment in the root region or rhizosphere. This is a very complex environment in which the effects of the plant on soil microorganisms and the effects of microorganisms on the plant are interacting, interdependent and immensely complex. Plant root exudates and breakdown products, derived from solar energy utilization, feed the microbes and the microbes in turn often benefit the plant or induce disease harmful to the plant. Much agricultural research in the past has treated the soil as a "black box" and not enough attention has been given to the interaction between roots, soil and soil organisms. An empirical approach to plant growth served well as long as fertilizers, water and land were abundant and inexpensive. However, we now recognize that the empirical approach is not sufficient as we consider how to maintain soil productivity and crop yields, as well as environmental quality.

Ecological studies of plant-microbe-soil relationships have often been descriptive and empirical and it is now recognized that microbe-microbe interactions in the soil are very important in determining how specific microbes interact with plants. Many soil microbes have been neglected and information on how they interact with other microbes and with the plant is needed.

Improved crop management and crop protection systems are required in order to promote sustainable agriculture and improve the quality of the environment. Cost-effective systems such as biological control and symbiotic nitrogen fixation, that would lessen the need for agricultural chemicals and decrease energy inputs, also would lessen environmental contamination and increase cost effectiveness. To design such systems the manipulation of soil microbes, both indigenous and introduced, is increasingly being advocated. Advances in plant and microbial molecular biology is offering possibilities for designing specific plant-microbe associations. Thus this symposium was conceived to explore the present state of rhizosphere research, emphasize current problems, explore potential solutions, and describe the potential for modern biotechnology applications to these problems.

This was the first symposium in the U.S. that brought together scientists working on problems in biocontrol of plant diseases, microbiology and ecology of the soil, symbiotic nitrogen fixation, plant-microbe interactions and plant growth promotion.

Donald L. Keister
Symposium Chairman
Soybean and Alfalfa
Research Laboratory
Beltsville Agriculture Research Center
Beltsville, Maryland, USA

Symposium organization

Beltsville Agriculture Research Center (BARC)
E.B. Knipling, Director, Beltsville Area
K.D. Murrell, Associate Director, Beltsville Area

Symposium Chairman
Donald L. Keister

Symposium Secretary
Margaret Blackwell

Program Committee
Donald L. Keister, Co-chairman
George C. Papavizas, Co-chairman
Donald D. Kaufman
Robert E. Davis
Daniel R. Shelton

Local Arrangements Committee
Robert C. Leffel, Chairman
Amanda M. Prigg-Bolgiano
Lawrence J. Sikora
Jack J. Meisinger
Gordon T. Carpenter
Daniel R. Shelton

Publicity
Deborah R. Fravel

Publications
Donald L. Keister
Perry B. Cregan

Finance
Frank J. Longen, Friends of Agricultural
Research – Beltsville

Part 1

Progress in rhizospere research

D. L. Keister and P. B. Cregan (Eds.), The rhizosphere and plant growth, 3–13.

PLSO BARC01

Rhizosphere research – 85 years of progress and frustration

A.D. ROVIRA
CSIRO, Division of Soils, Adelaide, South Australia

Key words: biology, biological control, models, plant growth promotion, rhizodeposition, rhizosphere, rhizoplane, root exudation

Abstract

This paper reviews the major steps which have contributed to our knowledge of rhizosphere biology since 1904. The progress made has been very considerable, and the frustrations limited. Prospects are bright for improving our understanding of the biology of the rhizosphere and for managing the rhizosphere microflora to increase plant growth.

Introduction

Progress in science is made in a series of steps with each step being the starting point for new developments in the subject. Research into the biology and chemistry of the rhizosphere is no exception and I will trace its progress through a series of such steps.

I plan to follow the history of the subject, examine the progress in the most important areas of rhizosphere biology, and examine those areas which have proven frustrating. Finally, I will look to the future in rhizosphere research to project some of the new developments which will arise from steps made to this point. As there are over 2000 publications on the rhizosophere, I shall select papers to highlight the steps which formed starting points for new developments. Readers are referred to three up to date books on the rhizosphere (Curl and Truelove, 1986; Lynch, 1990; Waisel and Eshel, 1990) and other reviews (Bowen, 1980; Foster and Bowen, 1982; Foster *et al.*, 1983; Foster, 1986; Lynch, 1988; Rovira *et al.*, 1983; Rovira, 1985; 1988; Whipps and Lynch, 1986).

Historical

The first step in the science of the rhizosphere was achieved by Lorenz Hiltner (1904), who published on his observation that the microbial activity in soil surrounding roots was greater than in soil remote from roots. Hiltner coined the term "rhizosphere" to described this zone of soil surrounding roots of legumes in relation to symbiotic nitrogen fixing bacteria. Interwoven with these studies on the stimulation of bacteria in the rhizosphere are the pioneering studies on legumes and Rhizobium.

Research on the influence of roots on soil bacteria between 1929 and 1938 was dominated by R.L. Starkey (1929) who reviewed the rhizosphere literature and set the stage for the next phase; he used light microscopy to observe the density and distribution of bacteria on buried glass slides over which roots had grown (Starkey, 1938).

Our understanding of the qualitative as well as the quantitative differences between rhizosphere soil and bulk soil (soil beyond the zone of root influence) was greatly advanced by the work of Lochhead and associates in Ottawa (Katznelson *et al.*, 1948). Gram negative, non-sporing rods with simple nutritional requirements and which respond in growth to amino acids were stimulated more by roots than were coccoid forms and Gram positive sporing rods (Lochhead and Chase, 1943; Lochhead and Thexton, 1947).

F.E. Clark, another pioneer, demonstrated that the R:S ratio (counts of organisms in rhizosphere soil to counts in bulk soil) was affected not only by the plant species and soil water content but also by the quantity of soil adhering to roots from which suspensions were prepared for counting. He proposed the term "rhizoplane" for the microbiology of the root surface (Clark, 1949).

In the USSR another school of soil microbiology which made the rhizosphere its major theme was developed by N.A. Krasilnikov. Its first publication (Krasilnikov, 1934) dealt with the effects of root excretions on Azotobacter and was followed by at least 26 publications on the interactions between plant roots and soil microorganisms (Krasilnikov, 1958). Much of this work related to the reports of Zinovyeka (1940) and Savostin (1948) that increased yields resulted from the treatment of seed with Azotobacter.

Progress

Quantification of the rhizosphere microflora

The R:S ratio has been used to compare the rhizosphere populations of different plant species, of the same plant species growing in different soils, of plants of different ages, and of plants growing at different soil moisture content. Ishizawa *et al.* (1957) and Louw and Webley (1959) defined the rhizosphere microflora as the soil and organisms removed when roots are shaken gently in water. In contrast, the rhizoplane microflora were removed only when roots were transferred to fresh diluent and shaken vigorously. Louw and Webley (1959) found that the ratio of direct to plate counts for soil, rhizosphere and rhizoplane suspensions, averaged 9.2, 3.2 and 3.0 respectively, which indicates that bacteria associated with roots were more capable of responding to the nutrients and conditions in the agar medium than those from bulk soil. The high populations of bacteria in rhizosphere favour the build up of protozoa which "graze" on these bacteria. This phenomenon is largely overlooked in rhizosphere studies and yet represents a dynamic aspect of this zone (Gerhardson and Clarholm, 1986).

Qualitative nature of the rhizosphere microflora

The higher proportions of non-sporulating rods among rhizosphere isolates and their ability to respond to nutrients such as amino acids reflects the mechanisms by which the roots and rhizospheres are colonized. The elongating tip of a root is sterile and as the root grows into the soil, contact is made with colonies of bacteria and fungal propagules distributed through the soil. Organisms which respond to the root excudates and with the shortest generation times will dominate.

Development of techniques for rhizosphere research

Cultural methods

Traditional techniques of serial dilutions and agar plate counts with minor variations have been the standard practice, even though such counts represent only 10 to 30% of the numbers recorded by direct microscopy (Louw and Webley, 1959; Rovira *et al.*, 1974). Selective media and isolation under special environmental conditions permit the growth of specific groups of microorganisms and widen the spectrum of isolates from the rhizosphere. However, many microorganisms occupying the rhizosphere and the rhizoplane will be competing with the culturable organisms and thus will not be accounted for with our current techniques.

Direct Light Microscopy

The use of glass slides buried in soil demonstrated greater populations in the vicinity of roots (Starkey, 1938). Rovira (1956) used phenolaniline blue-acetic acid (Jones and Mollison, 1948) to stain bacteria and fungi on roots and found bacteria concentrated along the junctions between root cells. A low occupancy (<10%) of the rhizoplane was found with eight plant species grown in soil by Rovira *et al.* (1974), but bacteria were not randomly spread over roots. They formed two patterns, viz. one on a microcolony scale and another of larger areas (Newman and Bowen, 1974). These patterns were probably related to the sites of major leakage of organic materials from roots and to the distribution of colonies of bacteria in soil which provide the inoculum for roots. Trolldenier (1973) and van Vuurde and Elenbaas (1978) used vital stains and U.V. fluorescence microscopy to distinguish living from dead bacteria on roots. Immunofluorescence techniques offer the possibility of detecting and counting specific bacteria and fungi and have been used in studies on the establishment of Rhizobium, Azospirillum and other organisms in the rhizosphere (Bohlool and Schmidt, 1980).

Such techniques of direct observation could contribute more to our understanding of the ecology of the rhizosphere than do culturing techniques, especially if combined with modern image analysing equipment.

Transmission Electron Microscopy (TEM)

Jenny and Grossenbacher's (1963) study of the root-soil interface by TEM represents another step in improving our understanding of the rhizosphere – they coined the term "mucigel" to describe the coating of mucilage, microbial cells and clay particles on roots. TEM studies by Dart and Mercer (1964), Greaves and Darbyshire (1972), Old and Nicholson (1975), Foster and Rovira (1978), Foster (1981) and Foster *et al.* (1983) provide an excellent visual portrayal of the micro-ecology of the rhizosphere and have enabled us to study the relationships between the root, the mucilage of plant and microbial origin, clay and sand particles. The photomicrographs show the intimate associations between colonies of bacteria with different morphology and structure (Fig. 1).

Scanning Electron Microscopy (SEM)

The application of SEM to roots by Gray (1967), Dart (1971), Rovira and Campbell (1975) and Umali-Garcia *et al.* (1980) represents another step in the development of our understanding of the spatial distribution of bacteria and fungi on roots. SEM has the advantages that it is simpler to prepare specimens and larger specimens can be examined than by TEM. A further improvement in SEM for rhizosphere studies was achieved by examining frozen specimens on a

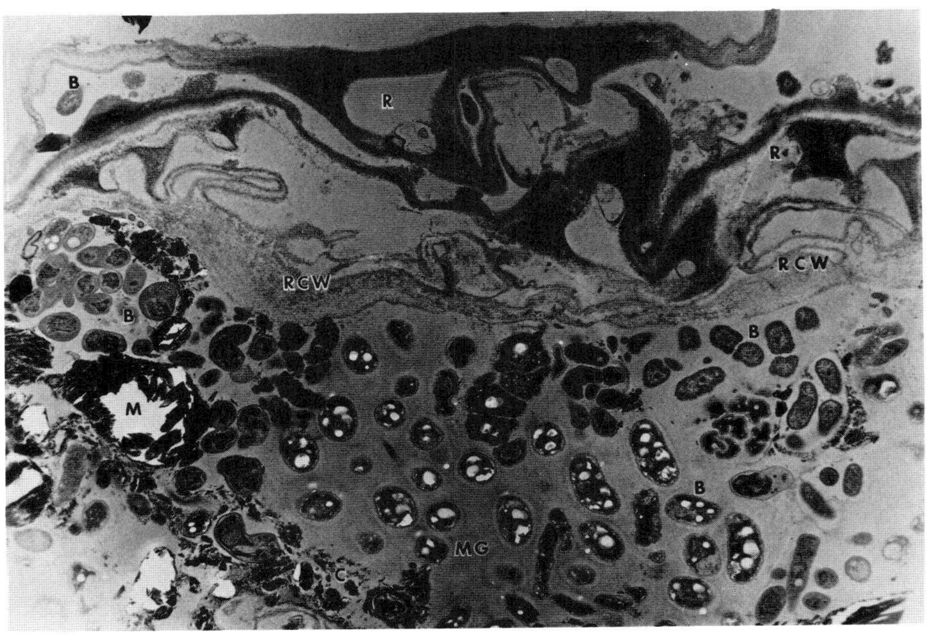

Fig. 1. TEM of root-soil interface of clover-bacterial colonies (B), mucigel (MG), clay (C), minerals (M), root cell wall (RCW) and root surface (R). (Foster and Rovira, 1978). ×6000

low temperature stage which reduces the collapse of cells of roots and fungi (Fig. 2) (Campbell and Porter 1982).

Release of organic compounds from roots

Healthy roots release many organic compounds (Rovira *et al.*, 1978). Our understanding of the carbonaceous substrates from roots depended upon the application of techniques from other fields. Paper chromatography enabled researchers to discover the wide range of amino acids, sugars and organic acids released from roots grown under axenic or sterile conditions (Katznelson *et al.*, 1954; Rivière, 1959; Rovira, 1969). A variety of techniques have been used to measure the release from roots of enzymes, vitamins, nematode hatching factors, fungal spore germinating factors and fungal zoospore attractants.

The observation by Rangaswami and Balasubramanian (1963) that roots of sorghum release sufficient hydrocyanic acid to modify the rhizosphere microflora and delay the germination of spores of root pathogenic fungi could be important in developing crop rotations to inhibit root pathogens.

The next advance in the research on microbial nutrients emmanating from roots came from the use of $^{14}CO_2$. Studies first were conducted with roots in solution culture (McDougall, 1970) followed by filter paper "sandwiches" to find the sites of exudation (McDougall and Rovira, 1970; Rovira, 1973). Sand or glass beads then were used to demonstrate that the exudation of soluble compounds increased as the resistance to root penetration increased (Barber and Gunn, 1974). Studies in sand and soil have shown that from 10 to 18% of the total carbon fixed as

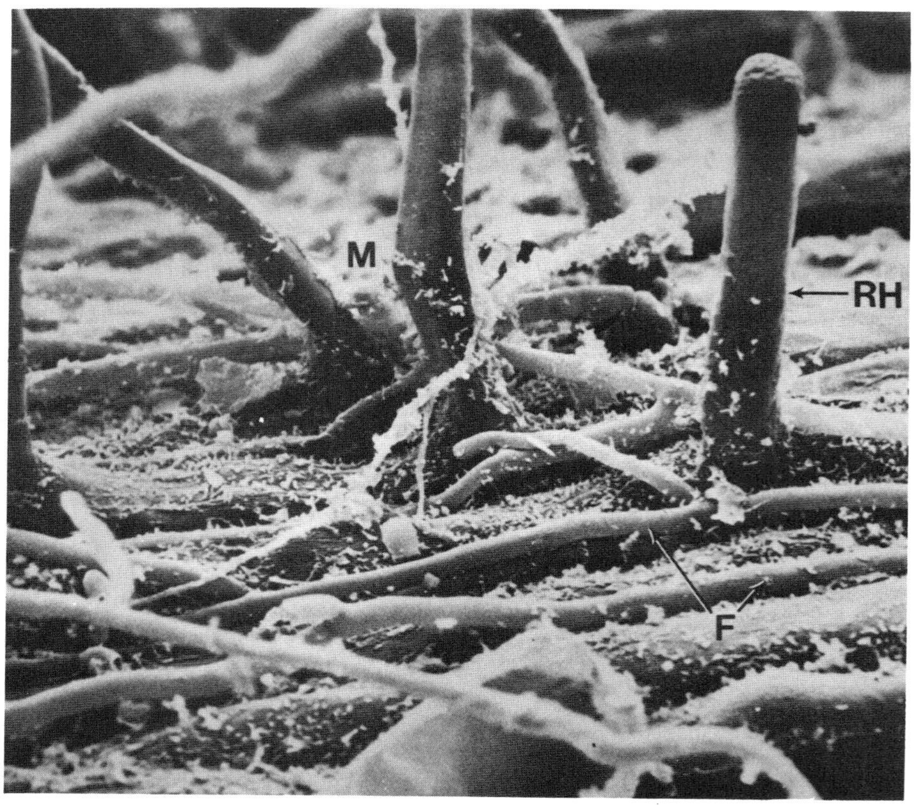

Fig. 2. SEM of wheat root showing hyphae (F) of *G. graminis* var. *tritici* bacteria, mucigel (M) and root hairs (RH) (photo by R C Campbell, University of Bristol, from Foster *et al.*, 1983). ×2000

photosynthate is released from roots into soil – possibly 1 to 2% as soluble exudates and the remainder from root cells as lysates and cell wall material (Balandreau and Knowles, 1978, Martin, 1975; Warembourg, 1975). Barber and Martin (1976) reported that with microorganisms present in soil, some 18% of the photosynthate was released from roots as compared with 10% from sterile roots. Whipps and Lynch (1983), however, found no difference in root products in sterile and non-sterile soils.

Effects of rhizosphere microorganisms on nutrient uptake, plant growth and plant health

Uptake and translocation of nutrients by roots

The discovery that rhizosphere microorganisms affect the uptake and translocation of plant nutrients by roots was another major step in rhizosphere research. Barber (1966; 1967) found that microbes competed with the roots for phosphate, resulting in reduced transport to the tops of barley seedlings. Krasilnikov (1958), Bowen and Rovira (1966; 1968), and Rovira and Bowen (1966) reported that more phosphate was taken-up, incorporated into "DNA"-phosphate and "RNA"-phosphate and translocated to the tops by non-sterile than sterile wheat roots. These observed differences were explained when Barber *et al.* (1976) found that increased or decreased uptake by non-sterile roots depended upon age, nutrition and species of plants, the origin of the microflora, the concentrations of phosphate and the duration of the experiment.

Loutit and Brooks (1970) demonstrated an interesting link between dental health in children and microbes on plant roots in two areas of New Zealand only 10 miles apart. They found that lower molybdenum levels in plants in one soil was due to the rhizosphere microorganisms which resulted in a marked reduction in uptake. The rhizosphere microflora has also been demonstrated to increase the uptake of manganese (Barber and Lee, 1974) and potassium (Williamson and Wyn-Jones, 1973) by cereal roots. These effects of microorganisms on root growth, root morphology, root hair development and the uptake and translocation of nutrients to plant tops highlight the importance of the rhizosphere microflora in plant nutrition and the need for multidisciplinary research.

Growth of roots and root hairs

Bowen and Rovira (1961) and Williamson and Wyn Jones (1973) demonstrated that the growth of roots and root hairs was greater in sterile agar and sand than the growth after the introduction of soil organisms.

Plant growth promotion

The reports from the USSR of large increases in yield from the treatment of seed with *Azotobacter chroococcum* ("Azotobacterin") led to the assumption that these increases were due to the non-symbiotic fixation of nitrogen in the rhizosphere. By 1958 some 10 million ha were sown with seed treated either with "Azotobacterin" or "Phosphobacterin" (phosphate-dissolving isolates of *Bacillus megaterium*). Following World War II, experiments conducted in the USA with the same cultures on a wide range of crops gave negative results (Allison, 1947a, b; Clark, 1948). Research on this topic ceased outside the USSR until Cooper (1959) reported seeing positive responses to "bacterization" in greenhouse crops during his visit to the USSR. This led to experiments with local isolates of Azotobacter in England (Brown *et al.* , 1964) and Australia (Rovira, 1963) which showed that tomato and wheat responded to seed treatment by flowering earlier and in some trials increasing yields. The responses ranged from 5 to 10% but were not consistent (Ridge and Rovira, 1968). However, field trials with carrots gave yield increases of 15 to 40% upon treatment of seed with *Bacillus subtilis* and *Streptomyces griseus* (Merriman *et al.*, 1974).

Mischoustin and Naumova (1962) published a critical analysis of experiments conducted by themselves and other Soviet scientists and concluded that seed treatment with Azotobacter increased yields significantly in one third of their trials, gave non-significant responses in another third and produced no responses in the remaining trials. The rationale behind the use of Azotobacter seed treatment was that it fixed atmospheric nitrogen in the rhizosphere and reduced the plants' dependence and soil and fertilizer nitrogen. Rubinchek (1963), Brown and associates (Brown, 1974), Vancura and Macura (1960), Vancura (1961), Elklund (1970) showed that plant growth promoting compounds, probably gibberellic acid and indole-acetic acid, are produced by a range of bacteria including species of Azotobacter, Bacillus and Pseudomonas and are responsible for the advanced development.

Dobereiner (1968) and Day *et al.* (1975) reported that roots of tropical grasses associated with Azotobacter and Azospirillum fixed large amounts of nitrogen. Kass *et al.* (1971), using acetylene reduction to estimate nitrogen fixation, concluded that in the field the association between *Azotobacter paspali* and Bahia grass (*Paspulum notatum*) could fix between 10 and 20 kg N ha^{-1} per year. However, other workers have arrived at lower figures for the amount of nitrogen fixed and the current thinking is that plant growth responses from Azotobacter and Azospirillum are probably due to plant growth promoting compounds (Schank *et al.*, 1981; Tien *et al.*, 1979).

Schroth and coworkers have demonstrated the widespread nature of plant growth promoting rhizobacteria (PGPR's) (Kloepper *et al.*, 1980) and emergence-promoting rhizobacteria (Kloepper *et al.*, 1986) and their effects on seed germination, plant growth and yield. Yield increases of 10 to 30% were obtained, but one feature was the unaccountable lack of response in some experiments. These effects are attributed primarily to siderophores and the inhibition of detrimental rhizobacteria rather than plant growth promotion (Bakker *et al.*, 1987; Kloepper *et al.*, 1988; Schippers *et al.*, 1987). Thus there are many mechanisms involved in plant growth promotion by rhizosphere bacteria, all of which must be considered when studying the effects of microorganisms on plants.

Biological control of root diseases

One area of rhizosphere microbiology which has attracted a great deal of attention has been on the use of bacteria and fungi to control root diseases (Chet, 1987; Cook and Baker, 1983, Weller, 1988). Due to space constraints only a few publications which highlight rhizosphere colonization and antibiotic production by biocontrol agents as the dominant theme will be quoted.

In the biological control of crown gall caused by the *Agrobacterium radiobacter* pv. *tumefaciens* with the non-pathogenic strain K84 we have an example where competition for sites (viz. wounds on roots and lower stem) and production of an antibiotic, "agrocin-84", are the major factors for the consistent effectiveness of K84 (Kerr, 1980; Thompson, 1987).

Pseudomonas fluorescens strain 2-79 (Weller and Cook, 1983), which controls take-all of wheat in greenhouse and field experiments, establishes itself in the rhizosphere and produces the antibiotic phenazine-1-carboxylic acid (PCA) in the wheat rhizosphere in sand culture (Brisbane and Rovira, 1988) and in soil (Thomashow and Weller, 1988). These are important landmarks in rhizosphere biology, because, while many isolates produce antibiotics in laboratory culture, production in the rhizosphere has seldom been demonstrated. The role antibiotics play in biological control of plant diseases was reviewed recently by Fravel (1988).

The movement of biocontrol bacteria from inoculated seed to roots is considered necessary for biocontrol. Although Loper *et al.* (1986) found that pseudomonads spread from coated seed potatoes to roots, this spread was facilitated by irrigation (Bahme *et al.*, 1988). Parke *et al.* (1986) demonstrated that the colonization of roots by *P. fluorescens* 2-79 was limited without the downward movement of water through soil. The improved "rhizosphere competence" of *Trichoderma harzianum*, combined with resistance to benlate (Papavizas *et al.*, 1982), presents a promising concept of combined biological and chemical control (Lifshitz *et al.*, 1985).

Mycophagous protozoa, nematodes and microarthropods have been demonstrated to interact with and in some cases control root pathogens, *e.g. Rhizoctonia solani* (Curl, 1979) and *Gaeumannomyces graminis* (Chakraborty and Warcup, 1983), and their potential as biocontrol agents is reviewed in Curl (1988).

Rhizobium as a rhizosphere colonist

Despite the economic importance of the legume-Rhizobium and the relevance of research on competition between effective and ineffective strains (as determined by nodule formation and nitrogen fixation) to other aspects of rhizosphere ecology, *e.g.* introduction of biological control agents, research on Rhizobium as a rhizosphere inhabitant has been limited; see reviews by Barnet *et al.* (1988) and Date (1988) and the paper on the populations and distribution of *R.leguminosarum* biovar *trifolii* in relation to the general soil microflora (Bottomley and Dughri, 1989).

Modelling the rhizosphere

Bowen and Rovira (1973) asked "Are modelling approaches useful in rhizosphere biology?" and concluded that modelling of some rhizosphere phenomena would be useful, but some of the microbial and plant parameters, *e.g.* sources of inocula for the rhizosphere, spatial distribution of microorganisms in soil, generation times of bacteria, growth rates of fungi, root extension rates, rates of cortical root-cell senescence, quantities of exudates and lysates, the soil conditions, need further study for a quantitative understanding of microbial colonization and growth in the rhizosphere. One value of an integrated conceptual model is that it exposes areas of ignorance requiring further study.

Newman and Watson (1977) produced a mathematical model which predicts the abundance of microbial biomass in the rhizosphere in relation to the distance out from the root and the amounts of soluble root exudate. The model has been validated to a certain extent by TEM studies of roots of wheat, clover, ryegrass and paspalum (Foster, 1981). The model can be used to predict the effects of microbial and substrate concentrations, changes in soil water, root density, root exudation rate, initial microbial population and microbial response to substrate concentration. Researchers interested in the ecology of rhizosphere microorganisms could benefit from applying the Newman-Watson model as a tool for testing ideas and identifying the key information needed to understand the rhizosphere.

Further developments in modelling have occurred through root pathology; these deal with root growth through soil, with the pathogens coming within the zone of influence of the root, growing towards the root, colonizing and infecting it (Gilligan, 1985; Huisman, 1982). Rhizosphere biologists would do well to study such models and apply them to studies on colonization of roots by non-pathogens, including microorganisms with potential for biological control of root diseases.

Frustrations of rhizosphere research

Our understanding of the biology of the rhizosphere has progressed a long way since 1904, to the point where the microorganisms in the rhizosphere are being managed to increase plant growth. However, this area has proven frustrating as results are variable. This frustration spreads across biological control of root diseases, promotion of plant growth, maintenance of high populations of effective rhizobia and is not surprising considering the complexity of the rhizosphere environment.

Our frustrations will continue unless more thought and effort is put into the micro-ecology of the rhizosphere. We need to understand interactions between the component microorganisms on a 1 to 100 micron scale rather than the "total root system" approach which forms the basis of most rhizosphere studies. Plate counting tells us something of the numbers and types of organisms which can grow on our media, but says nothing of how these and the large proportion which are not culturable on laboratory media are distributed in the rhizosphere. A further cause for frustration is that far too many reports deal with very specific conditions with little thought to the general principles of rhizosphere ecology.

This is a complex field of research and progress in managing the rhizosphere microflora will only come from asking the important questions on root colonization and in this respect the concept of serendipity cannot be ignored (Austin, 1977).

Future of rhizosphere research

Rhizosphere research over the past 85 years has made a great deal of progress. However, we need to develop better systems for tracking introduced micro-organisms in soil and on roots. The use of antibiotic resistance markers has been used widely for detecting introduced bacteria, but there is a potential problem that antibiotic resistant mutants may not perform as well as the parent strains.

The insertion of genes for β-galactosidase into fluorescent pseudomonads, has been used successfully to follow their movement in soil and along roots (Drahos *et al.*, 1986). This technique offers considerable promise for following pseudomonads because the members of this genus as a whole lack β-galactosidase and are incapable of using lactose.

Serological techniques offer the possibility of detecting organisms on the rhizoplane against the background microflora. Polyclonal antisera have been used to detect *Azospirillum brasilense* on grass roots (Schank *et al.*, 1979), *Rhizobium* spp. on legume roots (Schmidt *et al.*, 1968), and *Verticillium dahliae* in cotton roots (Gerik *et al.*, 1987). A refinement of this technique using monoclonal antibodies could provide for the location of specific bacteria in the rhizosphere (Wright *et al.*, 1986). Such serological methods also offer promise for microecology of the rhizosphere by tagging antibodies with gold and locating the reacting cells by TEM of ultrathin root-soil sections.

The application of probes based on DNA and ribosomal DNA, DNA restriction fragment length polymorphisms (RFLP) and enzyme-linked immunoabsorbant assay (ELISA) techniques which can detect low numbers of particular bacterial cells or fungal mycelia (Miller and Martin, 1988) in the rhizosphere to trace and monitor the growth and survival of introduced organisms and their product will provide the opportunity to greatly improve our understanding of the biology of the rhizosphere.

Conclusions

The 85 year history of the rhizosphere has been one of achievement and progress, so from this knowledge base we can look optimistically towards the goal of modifying the rhizosphere microflora to increase plant growth. I will conclude with a very pertinent statement by Charles Kettering – "A research problem is not solved by apparatus, it is solved in a person's head".

References

Allison F E 1947a Azotobacter inoculation of crops. I. Historical. Soil Sci. 64, 413–429.

Allison F E 1947b Azotobacter inoculation of crops. II. Effect on crops under greenhouse conditions. Soil Sci. 64, 489–497.

Austin J H 1978 Chase, Chance and Creativity. Colombia University Press, New York. 237 p.

Bahme J B, Schroth M N, Van Gundy S D, Weinhold A R and Tolentino D M 1988 Effect of inocula delivery systems on rhizobacterial colonization of underground organisms of potato. Phytopathology 78, 534–542.

Bakker P A H M, Bakker A W, Marugg J D, Weisbeck P J and Schippers B 1987. A bioassay for studying the role of siderophores in potato growth stimulation by *Pseudomonas* spp. in short potato rotation. Soil Biol. Biochem. 19, 443–450.

Balandreau J and Knowles R 1978 The rhizosphere. *In* Interactions Between Non-pathogenic Soil Micro-organisms and Plants. Eds Y R Dommergues and S V Krupa. pp 243–268. Elsevier, Amsterdam, New York.

Barber D A 1966 Effect of microorganisms on nutrient absorption by plants. Nature London 212, 638–640.

Barber D A 1967 Microorganisms and the inorganic nutrition of higher plants. Annu. Rev. Pl. Physiol. 19, 71–88.

Barber D A and Gunn K B 1974 The effect of mechanical forces on the exudation of organic substances by the roots of cereal plants grown under sterile conditions. New Phytol. 73, 39–45.

Barber D A and Lee R B 1974 The effect of microorganisms on the absorption of manganese by plants. New Phytol. 73, 97–106.

Barber D A and Martin J K 1976 The release of organic substances by cereal roots into soil. New Phytol. 76, 69–80.

Barber D A, Bowen G D and Rovira A D 1976 Effects of microorganisms on absorption and distribution of phosphate in barley. Aust. J. Pl. Physiol. 3, 801–808.

Barnet Y M, Trinick M J, Date R A and Roughley R J 1988 Ecology of the root nodule bacteria. *In* Microbiology in Action. Eds W G Murrell and I R Kennedy. pp 1–22. Research studies Press Ltd. Letchworth, England and John Wiley and Sons Inc., New York.

Bohlool B B and Schmidt E L 1980 The immunofluorescence approach in microbial ecology. Adv. Microbiol. Ecol. 4, 203–241.

Bottomley P J and Dughri M H 1989 Population size and distribution of *Rhizobium leguminosarum* bv *trifolii* in relation to total soil bacteria and soil depth. Appl. Environ. Microbiol. 55, 959–964.

Bowen G D 1980 Misconceptions, concepts and approaches in rhizosphere biology. *In* Contemporary Microbial Ecology. Ed. D G Ellwood. pp 283–304. Academic Press, London, England.

Bowen G D and Rovira A D 1961 The effects of microorganisms on plant growth. I. Development of roots and root hairs in sand and agar. Plant and Soil 15, 166–189.

Bowen G D and Rovira A D 1966 Microbial factor in short-term phosphate uptake studies with plant roots. Nature 211, 666–668.

Bowen G D and Rovira A D 1968 The influence of microorganisms on growth and metabolism of plant roots. *In* Root Growth. Ed. W J Whittington, pp 170–201. Butterworths, London, England.

Bowen G D and Rovira A D 1973 Are modelling approaches useful in rhizosphere biology? Bull. Ecol. Res. Comm. Stockholm 17, 443–450.

Brisbane P G and Rovira A D 1988 Mechanisms of inhibition of *Gaeumannomyces graminis* var. *tritici* by fluorescent pseudomonads. Plant Pathol. 37, 104–111.

Brown M E 1974 Seed and root bacterization. Annu. Rev. Phytopathol. 12, 181–197

Brown M E, Burlingham S K and Jackson R M 1964 Studies on Azotobacter species in soil. III. Effects of artificial inoculation on crop yields. Plant and Soil 20, 199–214.

Campbell R and Porter R 1982 Low temperature scanning electron microscopy of microorganisms in soil. Soil Biol. Biochem. 14, 241–245.

Chakraborty S and Warcup J H 1983 Soil amoebae and saprophytic survival of *Gaeumannomyces graminis* var. *tritici* in a suppressive pasture soil. Soil Biol. Biochem. 15, 181–185.

Chet I (Ed) 1987 Innovative Approaches to Plant Disease Control. John Wiley and Sons, New York. 322 p.

Clark F E 1948 Azotobacter inoculation of crops. III. Recovery of Azotobacter from the rhizosphere. Soil Sci. 65, 193–202.

Clark F F 1949 Soil microorganisms and plant growth. Adv. Agron. 1, 241–288.

Cook R J and Baker K F 1983 The Nature and Practice of Biological Control of Plant Pathogens. Am. Phytopath. Soc., St Paul, MN.

Cooper R 1959 Bacterial fertilizers in the Soviet Union. Soils Fertil. 22, 227–233.

Curl E A 1979 Effects of mycophagous Collembola on Rhizoctonia and cotton seedling disease. *In* Soil-Borne Plant Pathogens. Eds. B Schippers and W Gams. pp. 253–269, Academic Press, London, England.

Curl E A and Truelove B 1986 The Rhizosphere. Springer-Verlag, Berlin. 288 p.

Curl E A 1988 The role of soil microfauna in plant disease suppression. CRG Crit. Rev. Plant Sci. 7, 175–196.

Date R A 1988 Colonization of the rhizosphere by root nodule bacteria. *In* Microbiology in Action. Eds W G Murrell and I R Kennedy. pp 23–33. Research Studies Press Ltd, Letchworth, England and John Wiley and Sons, New York.

Dart P J 1971 Scanning electron microscopy of roots. J. Exp. Bot. 22, 163–168.

Dart P J and Mercer F V 1964 The legume rhizosphere. Arch. Mikrobiol. 47, 344–378.

Day J M, Neves M C P and Dobereiner J 1975 Nitrogen fixation on the roots of tropical forage grasses. Soil Biol. Biochem. 7, 107–112.

Dobereiner J 1968 Non-symbiotic nitrogen fixation in tropical soils. Pesq. Agropic. Brazil. 1, 357–365.

Drahos D J, Hemming B C and McPherson S 1986 Tracking recombinant organisms in the environment: β-galactosidase as a selectable non-antibiotic marker for fluorescent pseudomonads. Biotechnology 4, 439–444.

Eklund E 1970 Secondary effects of some pseudomonads in the rhizoplane of peat grown cucumber plants. Acta Agric. Scand. Suppl. 17. Stockholm. 57 p.

Foster R C 1981 The ultrastructure and histochemistry of the rhizosphere. New Phytol. 89, 263–273.

Foster R C 1986 The ultrastructure of the rhizoplane and rhizosphere. Annu. Rev. Phytopathol. 24, 211–234.

Foster R C and Bowen G D 1982 Plant surfaces and bacterial growth: The rhizosphere and rhizoplane. *In* Phytopathogenic Prokaryotes, Vol. 1. Eds M S Mount and G H Lacey. pp 159–185. Academic Press, New York.

Foster R C and Rovira A D 1978 The ultrastructure of the rhizosphere of *Trifolium subterraneum* L. *In* Microbiol. Ecology. Eds M W Loutit and J A R Miles. pp 278–290. Springer-Verlag, Berlin.

Foster R C, Rovira A D and Cock T W 1983 Ultrastructure of the Root-Soil Interface. The American Phytopathological Society, St Paul, MN. 157 p.

Fravel D R 1988 Role of antibiotics in the biocontrol of plant diseases. Annu. Rev. Phytopathol. 26, 75–91.

Gerhardson B and Clarholm M 1986 Microbial communities on plant roots. *In* Microbial Communities in Soil. Eds V Jensen, A Kjøller and L H Sørensen. pp 18–34. Elsevier, London, England.

Gerik J S, Lommel S A and Huisman O C 1987 A specific serological staining procedure for *Verticillium dahliae* in cotton root tissue. Phytopathology 77, 261–266.

Gilligan C A 1985 Probability models for host infection by soil-borne fungi. Phytopathology 75, 61–67.

Gray T R G 1967 Stereoscan electron microscopy of microorganisms. Science N.Y. 155, 1668–1670.

Greaves M P and Darbyshire J F 1972 The ultrastructure of the mucilaginous layer on plant roots. Soil Biol. Biochem. 4, 443–449.

Hiltner L 1904 Über neuere Erfahrungen und Probleme auf dem Gebiet der Bodenbakteriologie und unter besonderer Berücksichtigung der Gründüngung und Brache. Arb. Dtsch. Landwirt. Ges. 98, 59–78.

Huisman O G 1982 Interrelations of root growth dynamics to epidemiology of root-invading fungi. Annu. Rev. Phytopathol. 20, 303–327.

Ishizawa S T, Suzuki T, Sato O and Toyoda H 1957 Studies on microbiol population in the rhizosphere of higher plants with special reference to the method of study. Soil Plant Food (Tokyo) 3, 85–94.

Jenny H and Grossenbacker K 1963 Root-boundary zones as seen in this electron microscope. Soil Sci. Soc. Am. 27, 273–277.

Jones P C T and Mollison J E 1948 A technique for the quantitative estimation of soil microorganisms. J. Gen. Microbiol. 2, 54–69.

Kass D L, Drosdoff M and Alexander M 1971 Nitrogen fixation by *Azotobacter paspali* in association with Bahiagrass (*Paspalum notatum*). Soil Sci. Soc. Am. Proc. 35, 286–289.

Katznelson H, Lochhead A G and Timosin M I 1948 Soil microorganisms and the rhizosphere. Bot. Rev. 14, 487–543.

Katznelson H, Rouatt J V and Payne T M B 1954 Liberation of amino acids by plant roots in relation to desiccation. Nature 174, 1110–1111.

Kerr A 1980 Biological control of crown gall through production of agrocin-84. Plant Dis. 64, 25–30.

Kloepper J W, Lifshitz R and Schroth M N 1988 Pseudomonas inoculants to benefit plant production. ISI Altas of Science: Anim. Plant Sci. 1988, 60–64.

Kloepper J W, Schroth M N and Miller T D 1980 Effects of rhizosphere colonization by plant growth-promoting rhizobacteria on potato plant development and yield. Phytopathology 709, 1078–1082.

Kloepper J W, Scher F M, Leliberté M and Tipping B 1986 Emergence-promoting rhizobacteria: Description and implications for agriculture. *In* Iron, Siderophores and Plant Disease. Ed. T R Swinburne. pp 155–164. Plenum, New York.

Krasilnikov N A 1934 The effect of root excretions on the growth of Azotobacter and other soil microbes. Microbiology (USSR) 3, 3.

Krasilnikov N A 1958 Soil microorganisms and higher plants. Acad. Sci. USSR Moscow, 474 p. (Translated in 1961 for the NSF, USDA and The Israel Program for Scientific Translations.)

Lifshitz R, Liftshitz S and Baker R 1985 Decrease in the incidence of Rhizoctonia preemergence damping off by use of integrated chemical and biological controls. Plant Dis. 69, 431–434.

Lochhead A G and Chase F E 1943 Qualitative studies of soil microorganisms: V. Nutritional requirements of the predominant bacterial flora. Soil Sci. 55, 185–195.

Lochhead A G and Thexton R H 1947 Qualitative studies in soil microorganisms: VII The "rhizosphere effect" in relation to the amino acid nutrition of bacteria. Can J. Res. C, 25, 20–26.

Loper J E, Haack C and Schroth M N 1986 Population dynamics of soil pseudomonads in the rhizosphere of potato (*Solanum tuberosum* L.). Appl. Environ. Microbiol. 49, 416–422.

Loutit M W and Brooks R R 1970 Rhizosphere organisms and molybdenum concentrations in plants. Soil Biol. Biochem. 2, 131–135.

Louw H A and Webley D M 1959 The bacteriology of the root region of the oat plant grown under controlled pot culture conditions. J. Appl. Bacteriol. 22, 216–226.

Lynch J M 1988 Microbes are rooting for better crops. New Scientist 1610, 45–49.

Lynch J M (Ed.) 1990 The Rhizosphere. John Wiley and Sons, New York. 458 p.

Martin J K 1975 ^{14}C-labelled material leached from the rhizosphere of plants supplied continuously with $^{14}CO_2$. Soil Biol. Biochem. 7, 395–399.

McDougall B M 1970 Movement of ^{14}C-photosynthate in roots of wheat seedlings and exudation of ^{14}C from intact roots. New Phytol. 69, 37–46.

McDougall B M and Rovira A D 1970 Sites of exudation of ^{14}C-labelled compounds from wheat roots. New Phytol. 69, 999–1003.

Merriman P R, Price R D, Kollmorgen J F, Piggott T and Ridge E H 1974 Effect of soil inoculation with *Bacillus subtilis* and *Streptomyces graminis* on the growth of cereals and carrots. Aust. J. Agric. Res. 25, 219–226.

Miller S A and Martin 1988 Molecular diagnosis of plant disease. Annu. Rev. Phytopathol. 26, 409–432.

Mishoustin E H and Naumova A N 1962 Bacterial fertilizers, their effectiveness and mode of action. Mikrobiologiya 31, 543–555.

Newman E I and Bowen H J 1974 Patterns of distribution of bacteria on root surfaces. Soil Biol. Biochem. 6, 205–209.

Newman E I and Watson A 1977 Microbial abundance in the rhizosphere. Plant and Soil 48, 17–56.

Old K M and Nicholson T H 1975 Electron microscopical studies of the microflora of roots of sand dune grasses. New Phytol. 74, 51–58.

Parke J L, Moen R, Rovira A D and Bowen G D 1986 Soil water flow affects the rhizosphere distribution of a seed-borne biological control agent, *Pseudomonas fluorescens*. Soil Biol. Biochem. 18, 583–588.

Papavizas G C, Lewis J A and Abd-El-Moity T H 1982 Evaluation of new biotypes of *Trichoderma harzianum* for tolerance of benomyl and enhanced biocontrol capabilities. Phytopathology 72, 126–132.

Rangaswami G and Balasubramanian A 1963 Release of hydrocyanic acid by sorghum roots and its influence on the rhizosphere microflora and plant pathogenic fungi. Indian J. Exp. Biol. 1, 215–217.

Ridge E H and Rovira A D 1968 Microbial inoculation of wheat. Trans. 9th Congr. Int. Soc. Soil Sci. 3, 473–481.

Rivière J 1959 Contribution l'étude e la rhizosphère du blé. Ann. Agron. 45, 93–337.

Rovira A D 1956 A study of the development of the root surface microflora during the initial stages of plant growth. J. Appl. Bacteriol. 19, 72–79.

Rovira A D 1963 Microbial inoculation of plants. I. Establishment of free-living nitrogen fixers in the rhizosphere and their effects on maize, tomato and wheat. Plant and Soil 19, 304–314.

Rovira A D 1969 Plant root exudates. Bot. Rev. 35, 17–34.

Rovira A D 1973 Zones of exudation along plant roots and spatial distribution of micro-organisms in the rhizosphere. Pestic. Sci. 4, 361–366.

Rovira A D 1985 Manipulation of the rhizosphere microflora to increase plant production. *In* Reviews of Rural Science 6, Biotechnology and Recombinant DNA Technology in the Animal Production Industries. Eds. R A Leng, J S F Barker, D B Adams and K J Hutchinson. pp 185–197. Univeristy of New England, Armidale, Australia.

Rovira A D 1988 Ecology and management of the rhizosphere microflora. *In* Microbiology in Action. Eds. W G Murrell and I R Kennedy. pp 221–238. Research Studies Press Ltd, Letchworth, England and John Wiley and Sons Inc., New York.

Rovira A D and Bowen G D 1966 Phosphate incorporation by sterile and non-sterile plant roots. Aust. J. Biol. Sci. 19, 1167–1169.

Rovira A D and Campbell R 1975 A scanning electron microscope study of interactions between micro-organisms and *Gaeumannomyces graminis* (Syn. *Ophiobolus graminis*) on wheat roots. Microb. Ecol. 2, 177–195.

Rovira A D, Bowen G D and Foster R C 1983 The significance of rhizosphere microflora and mycorrhizas on plant

nutrition. *In* Encyclopedia of Plant Physiology, New Series Vol. 15. Eds. A Lauchli and R L Bieleski. pp 61–93. Springer-Verlag, Berlin, Heidelberg.
Rovira A D, Foster R C and Martin J K 1978 Origin, nature and nomenclature of the organic materials in the rhizosphere. *In* The Root-Soil Interface. Eds. J L Harley and R Scott-Russell. pp 1–4. Academic Press, London, England.
Rovira A D, Newman E I, Bowen H J and Campbell R 1974. Quantitative assessment of the rhizosphere microflora by direct microscopy. Soil Biol. Biochem. 6, 211–216.
Rubenchik L I 1963 Azotobacter and its use in agriculture. Jerusalem, Israel, Program for Scientific Translation, 179 p.
Savostin P V 1941 The nature of azotogen action. Microbiology (USSR) 10, 33–41.
Schank S C, Weir K L and MacRae I C 1981 Plant yield and nitrogen content of a digit grass in response to Azospirillum inoculation. Appl. Exp. Microbiol. 41, 342–345.
Schank S C, Smith R L, Weiser G C, Zuberer D A, Boulton J H, Quesenberry K H, Tyler M E, Milam J R and Littell R C 1979. Fluorescent antibody technique to identify *Azospirillum brasilense* associated with roots of grasses. Soil Biol. Biochem. 11, 287–295.
Schippers B, Bakker A W and Bakker A H M 1987 Interactions of deleterious and beneficial rhizosphere microorganisms and the effect of cropping practices. Annu. Rev. Phytopathol. 25, 339–358.
Schmidt E L, Bankole R O and Bohool B B 1968 Fluorescent antibody approach to study of rhizobia in soil. J. Bacteriol. 95, 1987–1992.
Starkey R L 1929 Some influences of the development of higher plants upon the microorganisms in the soil. I. Historical and introductory. Soil Sci. 27, 319–334.
Starkey R L 1938 Some influences of the development of higher plants upon the microorganisms in the soil. VI. Microscopic examination of the rhizosphere. Soil Sci. 45, 207–249.
Thomashow L S and Weller D M 1988 Role of antibiotics in root disease suppression by fluorescent pseudomonads. Abstr. 5th ICPP Congress, Kyoto, Japan II 7–4, 94.
Thompson J A 1987 The use of agrocin-producing bacteria in the biological control of crown gall. *In* Innovative Approaches to Plant Disease Control. Ed. I. Chet. pp 213–228. John Wiley and Sons, New York.
Tien T M, Gaskins M H and Hubbell D H 1979 Plant growth substances produced by *Azospirillum brasilence* and their effect on the growth of pearl millet (*Pennisetum americanum* L.). Appl. Exp. Microbiol. 37, 1016–1024.
Trolldenier G 1973 The use of fluorescence microscopy for soil microorganisms. Bull. Ecol. Res. Comm. Stockholm 17, 53–59.
Umali-Garcia M, Hubbell D H, Gaskins M H and Dazzo F B 1980 Association of Azospirillum with grass roots. Appl. Environ. Microbiol. 39, 219–226.
Vancura V 1961 Detection of gibberellic acid in Azotobacter cultures. Nature (Lond.) 192, 88–99.
Vancura V and Macura J 1960 Indole derivatives in Azotobacter cultures. Fol. Microbiol. 5, 293–297.
Vuurde J W L van and Elenbaas P E M 1978 Use of fluorochromes for direct observation of micro-organisms associated with wheat roots. Can. J. Microbiol. 24, 1272–1775.
Waisel Y and Eshel A 1990 Plant Roots: The Hidden Half. Marcel Dekker Inc. New York. *In press.*
Warembourg F R 1975 Le degagement de CO_2 dans la rhizosphere des plantes. Soc. Bot. Fr. Colloqu. Rhizosphere. 122, 77–100.
Weller D M 1988 Biological control of soilborne plant pathogens in the rhizosphere with bacteria. Annu. Rev. Phytopathol. 26, 379–407.
Weller D M and Cook R J 1983 Suppression of take-all by seed treatments with fluorescent pseudomonads. Phytopathology 73, 463–469.
Whipps J M and Lynch J M 1983 Substrate flow and utilization in the rhizosphere of cereals. New Phytol. 95, 605–623.
Whipps J M and Lynch J M 1986 The influence of the rhizosphere on crop productivity. Adv. Microb. Ecol. 9, 187–244.
Williamson F A and Wyn-Jones R G 1973 The influence of soil micro-organisms on growth of cereal seedlings and on potassium uptake. Soil Biol. Biochem. 5, 569–575.
Wright S F, Foster J G and Bennett O C 1986 Production and use of monoclonal antibodies for identification of strains of *Rhizobium trifolii*. Appl. Environ. Microb. 52, 119–123.
Zinovyeka H K 1940 On the method of Azotobacter preparation. Mikrobiol. Z. (Kiev) 7, 85–96. (Abst. in Biol. Abst. 16, 1210, 1942).

D. L. Keister and P. B. Cregan (Eds.), The rhizosphere and plant growth, 15–24.

PLSO BARC02

Substrate flow in the rhizosphere

J.M. LYNCH and J.M. WHIPPS
Microbiology Department, AFRC Institute of Horticultural Research, Littlehampton, West Sussex, BN17 6LP, UK

Key words: arrested metabolism, biomass, carbon budgets, cryptic growth, dormancy, oligotrophy, products, rhizodeposition, substrates

Abstract

The major source of substrates for microbial activity in the ectorhizosphere and on the rhizoplane are rhizodeposition products. They are composed of exudates, lysates, mucilage, secretions and dead cell material, as well as gases including respiratory CO_2. Depending on plant species, age and environmental conditions, these can account for up to 40% (or more) of the dry matter produced by plants. The microbial populations colonizing the endorhizosphere, including mycorrhizae, pathogens and symbiotic N_2-fixers have greater access to the total pool of carbon including that recently derived from photosynthesis. Utilization of rhizodeposition products induces at least a transient increase in soil biomass but a sustained increase depends on the state of the native soil biomass, the flow of other metabolites from the soil to the rhizosphere and the water relations of the soil. In addition, the phenomena of oligotrophy, cryptic growth, plasmolysis, dormancy and arrested metabolism can all influence the longevity of rhizosphere organisms. With this background, microbial growth in the rhizosphere will be discussed.

Introduction

Microbial growth in the rhizosphere is stimulated by the continual input of readily assimilable organic substrates from the root. In ecological terms, the substrate flowing from the roots is the product of photosynthesis and is thus primary productivity and the utilization of these organic substrates by rhizosphere micro-organisms results in secondary productivity. In turn, these micro-organisms can influence the plant as primary producer. Methodology is now available which enables substrate flow in the rhizosphere to be analysed in terms of the balance between the production and utilization of substrates. We will review current evidence on each of these components.

Rhizodeposition

This topic has been reviewed in depth previously (Whipps, 1990; Whipps and Lynch, 1985; 1986) and therefore only a summary will be given here. The term 'rhizodeposition' has been used to describe carbon *loss* from roots (Whipps and Lynch, 1985) and this material generates the rhizosphere effect. However, it should not be forgotten that the living root itself provides a substrate for endorhizosphere colonizers, a consideration often overlooked in the past.

The chemical components coming from roots are conveniently classified into the following four groups, depending on their mode of arrival: 1) water soluble *exudates*, such as sugars, amino acids, organic acids, hormones and vitamins which leak from the root without the involvement of metabolic energy; 2) *secretions*, such as polymeric carbohydrates and enzymes which depend upon metabolic processes for their release; 3) *lysates*, released when cells autolyse including cell walls, and, with time whole roots; 4) *gases* such as ethylene and CO_2. These groupings cover all stages of plant growth and development with the balance of these various processes changing with age of the plant. This can range from the release of simple sugars during membrane re-organization of the embryo during seed germination (Hale and Moore, 1979) through to root cortical death in the mature plant followed by whole root senescence. It also includes the sloughing-off of root cap cells; approximately 5000 cells per day are lost from maize (Moore and McClelen, 1983). Bearing in mind the complexity of substrate origin, the composition of rhizodeposition products can be expected to include all components of the plant cell.

A range of experiments has been carried out to measure rhizodeposition. Many have involved the growth of plants in nutrient solution culture under gnotobiotic conditions where exudates, sloughed cells and mucilage are collected. Addition of micro-organisms generally increases the carbon loss. These types of rhizodeposition have been analysed quantitatively and qualitatively by chemical analytical techniques (*e.g.* Kraffczyk *et al.*, 1984; Schönwitz and Ziegler, 1982) or by the use of $^{14}CO_2$ feeding experiments, mainly using pulse-chase techniques (*e.g.* Beck and Gilmour, 1983; Lee and Gaskins, 1982). Some experiments have been refined to examine the effect of a whole range of factors on rhizodeposition including textural quality (by addition of glass Ballotini or use of artificial soils), anoxia, water stress, cultivar, nutrient sources, environmental parameters and micro-organisms (see Hale and Moore, 1979; Whipps, 1990). Measured in this way, values for rhizodeposition do not exceed $600\,mg\,g^{-1}$ root dry weight (Newman, 1985; Whipps, 1990). These values must represent considerable underestimations of rhizodeposition as in many cases respiration from the root and accompanying microflora is ignored. In addition, nutrient solution culture is not suitable for the growth of many perennial plants and this, together with the difficulty of maintaining sterility in axenic cultures has given rise to much data on seedling rhizodeposition, mainly exudation and secretion. Similarly, use of artificial soils and addition of specific micro-organisms can bear little resemblance to the complex structural and functional relationships taking place in the soil.

Due to the complex nature of the soil, two types of tracer techniques have been developed to study rhizodeposition in this milieu. The first, pulse-chase technique, is relatively easy and rapid and is applicable to both laboratory and field experimentation. However, it results in a non-uniform distribution of label in the plant and can thus only provide information on movement and losses of recently fixed carbon and not on the degradation of whole cells or dying older roots. Even repeated pulse feeding cannot evenly label all pools of carbon in the plant. This has led to the development of the second system which is applicable only to the laboratory. Here, plants are grown in controlled environment growth chambers with shoots exposed to an atmosphere containing constant specific activity $^{14}CO_2$. Any $^{14}CO_2$ or ^{14}C-containing material found in the soil must then originate from rhizodeposition. The $^{14}CO_2$ detected in the soil can come from root respiration *per se* or from microorganisms utilizing root derived material. This technique cannot allow for carbon respired from the shoot and so distribution data can only be related to the net fixed carbon rather than total fixed carbon.

Using these constant feeding systems, ^{14}C derived from maize roots has been shown to spread 20 mm from the root after 25 days of growth and act as a primer for microbial degradation of the existing soil organic carbon (Helal and Sauerbeck, 1983; 1986). Except when activity is high during the early stages of decomposition of fresh organic matter, similar results have been found with wheat (Sallih and Bottner, 1988), resulting in an increase in soil biomass (Bottner *et al.*, 1988). Merckx *et al.* (1985) also showed that ^{14}C derived from wheat roots entered the soil microbial biomass and increased throughout a six week experimental period. However, the majority of these studies have examined the amounts and proportions of ^{14}C entering the roots and forming the rhizodeposition. Results from some of these studies are summarised in Table 1.

Table 1. Percentage of net fixed carbon moved to roots and the percentage of net fixed carbon moved to the roots which is lost when shoots of plants are exposed continually to $^{14}CO_2$ (Plants grown in soil). Conditions varied but were *c.* 18°C, 16 h day length, 0.03% CO_2 and none were sterilized

Plant	Age (d)	% of net fixed C transferred to root (A)	% of C transferred to root lost as: Respiration (B)	Rhizodeposition in soil (C)	Respiration and rhizodeposition in soil (B + C)	Reference
Wheat	21	38	25	17	42	Barber and Martin, 1976
	21	47	36	25	61	Whipps and Lynch, 1983
	21	59	39	29	68	Whipps, 1984
	28	44	39	11	50	Merckx *et al.*, 1985
	153	?	76	4	80	Johnen and Sauerbeck, 1977
Barley	21	43	21	23	44	Barber and Martin, 1976
	21	54	48	21	69	Whipps, 1984
Maize	14	28	16	52	67	Whipps, 1985
Mustard	73	?	66	13	78	Johnen and Sauerbeck, 1977
Tomato	14	43	20	70	90	Whipps, 1987
Pea	28	44	53	29	82	Whipps, 1987

$$A = \frac{\mu\text{Ci in root + rhizodeposition + } CO_2 \text{ from root and soil micro-organisms}}{\text{net } \mu\text{Ci fixed}} \times 100; \ \mu\text{Ci } \Omega \text{ amount of C}$$

$$B = \frac{\mu\text{Ci in } CO_2 \text{ from root and soil micro-organisms}}{\mu\text{Ci in root = rhizodeposition in soil + } CO_2 \text{ from root and soil micro-organisms}} \times 100$$

$$C = \frac{\mu\text{Ci in rhizodeposition in soil} \times 100}{\mu\text{Ci in root + rhizodeposition in soil + } CO_2 \text{ from root and soil micro-organisms}}$$

Overall, between 30–60% of net fixed carbon is transferred to roots in annual plants. These figures agree well with a range of nine different annual plants grown in nutrient solutions under a range of conditions (Lambers, 1987). Between 40 and 90% of the carbon transferred to the root is lost as rhizodeposition in the soil and respiration from the roots and associated micro-organisms. With wheat, and other annual plant species not shown the quantity of rhizodeposition in soil decreased with time, but the proportion lost as respiration increased. Similar experiments with perennial blue grama grass (*Bouteloua gracilis*) showed that over 70% of net fixed carbon was passed to roots during a simulated summer growing period (Dormaar and Sauerbeck, 1983). Subsequent loss of 30–60% of young, suberised roots occurs when root growth ceases (Ares, 1976).

These constant feeding techniques also suffer problems. For example, root growth may be physically restricted in pots and water and gas exchange and nutrient availability may all limit growth. The pots may leak and harvests have to be sequential and destructive. The seedlings often have considerable food reserves and it may take some time for all plant parts to become evenly labelled and, as shoot respiration cannot be measured *in situ*, distribution data always refers to net fixed carbon. Perhaps most frustrating is that it is impracticable to grow perennial species for long periods.

Estimates of rhizodeposition in field experiments have been obtained using $^{14}CO_2$ pulse-chase experiments and by analyzing components of crop growth and some results are given in Table 2. The value for percentage ^{14}C transferred below ground for wheat is lower than that determined for seedlings in constant feeding experiments (Table 1), but as rhizodeposition falls with time (Keith *et al.*, 1986), a lower annual figure is to be expected. The values of 1.3 t C ha^{-1} $annum^{-1}$ for rhizodeposition measured in these field experiments agrees well with the range of values for wheat and mustard (0.5–2.9 t C ha^{-1} $annum^{-1}$) obtained in $^{14}CO_2$ growth

Table 2. Carbon movement to and rhizodeposition from the roots in field experiments

	% net fixed C transferred to root	% of C transferred to root that is lost		Rhizodeposition (root C + soil C + CO_2C) (t C ha^{-1} $annum^{-1}$)	Reference
		Soil	CO_2		
Wheat[a]	15	25	50	1.3	Keith *et al.*, 1986
Agropyron-Koeleria dominated grass mixture	35–50	[b]	20–28	1.3	Warembourg and Paul, 1977
Douglas fir	73		40–47[d]	5.8–7.5[c]	Sanantonio, 1979; Fogel and Hunt, 1983

[a] Mean value of five sampling times (42–154 d)
[b] No value available
[c] Includes carbon in mycorrhizas
[d] Soil and CO_2 not separated

cabinet experiments (Johnen and Sauerbeck, 1977). This represents approximately 15% of total fixed carbon (Keith *et al.*, 1986). Field measurements of root production without allowance for respiration losses range between 0.2–1.6 t C ha^{-1} $annum^{-1}$ (Barber, 1971; Gregory *et al.*, 1978; Sivakumar *et al.*, 1977).

Perennial plants maintain roots throughout the year and rhizodeposition is known to be cyclical depending on periods of active growth. Using $^{14}CO_2$ pulse-feeding techniques, Singh and Coleman (1974) found that carbon was initially found in actively growing roots of blue grama in the top 20 cm of soil but it subsequently moved to lower regions of soil as roots grew deeper. Between 30–50% of net fixed carbon was transferred below gound in *Agropyron-Koeleria* perennial grassland (Table 2), less than that found in growth cabinet experiments with blue grama (Dormaar and Sauerbeck, 1983) and other field experiments (Kucera *et al.*, 1967; Sims and Singh, 1971). But as only 20–28% of the carbon moved to roots was respired, which is low compared with other grassland systems, the value of 1.3 t C ha^{-1} $annum^{-1}$ should be considered a minimum. Tree species pass on even greater quantities of carbon to their roots resulting in an annual rhizodeposition of 5.8–7.5 t C ha^{-1} (Table 2). In general, perennial plants invest more of their productivity in root material than annuals and this may be related to their need to withstand prolonged periods of stress.

Another aspect of rhizodeposition concerns the involvement of symbiotic root inhabitants such as Rhizobium and mycorrhizae. Legumes benefit from Rhizobium infection by improved N nutrition from N_2-fixation and the majority of plants benefit from mycorrhizae by improved mineral nutrition, particularly of P, and drought tolerance. In return, the heterotrophs gain organic material from the endorhizosphere. Measurement of total rhizodeposition from legumes in soil are scarce, the majority of studies being concerned with the efficiency and cost of N_2 fixation. Over the life of a range of legumes, between 30–50% of net fixed C enters the roots with 25–40% of net fixed C representing 63–79% of C entering the root being lost (Minchin *et al.*, 1981). Mean values of 13.4, 7.1 and 9.2 mg CO_2 respired g^{-1} root h^{-1} have been found for whole nodulated roots, roots (separated from nodules) and roots from non-nodulated nitrate-grown plants (Ryle *et al.*, 1983). The cost (CO_2 lost g^{-1} root) of N_2-fixation by nodulated roots was up to twice that when growing on nitrate-N (Pate *et al.*, 1979; Ryle *et al.*, 1979).

Studies on vesicular-arbuscular (VA) mycorrhizal infections indicated that conditions which increase permeability of roots or exudation increase infection (Graham *et al.*, 1981). Once synthesised, the cost of maintaining VA mycorrhizas in terms of extra carbohydrate moved to the roots is approximately 6–11% net fixed carbon (Snellgrove *et al.*, 1982; Koch and Johnson, 1984). In experiments with *Vicia faba* plants, it was shown that nodulated root systems of VA mycorrhizal plants fixed more N_2 than nodulated root systems of non-mycorrhizal plants. This was related to an increase in nodule biomass in doubly infected plants. The plants may compensate for the cost of these symbionts by increasing rates of photosynthesis or by changing morphological characteristics to increase specific leaf area (Bayne *et al.*, 1984; Bildusas *et al.*, 1986; Harris *et al.*, 1985).

Greater amounts of C are transferred to roots with ectomycorrhizae compared with non-mycorrhizal plants (12–22% total fixed C in pine) and this is also partially compensated for by increased rates of photosynthesis (Ried *et al.*, 1983). It has been estimated for Douglas fir that ectomycorrhizae comprise 6% of the total tree biomass and, that for a range of host trees, ectomycorrhizae have a biomass between 0.4–10.0 t C ha^{-1}, most of which must have come from the host (Fogel, 1985; Fogel and Hunt, 1983).

With such large inputs of C to the roots, associated micro-organisms and rhizodeposition may seem wasteful but in ecological terms it may enable the plants to grow where they otherwise could not. Photosynthetic compensation can occur, and, in situations where lack of nutrients or environmental conditions limit growth, plants may have excess source potential (Herold, 1980). Thus, when root growth is limited, excess carbohydrate and reducing power can build-up and the roots switch on an 'overflow metabolism' via an alternative, cyanide-resistant respiratory pathway (Barneix *et al.*, 1984; Lambers, 1980).

Carbon budgets

We have indicated previously that whereas the substrate made available by roots to the microbial population of the rhizosphere is apparently very large, there is still a larger microbial biomass present than can be accounted for theoretically (Barber and Lynch 1977; Whipps and Lynch, 1983). The results of this analysis are given in Table 3. The basis of the calculation was to (1) grow rhizosphere bacteria in pure shaken culture in liquid media; (2) take a sample to estimate the cell density microscopically; (3) harvest and weigh the culture; and (4) assess the dry weight per unit cell. The total number of bacteria colonizing the roots was assessed from viable count measurements of the surrounding solution, sand or the roots themselves. The latter estimate was obtained by shaking the roots with glass beads in a surfactant to ensure by microscopic observation that the roots were free from bacterial colonization. This procedure does not allow for bacteria in the endorhizosphere and therefore, if anything, underestimates rather than overestimates the microbial biomass associated with the root.

It is difficult to be certain about the efficiency at which the bacterial biomass associated with roots is produced but it seems reasonable to assume a yield factor of *c.* 0.35 g dry weight g^{-1} substrate (Barber and Lynch, 1977). From this assumption the substrate required to generate the biomass can be calculated and compared with measured carbon flow using the method of continuous $^{14}CO_2$ labelling of the plant referred to above. In the first barley study (Barber and Lynch, 1977), the apparent substrate supplied was only 9% of that required, and in the subsequent barley and wheat studies (Whipps and Lynch, 1983), the figures were 25% and 10% respectively. Unfortunately, it is difficult to comment on the generality of these calculations be-

Table 3. Substrate input and microbial biomass of gnotobiotic cereal plants

	Barley[a]	Barley[b]	Wheat[b]
Microbial biomass			
Mean cell wt ($\times 10^{-12}$ g)	1.9	3.2	3.2
Bacterial biomass ($\times 10^{-4}$ g mg^{-1} dry root)	2.6	2.0	3.4
Substrate input			
Calculated ($\times 10^{-4}$ g C mg^{-1} dry root)	3.7	2.8	7.8
Measured ($\times 10^{-5}$ g C mg^{-1} dry root)	3.4	7.5	7.9

[a] Barber and Lynch (1977)
[b] Whipps and Lynch (1983)

cause no other studies have been undertaken. Certainly the lack of measurement of the endorhizosphere population and the lack of consideration of the maintenance energy (Pirt, 1975) of the microbial population makes the actual discrepancy in the calculation even greater.

The discrepancies could in part be caused by underestimation of substrate production, but this seems rather unlikely because the figures are consistent with those referred to above by several other investigators. It is likely therefore that the microbial biomass has been overestimated in some way. The most obvious interpretation is that the bacteria growing on roots are very much smaller than those produced in pure liquid culture; indeed Arthrobacter cells can change shape at low growth rates (Chapman and Gray, 1981). This concept was investigated using two strains of *Enterobacter cloacae* which colonized lettuce roots readily (Lynch, 1989). There is often a tacit assumption that all Gram negative bacteria associated with roots are pseudomonads but *Enterobacter* spp. are also important colonists, especially in the endorhizosphere (Kleeberger *et al.*, 1983). They are of particular interest because various strains have the capacity to fix N_2 (Neilson and Sparell, 1976), to produce soil stabilizing agents (Chapman and Lynch, 1985) and to protect against damping-off diseases (Nelson *et al.*, 1986). Despite extensive studies of these bacteria in various conditions of starvation and water stress (osmotic and matric) associated with roots and in isolation, one of us (Lynch, unpublished) has found no evidence for a significant change in cell dimensions by light or electron microscopy. However, closer examination of thin sections by electron microscopy indicated that starved cells which, although viable after one year, exhibited considerable plasmolysis both around the periphery of the cell and towards the core. Such cell morphology has been found in other ecosystems, particularly in oil wells (Lappin-Scott *et al.*, 1988). It is unrealistic experimentally to quantify this phenomenon because it can only be observed in the thin sections but it could contribute to the underestimation of microbial biomass. Another consequence of this study was the observation that the Enterobacter cells survived without any apparent substrate input. This may mean that our ideas of maintenance, which were largely derived for pure microbial cultures in chemostats (Pirt, 1975), are not readily applicable to organisms in natural ecosystems. For example, Anderson and Domsch (1985) showed that such low levels of carbon input were required to sustain soil biomass that maintenance energy requirements had to be much smaller than previously thought. Much of the population introduced to roots and that colonizing the rhizosphere from the soil might be sustained in a viable state for long periods in a state of *arrested metabolism*. This could be considered as a state close to *dormancy* while not exhibiting sporulation or other morphological features commonly associated with the dormant state. However, it would be dangerous to extrapolate these ideas too far because other phenomena such as *cryptic growth* and *oligotrophy* may occur. In batch culture, cryptic growth takes place as organisms utilize the cellular contents of both living and dead cells so there is a very protracted death phase (Postgate and Hunter, 1962). It is expected that this state could occur to an extent in the rhizosphere. Oligotrophs as opposed to copiotrophs appear to be able to sustain themselves on very low inputs of carbon which can be derived from the atmosphere (Lynch, 1989). Wainwright (1988) has reviewed the evidence of this occurring in fungi and has even suggested that a part of this carbon could be derived from fixation of CO_2 by chemolithoheterotrophy (organisms can grow as heterotrophs but gain additional energy by the simultaneous oxidation of reduced forms of elements). However, it seems most likely, in our rhizosphere studies, where roots have been forced-aerated that some carbon substrates for bacteria could be derived from traces of organic carbon in the air supply.

It is therefore difficult to produce satisfactory carbon budgets of the rhizosphere until more extensive and precise experimentation in the laboratory has been carried out. It is even more difficult to produce budgets of rhizospheres under field conditions. Lynch and Panting (1980a) have modified the fumigation-respiration technique of biomass determination (Jenkinson and Powlson, 1976) for use with instant cores of soil. Using this method, it was demonstrated that the soil biomass increased with increased rooting density, brought about by crop growth in the spring and by the greater proliferation of roots in the surface layers of direct-drilled crops as against those seeded after first ploughing the soil. Similarly, the increased root growth in grassland versus arable cropping in the same soil series caused a three-fold increase in soil microbial biomass (Lynch and Panting, 1980b). However, this information can only be obtained by using intact cores of soil because sieving decreases the soil biomass in winter and increases it in summer (Lynch and Panting, 1981). Sieving breaks up roots and allows a rapid proliferation of bacteria on the readily assimilable substrates so released. Although this concept has been questioned by Jenkinson and Powlson (1980), homogenization of soils by sieving must disrupt the soil ecophysiology to a considerably greater degree than the passage of the plough through the soil which has been shown by others to modify the size of soil biomass (Doran, 1987). The limitation therefore in attempting to produce energy budgets in the rhizosphere of field soils seems to be one of reliability of the methodology used. Far from being a problem in estimates of biomass (Martens, 1985; Martin and Foster, 1985; Merckx and Martin, 1987; Sparling *et al.*, 1985) the presence of roots in soil should be a stimulus to the understanding of substrate flowing in the soil. With suitable methodology, it should also prove possible to develop dynamic models of microbial activity in the rhizosphere (Bazin *et al.*, 1990).

Products

As indicated above, the yield of micro-organisms utilizing the products of rhizodeposition can be assumed to be *c.* 0.35 g dry weight g^{-1} substrate consumed. A proportion of remaining substrate consumed will appear as CO_2 from microbial respiration and another as microbial products. A further proportion of these microbial products will be consumed by other microorganisms or be absorbed by the root system, but some products will be recalcitrant and have long half-lives in the soil/plant ecosystem. The range of products has been reviewed (Lynch, 1976; 1990) and the categories of products are summarised in Table 4.

Some of the products formed as primary metabolites make a substantial contribution to the carbon budget in the rhizosphere. Other secondary metabolites are formed in low concentrations whereas they may not make a major direct impact of the carbon budget, they may regulate rhizodeposition.

Methods have only been developed relatively recently to definitively characterize the categories of products. However, it should now be possible to carry out experiments to determine the dynamics of product formation and the effect of environmental and chemical perturbations on steady-state processes which exist. Such analyses are available for pure cultures of micro-organisms (Pirt, 1971) and the complexity

Table 4. Categories of microbial products in the rhizosphere

Class of compound	Function
(*a*) *Likely absorbed by roots*	
Plant growth regulators	*Potentially* growth stimulatory at low concentrations but inhibitory in higher concentrations
Organic acids	Phytotoxic
H_2S, HCN	Phytotoxic
Antibiotics	Disease control but phytotoxic if absorbed
Ionophores	Promotion of nutrient uptake and possible disease control by making Fe unavailable to pathogens
(*b*) *Likely bound by roots*	
Lectins/Agglutinins (Proteins)	Promote specific binding of micro-organisms to roots and other micro-organisms
Polysaccharides	Non-specific microbial binding, stabilization of soil structures
Free enzymes	Various – oxidoreductases, transferases, hydrolases, lysases including involvement in nutrient cycling, pathogenesis and disease control

of mixed cultures in a complex environment is obviously far more of a challenge. In the first instance, it would seem sensible to develop experimental procedures based on microcosms and increase the complexity of experiment thereafter.

A major opportunity is to change the quality and quantity of chemicals produced in the rhizosphere. This could be done either by inoculating the roots with organisms synthesising such compounds or by genetic modification of known rhizosphere component microorganisms. Direct chemical treatment or genetic modification of the plant are also possibilities. However, genetic manipulation should only be pursued in detail after the ecophysiology of the rhizosphere ecosystem is better understood. Only then will the molecular biologists have more precise targets on which to use their skills.

Conclusions

Central to the understanding of substrate flow in the rhizosphere is the analysis of rhizodeposition. A substantial amount of information has been accumulated especially from the continuous ^{14}C labelling experiments. However, most of the studies were carried out with young plants and there is a need to carry the analyses through to maturity. There is also a need to extend the continuous labelling experiments to host/symbiont and host/pathogen interactions. Unfortunately, with the possible exception of the excellent facilities which are retained for these studies in Wageningen, there is very little current experimental work being carried out internationally in this field. The emphasis of rhizosphere studies has shifted strongly towards rhizosphere manipulation and whereas this can be fully justified in strategic and applied science areas, a better understanding of the underlying physiological energetics involved would provide a much better framework on which to base manipulation.

References

Anderson T-H and Domsch K H 1985 Determination of ecophysiological maintenance requirements of soil microorganisms in a dormant state. Biol. Fertil. Soils 1, 81–89.

Ares J 1976 Dynamics of the root system of blue grama (*Bouteloua gracilis* (H.S.K.) Lag.). J. Range Manage. 29, 208–213.

Barber D A and Lynch J M 1977 Microbial growth in the rhizosphere. Soil Biol. Biochem. 9, 305–308.

Barber D A and Martin J K 1976 The release of organic substances by cereal roots in soil. New Phytol. 76, 69–80.

Barber S A 1971 Effect of tillage practice on corn (*Zea mays* L.) root distribution and morphology. Agron. J. 63, 724–726.

Barneix A J, Breteler H and Van de Geijn S C 1984 Gas and ion exchanges in wheat roots after nitrogen supply. Physiol. Plant 61, 357–362.

Bayne H G, Brown M S and Bethlenfalvay G J 1984 Defoliation effects on mycorrhizal colonization, nitrogen fixation and photosynthesis in the *Glycine-Glomus-Rhizobium* symbiosis. Physiol. Plant. 62, 576–580.

Bazin M J, Markham P, Scott E and Lynch J M 1990 Microbial interactions in the rhizosphere. *In* The Rhizosphere. Ed. J. M Lynch. pp 99–127. John Wiley, Chichester.

Beck S M and Gilmour C M 1983 Role of wheat root exudates in associative nitrogen fixation. Soil. Biol. Biochem. 15, 33–38.

Bildusas I J, Dixon R K, Pfleger F L and Stewart E L 1986 Growth, nutrition and gas exchange of *Bromus inermis* inoculated with *Glomus fasciculatum*. New Phytol. 102, 303–311.

Bottner P, Sallih Z and Billies G 1988 Root activity and carbon metabolism in soils. Biol. Fertil. Soils 7, 71–78.

Chapman S J and Gray T R G 1981 Endogenous metabolism and macromolecular composition of *Arthobacter globiformis*. Soil Biol. Biochem. 13, 11–18.

Chapman S J and Lynch J M 1985 Some properties of polsaccharides of microorganisms from degraded straw. Enzyme Microb. Technol. 7, 161–163.

Doran J W 1987 Microbial biomass and mineralizable nitrogen distributions in no-tillage and plowed soils. Biol. Fertil. Soils 5, 68–75.

Dormaar J F and Sauerbeck D R 1983 Seasonal effects of photoassimilated carbon-14 in the root system of blue grama and associated soil organic matter. Soil Biol. Biochem. 15, 475–479.

Fogel R 1985 Roots as primary producers in below-grown ecosystems. *In* Ecological Interactions in Soil. Plants, Microbes and Animals. Eds. A H Fitter, D Atkinson, D J Read and M B Usher. pp 23–26. Blackwell Scientific Publications, Oxford.

Fogel R and Hunt G 1983 Contribution of mycorrhizae and soil fungi to nutrient cycling in a Douglas-fir ecosystem. Can. J. For. Res. 12, 219–232.

Graham J T , Leonard R T and Menge J A 1981 Membrane-mediated decrease in root exudation responsible for phosphorus inhibition of vesicular-arbuscular mycorrhiza formation. Plant Physiol. 68, 548–552.

Gregory P J, McGowan M, Biscoe P V and Hunt B 1978 Water relations of winter wheat. 1. Growth of the root system. J. Agric. Sci. 91, 91–102.

Hale M G and Moore L D 1979 Factors affecting root exudation. II: 1970–1978. Adv. Agron. 31, 93–124.

Harris D, Pacovsky R S and Paul E A 1985 Carbon economy of soybean-*Rhizobium-Glomus* association. New Phytol. 101, 427–440.

Helal H M and Sauerbeck D R 1983 Method to study turnover processes in soil layers of different proximity to roots. Soil Biol. Biochem. 15, 223–225.

Helal H M and Sauerbeck D R 1986 Effect of plant roots on carbon metabolism of soil microbial biomass. Z. Pflanzenernaehr. Bodenkd. 149, 181–188.

Herold A 1980 Regulation of photosynthesis by sink activity – the missing link. New Phytol. 86, 131–144.

Jenkinson D S and Powlson D S 1976 The effects of biocidal treatments on metabolism in soil. V. A method for measuring soil biomass. Soil Biol. Biochem. 8, 209–213.

Jenkinson D S and Powlson D S 1980 Measurement of microbial biomass in intact soil cores and in sieved soil. Soil Biol. Biochem. 12, 579–581.

Johnen B G and Sauerbeck D R 1977 A tracer technique for measuring growth, mass and microbial breakdown of plant roots during vegetation. *In* Soil Organisms as Components of Ecosystems. Eds. V Lohm and T Persson. Proc. VI Int. Soil. Zoo. Coll. Ecol. Bull. Stockholm 25, 366–373.

Keith H, Oades J M and Martin J K 1986 Input of carbon to soil from wheat plants. Soil Biol. Biochem. 18, 445–449.

Kleeberger A, Castorph H and Klingmuller W 1983 The rhizosphere microflora of wheat and barley with special reference to Gram negative bacteria. Arch. Microbiol. 136, 306–311.

Koch K E and Johnson C R 1984 Photosynthate partitioning in split-root citrus seedlings with mycorrhizal and non-mycorrhizal root systems. Plant Physiol. 75, 26–30.

Kraffczyk I, Trolldenier G and Beringer H 1984 Soluble root exudates of maize: influence of potassium supply and rhizosphere micro-organisms. Soil Biol. Biochem. 16, 315–322.

Kucera C L, Dahlman R C and Koelling M R 1967 Total net productivity and turnover on an energy basis for tallgrass prairie. Ecology 48, 536–541.

Lambers H 1980 The physiological significance of cyanide-resistant respiration in higher plants. Plant Cell Environ. 3, 293–302.

Lambers H 1987 Growth, respiration, exudation and symbiotic associations: the fate of carbon translocated to the roots. *In* Root Development and Function. Eds. P J Gregory, J V Lake and D A Rose. pp 125–145. Cambridge University Press, Cambridge.

Lappin-Scott H M, Cusack F, Macleod A and Costerton J W 1988 Starvation and nutrient resuscitation of *Klebsiella pneumoniae* isolated from oil well waters. J. Appl. Bacteriol. 64, 541–549.

Lee K J and Gaskins M H 1982 Increased root exudation of ^{14}C-compounds by sorghum seedlings inoculated with nitrogen-fixing bacteria. Plant and Soil 69, 391–399.

Lynch J M 1976 Products of soil micro-organisms in relation to plant growth. CRC Crit. Revs Microbiol. 5, 67–107.

Lynch J M 1986 Rhizosphere microbiology and its manipulation. Biol. Agric. Hort. 3, 143–152.

Lynch J M 1989 Development and interactions between microbial communities on the root surface. *In* Interrelations Between Micro-organisms and Plants in Soil. Eds. V Vančura and F Kunc. pp 5–12. Academic, Prague and Elsevier, Amsterdam.

Lynch J M 1990 Microbial metabolites. *In* The Rhizosphere. Ed. J M Lynch. pp 177–206. John Wiley, Chichester.

Lynch J M and Panting L M 1980a Cultivation and the soil biomass. Soil Biol. Biochem. 12, 29–33.

Lynch J M and Panting L M 1980b Variations in the size of the soil biomass. Soil Biol. Biochem. 12, 547–550.

Lynch J M and Panting L M 1981 Measurement of the microbial biomass in intact cores of soil. Microbial Ecol. 7, 229–234.

Martens R 1985 Limitations in the application of the fumigation technique for biomass estimations in amended soils. Soil Biol. Biochem. 17, 57–63.

Martin J K and Foster R C 1985 A model system for studying the biochemistry and biology of the root-soil interface. Soil Biol. Biochem. 17, 261–269.

Merckz R and Martin J K 1987 Extraction of microbial biomass components from rhizosphere soils. Soil Biol. Biochem. 19, 371–376.

Merckz R, Den Hartog A and Van Veen J A 1985 Turnover of root-derived material and related microbial biomass formation in soils of different texture. Soil Biol. Biochem. 17, 565–569.

Minchin F R, Summerfield R J, Hadley P, Roberts E H and Rawsthorne S 1981 Carbon and nitrogen nutrition of nodulated roots of grain legumes. Plant Cell. Environ. 4, 5–26.

Moore R and McClelen C E 1983 A morphometric analysis of cellular differentiation in the root cap of *Zea mays*. Am. J. Bot. 70, 611–617.

Nelson A H and Sparell L 1976 Acetylene reduction (nitrogen fixation) by Enterobacteriacae isolated from paper mill process waters. Appl. Environ. Microbiol. 32, 197–205.

Nelson E B, Chao W-L, Norton J M, Nash G T and Harman G E 1986 Attachment of *Enterobacter cloacae* to *Pythium ultimum* hyphae: possible role in the biological control of *Phythium* pre-emergence damping-off. Phytopathology 76, 327–335.

Newman E I 1985 The rhizosphere: carbon sources and microbial populations. *In* Ecological Interactions in Soil: Plants, Microbes and Animals. Eds. A H Fitter, D Atkinson, D J Read and M B Usher. pp 107–121. Blackwell Scientific Publications, Oxford.

Pate J S, Layzell D B and Atkins C A 1979 Economy of carbon and nitrogen in a nodulated and non-nodulated (NO_3-grown) legume. Plant Physiol. 64, 1083–1088.

Pirt S J 1975 Principles of Microbe and Cell Cultivation. Blackwell Scientific Publications, Oxford, 274 p.

Postgate J R and Hunter J R 1962 The survival of starved bacteria. J. Gen. Microbiol. 29, 233–263.

Reid C P P, Kidd F A and Ekwebelam S A 1983 Nitrogen nutrition, photosynthesis and carbon allocation in ectomycorrhizal pine. Plant and Soil 71, 415–431.

Ryle G J A, Arnott R A, Powell C E and Gordon A J 1983 Comparisons of the respiratory effluxes of nodules and roots in six temperate legumes. Ann. Bot. 52, 469–477.

Ryle G J A, Powell C E and Gordon A J 1979 The respiratory costs of nitrogen fixation in soyabean, cowpea and white clover. II. Comparisons of the cost of nitrogen fixation and the utilization of combined nitrogen. J. Exp. Bot. 30, 145–153.

Sallih Z and Bottner P 1988 Effect of wheat (*Triticum*

aestivum) roots on mineralization rates of soil organic matter. Biol. Fertil. Soils 7, 67–70.

Santantonio D 1979 Seasonal dynamics of fine roots in mature stands of Douglas-fir of different water regimes: a preliminary report: *In* Root Physiology and Symbiosis. Eds. A Reidacker and J Gagnaie-Michard. pp 90–203. Proceedings IUFRO Symposium on Root Physiology and Symbiosis, Nancy France, 1978, Vol. 6. CNRF Champenoux, France.

Schönwitz R and Ziegler H 1982 Exudation of water-soluble vitamins and of some carbohydrates by intact roots of maize seedlings (*Zea mays* L.) into a mineral nutrient solution. Z Pflanzenphysiol. 107, 7–14.

Sims P L and Singh J S 1971 Herbage dynamics and net primary production in certain ungrazed and grazed grasslands in North America. *In* Preliminary Analysis of Structure and Function in Grasslands. Ed. N R French. pp 59–124. Range Science Department Science Series No. 10. Colorado State University, Fort Collins, Colorado.

Singh J S and Coleman D C 1974 Distribution of photo-assimulated ^{14}carbon in the root system of a shortgrass prairie. J. Ecol. 62, 359–365.

Sivakumar M V K, Taylor H M and Shaw R H 1977 Top and root relations of field-grown soybeans. Agron. J. 69, 470–473.

Snellgrove R C, Splittstoesser W E, Stribley D P and Tinker P B 1982 The distribution of carbon and the demand of the fungal symbiont in leek plants with vesicular-arbuscular mycorrhizas. New Phytol. 92, 75–87.

Sparling G P, West A W and Whale K N 1985 Interference from plant roots in the estimation of soil microbial ATP, C, N and P. Soil Biol. Biochem. 17, 275–278.

Wainwright M 1988 Metabolic diversity of fungi in relation to growth and mineral cycling in soil – a review. Trans. Br. mycol. Soc. 90, 159–170.

Warembourg F R and Paul E A 1977 Seasonal transfers of assimilated ^{14}C in grassland: plant production and turnover, soil and plant respiration. Soil Biol. Biochem. 9, 295–301.

Whipps J M 1984 Environmental factors affecting the loss of carbon from the roots of wheat and barley seedlings. J. Exp. Bot. 35, 767–773.

Whipps J M 1985 Effect of CO_2 concentration on growth, carbon distribution and loss of carbon from the roots of maize. J. Exp. Bot. 36, 644–651.

Whipps J M 1987 Carbon loss from the roots of tomato and pea seedlings grown in soil. Plant and Soil 103, 95–100.

Whipps J M 1990 Carbon economy. *In* The Rhizosphere. Ed. J M Lynch. pp 59–97. John Wiley, Chichester.

Whipps J M and Lynch J M 1983 Substrate flow and utilization in the rhizosphere of cereals. New Phytol. 95, 605–623.

Whipps J M and Lynch J M 1985 Energy losses by the plant in rhizodeposition. Ann. Proc. Phytochem. Soc. Eur. 26, 59–71.

Whipps J M and Lynch J M 1986 The influence of the rhizosphere on crop productivity. Adv. Microb. Ecol. 9, 187–244.

D. L. Keister and P. B. Cregan (Eds.), The rhizosphere and plant growth, 25–32.

PLSO BARC03

Microbial dynamics in the rhizosphere: Possible strategies in managing rhizosphere populations

GLYNN D. BOWEN
Joint FAO/IAEA Division, International Atomic Energy Agency P.O. Box 100, Vienna, Austria

Key words: competition, microbial dynamics, microbial probes, migration, recognition, rhizosphere

Abstract

Components of growth and population composition of micro-organisms in the rhizosphere are examined. The value of experimental approaches *e.g.* using single gene mutants is indicated in examining individual components of growth, migration, and competition. Flagella and agglutinin properties may be of minor importance, although some rhizosphere bacteria may have special recognition relations with the root. The use of microbial probes is discussed.

Introduction

The benefits to be derived from micro-organisms, be they infective (such as Rhizobium, *Frankia* or mycorrhizal fungi) or non-infective (such as plant growth promoting bacteria and bio-control organisms) rest on the successful establishement of the organisms in the rhizosphere. This usually must be achieved in non-sterile soil and often under non-ideal environmental conditions *e.g.* highly acid soils, low soil moisture *etc.* Soil conditions, plant and micro-organisms can all be managed but by far the simplest is management of the plant and/or the micro-organism. This paper examines therefore what we know about the determinants of microbial growth and composition of the rhizosphere and seeks to define the characteristics we could possibly select for (or 'engineer' for) in micro-organisms and plants. A large number of tools are now available for the study of rhizosphere microbial ecology, an exciting field.

Most of us intuitively focus on inoculation of a particular micro-organism (of a suite of micro-organisms). However, it may be equally important to manage agricultural systems to increase the population of existing beneficial organisms *e.g.* vesicular arbuscular mycorrhizal fungi or to ensure that the population of an introduced micro-organism stays high from season to season.

Microbial substrates in the rhizosphere

Rovira *et al.* (1983) listed five main sources of rhizosphere substrates ranging from exudates and secretion from intact cells (in a strict biophysical sense the status of any particular substrate in these two categories has rarely been examined) to mucilages of plant origin, to lysates from senescing cells. While there has been emphasis on losses of exudates/secretions from the apical few centimeters of the root, this may be quantitatively misleading: (i) The distribution of translocated assimilate along the root, some of which is reflected in exudates or secretions, differs markedly with soil temperature. In wheat only 30% and 46% of ^{14}C assimilates occurred in the apical 3 cm at 5°C and 10°C respectively, compared with 64% at 20°C (Rovira and Bowen, 1973). It is probable that exudation from older parts of root may be significant under many field soil temperatures. (ii) Lysates coming from senescing cells, particularly in the more mature parts of roots (but by no means restricted to them, Henry and Deacon, 1981) can be significant soures of substrate for microbial growth.

Bowen and Rovira (1973) found the growth dynamics of the soil microflora on the 9–10 cm portion of washed, sterile roots, exposed to soil micro-organisms for the first time, were almost identical to those in the apical cm of root and hypothesized that lysates form much of the substrate available to micro-organisms in the older parts of the root.

There has been considerable concern about the effects on plant growth of the large losses of substrates from roots *e.g.* up to some 10% of the photosynthate under sterile conditions and up to some 20–30% under non-sterile conditions and up to some 20–30% under non-sterile conditions (see Rovira *et al.*, 1983). However it is critical to examine the sources of the losses: if they are largely from intact cells (*e.g.* exudates) that may be of importance, but if the main loss is as lysates from senescing cells and sloughed root material (and from root turnover) it may be of little real consequence for plant growth. However, varietal differences may be important.

Microbial probes

Historically, there has been a tendency to collect root 'exudates', analyze them and (sometimes) speculate on their capacity to support microbial growth. Such studies may be useful in relating changes in root exudation to the physiology of the plant and in predicting the effect of an environmental factor on microbial populations. However, a *microbial probe* approach may be just as effective. If we wish to relate losses to microbial growth why not measure microbial growth? *e.g.* one could inoculate a root uniformly with a known general population and study its growth dynamics. Experimental study of microbial growth on different parts of the root (Bowen and Rovira, 1973; Bowen, 1979) indicated the amount of substrate for microbial growth at different sites along the root. However, a limitation is that some substrates will diffuse into soil, the extent depending on the soil and its moisture. It may be interesting to compare the exudate analysis approach with the microbial probe approach.

Examinations of microbial growth curves on roots in soil indicate a rapid logarithmic increase in microbial population over the first 3–4 days followed by a levelling of the curve. The rapid phase has been equated (Bowen, 1979) with an intensity factor *i.e.* the richness of the substrate, and the later period has been equated with the carrying capacity of the system, *i.e.* the amount of energy able to support the population which has developed. In fact the second phase is not stationary but shows a slow rise due to growth of micro-colonies along the root. Generation times in the slow phase may be 100 hrs or more, but only some 5–6 hrs in the rapid phase.

Microbial probes for detecting the location of *specific* factors with great resolution were first employed by Frenzel (1960) to determine sites of loss of particular amino acids along roots; roots were inoculated with amino acid requiring mutants of *Neurospora crassa* and the sites of spore germination were observed. Studies of sites of growth of obligate anaerobic organisms could be potentially useful in defining anaerobic sites on roots and so on. All that is needed is that one can recognize the test organism microscopically (microsite analysis on the root) *e.g.* by fluorescence antibody techniques or can isolate it from particular segments of roots on selective media.

Microbial probes have recognized specific basidiospore germination factors (Theodorou and Bowen, 1986) and the localization of exudation largely in junctions along roots (Bowen, 1979). The inability of two out of five strains of an ectomycorrhizal fungus to colonize sterile roots in soil at 16°C, while all five strains colonized equally well at 20°C (Theodorou and Bowen, 1971) indicates changes in specific root substrate losses, probably undetectable by routine analyses.

Substrates and microbial composition of the rhizosphere

Four basic considerations arise in relating substrate loss to microbial growth and composition in the rhizosphere.

(i) *Substrates used by a broad range of micro-organisms.* These substances are the ones usually assayed for in general root exudate studies. Micro-organisms with rapid growth rates (and with short response times) and with rapid germination of propagules (fungi) have an obvious advantage in using these for colonization. However, subtle differences in substrate composition between closely related plant species and changes in growth conditions may have large effects on the relative growth rates of even closly related bacterial strains, *e.g.* growth of three strains of *Pseudomonas fluorescens* differed 100 fold on roots of *Eucalyptus seiberi*, the ranking of which was changed on roots of *E. globus* (Bowen, 1979). However, while fast growth rate is an important factor, the ability to capitalize on these in competition with other soil-organisms is very dependent on proximity to the source of the substrate (Bowen, 1979; 1980).

(ii) *Specific substrates little used by other micro-organisms* obviously gives organisms a high competitive advantage. It would therefore be a highly appropriate strategy to engineer plants to secrete substrates, specifically and uniquely used by the 'desired' organisms. This is a strategy demanding much more attention. The selective stimulation of some rhizobia by some isoflavonoids (Lameta and Jay, 1987) may be an example of this.

(iii) *Localized enhancement of loss.* There would be very considerable competitive advantage to to an organism, especially a slow grower or a late arrival, to locally increase the permeability of the host cell plasmalemma, thus giving a specialized 'pipeline' to the plant. For example, indole acetic acid (IAA) increases plasmalemma permeability and is produced by some ectotrophic mycorrhizal fungi. This hypothesis appears not to have been experimentally tested, although this is possible using a ^{14}C microautoradiographic approach or, more specifically, by comparing the growth of auxin-positive and auxin-negative mutants, a approach little used so far. However, Loper and Schroth (1986) showed no differences in colonization of sugar beet roots by IAA + ve and IAA − ve strains of *P. syringae* pv. *savastonoi*. The roots were grown in the absence of microbial competition but nevertheless one might have expected some increases in colonization by the IAA + ve strains. There are, of course, many other possible compounds with an effect on plant permeability.

(iv) *Translocation.* Organisms, such as translocating fungi, once established at a position, have great advantages in avoiding competition both for energy sources and associated spaces, whereas bacteria grow preferentially along the junctions of cells, this impeding the movement of other bacterial on the root.

Recognition

Chao *et al.* (1988) found 30% of bacteria from non-rhizosphere soil showed positive agglutination to toward pea root compared with 82% of bacteria isolated from the rhizosphere. However, is agglutination an *essential* part of rhizosphere competence or are we dealing with a correlated character of rhizosphere micro-organisms? The agglutinin response, greatly conditioned by the ionic composition of the medium (James *et al.*, 1985) appears to be a very widely occuring property and with low specificity. Many tests of agglutination employ cell concentration of 10^5 – 10^8 CFU but the adsorption is 3–4 orders of magnitude lower than this, *i.e.* 10–100 CFU absorbed per million CFU exposed to the root. One is forced to ask the ecological reality and importance of this. Anderson *et al.* (1988) examined root colonization by single gene mutants of *P. putida* differing in agglutination ability. The results, indicating some 40 to 60×10^5 CFU/g root 8 days after an initial inoculation of 2×10^3 CFU/used (Table 1), suggests (to me) little real role of the agglutination reaction as *primary* factor in root colonization.

However, I suggest that some non-invasive organisms may well have a special recognition relationship of which lectin/agglutinin reactions are only one phase. Dazzo *et al.*, (1984) distinguished between a non-specific attachment be-

Table 1. Colonization of bean roots by *Pseudomonas putida* with different agglutinability (from Anderson *et al.*, 1988)

	Population CFU × 10^5/g root[a]	
	Around root	Firmly adhering
Agglutinin-positive	~60	(5.5)
Agglutinin-negative	~40	(1.0)
Agglutinin-enhanced	~50	(1.4, 3.4)

[a] Population after 8 days in moist vermiculite. Seed was inoculated with 2×10^3 CFU per seed. Data indicate growth around the root while the data in parenthesis indicate the population firmly adhering to the root.

tween Rhizobium and clover root hairs (and root hairs of heterologous legumes) and symbiotic specificity accompanied by extracellular microfibrils formed 4–12 hr after initial confrontation ("phase 1C, phase 2 adhesion"), attachment/recognition processes probably precede this. Similar microfibrilar attachments are also observed with several rhizosphere micro-organisms (Foster and Bowen, 1981). In rhizobia there is a change of host/bacteria physiology (Dazzo *et al.*, 1988), including increases in porosity of root hair walls, and alteration of plasmalemma and of proton flux. Non-homologous bacteria can also cause somewhat similar change (Dazzo *et al.*, 1988; Rolfe *et al.*, 1988) but differing in intensity. Some of these changes may be antibiotic in nature *e.g.* phenols (Rolfe *et al.*, 1988). It would be interesting to study if such non-homologous, non-infecting rhizobia proliferate in the rhizosphere *i.e.* become 'rhizosphere' micro-organisms. It is a strong possibility that a number of rhizosphere bacteria can change the host root metabolism by similar recognition factors and substrates for the micro-organisms, without subsequent infection.

Colonization of new root surfaces

Many micro-organisms will completely colonize the root in sterile soil but in non sterile soil colonization is often severely restricted, *e.g.* with ectomycorrhizal fungi (Bowen, 1979) and with *Enterobacter cloacae* (Chao *et al.*, 1986).

Two cardinal factors in colonization of new root surfaces are involved: (i) Conditions for movement of micro-organisms along roots and their possession of new sites before their occupancy by organisms in the soil. and (ii) Interactions with other micro-organisms.

Usually, but not always, there will be a need to colonize a significant percentage of the root system. The main exception is where the inoculated micro-organism exerts its major effect *via* young roots *e.g.* growth promoting factors or protection of crown roots against disease organisms. Even here there are good reasons for establishment of organisms over larger parts of the root system, because of their impact on root colonization in subsequent seasons. Two examples of the importance of establishment along the root are given below:

(1) Despite some claims to the contrary, there is strong evidence that, except probably in soils with free water, movement of bacteria in soils is extremely limited (Bowen and Rovira, 1976). Fungi, because of their larger reserves, can move to the root over much larger distances and thus the soil population needed to produce high infection/root colonization is much lower (see Bowen, 1979; 1980). With small organisms such as biocontrol bacteria and rhizobia and some small progagule fungi used in biocontrol, *e.g.* Trichoderma the root must almost touch the microcolonies of the propagules in the soil. The opportunity of colonization is therefore very much a statistical matter and organisms giving high populations over the whole of a root system in one season, will be advantaged in colonizing the next season's roots significantly, provided they survive from season to season. There is increasing evidence, that rhizobia migrate very poorly along roots in non-sterile soils (Danso and Bowen, 1988). Thus, in a self sowing, pasture species the major concentration of inoculated rhizobia for nodulation in the second season will be restricted to the position of the first season's seeding. In the first season, however, poorly effective indigenous rhizobia may have formed nodules over large parts of the root, thus building up a high population throughout the soil which can dominate the next season's nodulation.

Where biocontrol of pathogens occurs in the *soil* phase as well as the rhizosphere, *e.g.* *Gauemannomyces graminis* var. *tritici* (Wildermuth *et al.*, 1979), high populations of the biocontrol organisms over the whole root system, and subsequent high numbers through soil, must optimize contact with the pathogen propagules.

Similarly, intermediate crops should be selected to increase natural populations of beneficial organisms, *e.g.* to increase populations of vesicular arbuscular mycorrhizal fungi.

(ii) Recent studies on Rhizobium migration suggest that the restriction of nodulation to the crown/top root in non-sterile soil may be little more than a reflection of poor migration of rhizobia along the root. Hardarson *et al.*, (1989) found that when soybean was inoculated to a non-sterile soil, 85% of the nodules were in the surface 5 cm, in the tap root and crown area. By contrast, when the soil was uniformly inoculated only 28% were in that area and there was a four fold increase in nitrogen fixation. This is a rich area for further research.

Movement along the root

Several modes of movement along the root, all capable of simple experimental resolution, have been suggested.

(i) *Motility*. As discussed by Howie *et al.*, (1987), at soil water matrices lower than approx. −0.5 bar, water filled pores are probably too small for flagellated bacteria to exhibit motility. Howie *et al.*, showed that flagella negative mutants of *P. fluorecens* colonized wheat roots as well as flagella positive even to 7–9 cm in soil.

(ii) *By downward flow of water*. Parke *et al.* (1986) experimentally showed a major effect of downward water movement along the roots in distributing micro-organisms to new sites for subsequent growth. With inoculation of wheat roots with *P. fluorescens* in soil at −0.1 MPa no growth could be detected 0.75 cm below the inoculation point after 24 hrs but when water was applied (=22 mm rainfall) there was significant movement, and growth on the root occurred 3.5 cm below the inoculation point within 24 hrs. Similarly, Chao *et al.*, (1986) found percolating water enhanced downward movement of bacteria and (small) fungal propagules. Where irrigation is available this could be used to enhance the rhizosphere colonization at depths by an inoculated organism. There is a great need for more studies to quantitatively relate water application to water flow and bacterial movement in sufficient quantity to establish themselves on the new root surface.

(iii) *Bacterial distribution by the growing root*. In a system with no downward water movement, Howie *et al.*, (1987) found colonization by introduced *P. fluorescens* to 9 cm below the seed in 6–7 days in (see also Parke *et al.*, 1986). They suggested a passive carrying and deposition of bacteria by elongating tips followed by their multiplication. Distribution of bacteria along the root could occur by colonization of the root cap, which is then sloughed as the root moves through soil, and by limited extension of the root (for a very small distance) behind the cap – often much less than a cm. It would be possible to separate the movement factor from subsequent growth with great precision by applying heavily radioactively labelled anion exchange resins or killed rhizobacteria (isotope labelled or using fluorescent antibody methods) to the root tip and following their subsequent movement. Movement on the root tip and subsequent growth will depend on the growth rate of the bacteria *e.g.* on the root cap, and the extension rate of that area. A rapidly growing root and slowly-growing bacteria would lead to a sharp profile of dilution, the extent of which will depend on the initial bacterial population. The reverse will occur with a slowly growing root and rapidly growing bacteria. Sloughed root cap cells may be especially important as these would provide an immediate and continuing source of nutrient for the associated bacteria; Vermeer and McCully (1982), found isolated corn root-cap cells living within the rhizosphere, far from the root tip.

(iv) *Translocation*. As indicated above, translocating fungi can spread along a root without having to compete with other organisms all along the root.

Interactions in the rhizosphere

Experimental studies in rhizosphere colonization have shown that interactions between rhizosphere bacteria on laboratory media can be quite misleading and that rhizosphere interactions are best studied in a plant/soil environment (Bowen, 1979; 1980). Experimental studies have shown that other organisms, including protozoa, can reduce growth of the test organism, have no effect or even stimulate them (Bowen and Theodorou, 1979; Chakraborty *et al.*, 1985) (Table 2). Stimulation has received far less study than depression.

It is important to assess when a depression of microbial growth is due to a *positive* process such as antibiotic production or production of factors giving the producing organisms an advantage *e.g.* siderophores, change of pH *etc.*, or to competition for substrate. Competition with the target organisms for substrates can often be dis-

Table 2. Rhizoplane interactions between mycorrhizal fungi, bacteria, and protozoa (Bowen and Theodorou, 1979, Chakraborty *et al.*, 1985)

Test fungi	Root length of *Pinus radiata* colonized by fungi (mm) after 4 weeks at 20°C				
	Added organisms				
	Exp. 1			Exp. 2	
	Control	*P. fluorescens*	*Bacillus* sp. 1	Control	*Bacillus* sp. 2
Rhizopogon luteolus	27.6*[a]	8.3[b]	23.7[a]	20.7[a]	34.1[c]
Corticum bicolor	17.2[a]	3.2[b]	15.3[a]	15.8[a]	27.1[c]
		Saccamoeba sp.		Unidentified leptomyxid	
R. luteolus	30.5[a]		5.3[b]	17.8[a]	

* Along rows, data followed by the same superscript are not significantly different.

tinguished from the positive processes above by comparing the effect of two closely related organisms, presumably with similar nutritional requirements. Bowen and Theodorou (1979) found two strains of *P. fluorescens* grew to the same extent in the rhizosphere but only one depressed the mycorrhizal test fungus. In studies by Thomashow and Weller (1988), a number of strains of *P. fluorescens* all colonized the rhizosphere to the same extent but ones which produced phenazine gave considerably more inhibition of *G. graminis* var. *tritici* (Table 3). The other pseudomonads however reduced the disease to some extent – competition or antagonism?

The success of a rhizosphere micro-organism would be enhanced by a rapid growth rate on a wide range common exudates, production of antibiotics or other advantage producing compounds, such as siderophores, and tolerance of such products formed by other micro-organisms. There appears to be wide variation in generation time of organisms in the rhizosphere, *e.g.* 5 hrs for some pseudomonads, 39 hours for some *Bacillus* spp. (Bowen and Rovira, 1976). Generation times for rhizobia on young roots appear to have been studied little. A long generation time may be a reason for poor migration in the face of competition. Root microbiologists have largely ignored the study of generation times but this may be an extremely important factor to screen for.

Conclusions

I have indicated some of the possible strategies in rhizosphere management: *plant* breeding/engineering to provide unique substrates specific for the introduced organism, *microbial* selection/engineering for rapid growth rates on a range of substrates, use of a specific substrate, local increase in the plasmalemma permeability, produc-

Table 3. Colonization and take-all control by strains of *P. fluorescens* (Thomashow and Weller, 1988)

Strain	Phenazine Production	Root disease Rating[b]	Colonization log CFU/cm root 3.5 cm	Colonization log CFU/cm root 5–7 cm
2.79 RN_{10}	+	4.89	2.97	2.55
2-79-21445	–	5.71	2.96	2.20
2-79-611	–	6.25	2.70	2.34
0		7.19		

[a] After 11–13 days in soil at 15–18°C. Seed inoculation $1–2 \times 10^8$ CFU/seed.
[b] All data in this column differ significantly. ($P = 0.05$)

tion and tolerance of antibodies or local change of environment (such as pH change on siderophore production), ready movement to new surfaces (possibly related to growth rate) and seasonal persistence, and *management*, including choice of fertilizer (*e.g.* $NO_3^- - NH_4^+$) choice of crop sequence and irrigation. A great deal has been achieved to now, more or less by *ad hoc* testing. However, it is appropriate to briefly look forward to a better development of the science of ecology of the rhizosphere.

First, I believe we must have a conceptual framework of the systems we are dealing with, recognizing spatial and temporal variations along the root and in soil. We need to recognize that colonization from soil, by pathogens or symbiotic organisms, has three phases: propagule germination, movement to the root and growth on the root in competition with microflora (and microfauna) on the root (Bowen, 1979, 1980).

Having recognized these separate components it is usually a simple matter to *experimentally* determine their relative importance, in an ecologically real situation *e.g.* plants growing in soil rather than organisms interacting in laboratory media. Initially it will help to have simplified systems *e.g.* using sterile soil indicates the *potential* of an organism to grow under various rhizosphere conditions (*e.g.* Theodorou and Bowen, 1971) but eventually these systems must be made more complex *e.g.* comparing performance in sterile and non-sterile soil and by addition of other (known) organisms. One must then examine, experimentally, the *mechanisms* for the factors which are shown to be important. This requires hypotheses which can be tested experimentally, *e.g.* the role of water flow in movement, the importance of agglutinin reactions, auxin and antibiotic production. The use of single gene mutants is invaluable in such studies.

All of this is in the category of experimental rhizosphere dynamics; it's development will be helped tremendously by a creative use of existing and emerging techniques. One such is the use of single gene mutants. The use of selective media has been proven in many studies, particularly with pseudomonads or double antibiotic resistant mutants of rhizobia. However for other organisms it may be much more practicable to use specific DNA probes (Tiedje, this volume) or monoclonal antibodies (Wright and Morton, 1989) possibly linked with ELISA techniques. Such methods will need to be quantitative and will be indispensable in measuring the rhizosphere population of important organisms which are difficult to culture quantitatively *e.g.* Frankia. It is possible that the use of restriction fragment length polymorphism (RFLP) technology will be able to distinguish correlated seets of genes important in root colonization under particular conditions, or high competitiveness in root infection *e.g.* by rhizobia but this seems further away at present.

In conclusion, I should point out that rhizosphere and soil microbiologists seem determined to ignore the vast reservoir of useful concepts developed in population biology in other fields. True, they aren't always directly applicable to microbial systems, but many analogous situations do occur. It is incredible that the power of generation time concepts have been almost ignored in rhizosphere microbiology yet it is a fundamentally more sound approach than R:S ratios (Bowen and Rovira, 1976). The relevance of the concept of r/K strategies to biocontrol is starting to be seen (Bowen, 1980). "Traditional" population ecology in other fields contains valuable analyses of migration coupled with multiplication, concepts of diversity/stability, and niche concepts/stability of ecological systems. All of these are applicable to rhizosphere population biology. Microbiology was to the fore in early experimental studies of population biology (*e.g.* Gause, 1934) but we seem to have lost the initiative. Therein also lies the challenge, and the opportunities, because microbial systems surely offer great experimental advantages over macro systems, especially with recently emerging techniques.

References

Anderson A J, Habibzadegah-Tari P and Tepper C S 1988 Molecular studies on the role of a root surface agglutinin in adherence and colonization by *Pseudomonas putida*. Appl. Environ. Microbiol. 54, 375–380.

Bowen G D 1979 Integrated and experimental approaches to the growth of organisms around roots. *In* Soil Borne Plant Pathogens. Eds. B Schippers and W. Gams. pp 209–227. Academic Press, London.

Bowen G D 1980 Misconceptions, concepts and approaches in rhizosphere biology. *In* Contemporary Microbial Ecology. Ed. D G Ellwood. pp 283–304. Academic Press, London.

Bowen G D and Rovira A D 1973 Are modelling approaches useful in rhizosphere biology? Bull. Ecol. Res. Comm. Stockholm 17, 443–450.

Bowen G D and Rovira A D 1976 Microbial colonization of plant roots Annu. Rev. Phytopathol. 14, 121–144.

Bowen G D and Theordorou C 1979 Interactions between bacteria and ectomycorrhizal fungi. Soil Biol. Biochem. 11, 119–126.

Chakraborty S, Theodorou C and Bowens G D 1985 The reduction of root colonziation by mycorrhizal fungi by mycopohagous amoebae. Can. J. Microbiol. 31, 295–297.

Chao E L, Nelson E B, Harman E E and Hoch H C 1986 Colonization of the rhizosphere by biological control agents applied to seeds. Phytopathology 76, 60–65.

Chao E L, Li R K and Chang W T 1988 Effect of root agglutinin on microbial activities in the rhizosphere. Appl. Environ. Microbiol. 54, 1838–1841.

Danso S K A and Bowen G D 1989 Methods of inoculation and their influence on nodulation pattern and nitrogen fixation. Soil Biol. Biochem. 21, 1053–1058.

Dazzo F B, Truchet G L, Sherwood J E, Hrabak F M, Abe M and Parkratz S H 1984 Specific phases of root hair attachment in the *Rhizobium trifolii* – clover symbiosis. Appl. Envir. Microbiol. 48, 1140–1150.

Dazzo F B, Hollingsworth R I and (16 others) 1988. Signal recognition responses in the *Rhizobium trifolii* – white clover symbiosis. *In* Molecular Genetics of Plant Microbe Interactions. Eds. R Palacios and D P S Verma. pp 35–40. APS Press, St. Paul.

Foster R C and Bowen G D 1981 Plant surfaces and bacterial growth. *In* Prokaryotic Pathogens. Eds. R Mount and C Lacey. vol. 1 pp 159–185. Academic Press, New York.

Frenzel B 1960 Zur Atiologie der Anreicherung von Aminosäuren and Amiden im Wurzelraum von *Helianthus annuus L.*: Ein Beitrag zu Klärung der Problem der Rhizosphere. Planta 55, 169–207.

Gause G F 1934 The Struggle for Existence. Hafner, New York.

Hardarson G, Gobls M and Danso S K A 1989 Effect of nodulation patterns on nitrogen fixation in soybean (*Glycine max* (L) Merrill). Soil Biol. Biochem. 21, 783–787.

Henry C M and Deacon J W 1981 Natural (non-pathogenic) death of the cortex of wheat and barley seminal roots as evidenced by nuclear staining with acridine orange. Plant and Soil 60, 255–274.

Howie W J, Cook R J and Weller D M 1987 Effects of soil matric potential and cell motility on wheat root colonization by fluorescent pseudomonads suppressive to take-all. Phytopathology 77, 286–292.

James D W, Suslow T V and Steinback K E 1985 Relationship between rapid firm adhesion and long term colonization of roots by bacteris. Appl. Environ. Microbiol. 50, 392–397.

Lameta A A and Jay M 1987 Study of soybean and lentil root exudates Plant and Soil 101, 267–272.

Loper J E and Shroth M N 1986 Influence of bacterial sources of indole-3-acetic acid on root elongation of sugar beet. Phytopathology 76, 386–389.

Parke J L, Moen R, Rovira A D and Bowen G D 1986 Soil water flow affects rhizosphere distributions of a seed born biological control agent, *Pseudomonas fluorescens*. Soil Biol. Biochem. 18, 583–88.

Rolfe B G, Bathley M, Redmond J W (and 7 others) 1988 Phenolic compounds secreted by legumes. *In* Nitrogen Fixation: Hundred Years After. Eds. H Bothe, F J de Bruin and W E Newton. pp 405–409. Gustav Fischer. Stuttgart.

Rovira A D and Bowen G D 1973 The influence of root temperature on ^{14}C assimilate profiles in wheat roots. Planta 114, 101–107.

Rovira A D, Bowen G D and Foster R C 1983 The nature of the rhizosphere and the influence of the rhizosphere microflora and mycorrhizas on plant nutrition. *In* Encyclopedia of Plant Physiology. New series. Eds. A Läuchli and R L Bieleske. Vol. 15A, pp 61–93 Springer-Verlag, Heidelberg.

Theodorou C and Bowen G D 1971 Influence of temperature on the mycorrhizal associations of *Pinus radiata*. Aust. J. Bot. 19, 13–20.

Theodorou C and Bowen G D 1986 Germination of basidiospores of mycorrhizal fungi in the rhizosphere of *Pinus radiata*, D Don. New Phytol. 106, 217–223.

Thomashow L S and Weller D M 1988 Role of a phenazine antibiotic from *Pseudomonas fluorescens* in biological control of *Gauemannom yces graminis* var. *tritici*. J. Bacteriol. 170, 3499–3508.

Vermeer J and McCully M E 1982 The rhizosphere in *Zea*: new insight into its structure and development. Planta 156, 45–61.

Wildermuth G B and Rovira A D and Warcup J H 1979 Mechanism and site of suppression of *Gaeumannomyces graminis* var. *tritici* in soil. *In* Soil-Borne Plant Pathogens. Eds. B Schippers and W Gams. pp 157–164. Academic Press, London.

Wright S F and Morton J B 1989 Detection of vesicular arbuscular mycorrhizal fungus colonization of roots by using a dot-immunoblot assay. Appl. Environ. Microbiol. 55, 761–763.

D. L. Keister and P. B. Cregan (Eds.), The rhizosphere and plant growth, 33–42.

PLSO BARC04

Root colonization by indigenous and introduced microorganisms

J.L. PARKE
Department of Plant Pathology, University of Wisconsin-Madison, Madison, WI 53706, USA

Key words: bacteria, fungi, population dynamics, quantitative recovery, rhizosphere competence, soil factors

Abstract

Root colonization is defined as the proliferation of microorganisms in, on, or around roots. It includes dispersal of microorganisms from a source of inoculum to the actively growing root, and multiplication or growth in the rhizosphere. Soil physical, chemical, and biological factors have been shown to affect root colonization, but few phenotypic attributes of plants and microorganisms which contribute to successful root colonization have been identified. Quantitative studies on the distributon of root colonists in time and space are needed to develop mathematical models that describe and predict the root colonization process. This would enable more effective management of rhizosphere populations to achieve biological control of soilborne disease or to enhance plant growth.

Introduction

Certain root-associated microorganisms increase nutrient availability, produce growth hormones, or deter root pathogens. For many of these beneficial effects to occur, the microorganism must first become established in or near the roots. Attempts to manage the rhizosphere microflora to achieve these beneficial effects often fail because the process of root colonization is poorly understood.

I will attempt to provide a general overview of root colonization by indigenous and introduced microorganisms, and to describe some attributes of plants, microorganisms, and the soil environment that appear to contribute to successful establishment of microbial populations in the rhizosphere. Finally I will consider some of the difficulties associated with quantitative studies on root colonization, and suggest some approaches for future research.

The root colonization process

Root colonization is defined here as the proliferation of microorganisms in, on, and around the growing root. For many microorganisms, the zone they occupy is a function of root age, so I will not attempt to distinguish between root colonists and rhizosphere colonists. Colonization begins with a dispersal phase, which can result either from growth or movement of the microorganism towards the root, or from contact of a stationary propagule by a growing root. This is followed by a growth phase, in which the organism multiplies or grows. Root colonization by rhizobia, Agrobacterium, *Frankia* and ectomycorrhizal fungi also involve host-specific recognition events that will not be addressed here.

As the root grows, cell division in the apical meristem gives rise to root cap cells and copious quantities of mucilage are produced. Root cap cells and mucilage are abraded by soil particles and are then utilized by microorganisms able to degrade cellulose and pectin (Foster, 1986). The root cap itself is generally sterile (Rovira and Campbell, 1974), probably because cell division and elongation result in root growth rates (>1 cm per day) too rapid for microbial colonization to occur. The zone of elongation is also relatively free of microorganisms, but the root hair zone is a site of intense microbial activity. Microcolonies of bacteria become embedded in the mucigel surrounding the root, in the grooves between epidermal cells and on the surface of the epidermal cells (Bennett and Lynch, 1981). Root hairs of legumes are infected by rhizobia and nodules form. Mycorrhizal fungi colonize the root cortex, forming arbuscules, vesicles, and extramatrical hyphae (vesicular-arbuscular mycorrhizae), or ensheathing short roots (ectomycorrhizae). Ageing epidermal and cortical cells become invaded by bacteria and fungi (Henry and Deacon, 1981). The root is a dynamic substrate, changing in space, as new cells are produced by the root apex, as well as in time, as differentiation of root tissues and senescence proceed.

Root colonization by indigenous microorganisms occurs when a root contacts bacteria and fungi associated with decaying organic matter and colloidal material scattered through the soil (Foster, 1986). As the root grows, it continues to contact additional inocula. In contrast, organisms introduced on seeds or other plant propagative materials must disperse along the growing root from a single source of inoculum, and colonize the root in competition with indigenous microorganisms already established there.

Dispersal in the rhizosphere

How do microorganisms colonize regions of new root growth? Bowen and Rovira (1976) postulated three means of dispersal for root-colonizing bacteria: motility, passive movement in a film of water on the root surface, and passive transport on the root apex or fungal hyphae. Few studies, however, have sought to provide experimental evidence for the mechanism of dispersal.

Howie *et al*. (1987) suggested that root colonization by introduced bacteria occurs by a process of passive transport of the bacteria on the root apex (Phase 1). As the bacteria multiply, some are left behind or shed along with root cap cells to serve as a source of inoculum for older sections of roots. The dispersal phase is then followed by proliferation of the bacteria where they have been deposited (Phase 2). Both motile and nonmotile strains of *Pseudomonas fluorescens* 2-79 were detected on wheat roots 7–9 cm below the coated seed at populations of 10^1 to 10^3 colony-forming units (cfu)/cm root in the absence of percolating water. From this, Howie *et al*. (1987) concluded that passive carriage with the root apex was the most likely mechanism of dispersal.

However, Chao *et al*. (1986) found that fungi and bacteria introduced on pea seeds including *Trichoderma harzianum*, *T. koningii*, *Gliocladium virens*, *Penicillium funiculosum*, *Enterobacter cloacae*, *Pseudomonas fluorescens*, and *P. putida*, failed to colonize roots more than 3 cm below the seed, but dispersal was increased by the application of water to the soil. Similarly, *Bradyrhizobium japonicum* and *P. putida* were restricted to a zone 0–2.7 cm below the seed on roots of soybean and snap bean unless water was allowed to infiltrate the soil (Madsen and Alexander, 1982). Parke *et al*. (1986) found that although *P. fluorescens* strain 2-79 was initially carried on the root tip, the bacterial population did not appear to keep pace with the root, and was left behind after 4 days. These studies suggest that passive carriage by the root is not always an effective means of dispersal.

In nearly all root colonization studies, the application of water to the root zone is not controlled. Thus, dispersal of the bacteria may result simply from redistribution of the inoculum by percolating water. Bahme and Schroth (1978) and Liddell and Parke (1989) both found that the spatial distribution of *P. fluorescens* in the rhizosphere was substantially different before and after irrigation.

The role of motility in root colonization by rhizosphere bacteria has not been resolved. In any case, motility is limited to soil moisture conditions that provide a continuity of water-filled pores that are at least as large as the bacterial cells (Griffin and Quail, 1968; Papendick and Campbell, 1981). Motile bacteria can apparently move considerable distances through wet soil towards seeds and roots. *Pseudomonas fluorescens* migrated 2 cm in 48 h towards imbibed soybean seeds (Scher *et al*., 1985) and *Azospirillum brasiliense* migrated up to 4.5 cm in 72 h towards wheat roots (Bashan, 1986).

The use of non-motile mutants has been helpful in determining the importance of motility to root colonization and nodulation. Co-inoculation of roots with equal numbers of wild-type motile and non-motile mutants of Rhizobium indicated that motile strains form a disproportionate number of nodules (Mellor *et al.*, 1987). Tn*5*-induced non-motile mutants of Pseudomonas were impaired in their ability to move towards seed (Scher *et al.*, 1985) or colonize roots (De Weger *et al.*, 1987) when compared to their wild-type parent strains.

A problem with many of these motility studies is that soil matric potential was not carefully controlled to ensure that soil drainage, and thus passive movement of the bacteria, did not occur. The addition of liquid inoculum to soil in which moisture is controlled gravimetrically can result in rapid redistribution of soil water and microorganisms. Finally, in experiments involving non-motile mutants, it is important to determine that differences in migration are not due to pleitropic effects on growth rate, antibiotic production, or attachment.

The importance of bacterial motility to root colonization mediated by percolating water, similar to what would happen during rain events or irrigation, has not been evaluated. Lawrence *et al.* (1987) speculated that motility of *P. fluorescens* is important in overcoming attractions between bacteria and solid surfaces, and may be necessary for the bacteria to detach from roots or soil particles and to enter into the lanes of fast-moving water beyond the quiescent boundary layer. A similar phenomenon was described by Duniway and McKeen (1987) for zoospores of Phytophthora.

Taken together, these studies demonstrate that passive transport on the root apex, passive movement with percolating soil water, and motility all may contribute to dispersal of rhizosphere microorganisms. The relative contribution of each component may depend on variables including the rate of root growth, soil texture and moisture characteristics, and attributes of the microorganisms such as growth rate, motility, chemotaxis, and cell surface properties.

Factors affecting root colonization: the soil environment, the microorganism and the plant

The soil environment

Soil matric potential

Soil matric potentials reported as optimal for root colonization by Pseudomonas range from −1 kPa to −140 kPa (−0.01 to −1.4 bars) (Howie *et al.*, 1987; Liddell and Parke, 1990; Parke *et al.*, 1984). This variation may reflect differences in the metabolic requirements of these strains, or in the ability of the bacteria to compete with indigenous microorganisms in the different soils (Parke *et al.*, 1984). At high soil matric potentials when a large percentage of pore space is water-filled, oxygen may be inadequate, whereas at low soil matric potential, motility, cell turgor, and diffusion of nutrients could be reduced (Howie *et al.*, 1987).

Soil texture

Soil texture influences both passive transport (Bitton *et al.*, 1974) and root colonization (Bahme and Schroth, 1987) by bacteria. In general, bacteria move greater distances in coarse-textured soils than in fine-textured soils (Wilkinson *et al.*, 1981).

Temperature

Loper *et al.* (1985) found that colonization of potato roots by two Pseudomonas strains was best at 12 or 18°C, despite optimal temperatures for bacterial growth *in vitro* of 28 or 30°C. Competition from the indigenous microflora may have been reduced at the lower temperatures, or roots elongating more slowly may have been more effective in transporting the bacteria. Root colonization by indigenous VA and ectomycorrhizal fungi, and species composition of ectomycorrhizae that formed, were also affected by root zone temperature (Parke *et al.*, 1983).

pH

Soil pH affects survival and nodulation by Rhizobium (Dowling and Broughton, 1986) and manipulation of soil pH can alter the relative competitive abilities of indigenous strains (Dughri and Bottomley, 1984). Soil pH also has a significant affect on disease suppression and rhizosphere colonization by Trichoderma (Ahmad and Baker, 1987a; Chet and Baker, 1980).

Biological properties of soils

Root colonization by introduced mircoorganisms is greatly restricted in natural soil as compared to sterilized soil (Chao *et al.*, 1986; Scher *et al.*, 1984; Wong and Griffin, 1976) which indicates that antagonism or competition by the indigenous microflora are likely to be major determinants of successful establishment in the rhizosphere.

As the total population of rhizosphere bacteria does not change as a result of introducing bacteria (Iswandi *et al.*, 1987; Weller, 1983), it appears that introduced bacteria may reduce the resources available in the rhizosphere. The same principle may apply to other root colonists. Prior occupancy of roots by any species of VA mycorrhizal fungus reduced colonization by other VA fungi (Abbott and Robson, 1984).

Many rhizosphere microorganisms produce antibiotics or siderophores with a broad spectrum of activity; in addition to affecting plant pathogens (Fravel, 1988; Weller, 1988) we would expect them to reduce competition from subclinical pathogens or deleterious rhizobacteria (Schippers *et al.*, 1987; Suslow and Schroth, 1982) or even reduce populations of beneficial organisms such as Trichoderma (Hadar *et al.*, 1984).

Competition can be reduced to favor root colonization by indigenous or introduced microorganisms by selective treatment of the soil or application of pesticides or antimicrobial agents (Li and Alexander, 1986). Soil fumigation has been used to rid citrus nurseries of indigenous VA mycorrhizae to facilitate introduction of isolates more effective in phosphorus uptake (Menge, 1983).

Predation by soil macro- or microorganisms can also affect root colonization. Colonization of pine roots by ectomycorrhizal fungi was reduced by mycophagous amoebae (Chakraborty *et al.*, 1985). Bacteriophage are plentiful in soil and may reduce rhizosphere populations of many bacteria, including Rhizobium (Patel and Craig, 1984) and Pseudomonas (Stephens *et al.*, 1987).

Some interactions among root colonists are synergistic rather than antagonistic. These include enhanced nodulation and N_2 fixation in Rhizobium-legume symbioses as a result of colonization with VA mycorrhizal fungi (Bagyaraj, 1984). The combination of indigenous rhizobia, VA mycorrhizal fungi, and a plant growth-promoting strain of *Pseudomonas putida* resulted in increased nodulation and root growth compared to treatments where either the VA fungi or the introduced strain were not included (Meyer and Linderman, 1986). It is likely that additional examples of complex interactions among root colonists will become apparent with further research in this area.

Properties of the microorganism

Various terms used to describe the potential of an organism to colonize roots include competitiveness (Dowling and Broughton, 1986), infectivity (Abbott and Robson, 1984), rhizosphere competence (Ahmad and Baker, 1987a, Schmidt, 1979) and colonization potential (Bennett and Lynch, 1981). Phenotypic attributes that appear to contribute to root colonization by certain microorganisms are considered below.

Cell surface properties

Cell surface properties of bacteria are important in adsorption and permanent adhesion to soil colloids (Marshall, 1980) as well as specific host-bacterium recognition events involving rhizobia and Agrobacterium (Dazzo, 1980). However, the importance of surface properties to root colonization by other rhizosphere bacteria is poorly understood. In short-term binding assays, James *et al.* (1985) found that Pseudomonas and other bacteria attached to radish roots largely by means of reversible sorption. Attachment did not appear to involve specific recognition events, nor was attachment a good predictor of root-colonizing ability. In contrast, Anderson *et al.* (1988) suggested that specific recognition events promoted attachment and facilitated root colonization by pseudomonads in the bean rhizosphere. Pili are involved in attachment of Rhizobium, Klebsiella, and Enterobacter to plant roots (Korhonen *et al.*, 1986; Vesper and Bauer, 1986), and *Pseudomonas fluorescens* strain 2-79 piliated cells appear to attach to corn roots with a greater frequency than cells lacking pili (Vesper, 1987). Cell surface properties could greatly influence the mobility of bacteria with infiltrating water, since only those bacteria reversibly sorbed, and not those attached to roots or soil colloids by permanent adhesion, are likely to be removed by infiltrating water.

Substrate utilization
The ability of root-colonizing microorganisms to utilize root exudates and root secretions may be related to root-colonizing ability. There is one example in which rhizosphere competence of an organism was increased by genetic changes that influenced substrate utilization (Ahmad and Baker, 1987a,b). *Trichoderma harzianum* mutants, selected on the basis of resistance to benomyl, colonized roots to a greater extent than the wild-type. Rhizosphere competence was correlated with increased production of cellulase by these mutants.

As techniques for genetic modification of fungi and bacteria continue to improve, it should be possible to modify nutrient utilization of rhizosphere organisms to increase rhizosphere competence more specifically.

Osmotolerance
Loper *et al.* (1985) found there was a good correlation between rhizosphere competence and osmotolerance *in vitro* for eight Pseudomonas strains. It is unclear how osmotolerance might relate to growth in the rhizosphere, since bacteria are believed to be more sensitive to low matric potential than to low osmotic potential.

Growth rate
The rate of growth is likely to affect root colonization, and would certainly be useful in selecting strains and predicting their fate in the rhizosphere, but estimates of growth rate in the rhizosphere are difficult to determine.

The plant

Varietal differences
Rhizosphere organisms are largely dependent on root exudates and secretions as a substrate. Both the type and the amount of exudation are influenced by environmental conditions, but are largely under genetic control (Hale *et al.*, 1978). Different lines of wheat and potatoes have been shown to sustain different rhizosphere microfloras (Azad *et al.*, 1985; Weller, 1986). Substitution of chromosome pairs resulted in differences in the rhizosphere microflora that influenced susceptibility to Cercospora foot rot of wheat (Neal *et al.*, 1970). The concept of breeding plants for their ability to sustain particular rhizosphere microfloras opens some intriguing possibilities. But in fact, this may have occurred unwittingly as field-grown disease-resistant phenotypes, possibly supporting rhizosphere communities suppressive to pathogens, have been selected.

Nutrient status of host
Mycorrhizal colonization is generally suppressed by application of phosphorus to soil (Jasper *et al.*, 1979). Recent evidence suggests that phosphorus effects may be mediated by the host due to suppression of mycorrhiza-specific phosphatase activity, reduction in soluble carbohydrates available to the fungus, or changes in root cell membranes (Cooper, 1984). Proponents of the 'leaky membrane' theory suggest that, compared to P-sufficient plants, P-deficient plants have fewer root cell membrane phospholipids resulting in increased cell membrane permeability, increased leakage of reducing sugars and amino acids, and therefore greater mycorrhizal colonization (Graham *et al.*, 1981).

Problems associated with quantitative studies on root colonization

Recovery methods
There are several inherent difficulties in studying root colonization. Soil is an opaque medium which obscures direct observation of microorganisms using light microscopy. Washing techniques and many fixing procedures disrupt the rhizosphere, confounding observation. Electron microscopy has provided for keen insight into the ultrastructure of the rhizosphere (Foster *et al.*, 1983), but does not allow quantitative determinations of particular strains of microorganisms.

Therefore, indirect techniques such as dilution plate counts or most-probable-number determinations, often in combination with antibiotic-resistance markers and selective media, are widely used to study the populations of organisms in the rhizosphere. However, estimates of microbial populations based on plate counts do not agree with estimates based on direct microscopy (Rovira and Davey, 1974). In addition, strains marked with antibiotic resistance may be less fit than their wild-type parent strains (Compeau *et al.*, 1988) and markers may be lost, leading to underestimates of population size (Halverson and Handelsman, 1989). The use of immunological techniques (*e.g.* agglutination reactions, fluorescent antibody techniques, and ELISA) used in studies on rhizobia (Dowling and Broughton, 1986), have only recently been used for other rhizosphere organisms (Levanony *et al.*, 1987). Chromosomal markers using the *lac*ZY genes (Drahos *et al.*, 1986), the *lux* gene (Howie *et al.*, 1988), or an ice-nucleation-active gene (Suslow *et al.*, 1987) potentially provide more specific and more stable markers for quantitative studies of bacterial strains introduced into the rhizosphere.

Defining the sampling unit

Another difficulty in quantifying rhizosphere microorganisms is in defining the sampling unit and determining the appropriate units of expression. The rhizosphere is defined as that zone of soil in which the microflora is influenced by the root (Hiltner, 1904). From a practical standpoint, sampling of root colonists usually involves suspension of the root plus 'firmly adhering soil' (Rovira and Davey, 1974) in a known volume of diluent, followed by dilution-plating. Obviously, the amount of adhering soil can vary greatly depending on soil type, soil moisture content, and the vigor with which root samples are shaken to remove 'loosely adhering' soil. Because the density of microorganisms is greatest near the root, with the greatest concentration generally within 0–20 μm of the root surface (Foster, 1986), inclusion of different amounts of soil can dramatically affect population determinations expressed per unit soil volume or soil mass (Parke *et al.*, 1990). For this reason, populations of root colonists should be expressed per unit root length or root mass.

Variability in rhizosphere populations

Populations of introduced rhizobacteria among individual plants are generally highly variable (Loper *et al.*, 1984). A few plants sustain high populations of bacteria, while most plants support low populations. Data based on population means alone may be misleading in that they overestimate population size. Frequency distributions would also be very informative. For example, mean populations of *Pseudomonas fluorescens* strain 2-79 sampled from wheat roots 3–5 cm below the seed were 2.6 log cfu cm^{-1} root, but the bacterium was detected on only 56% of the roots emerging from seeds (Bull, 1987). Populations of rhizobacteria are often determined from whole root systems. However, populations of introduced strains are generally lognormally distributed within the root system, with highest populations occurring near the seed, and lower populations near the root tip (Weller, 1983). Studies in which populations are determined from discrete regions of the root, and at several time points, are most useful for determining population dynamics in the rhizosphere (Bahme and Schroth, 1987). Unfortunately, these studies are extremely labor-intensive.

Measuring growth of rhizosphere populations

It is relatively easy to demonstrate persistence of rhizosphere populations (Marois *et al.*, 1984) or changes in population distribution (Liddell and Parke, 1989; Parke *et al.*, 1986), but it is difficult to demonstrate that growth has occurred. In fact, most studies report a decline in the recoverable populations of introduced organisms over time, even if the organisms are considered to be good root colonists (Bahme and Schroth, 1987; Juhnke *et al.*, Weller *et al.*, 1983; Weller, 1984). This may be because large initial populations are added, far greater than the carrying capacity of the root (Bennett and Lynch, 1981). Caution should be used in determining that growth, and not simply re-distribution of the initial inoculum, has occurred.

One of the most difficult things to measure in the rhizosphere is generation time (doubling time) of bacteria. Because of the high variability in rhizosphere populations among plants, it is difficult to measure changes in populations less than 0.5–1.0 log unit with any degree of accuracy. One method with potential for more accurately determining generation time *in situ* is based on the incorporation of tritiated-thymidine incorporated into cellular DNA (Christensen and Funck-Jensen, 1989).

Density-dependent relationships
Persistence and high populations are considered important attributes of root colonists (Weller, 1988). However, a density-dependent relationship between inoculum level and rhizosphere population, and between rhizosphere population and plant response, are assumed but rarely demonstrated.

Bacteria introduced onto barley roots rapidly achieved populations of 10^7 colony-forming units mg^{-1} root regardless of initial inoculum levels (10^3, 10^5, or 10^7 cfu mg^{-1}) (Bennett and Lynch, 1981). Declining populations of Pseudomonas on treated potato tubers increased 700-fold within 2 days once the tubers were planted, suggesting that initial inoculum density was not as important as conditions favorable for rapid bacterial growth (Geels and Schippers, 1983). In Rhizobium, it is difficult to predict success in nodulation based solely on abundance of different strains in the rhizosphere, since both relative abundance and competitiveness determine nodulation frequency (Dowling and Broughton, 1986; Weaver and Frederick, 1974).

In some host-microbe systems, a 'critical mass' of microorganisms may be necessary to achieve plant growth effects. For instance, Iswandi *et al.* (1987) showed that Pseudomonas strain 7NSK2 populations must be at least 10^6 CFU per seed in order to increase maize growth significantly. A strain present for only a brief period in a localized zone on the root may be just as effective as a strain consistently recovered from the entire root. This is especially true of microorganisms effective in biological control of damping-off diseases, where opportunity for root infection may be limited to a few days after the seed has germinated.

However, there are examples in which plant growth response or disease suppression is related to inoculum density. In an elegant study by C.T. Bull, D.M. Weller, and L.S. Thomashow (personal communication), the population of *P. fluorescens* strain 2-79 on wheat roots was directly related to initial populations on wheat seeds; furthermore, there was an inverse relationship between populations of 2-79 on the roots and the number of lesions made by *Gauemannomyces graminis* var. tritici. Interestingly, there was no such relationship when a Tn*5* mutant of 2-79, deficient in production of the phenazine antibiotic, was applied to wheat seeds, but populations of a mutant restored for phenazine production were also inversely related to the number of root lesions. This is strong evidence for the involvement of phenazine in biological control, and clearly demonstrates the importance of root colonization to disease suppression in this system.

Modeling root colonization

Newman and Watson (1977) provided a model relating microbial abundance in the rhizosphere to root age, root exudation rate, soil water content, and initial populations. From this they predicted that the greatest density of microorganisms would occur near the root surface. This model has since been validated, to a certain extent, by electron microscopy (Foster, 1986). The future challenge is to expand the model to describe root colonization (dispersal and growth) in relation to root growth and plant response. Such an undertaking is necessarily very complex, as it will involve coupling the Newman and Watson model with models for root growth, and, perhaps, infection by pathogens. Significant progress has been made in modeling infection by mycorrhizal fungi (Walker and Smith, 1984) and soilborne pathogens (Gilligan, 1983).

The classic view of the rhizosphere is represented by an individual root growing in soil. A more realistic view in some soils would consist of overlapping rhizospheres as a result of high rooting density (Newman, 1985). Individual plants may even be linked by fungal networks (Francis and Read, 1984), and rhizobacteria apparently can spread from plant to plant (Bashan and Levanomy, 1987), further complicating root colonization models.

Approaches for future studies

Significant progress has been made in root colonization research, but new techniques for marking organisms and genetically modifying them offer tremendous potential for improving quantitative studies. Some priorities and experimental approaches include the following:

1. Identify phenotypic characteristics of microorganisms and plants that are important to root colonization.
2. Increase our understanding of microbial ecology and physiology, and use this information to develop more effective strategies for managing the rhizosphere microflora.
3. Rigorously control soil factors, such as soil matric potential, in studies on root colonization.
4. Expand our view of the rhizophere from the zone surrounding an individual root to overlapping rhizospheres penetrated by many roots.
5. Determine the importance of soil fauna as vectors and predators of rhizosphere microorganisms.
6. Develop better methods to measure microbial competition in the rhizosphere.
7. Develop methods to link root colonization with microbial activity, such as the production of antibiotics, so that population measurements have some biological relevance.
8. Improve quantitation of microbial population dynamics in relation to root growth, plant age, and infection by pathogens using improved techniques for marking organisms. Use this data to develop mathematical models to describe and predict root colonization.

References

Abbott L K and Robson A D 1984 Colonization of the root system of subterranean clover by three species of vesicular-arbuscular mycorrhizal fungi. New Phytol. 96, 274–281.

Ahmad J S and Baker R 1987a Rhizosphere competence of *Trichoderma harzianum*. Phytopathology 77, 182–189.

Ahmad J S and Baker R 1987b Competitive saprophytic ability and cellulolytic activity of rhizosphere competent mutants of *Trichoderma harzianum*. Phytopathology 77, 358–362.

Anderson A J, Habibzadegah-Tari P and Tepper C S 1988 Molecular studies on the role of a root surface agglutinin in adherence and colonization by *Pseudomonas putida*. Appl. Environ. Microbiol. 54, 375–380.

Azad H R, Davis J R, Schnathorst W C and Kado C I 1985 Relationship between rhizoplane and rhizosphere bacteria and Verticillium wilt resistance in potato. Arch. Microbiol. 140, 347–351.

Bagyaraj D J 1984 Biological interactions with VA mycorrhizal fungi. *In* VA Mycorrhiza. Eds. C L Powell and D J Bagyaraj. pp 131–153. CRC Press, Boca Raton.

Bahme J B and Schroth M N 1987 Spatial-temporal colonization patterns of a rhizobacterium on underground organs of potatoes. Phytopathology 77, 1093–1100.

Bashan Y 1986 Migraton of the rhizosphere bacteria *Azospirillium brasilense* and *Pseudomonas fluorescens* toward wheat roots in soil. J. Gen. Microbiol. 132, 3407–3414.

Bashan Y and Levanony H 1987 Horizontal and vertical movement of *Azospirillum brasiliense* Cd in the soil and along the rhizosphere of wheat and weeds in controlled and field environments. J. Gen. Microbiol. 133, 3473–3480.

Bennett R A and Lynch J M 1981 Colonization potential of bacteria in the rhizosphere. Curr. Microbiol. 6, 137–138.

Bitton G, Lahav N and Henis Y 1974 Movement and retention of *Klebsiella aerogenes* in soil columns. Plant and Soil 40, 373–380.

Bowen G D and Rovira A D 1976 Microbial colonization of plant roots. Annu. Rev. Phytopathol. 14, 121–144.

Bull C T 1987 Wheat root colonization by disease-suppressive or nonsuppressive bacteria and the effect of population size on severity of take-all caused by *Gaeumannomyces graminis* var *tritici*. MS thesis. Wash. State Univ., Pullman. 75 p.

Chakraborty S, Theodorou C and Bowen G D 1985 The reduction of root colonization by mycorrhizal fungi by mycophagous amoebae. Can. J. Microbiol. 31, 295–297.

Chao W L, Nelson E B, Harman G E and Hoch H C 1986 Colonization of the rhizosphere by biological control agents applied to seeds. Phytopathology 76, 60–65.

Chet I and Baker R 1980 Introduction of suppressiveness to *Rhizoctonia solani* in soil. Phytopathology 70, 994–998.

Christensen H and Funck-Jensen D 1989 Growth rate of rhizosphere bacteria measured directly by the tritiated thymidine incorporation technique. Soil Biol. Biochem. 21, 113–117.

Compeau G, Al-Achi B J, Platsouka E and Levy S B 1988 Survival of rifampin-resistant mutants of *Pseudomonas fluorescens* and *Pseudomonas putida* in soil systems. Appl. Environ. Microbiol. 54, 2432–2438.

Cooper K M 1984 Physiology of VA mycorrhizal associations. *In* VA Mycorrhiza. Eds. C L Powell and D J Bagyaraj. pp 156–186. CRC Press, Baton Rouge.

Dazzo P 1980 Adsorption of microorganisms to roots and other plant surfaces. *In* Adsorption of Microorganisms to Surfaces. Eds. G Bitton and K C Marshall. pp 253–316. Wiley and Sons, New York.

De Weger L A, van der Vlugt C I M, Wijfjes A H M, Bakker P A H M, Schippers B and Lugtenberg B 1987 Flagella of a plant-growth-stimulating *Pseudomonas fluorescens* strain are required for colonization of potato roots. J. Bacteriol. 169, 2769–73.

Dowling D N and Broughton D N 1986 Competition for nodulation of legumes. Annu. Rev. Microbiol. 40, 131–157.

Drahos D J, Hemming B C and McPherson S 1986 Tracking recombinant organisms in the environment: B-galactosidase as a selectable non-antibiotic marker for fluorescent pseudomonads. Bio/Technology 4, 439–444

Dughri M H and Bottomley P J 1984 Soil acidity and the composition of an indigenous population of *Rhizobium trifolii* in nodules of different cultivars of *Trifolium subterraneum* L. Soil Biol. Biochem. 16, 405–411.

Duniway J M and McKeen C D 1987 Dispersal of *Phytophthora cryptogea* zoospores in soils and glass microbeads by water flow. (Abstr.) Phytopathology 77, 1744.

Foster R C 1986 The ultrastructure of the rhizoplane and rhizosphere. Annu. Rev. Phytopathol. 24: 211–34.

Foster R C, Rovira A D and Cock T W 1983 Ultrastructure of the Root-Soil Interface. Am. Phytopathol. Soc., St. Paul, 157 pp.

Francis R and Read D J 1984 Direct transfer of carbon between plants connected by vesicular-arbuscular mycorrhizal mycelium. Nature 307, 53–65.

Fravel D R 1988 Role of antibiosis in the biocontrol of plant diseases. Annu. Rev. Phytopathol. 26, 75–91.

Geels F P and Schippers B 1983 Selection of antagonistic fluorescent *Pseudomonas* spp. and their root colonization and persistence following treatment of seed potatoes. Phytopathol. Z. 108, 193–206.

Gilligan C A 1983 Modeling of soilborne pathogens. Annu. Rev. Phytopathol. 21, 45–64.

Graham J H, Leonard R T and Menge J A 1981 Membrane-mediated decrease in root exudation responsible for phosphorous inhibition of vesicular-arbuscular mycorrhiza formation. Plant Physiol. 68, 548–1981.

Griffin D M and Quail G 1968 Movement of bacteria in moist, particulate systems. Aust. J. Biol. Sci. 21, 579–82.

Hadar Y, Harman G E and Taylor A G 1984 Evaluation of *Trichoderma koningii* and *T. harzianum* from New York soils for biological control of seed rot caused by *Pythium* spp. Phytopathology 74, 106–110.

Hale M G, Moore L D and Griffin G J 1978 Root exudates and exudation. *In* Interactions Between Non-pathogenic Soil Microorganisms and Plants. Eds. Y R Dommergues and S V Krupa. Elsevier, North Holland Biomedical Press, Amsterdam.

Halverson L J and Handelsman J 1990 Stability of antibiotic-resistance markers in *Bacillus cereus* UW85. *In* The Rhizosphere and Plant Growth. Eds. D L Keister and P B Cregan. p 107. Kluwer Academic Publishers, Dordrecht, The Netherlands.

Henry C M and Deacon J W 1981 Natural (non-pathogenic) death of the cortex of wheat and barley seminal roots, as evidenced by nuclear staining with acridine orange. Plant and Soil 60, 255–274.

Hiltner L 1904 Über neuere Erfahrungen und Probleme auf dem Gebiet der Bodenbakteriologie und unter besonderer Berücksichtigung der Gründüngung und Brache. Arb. Deut. Landw. Ber. 98, 59–78.

Howie W J, Cook R J and Weller D M 1987 Effects of soil matric potential and cell motility on wheat root colonization by fluorescent pseudomonads suppressive to take-all. Phytopathology 77, 286–92.

Howie W, Correll M, Gutterson N and Suslow T 1988 Indirect evidence for Oomycin A *in situ*: effect of soil temperature, moisture, and texture. Phytopathology 78, 1558 (Abstr.).

Iswandi A, Bossier P, Vandenabeele J and Verstaete W 1987 Influence of the inoculation density of the rhizopseudomonad strain 7NSK2 on the growth and the composition of the root microbial community of maize (*Zea mays*) and barley (*Hordeum vulgare*). Biol. Fertil. Soils 4, 119–123.

James D W Jr., Suslow T V and Steinback K E 1985 Relationship between rapid, firm adhesion and long-term colonization of roots by bacteria. Appl. Environ. Microbiol 50, 392–97.

Jasper D A, Robson A D and Abbott L K 1979 Phosporous and the formation of vesicular-arbuscular mycorrhizas. Soil Biol. Biochem. 11, 501–505.

Juhnke M E, Mathre D E and Sands D C 1987 Identification and characterization of rhizosphere-competent bacteria of wheat. Appl. Environ. Microbiol. 53, 2793–99.

Korhonen T K, Nurmiaho-Lassila E L, Laakso T and Haahtella K 1986 Adhesion of fimbriated nitrogen-fixing enteric bacteria to roots of grasses and cereals. Plant and Soil 90, 59–69.

Lawrence J R, Delaquis P J, Korber D R and Caldwell D E 1987 Behavior of *Psuedomonas fluorescens* within the hydrodynamic boundary layers of surface microenvironments. Microb Ecol. 14, 1–14.

Levanony H, Bashin Y and Kahana Z E 1987 Enzyme-linked immunosorbent assay for specific identification and enumeration of *Azospirillum brasilense* Cd. in cereal roots. Appl. Environ. Microbiol. 53, 358–364.

Li D and Alexander M 1986 Bacterial growth rates and competition affect nodulation and root colonization by *Rhizobium meliloti*. Appl. Environ. Microbiol. 52, 807–811.

Liddell C M and Parke J L 1989 Enhanced colonization of pea taproots by a fluorescent pseudomonad biocontrol agent by water infiltration into soil. Phytopathology 79, 1327–1332.

Loper J E, Haack C and Schroth M N 1985 Population dynamics of soil pseudomonads in rhizosphere of potato (*Solanum tuberosum* L.) Appl. Environ. Microbiol. 49, 416–422.

Loper J E, Suslow T V and Schroth M N 1984 Lognormal distribution of bacterial populations in the rhizosphere. Phytopathology 74, 1454–1460.

Madsen E L and Alexander M 1982 Transport of *Rhizobium* and *Pseudomonas* through soil. Soil Sci. Soc. Am. J. 46, 557–60.

Marois J J, Fravel D R and Papavizas G C 1984 Ability of *Talaromyces flavus* to occupy the rhizosphere and its interaction with *Verticillium dahliae*. Soil Biol. Biochem. 16, 387–390.

Marshall K C 1980 Adsorption of microorganisms to soils and sediments. *In* Adsorption of Microorganisms to Surfaces. Eds. G Bitton and K C Marshall. pp 317–329. Wiley and Sons, New York.

Mellor H Y, Glenn A R, Arwas R and Dilworth M J 1987 Symbiotic and competitve properties of motility mutants of *Rhizobium trifolii* TAI. Arch. Microbiol. 148, 34–39.

Menge J A 1983 Utilization of vesicular-arbuscular mycorrhizal fungi in agriculture. Can. J. Bot. 61, 1015–1024.

Meyer J R and Linderman R G 1986 Response of subterranean clover to dual inoculation with vesicular-arbuscular mycorrhizal fungi and a plant-growth promoting bacterium, *Pseudomonas putida*. Soil Biol. Biochem. 18, 185–190.

Neal J L Jr., Atkinson T G and Larson, R I 1970 Changes in the rhizosphere microflora of spring wheat induced by disomic substitution of a chromosome. Can J. Microbiol. 16, 153–158.

Newman E I 1985 The rhizosphere: carbon sources and microbial populations. *In* Ecological Interactions in Soil. Ed. A H Fitter. pp 107–121.

Newman E I and Watson A 1977 Microbial abundance in the rhizosphere: a computer model. Plant and Soil 48, 17–56.

Papendick R I and Campbell G S 1975 Water potential in the rhizosphere and plant and methods of measurement and experimental control. *In* Water Potential Relations in Soil Microbiology. pp 39–49. Soil Sci Soc. Am. Special Pub. No. 9. Madison, WI.

Parke J L, Liddell C M and Clayton M K 1990 Relationship between soil mass adhering to pea taproots and recovery of *Pseudomonas fluorescens* from the rhizosphere. Soil Biol. Biochem. 22, 495–499.

Parke J L, Linderman R G and J M Trappe 1983 Effect of root zone temperatures on ectomycorrhiza and vesicular-arbuscular mycorrhiza formation in disturbed and undisturbed forest soils of southwest Oregon. Can. J. For. Res. 13, 657–665.

Parke J L, Moen R, Rovira A D and Bowen G D 1986 Soil water flow affects the rhizosphere distribution of a seedborne biological control agent, *Pseudomonas fluorescens*. Soil Biol. Biochem. 18, 583–588.

Parke J L, Rovira A D and Bowen G D 1984 Soil matric potential affects colonization of wheat roots by a pseudomonad suppressive to take-all. (Abstr.) Phytopathology 74, 806.

Patel J J and Craig A S 1984 Isolation and characterization of bacteriophages active against strains of *Rhizobium trifolii* used in legume inoculants in New Zealand. NZ J. Sci. 27, 81–86.

Rovira A D and Campbell R 1974 Scanning electron microscopy of microorganisms on the roots of wheat. Microb. Ecol. 1, 15–23.

Rovira A D and Davey C B 1974 Biology of the rhizosphere. *In* The Plant Root and its Environment. Ed. E W Carson. pp 153–204. Univ. Press, Charlottesville, VA.

Scher F M, Ziegle J S and Kloepper J W 1984 A method for assessing the root-colonizing capacity of bacteria on maize. Can. J. Microbiol. 30, 151–57.

Scher F M, Kloepper J W and Singleton C A 1985 Chemotaxis of fluorescent *Pseudomonas* spp. to soybean seed exudates in vitro and in soil. Can. J. Microbiol. 31, 570–574.

Schippers B, Bakker A W and Bakker P A H M 1987 Interactions of deleterious and beneficial rhizosphere microorganisms and the effect of cropping practices. Annu. Rev. Phytopathol. 25, 339–358.

Stephens P M, O'Sullivan M and O'Gara F 1987 Effect of bacteriophage on colonization of sugarbeet roots by fluorescent *Pseudomonas* spp. Appl. Environ. Microbiol. 53, 1164–1167.

Schmidt E L 1979 Initiation of plant root microbe interactions. Annu. Rev. Microbiol. 33, 355–379.

Suslow T V, Matsubara D and Davies M 1987 Application of tube-nucleation assays to rapid population estimates of rhizobacteria expressing novel ice-nucleation activity. Curr. Plant Sci. Biotechnol. Agric. 4, 1018–1024.

Suslow T V and Schroth M N 1982 Role of deleterious rhizobacteria as minor pathogens in reducing crop growth. Phytopathology 72: 111–15.

Suslow T V and Schroth M N 1982 Rhizobacteria of sugarbeets: effects of seed application and root colonization on yield. Phytopathology 72, 199–206.

Vesper S J 1987 Production of pili (fimbriae) by *Pseudomonas fluorescens* and correlation with attachment to corn roots. Appl. Environ. Microbiol. 53, 1397–1403.

Vesper S J and Bauer W D 1986 Role of pili (fimbriae) in attachment of *Bradyrhizobium japonicum* to soybean roots. Appl. Environ. Microbiol. 52, 134–141.

Weaver R W and Frederick L R 1974 Effect of inoculum rate on competitive nodulation of *Glycine max* L. Merrill. I. Greenhouse studies. Agron. J. 66, 229–232.

Walker N A and Smith S E 1984 The quantitative study of mycorrhizal infection. II. The relation of rate of infection and speed of fungal growth to propagule density, the mean length of the infection unit and the limiting value of the fraction of the root infected. New Phytol. 96, 55–69.

Weller D M 1983 Colonization of wheat roots by a fluorescent pseudomonad suppressive to take-all. Phytopathology 73, 1548–1553.

Weller D M 1984 Distribution of a take-all suppressive strain of *Pseudomonas fluorescens* on seminal roots of winter wheat. Appl. Environ. Microbiol. 48, 897–899.

Weller D M 1986 Effects of wheat genotype on root colonization by a take-all suppressive strain of *Pseudomonas fluorescens*. Phytopathology 76, 1059 (Abstr.).

Weller D M 1988 Biological control of soilborne plant pathogens in the rhizosphere with bacteria. Annu. Rev. Phytopathol. 26, 379–407.

Wilkinson H T, Miller R D and Millar R L 1981 Infiltration of fungal and bacterial propagules into soil. Soil Sci. Soc. Am. J. 45, 1034–1039.

Wong P T W and Griffin D M 1976 Bacterial movement at high matric potentials-I. In artificial and natural soils. Soil Biol. Biochem. 8, 215–218.

D. L. Keister and P. B. Cregan (Eds.), The rhizosphere and plant growth, 43–50.

PLSO BARC05

Genetic approaches for studying rhizosphere colonization

S.T. LAM, D.M. ELLIS and J.M. LIGON
CIBA-Geigy Biotechnology Research, P.O. Box 12257, RTP, NC 27709, USA

Key words: β-galactosidase, plant-inducible promoter, Pseudomonas, rhizobacteria, root colonization, transposon mutagenesis

Abstract

Most bacterial traits involved in colonization of plant roots are yet to be defined. Studies were initiated to identify genes in Pseudomonas which play significant roles in this process. The general approach is to use transposons to construct collections of insertion mutants, each of which is then screened for alterations in its interactions with the host plant. In one study a Tn*5* derivative containing a constitutively expressed β-galactosidase (*lacZ*) gene was used to generate a collection of insertion mutants which could be distinguished from the wild-type parent on X-gal plates. Each mutant was examined for its ability to colonize wheat seedlings in the presence of the wild-type parent. Mutants which gave wild-type:mutant ratio of 20:1 or greater were obtained. In a second study a Tn*5* derivative which carries a promoterless *lacZ* gene located near one end of the transposon was constructed. Expression of the *lacZ* gene depends on the presence of an active promoter outside of the transposon in the correct orientation. Insertion mutants generated with this transposon were examined for changes in β-galactosidase expression in the presence and absence of plant root exudate. A number of mutants which showed differential *lacZ* expression have been identified.

Introduction

Rhizobacteria have the potential to be useful in agriculture for plant disease control and growth promotion. Successful root colonization by these organisms is generally accepted as a prerequisite for such applications, and it has been hypothesized that variable root colonization is probably a main reason for inconsistent control (Weller, 1988). To realize this potential we need a better understanding of the process of root colonization. Most bacterial traits involved in root colonization are yet to be defined. In this paper we discuss some recent results in this area, with the emphasis on Pseudomonas, including our efforts to idenfity bacterial genes which play significant roles in this process.

Root colonization

Bacterial root colonization comprises a series of steps: migration towards plant roots, attachment, distribution along the root, and finally growth and establishment of the population. Research efforts have focused on the determination of the roles, if any, of known bacterial structures and behavioral traits in these steps.

Chemotaxis has been demonstrated in a number of plant associated bacteria (Aguilar *et al.*, 1988; Loake *et al.*, 1988). Scher *et al.* (1985) showed that Pseudomonas strains actively migrated towards soybean seeds in soil, and hypothesized that this event may be the first step in establishment of seed and root colonization. On the other hand, no correlation was observed between chemotactic ability and root colonization in pot and field assays (Scher *et al.*, 1988). In the latter study, the introduced bacteria were directly inoculated on to seeds. This may have circumvented the need for chemotaxis.

A root-colonizing *Pseudomonas putida* strain was shown to be agglutinated by a glycoprotein complex present in bean and cucumber root washes (Anderson *et al.*, 1988; Tari and Anderson, 1988). Mutants were isolated which were no longer agglutinable (Agg-). Compared to the wild-type parent, Agg- mutants adhered to roots at lower levels in short-term binding studies, colonized roots to a lesser extent, and also provided less protection to cucumber plants against *Fusarium oxysporum* (Tari and Anderson, 1988). In another study, pea root exudate was found to agglutinate a higher percentage of bacterial isolates from pea rhizosphere compared to those from nonrhizosphere samples (Chao *et al.*, 1988). Two agglutination-positive strains, when inoculated on to seeds, could be recovered from the entire rhizosphere (up to 7 cm below the seed) of the resulting seedlings. An agglutination-negative strain, treated similarly, could not be recovered from the rhizosphere 3 cm below the seed. The agglutination-positive strains also yielded higher rhizosphere/soil (R/S) ratios. Thus it appears that cell surface molecule interactions between bacteria and host plant play an important role in attachment to, and eventual colonization of, the host plant root.

Cell surface polysaccharide molecules in Agrobacterium and Rhizobium mediate their attachment to, and subsequent interaction with, plant cells (Halverson and Stacey, 1986). The root-colonizing Pseudomonas strains WCS358 and WCS374 have lipopolysaccharides with long O-antigenic side chains (De Weger *et al.*, 1987a). From these strains, De Weger constructed mutants which no longer produce the O-antigenic side chains. Such mutants did not differ from their wild-type parents with respect to attachment to Sephadex beads or sterile potato roots (De Weger, 1988). However, colonization studies with one of the mutants indicated that it was less capable of colonizing distal roots as compared to the wild type (De Weger, pers. comm.).

The role of bacterial motility in their distribution along plant roots remains unresolved. In two studies (Howie *et al.*, 1987; Scher *et al.*, 1988), nonmotile mutants colonized wheat and soybean roots, respectively, to the same extent as their wild-type parents. In a third study (De Weger *et al.*, 1987b), nonmotile mutants were severely impaired in their ability to colonize lower parts of potato roots when compared to their wild-type parents. The discrepancy serves to illustrate the point that whether a trait is important may depend on the system one is working with.

In a series of studies, Ahmad and Baker found that among Trichoderma mutants selected for resistance to benomyl, many showed increased rhizosphere competence (Ahmad and Baker, 1987a). These rhizosphere competent mutants have increased cellulase activity and competitive saprophytic ability (Ahmad and Baker, 1987b; Ahmad and Baker, 1988a; b). The authors hypothesize that these characteristics allow the mutants to utilize the components of root mucigel and other complex carbohydrate substrates, resulting in more efficient colonization of the rhizoplane.

Genetic approaches

Analysis of mutant phenotype is a powerful way to reveal the molecular basis of a complicated process. In general, one postulates that a particular genetic trait may play a significant role in the process of interest. Mutants are isolated which are altered in that trait. One then compares the behavior of such mutants and the wild-type parent with respect to the process of interest. If a significant difference can be observed between a mutant and the wild-type parent, and the mutant can be shown to differ from the wild-type parent only in the selected trait, one may infer that the genetic trait indeed plays a role in the process. This approach has been applied successfully in a number of studies; the isolation and analysis of the Agg- mutants discussed in the last section is a good example. The approach has the advantage that clear answers can be obtained for the trait in question, but the disadvantage that one can only examine one trait at a time.

Alternatively, one may start with a random collection of mutants, and directly examine the phenotype of each mutant with respect to the process of interest. This strategy has rarely been used because, for complicated processes, it is usually difficult to devise meaningful and unambiguous assays, the assays are generally laborious, and one must have a very efficient way to generate mutants. The development of transposons as mutagens has made it feasible to apply this strategy in some cases. If carried out successfully, this approach can potentially reveal all important traits (as limited by the design of the assay), including ones which may not have been suspected otherwise.

We have initiated two studies using this approach: 1) to identify bacterial genes which play significant roles in competitive colonization, and 2) to identify bacterial genes which respond to host plant substances. The system we are study-

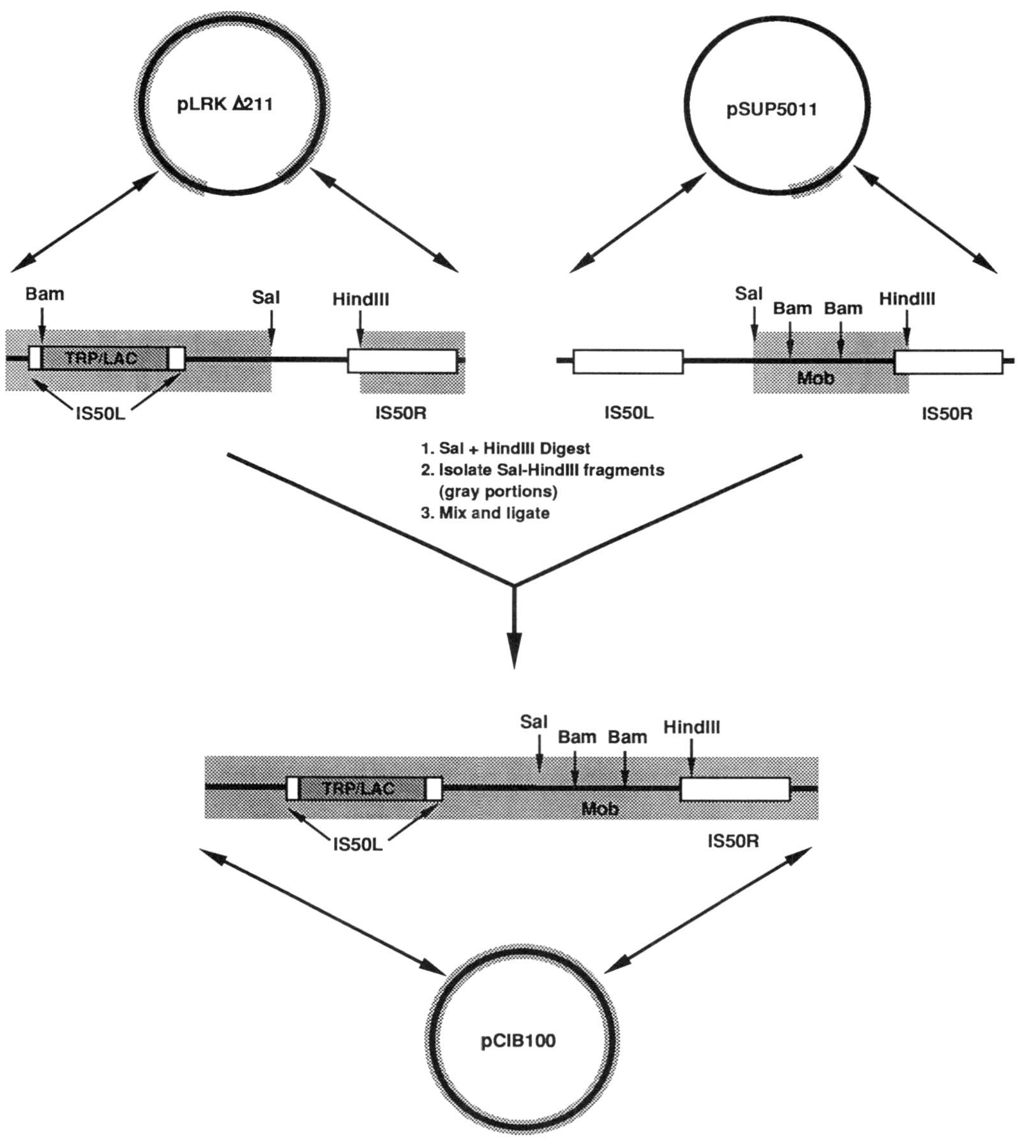

Fig. 1. Construction of pCIB100. MOB, plasmid mobilization site.

ing is the colonization of wheat roots by the *P. fluorescens* strain 2-79 (Weller, 1983).

Competitive colonization

pCIB100

We introduced the plasmid mobilization segment of pSUP5011 (Simon *et al.*, 1983) into pLRKΔ211 (Kroos and Kaiser, 1984) to form pCIB100 (Fig. 1). This plasmid carries a Tn*5*-*lac* transposon which confers kanamycin resistance and contains a *trp-lac* fusion fragment without an *E. coli* promoter. The *trp-lac* coding sequence is located at one end of the transposon and oriented such that, in *E. coli*, the *lacZ* gene can only be expressed from a functional and correctly oriented promoter outside that end of the transposon. The plasmid has a ColE1 replicon and cannot be maintained in Pseudomonas. It can, however, be mobilized into Pseudomonas and can therefore be used as a suicide vector for delivering the transposon. *E. coli* strain S17-1 (Simon *et al.*, 1983) is sensitive to ampicillin and has an IncP1 group conjugative plasmid integrated into the chromosome. Pseudomonas strain 2-79 is sensitive to kanamycin and resistant to ampicillin. Using strain S17-1 containing pCIB100 as donor and strain 2-79 as recipient in conjugation experiments, we were able to obtain stable kanamycin-resistant transconjugants in strain 2-79 by selecting for both kanamycin and ampicillin resistance. These transconjugants were shown by Southern hybridization experiments to contain the transposon sequence on different restriction fragments. To our surprise, all transconjugants of strain 2-79, which is naturally Lac$^-$, expressed the *lacZ* gene constitutively (blue on media containing the chromogenic substrate X-gal). It appears that some sequence within the transposon 5′ of the *lacZ* coding region is recognized as a functional promoter in strain 2-79 (see next section). Thus, with pCIB100, we are able to introduce into strain 2-79 an insertion mutation, an antibiotic resistance marker, and a visualizable marker in one step.

Competitive colonization assays

Over 1,200 insertion mutants were screened in a pairwise competitive colonization assay to identify mutants defective in the ability to colonize host plant roots in the presence of the wild-type parent. Equal numbers of mutant and wild-type bacteria were used to co-inoculate wheat seeds. The seeds were allowed to germinate, and after 5 to 7 days bacteria were recovered from the seedling roots. The recovered bacteria were plated on media containing X-gal. The mutants were distinguished from the wild-type parent by their blue color. We found that the ability of these mutants to compete varied over a wide range (Fig. 2). The majority are slightly less able to compete with the wild-type parent. This may indicate that many genes contribute to the process of competitive colonization in a minor way. Alternatively, it may reflect a small genetic load imposed by the presence of the transposon in the mutants. In either case, it is clear that the transposon in general did not cause a drastic reduction in the mutants' ability to compete. Severe competition-defective mutants resulted from disruption in critical genes, presumably caused by the insertion of the transposon.

Competition-defective mutants

Five competition-defective mutants (62, 94, 149, 383, 772) were analyzed further. They were shown reproducibly to be deficient in their ability to colonize wheat roots in the presence of the wild-type parent (Fig. 3). However, when inoculated alone, they were able to colonize wheat roots under gnotobiotic conditions at levels similar to the wild type parent. These mutants were prototrophic and motile. Two other insertion mutants, 20 and 604, were shown to compete favorably with the wild type parent (Fig. 3), confirming that the presence of the transposon alone did not necessarily diminish competitiveness.

Plant-responsive genes

In plant-associated bacteria, the nodulation genes in Rhizobium are induced by flavones and related compounds (for review see Djordjevic *et al.*, 1987), and the *vir* genes in Agrobacterium are induced by acetosyringone and related compounds (Stachel *et al.*, 1985). Both types of compounds are released into the rhizosphere by the host plant, and induction of the bacterial

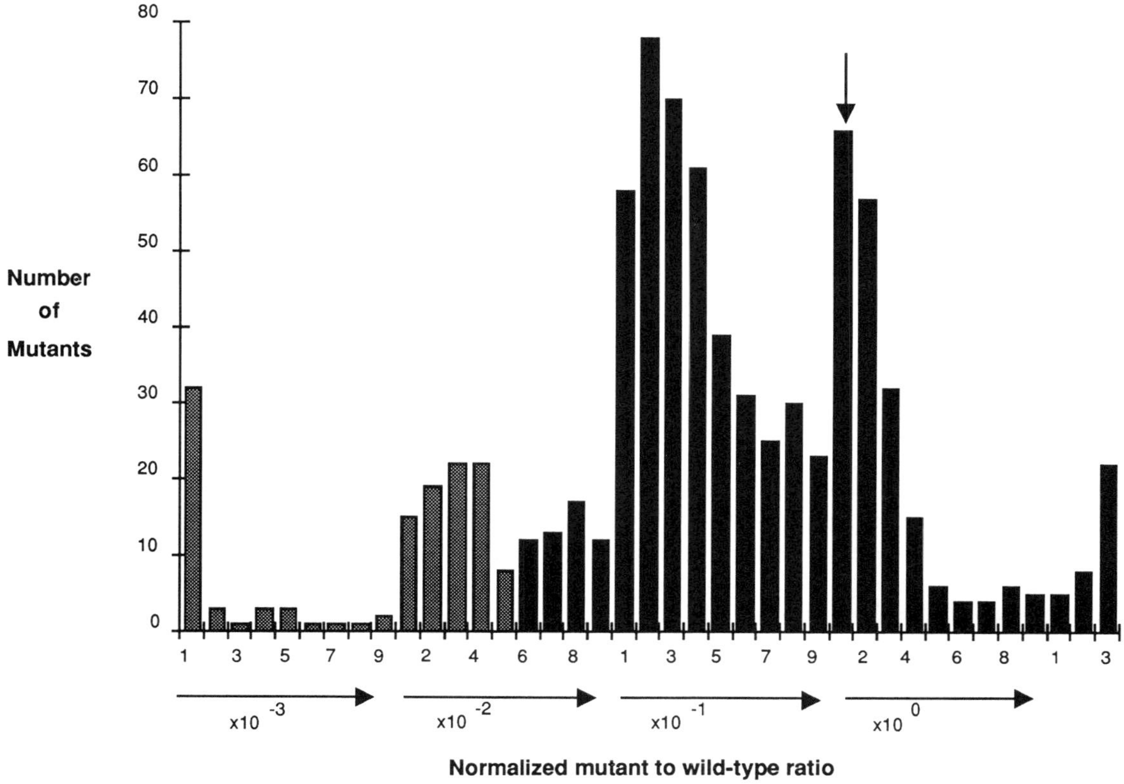

Fig. 2. Distribution profile of the 1200 insertion mutants tested. The x-axis indicates the normalized mutant to wild-type ratio, which is recovered mutant number/wild-type number divided by inoculated mutant number/wild-type number. The arrow in the graph indicates the mutants which had a ratio of 1.0, *i.e.* they were equal in colonization ability to the wild-type. The shaded bars indicate those mutants which had ratios ≤0.05 and were reexamined.

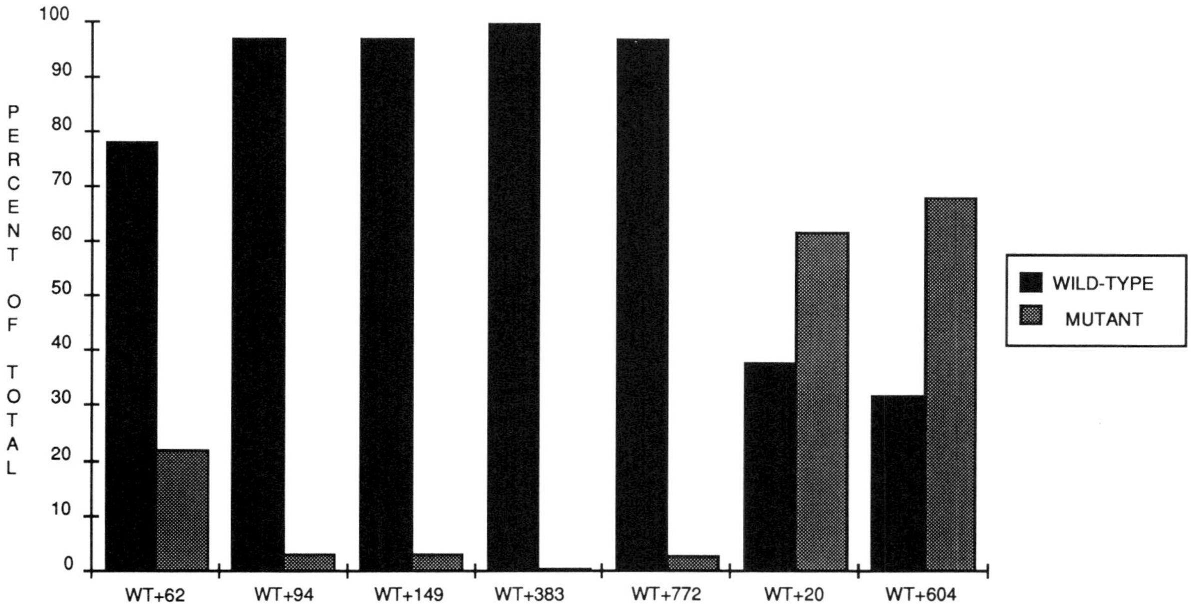

Fig. 3. Composition of rhizosphere populations 7 days after planting.

genes is critical for subsequent infections. Similar interactions have not been demonstrated in non-pathogenic Pseudomonas. Nevertheless, plant-associated pseudomonads do interact with their host plant, and the presence of a root creates a large environmental change to pseudomonads in the soil. We are interested in identifying Pseudomonas genes which are regulated in response to this change. Some of these genes may play significant roles in the interaction between the bacteria and the host plant.

pCIB116

Our approach is to use a transposon with a promoterless *lacZ* gene (promoter-probe transposon) to construct a collection of insertion mutants, then to screen for mutants which show differential *lacZ* expression in the presence and absence of root exduate. As discussed above, the promoterless *trp-lac* fusion in pCIB100 is expressed constitutively in strain 2-79. The fusion contains more than 1 kb of *trp* sequence 5′ of the *lacZ* coding region. We suspected that some sequence within the transposon 5′ of the *lacZ* coding region was recognized as a functional promoter in strain 2-79. A different promoterless *lacZ* construct without any *trp* sequence was made and was used to replace the *trp-lac* fusion in pCIB100. The resulting plasmid, pCIB116 (Fig. 4), was shown to behave as a promoter-probe in strain 2-79. Using strain S17-1 containing pCIB116 as donor and strain 2-79 as recipient in conjugation experiments, over 10,000 insertion mutants were made.

β-galactosidase assay

Individual mutants were grown in 96-well microtiter plates. Pairs of plates were set up: one containing root exudate, the other not. Root exudate was collected from sterile wheat seedlings and stored as lyophilized powder. After overnight growth, β-galactosidase activity of the

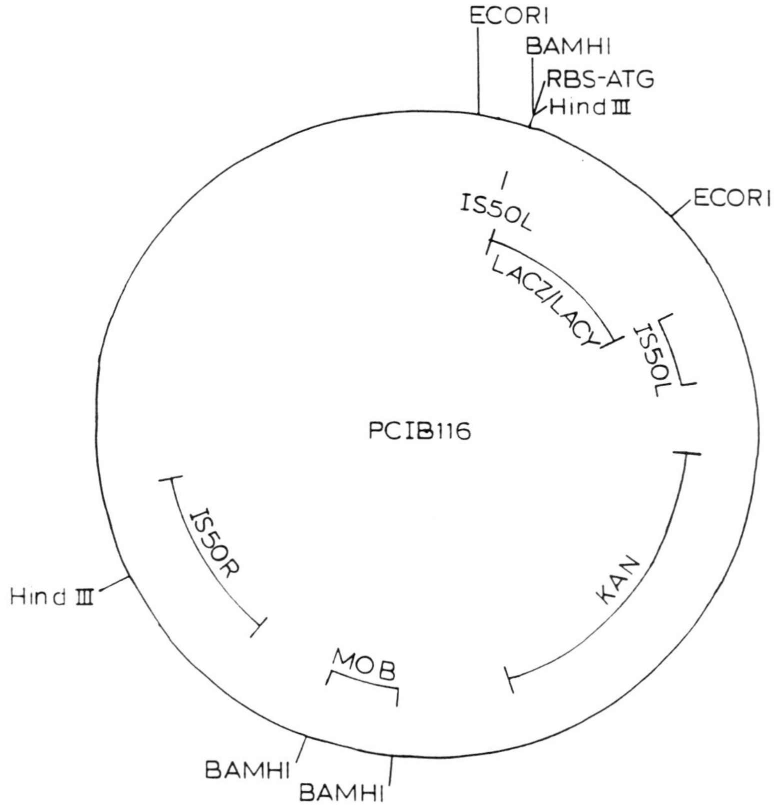

Fig. 4. pCIB116. RBS, ribosome binding site; KAN, kanamycin resistance.

cultures was measured according to a modification of the procedure of Miller (1972). Enzyme activity produced in the control (no root exudate) and in the root exudate supplemented cultures were compared (Fig. 5). Of ten thousand mutants screened, approximately 5% showed increases in β-galactosidase activity of 2-fold or greater in the presence of root exudate. Upon re-testing, 50 mutants which gave the largest and most reproducible responses to root exudates were selected. The three mutants which showed the strongest responses (67G6, 70C10, 98D12) are being studied further.

Plant assay

To test whether the induction of gene expression occurs on roots, the three mutants were examined in a plant assay. Sterilized wheat seeds were inoculated with individual mutants and allowed to germinate and grow under gnotobiotic conditions. After 7 days, bacteria were recovered from the seedling roots and immediately assayed for β-galactosidase activity. The enzyme activity per cell recovered from roots was elevated when compared to that obtained from overnight cultures grown in minimal medium (Table 1), indicating that the genes at the insertion sites in these mutants were indeed induced by the host plant.

Table 1. Direct verification of inductibility on plant roots

Strains	LacZ activity (modified Miller units)		
	Control	Exudate	On plant
Wild-type	0.00	0.00	0.00
67G6	1.00	7.50	3.67
70C10	0.93	12.45	9.88
98D12	0.75	13.58	9.96

Outlook

At the moment we do not know the nature of the alterations in these mutants. We do know, however, that they are affected in their interactions with the host plant. A number of questions

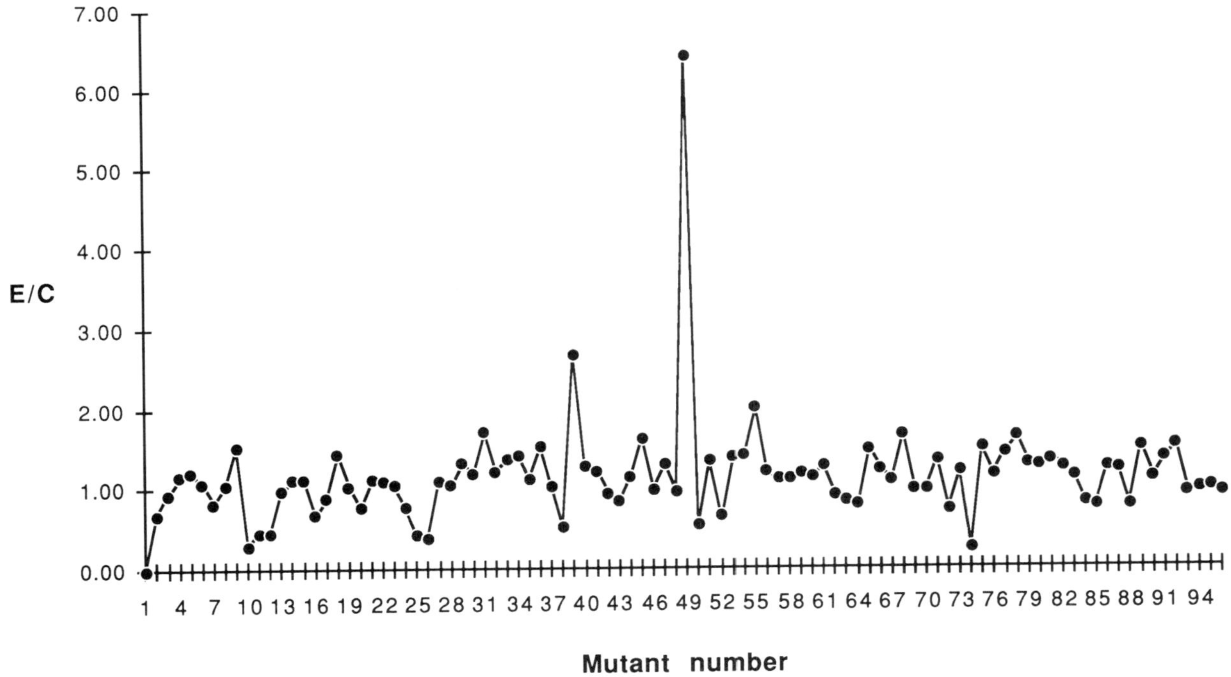

Fig. 5. Differential *lacZ* gene expression in the insertion mutants in the presence and absence of root exudate. E/C, *lacZ* activity in the presence of root exudate divided by that in the absence of root exudate.

are being pursued concerning these mutants. Do the competition-defective mutants show other defects in the absence of the plant host? What are the inducing substances in the root exudate? Do the insertions in the plant-responsive genes affect the ability of the bacteria to colonize the plant host? We hope that, by studying the changes in these mutants, we can gain a better understanding of the process of colonization.

References

Aguilar J M M, Ashby A M, Richards A J M, Loake G J, Watson M D and Shaw C H 1988 Chemotaxis of *Rhizobium leguminosarum* biovar *phaseoli* towards flavonoid inducers of the symbiotic nodulation genes. J. Gen. Microbiol. 134, 2741–2746.

Ahmad J S and Baker R 1987a Rhizosphere competence of *Trichoderma harzianum*. Phytopathology 77, 182–189.

Ahmad J S and Baker R 1987b Competitive saprophytic ability and cellulytic activity of rhizosphere-competent mutants of *Trichoderma harzianum*. Phytopathology 77, 358–362.

Ahmad J S and Baker R 1988a Growth on carbon substrates of rhizosphere-competent mutants of *Trichoderma harzianum*. Can. J. Microbiol. 34, 807–814.

Ahmad J S and Baker R 1988b Implications of rhizosphere-competence of *Trichoderma harzianum*. Can. J. Microbiol. 34, 229–234.

Anderson A J, Habibzadegah-Tari P and Tepper C S 1988 Molecular studies on the role of the root surface agglutinin in adherence and colonization by *Pseudomonas putida*. Appl. Environ. Microbiol. 54, 375–380.

Chao W L, Li R K and Chang W T 1988 Effect of root agglutinin on microbiol activities in the rhizosphere. Appl. Environ. Microbiol. 54, 1838–1841.

De Weger L A 1988 Role of the bacterial cell surface in potato growth-stimulation by fluorescent pseudomonas bacteria. Ph. D. diss. Leiden University, Netherlands, 91 p.

De Weger L A, Jann B, Jann K and Lugtenberg B 1987a Lipopolysaccharides of *Pseudomonas* spp. that stimulate plant growth: composition and use for strain identification. J. Bacteriol. 169, 1441–1446.

De Weger L A, van der Vlugt C I M, Wijfjes A H M, Bakker P A H M, Schippers B and Lugtenberg B 1987b Flagella of a plant-growth-stimulating *Pseudomonas fluorescens* strain are required for colonization of potato roots. J. Bacteriol. 169, 2769–2773.

Djordjevic M A, Gabriel D W and Rolfe B G 1987 *Rhizobium* – the refined parasite of legumes. Annu. Rev. Phytopathol. 25, 145–168.

Halverson J L and Stacey G 1986 Signal exchange in plant-microbe interactions. Microbiol. Rev. 50, 193–225.

Howie W J, Cook R J and Weller D M 1987 Effects of soil matric potential and cell motility on wheat root colonization by fluorescent pseudomonads suppressive to take-all. Phytopathology 77, 286–292.

Kroos L and Kaiser D 1984 Construction for Tn*5 lac*, a transposon that fuses *lacZ* expression to exogenous promoters, and its intoroduction into *Myxococcus xanthus*. Proc. Natl. Acad. Sci. USA 81, 5816–5820.

Loake G J, Ashby A M and Shaw C H 1988 Attraction of *Agrobacterium tumefaciens* C58C1 towards sugars involves a highly sensitive chemotaxis system. J. Gen. Microbiol. 134, 1427–1432.

Miller J H 1972 Experiments in Molecular Genetics. Cold Spring Harbor Laboratory, New York, 466 p.

Scher F M, Kloepper J W and Singleton C A 1985 Chemotaxis of fluorescent *Pseudomonas* spp. to soybean seed exudates *in vitro* and in soil. Can. J. Microbiol. 31, 570–574.

Scher F M, Kloepper J W, Singleton C, Zaleska I and Laliberte M 1988 Colonization of soybean roots by *Pseudomonas* and *Serratia* species: relationship to bacterial motility, chemotaxis, and generation time. Phytopathology 78, 1055–1059.

Simon R, Priefer U and Puhler A 1983 Vector plasmids for *in-vivo* and *in-vitro* manipulations of gram-negative bacteria. *In* Molecular Genetics of Bacteria-Plant Interaction. Ed. A Puhler. pp 98–106. Springer-Verlag, Berlin.

Stachel S E, Messens E, Van Montagu M and Zambryski P 1985 Identification of the signal molecules produced by wounded plant cells that activate T-DNA transfer in *Agrobacterium tumefaciens*. Nature 318, 624–629.

Tari P H and Anderson A J 1988 Fusarium wilt suppression and agglutinability of *Pseudomonas putida*. Appl. Environ. Microbiol. 54, 2037–2041.

Weller D M 1983 Colonization of wheat roots by a fluroescent pseudomonad suppressive to take-all. Phytopathology 73, 1548–1553.

Weller D M 1988 Biological control of soilborne plant pathogens in the rhizosphere with bacteria. Annu. Rev. Phytopathol. 26, 379–407.

D. L. Keister and P. B. Cregan (Eds.), The rhizosphere and plant growth, 51–59.

PLSO BARC06

Living plant cells released from the root cap: A regulator of microbial populations in the rhizosphere?

M.C. HAWES
Departments of Plant Pathology and Molecular and Cellular Biology, University of Arizona, Tucson, AZ 85721, USA

Key words: *Agrobacterium tumefaciens*, binding, chemotaxis, crown gall, *Pythium dissotocum*, *Pythium catenulatum*, root cap cells, rhizosphere colonization, zoospores

Abstract

A group of cells at the periphery of the root cap separate from plant roots as they move through the soil. Experimentally, these cells can be separated from the root by gentle agitation in water. Detached root cap cells provide a convenient system for comparing cellular and whole-plant responses to fungal and bacterial pathogens. Results with several host-parasite combinations indicate that these cells express host-specific traits with respect to chemotaxis, binding, and infection. Mutants of *Agrobacterium tumefaciens* were used to test the hypothesis that recognition of the cells plays a role in establishment of microbial-plant associations. Tn*5* mutants deficient in chemotaxis to the cells exhibit reduced ability to colonize the rhizosphere and to induce crown gall tumorigenesis on pea plants grown in soil. The discovery that cells shed from the cap express host-specific genes suggests that some microorganisms may have greater access to constituents of the cells than others. This cellular selectivity could have a significant impact on rhizosphere populations.

Sloughed root cap cells

In this review I will discuss new evidence about the properties of cells that are shed from root cap, and speculate about their possible impact on rhizosphere biology. Once these cells are released from the root, they are developmentally and morphologically distinct from the root cap. Because the cells then form a physical and biological interface between the root and the soil, we refer to them as 'root border (BRD) cells' (Hawes and Lin, 1990).

BRD cells are the product of a meristematic region distal to the apical meristem of the root (Barlow, 1975; Feldman, 1984). New cells arise by mitosis, proceed through a series of developmental stages, and then separate from the tip and ensheath the root as it grows through the soil (Guinel and McCully, 1987; Hawes and Lin, 1990; Vermeer and McCully, 1982). BRD cells constitute a group that is distinct from peripheral cells described by Moore and McClelan (1983); they can be experimentally defined as those which separate from the root on brief contact with water followed by gentle agitation (Fig. 1A-C) (Hawes and Wheeler, 1982). The cells are frequently not included in photographs of root caps (for example, Bennet et al., 1987; Hinch and Clarke, 1980; Moore and McClelan, 1983; Rost and Jones, 1988) because of the ease with which they do separate from the root: most germination procedures and all washing methods remove virtually all of them. Although outer root cap cells have long been known to slough from the root, they have been viewed as a passive and nonspecific source of carbon and nitrogen compounds that are released as the cells deteriorate (Voeller *et al.*, 1964). Recent studies suggest that BRD cells have the potential to play a more specific role in regulating the composition of microorganism populations in the rhizosphere.

First, Vermeer and McCully (1982) found that

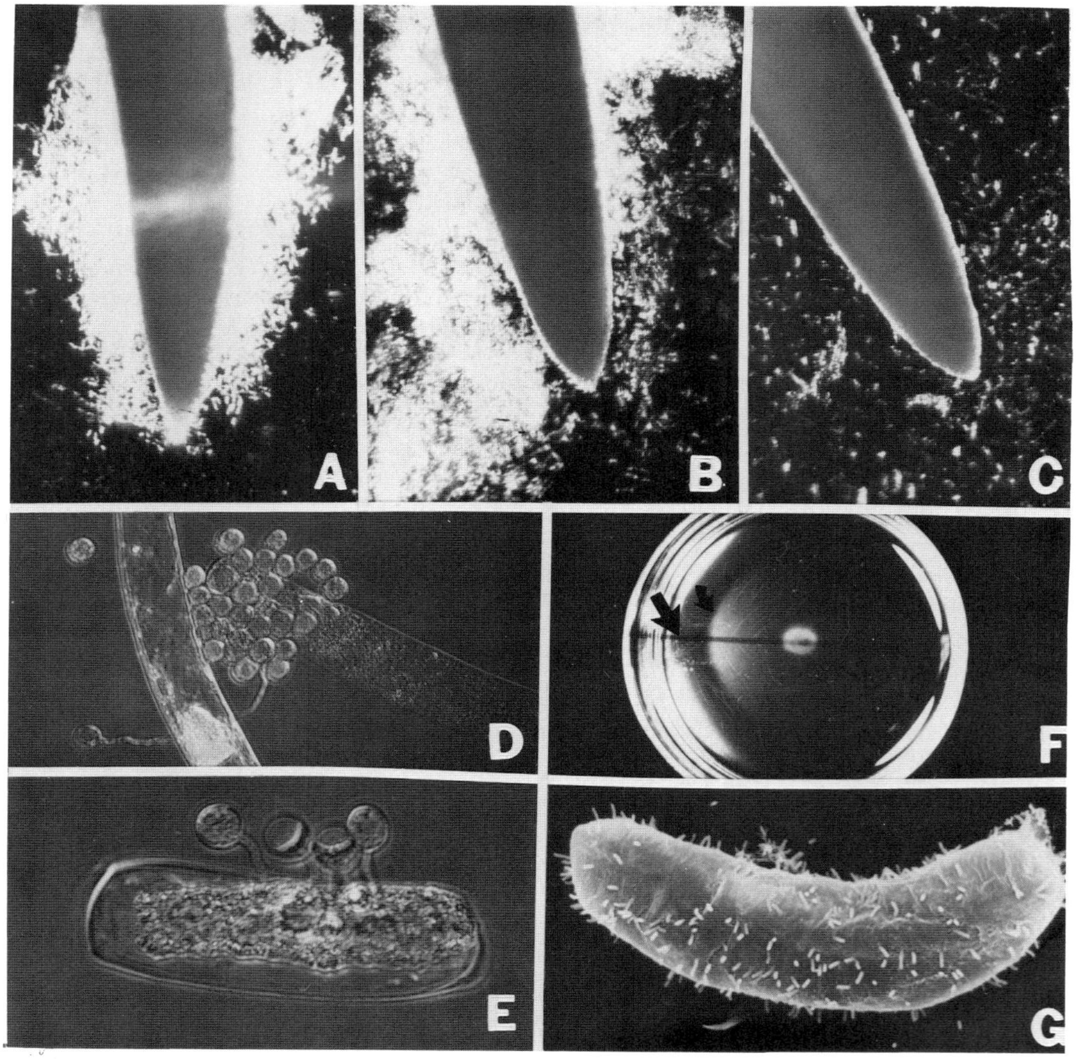

Fig. 1. Properties of BRD cells. **A-C.** Isolation of BRD cells form the tip of a cotton seedling germinated for 48 h on water agar overlaid with filter paper. (A) Swelling of mucilage containing masses of BRD cells 30 seconds after immersion into distilled water; (B) Spreading of cells away from the top 60 sec later; (C) Complete separation of BRD cells from root tip after agitating the water by taking it up and releasing it with a pasteur pipette once. **D-E.** Interactions of *Pythium dissotocum* zoospores with cotton BRD cells. (D) Attraction to and accumulation of zoospores on BRD cells; and (E) collapse of the BRD cell cytoplasm in response to penetration by germinating zoospores of *P. dissotocum*. **F-G.** Interactions of *Agrobacterium tumefaciens* with pea BRD cells. (F) Directional swarm assay of chemotaxis. *A. tumefaciens* cells in a 10 μl droplet placed in the center of a plate of semisolid water agar move in a swarm (small arrow) toward a 10 μl droplet containing BRD cells (large arrow); (G) Cells of *A. tumefaciens* bound to the surface of a pea BRD cell.

under natural conditions BRD cells of maize are capable of surviving in the rhizosphere independently of the root. Although the long-term viability and growth of sloughed cells in culture was reported many years ago (Gautheret, 1933; Knudson, 1919) and was confirmed recently (Guinel and McCully, 1987; Hawes and Wheeler, 1982; Hawes and Pueppke, 1986), this was the first report that root cap cell integrity can be maintained in the soil. Second, large numbers of viable BRD cells are lost from roots of some crop species (Hawes and Pueppke,

1986). The numbers of cells from aseptically grown seedlings can vary by orders of magnitude among different families, from more than 10,000 per root tip from cotton to fewer than ten from tobacco.

Root caps of *Arabidopsis thaliana* seedlings do not appear to release any BRD cells (Hawes, unpubl.).

It is interesting to note that the number of BRD cells from individual roots is a property that is highly conserved at the species and family levels. For example, the mean values for five legume species range from 2800 to 3100, and are not statistically distinct from each other (Hawes and Pueppke, 1986). The numbers do not appear to be correlated with root size: the small roots of eggplant, *Amaranthus tricolor*, and carrot seedlings yield mean values of 15, 126, and 1000 cells per root, respectively (Hawes and Pueppke, 1986). Viability of BRD cells also can vary among species in different families. While populations of cells from most species in ten families exhibit viability of 90% or higher, values for *Solanum* species are significantly lower. In fact, in two species of the Compositae family, sunflower and zinnia, most BRD cells are dead when they separate from the root (Hawes, unpublished).

At this time there is no obvious association between plant growth properties among species that release large numbers of viable BRD cells and those that do not. However, it is clear that plants of many species commit a large amount of energy to the daily production of thousands of healthy cells that are destined to be lost in to the soil. It is likely that this expensive process confers some selective advantage to the species that maintain it.

Proposed functions of the root cap

Perspectives about the root cap have changed little since 1914, when Haberlandt suggested that the primary functions of the root cap include gravity perception and physical protection of the meristem (reviewed in Barlow, 1975; Lucas, 1987). Although mechanistic details continue to be subject to debate, evidence for a role in gravitropism is good (reviewed in Moore and Evans, 1986). Physical protection was thought to be provided by two means: The polysaccharide slime produced by the root cap was believed to lubricate passage of the root through soil. In addition, dying cells released from the cap were thought to provide mechanical protection against abrasion of the tip (Voeller *et al.*, 1964). However, studies from McCully's laboratory have raised questions about the generally accepted picture of dying root cap cells and the slime that encase them as serving a rather passive function in assisting plant root growth. Apart from the discovery that BRD cells are viable (see above), Guinel and McCully (1986) recently reported that maize root cap mucilage actually has limited water-holding capacity and would not be particularly effective as a lubricant under most conditions. Although lubricative capacity of the mucilage is open to question as a result of those data, the physical shedding of BRD cells is likely to reduce the friction of root passage to some extent. Nevertheless, it is clear that the release of large numbers of cells of living cells is not required from the fact that species which slough fewer than ten cells a day exhibit normal root development (Hawes and Pueppke, 1986). A correlation between root size and cell number might be expected if cells were required for physical protection against abrasion but, as previously mentioned, such relationship was not observed in studies with seedlings. An understanding of the role of BRD cells in plant growth and development warrant more research in the light of recent studies.

It is important to point that little is known about the state of BRD cells, or indeed any root cells, under natural conditions (McCully, 1987). The variation in numbers and viability of BRD cells even *in vitro* means that generalizations about BRD cells have to be made with caution.

Root cap cells as a source of root exudates

A significant amount of fixed carbon is lost into the rhizosphere of plants; estimates of net carbon lost as exudates range from 1% to 40% (for example Barber and Martin, 1976; Haller and

Stolp, 1985; Merckx *et al.*, 1986; Prikryl and Vancura, 1980; Whipps and Lynch, 1985; reviewed in Curl and Truelove, 1986). One biological effect of this material is clear: numerous papers over the last 100 years have documented that root exudates can influence microbial populations in the vicinity of the root (reviewed in Curl and Truelove, 1986; Rovira, 1969; Rovira, 1985; Weller, 1988). Population levels of soilborne microorganisms routinely are found to be much higher in the rhizosphere than in bulk soil, because of the stimulating effects of root exudates. Recently, components of root exudates actually have been shown to directly affect the expression of genes involved in symbiosis and pathogenesis (Peters *et al.*, 1986; Redmond *et al.*, 1986; Shearman *et al.*, 1986; Stachel *et al.*, 1985; Van Brussel *et al.*, 1986; Zaat *et al.*, 1987). For example, acetosyringone and other phenolic compounds induce expression of virulence genes of *Agrobacterium tumefaciens* (Spencer and Towers, 1988; Stachel *et al.*, 1985).

Secretory cells of the root cap, together with BRD cells, are major contributors of root exudates (Bacic *et al.*, 1986; Chaboud, 1983; Griffin et al., 1976; Miki *et al.*, 1980; Morre *et al.*, 1967; reviewed in Oades, 1978). Although few attempts have been made to assess the quantitative and qualitative contribution of BRD cells, living or dead, there can be no doubt that much of the microbial growth-stimulating photosynthate lost from the roots of many species enters the rhizosphere in the form of living cells. I will review evidence for several host-pathogen combinations which I believe suggests that the nutrients in BRD cells may be preferentially more accessible to some microorganisms than others.

BRD cells as a system for evaluation of cell-cell interactions

To my knowledge, nothing is known about the physiological state and metabolic activity of living BRD cells in the soil. However, the cells happen to be uniquely suited to the study of cell surface properties and their role in microbial-plant interactions under controlled conditions (Hawes and Wheeler, 1982; Sherwood, 1987). The evaluation of cellular responses *in vitro* is made simple by the normal way the cells are released: because they are programmed to separate from the root, they can be isolated nondestructively by dipping the root into a small volume of liquid and agitating gently. The plant is undamaged by the procedure, and the cells retain surfaces that are intact and in the form in which they are naturally encountered by microorganisms. With many species yields from single roots are sufficient for physiological and biochemical studies, and cell populations can be serially diluted to determine dosage response effects (Hawes *et al.*, 1988). Significantly, the same plant from which the cells are obtained can then be inoculated to test for whole-plant response to the microorganism, or it can be grown to maturity and progeny can be harvested to evaluate inheritance of phenotypes of interest (Hawes and Pueppke, 1985; 1987; 1989). Because the single cells have walls, no osmoticum is required for stability, and the cells can survive in distilled water for a day or more (Hawes, 1983a, b). The cells can be induced to divide and grow into callus, or they can be retained in their original developmental state for weeks or months (Caporali, 1983; Hawes and Pueppke, 1986; Hawes and Wheeler, 1982; Knudson, 1919).

Properties of BRD cells with respect to pathogens

Genotype dependent responses of BRD cells to fungal pathogens

The first hint that BRD cells have selective properties was provided by studies with host-specific fungal toxins, whose activity can be conveniently assayed using isolated root cap cells (Hawes, 1983). Toxins from a number of leaf-invading fungal pathogens such as *Cochliobolus victoriae* and *Helminthosporium maydis*, race T severely damage or kill plants of genotypes that are susceptible to the fungus, but they cause little or no damage to resistant genotypes. Toxin assays revealed that the phenotype of the whole plant is expressed in populations of root cap cells: Cells from plants that are highly susceptible to the fungus are killed in minutes, whereas cells from resistant genotypes are insensitive to the toxin (Hawes and Wheeler, 1982; 1984; Hawes, 1983a, b). BRD cells are also capable of induced defensive responses, and can be invaded by fungal hyphae in a genotype-dependent manner (Sherwood, 1987). In response to inoculation of maize BRD cells with *Colletotrichum graminicola*, papilla formation and infection among different genotypes are correlated with lesion development in inoculated leaves of intact plants. Conclusions that can be made about BRD cells from these studies are limited by the artificial nature of these host-pathogen combinations: roots are not the natural sites of infection of *C. graminicola* and toxin-producing fungi. Nevertheless, the discovery that BRD populations mimic whole-plant responses in a genotype-dependent manner suggests that they can potentially exert selective effects on microbial populations.

BRD cell-specific fungal chemotaxis, binding, and infection

To approach a more realistic evaluation of BRD cell properties, zoospores of soilborne fungi were used to assess the role of BRD cells in recognition of host roots (Goldberg *et al.*, 1989). Chemoattraction, attachment, and penetration of BRD cells by zoospores, before and after separation of the plant cells from the root, were evaluated. Care was taken to ensure that BRD cells were still attached when the root was immersed into a suspension of zoospores. Zoospores of *Pythium dissotocum* were found to be instantly attracted to the root cap region of cotton seedlings, which extends approximately 1-2 mm from the tip, in such large numbers that the natural BRD cell separation process is inhibited. The response is tissue-specific, with a clear line demarcation above the point where BRD cells are attached to the root. Zoospores are not attracted to sites above this line, even when the root is wounded. Attachment and penetration of the cells occurs rapidly, and root necrosis follows within 24 hours. Zoospores of a second species, *P. catenulatum*, are not attracted to BRD cells on the roots of cotton seedlings and do not infect the plants.

Of particular interest was the finding that when BRD cells are washed from the root, attraction of *P. dissotocum* zoospores to the root is almost entirely eliminated (Goldberg *et al.*, 1989). Only a small, highly localized group of BRD cells at the very tip of the root that are still in the process of sloughing, remain attractive. In contrast, BRD cells washed from the root and suspended in water remain attractive to *P. dissotocum* zoospores, which attach, penetrate and kill the cells within minutes (Fig. 1D, F). On the other hand, *P. catenulatum* zoospores do not infect the cells and do not appear to utilize them for growth. It is important to note that BRD cells themselves, not just the soluble materials lost from them, are crucial to the interaction. Indeed, *P. catenulatum* zoospores are attracted to components of root and BRD cell extracts, even though they are not attracted to the intact cells (Goldberg, Timmermann and Hawes, unpublished).

Agrobacterium tumefaciens *binding and chemotaxis to BRD cells*

The above described studies of BRD cell interactions with a variety of fungi demonstrated that plant genes that encode responses to pathogens are expressed in populations of BRD cells. Because the cells can in effect expand the physical range of host-specific responses beyond the boundary of the root, this discovery has potentially important implications for the biology of the rhizosphere. I have used *A. tumefaciens* as a model system to test the hypothesis that recognition of BRD cells by soilborne bacteria may influence colonization (Hawes and Pueppke, 1985; 1987; 1989a; 1989b; Hawes *et al.*, 1989). As in studies with zoospores, I used chemotaxis and binding as indicators of recognition because these phenomena may be important preludes to rhizosphere colonization and root infection (James *et al.*, 1985, Vesper et al., 1987). Initial studies showed that variation in surface properties of BRD cells from different plant species result in a dramatic selectivity in ability to bind *A. tumefaciens* (Hawes and Pueppke, 1985; 1987). Thus, BRD cells from some plant species bind bacteria in large numbers (Fig. 1G), whereas BRD cells from other species bind few or no bacteria. Binding responses are strongly correlated with susceptibility of plants to crown gall tumorigenesis.

BRD cells from different plants also vary in their ability to act as chemoattractants for the bacteria. For example, BRD cells from pea and corn are attractants for *A. tumefaciens* cells, but cells from bean and cotton (which are attractive to *P. dissotocum* zoospores) do not attract the bacteria (Hawes *et al.*, 1988). I set out to use mutants of *A. tumefaciens* to exploit such differences in order to determine if recognition of BRD cells is important in colonization and disease. In preliminary experiments, chromosomal virulence mutants which are deficient in both binding and in motility were found to have reduced ability to colonize the rhizosphere of pea (Hawes and Pueppke, 1989a). In subsequent experiments, a directional swarm assay (Fig. 1F) was used to characterize chemotaxis to suspensions of BRD cells and to select mutants with more specific defects in recognition of roots and root cap cells (Hawes *et al.*, 1988). Ten mutants with several distinct phenotypes, including nonmotile, slow-migrating, and chemotaxis-deficient mutants, were isolated. All exhibit wild type levels of binding to root cap cells, but all are deficient in chemotaxis to root cap cells. One of these mutants retained an attraction to excised root tips, even though it is not attracted to isolated BRD cells. The phenotype suggests that this "wound-specific" mutant is attracted to substances that are released from cells damaged when the tip is excised from the root.

When virulence of the mutants on pea plants was tested in an *in vitro* growth pouch assay, their tumor-inducing capacities were indistinguishable from that of the parent strain (Hawes *et al.*, 1989; Hawes and Smith, 1989). The mutants were also fully virulent on plants grown in sand. However, they were avirulent on plants grown in silty loam soil. The date indicate that chemotaxis is important to the pathogenicity of *A. tumefaciens* under some environmental conditions, but not in others. This finding may explain contradictory reports of the relative significance of chemotaxis and motility in the establishment of microbial-plant associations. Thus, Scher *et al.* (1984), Scher *et al.* (1988) and Howie *et al.* (1987) reported that nonmotile mutants or strains of bacteria colonize plant roots as effectively as motile strains. In contrast, deWeger *et al.* (1987) found that colonization by nonmotile mutants of *Pseudomonas fluorescens* was drastically reduced, by several orders of magnitude, compared with colonization by the motile parent strain. Others have reported that motility and chemotaxis in *Rhizobium meliloti* facilitates competitiveness in the establishment of symbiosis (Ames and Berman, 1981; Caetano-Anolles *et al.*, 1988).

It is of particular interest that the "wound-specific" *A. tumefaciens* mutants, which is attracted to excised root tips but not isolated root cap cells, is no more pathogenic than nonmotile mutants. Although wild type *A. tumefaciens* is attracted to both BRD cells and excised root tips, these data suggest that, under the test conditions, attraction to BRD cells is more important than attraction to substances from wounds: loss of attraction to BRD cells always results in reduced virulence but loss of attraction to excised root tips does not. Presumably chemotaxis to BRD is required in order for bacteria to get into the vicinity in sufficient numbers to cause infection. Other studies have indicated that attraction of *A. tumefaciens* to substances that induce virulence genes is effective at extremely low concentrations, and it has been suggested that chemotaxis to such compounds from wounds is a crucial early step leading to infection (Ashby *et al.*, 1988). Perhaps attraction to wounds facilitates infection once bacteria are in the rhizosphere.

Speculation about BRD cell impact on rhizosphere biology

What selection pressure exists for some plant species to maintain a system that must require the coordination of a number of genes, and the expense of a great deal of energy? One possibility is that the ability of BRD cells to carry host-specific traits into the rhizosphere acts to regulate levels of microbial populations. Others have speculated that components of root exudates may function in extracellular recognition by microorganisms (for example, Oades, 1978), and recent reports that specific molecules can exert direct effects on gene expression, mentioned previously, support this long-standing hypothesis. I believe that the expression of selectivity at the cellular level may act to regulate populations, particularly at the root tip, in ways that cannot be implemented through the chemical structure of soluble root exudates alone. Selective attraction and binding of beneficial microorganisms, such as nitrogen fixing or plant growth promoting bacteria, to BRD cells could increase their access to nutrients in the cells and stimulate their growth at the expense of pathogens. Alternatively, the expression of resistance genes in BRD cells in the rhizosphere might directly suppress pathogen populations before they reach the root.

A third possibility is that the cells could provide a protective edge for the growing root tip by acting as a natural "trap" for pathogenic organisms. Under the conditions of constrained motility that are operative in most soils, chemotaxis and binding to BRD cells that are separated from the root could immobilize a pathogen while the root tip passes safely by. This could be especially effective for slow-growing bacteria or for zoospores which lose their motility upon contact with host signals. If nutrients in the invaded cell were insufficient to support growth or movement toward the root, the infection would be effectively halted. This hypothesis is untested. However, it is interesting to point out that when roots harvested from nonsterile soil are washed to remove unbound or loosely bound organisms, the tips are frequently found to be free of microbial colonization (Curl and Truelove, 1986). Similar results have occurred in *in vitro* assays of fungal chemotaxis in which roots are washed either before or after exposure to zoospores. For example, Longman and Callow (1987) found that when cress seedlings were grown on moist filter paper, incubated with *Pythium aphanidermatum* zoospores and then washed thoroughly, zoospore attachment was not observed in the root tip region. Similar results were obtained with *Phytophthora cinnamoni* zoospores and maize roots (Hinch and Clarke, 1980). Such findings have led to an assumption that microorganisms do not colonize the root tip as readily as other parts of the root. An alternative explanation is that microorganisms colonize BRD cells which are removed together with attached bacteria when the roots are washed prior to experimental observation.

The only genetic evidence that recognition of BRD cells has an impact on colonization, from studies with *A. tumefaciens* mutants, suggests that chemotaxis to the cells can be important. However, it is impossible to draw conclusions about the relative benefits of the phenomenon to the plant. Thus, *A. tumefaciens* – with a tumor-inducing plasmid – is a serious pathogen of horticultural crops, whereas *A. radiobacter* – without a Ti plasmid – is the most effective biological control agent currently available. Because the genes for binding and chemotaxis to root cap cells are encoded by chromosomal genes that are present in both *A. tumefaciens* and *A. radiobacter*, the effect could either be beneficial or detrimental to the plant. If recognition of BRD cells does play a role in colonization of the rhizosphere in nature, the relative benefits or detriments to the plant are likely to depend on the environment and on the particular host-microbe combination.

References

Ames P and Bergman K 1981 Competitive advantage provided by bacterial motiliy in the formation of nodules by *Rhizobium meliloti*. J. Bacteriol. 148, 728–729.

Ashby A M, Watson M D, Loake G J and Shaw C H 1988 Ti plasmid specified chemotaxis of *Agrobacterium tumefaciens* C58C toward vir inducing phenolic compounds and soluble factors from monocotyledonous and dicotyledonous plants. J. Bacteriol. 170, 4181–4187.

Bacic A. Moody SF and Clarke A E 1986 Structural analysis of secreted root slime from maize. Plant Physiol. 80, 771–777.

Barber D A and Martin J K 1976 The release of organic substances by cereal roots into soil. New Phytol. 76, 69–80.

Barlow P W 1975 The root cap. *In* The Development and Function of Roots, Eds. J G Torrey and D T Clarkson. pp 21–54. Academic Press London, New York and San Francisco.

Bennet R J, Breen C M and Fey M V 1987 The effects of aluminum on root cap function and root development in *Zea mays*. Environ. Exp. Bot. 27, 91–104.

Caetano-Anolles G, Wall L G, De Micheli A T, Macchi E M, Bauer W D and Favelukes G 1988 Role of motility and chemotaxis in efficiency of nodulation by *Rhizobim meliloti*. Plant Physiol. 86, 1228–1235.

Caporali L 1983 Cytological study of cultured cells from maize root cap. Plant Sci. Lett. 31, 231–236.

Chaboud A 1983 Isolation, purification and chemical composition of maize root cap slime. Plant and Soil 73, 395–402.

Curl E A and Truelove B 1986 The Rhizosphere. Berlin: Springer-Verlag. 288 p.

DeWeger L A, Van Der Vlugt C I M, Wijfjes A H M, Bakker P A H M, Schippers B and Lugtenberg B 1987 Flagella of a plant-growth-stimulating *Pseudomonas fluorescens* strain are required for colonization of potato roots. J. Bacteriol. 169, 2769–2773.

James D W, Suslow T V and Steinback K E 1985 Relationship between rapid, firm adhesion and long term colonization of roots by bacteria. Appl. Environ. Microbiol. 50, 392–397.

Feldman L J 1984 The development and dynamics of the root apical meristem. Am. J. Bot. 71, 1308–1314.

Gautheret R 1933 Cultures de cellules detachees de la coiffe. C.R. Acad. Sci. 186, 638–640.

Goldberg N P, Hawes M C and Stanghellini M E 1989 Specific atraction to and infection of isolated cotton root cap cells by zoospores of *Pythium dissotocum*. Can. J. Bot. 67, 1760–1767.

Griffin G J, Hale M G and Shay F J 1975 Nature and quantity of sloughed organic matter produced by roots of axenic peanut plants. Soil Biol. Biochem. 8, 29–32.

Guinel F C and McCully M E 1986 Some water-related physical properties of maize root cap mucilage. Plant, Cell Environ. 9, 657–666.

Guinel F C and McCully M E 1987 The cells shed by the root cap of Zea: their origin and some structural and physiological properties. Plant, Cell Environ. 10, 565–578.

Haller T and Stolp H 1985 Quantitative estimation of root exudation of maize plants. Plant and Soil 86, 217–216.

Hawes M C 1983a Use of isolated root cap cells in a simple assay for HMT toxin. Phytopathology 73, 1184–1187.

Hawes M C 1983b Sensitivity of isolated oat root cap cells and protoplasts to victorin. Physiol. Plant Pathol. 22, 65–76.

Hawes M C and Lin H J 1990 Correlation of pectolytic enzyme activity with the programmed release of cells from the root cap of *Pisum sativum*. Plant Physiol. (*in press*).

Hawes M C and Pueppke S G 1986 Isolated peripheral root cap cells: Yield from different plants, and callus formation from single cells. Am. J. Bot. 73, 1466–1473.

Hawes M C and Pueppke S G 1987 Correlation between binding of *A. tumefaciens* by root cap cells and susceptibility of plant to crown gall. Plant Cell Reports 6, 287–290.

Hawes M C and Puepke S G 1985 Relationship between binding of root cap cells to *Agrobacterium tumefaciens* and susceptibility of plants to crown gall. *In* Current Topics in Plant Biochemistry and Physiology 4, 244–245.

Hawes M C and Pueppke S G 1989 *Agrobacterium tumefaciens* chromosomal virulence mutants are deficient in rhizposhere colonization ability. Plant and Soil 113, 129–132.

Hawes M C and Pueppke S G 1989 Variation in binding and virulence of *Agrobacterium tumefaciens* chromosomal virulence mutants on different plants. Plant Physiol. 90, 113–118.

Hawes M C, Robbs S L and Pueppke S G 1989 Use of a growth pouch assay to measure genotypic variation in susceptibility of *Pisum sativum* to crown gall. Plant Physiol. 90, 180–184.

Hawes M C and Smith L Y 1989 Requirement for chemotaxis in pathogenicity of *Agrobacterium tumefaciens* on roots of soil-grown pea plants. J. Bacteriol. 171, 5668–5671.

Hawes M C, Smith L Y and Howarth A J 1988 Selection and characterization of *A. tumefaciens* mutants deficient in chemotaxis to root exudates. Molec. Plant Microbe Interactions 1, 182–186.

Hawes M C and Wheeler H 1982 Factors affecting victorin-induced cell death: temperature and plasmolysis. Physiol. Plant Pathol. 20, 137–144.

Hawes M C and Wheeler H 1984 Detection of effects of nuclear genes on susceptibility to *Helminthosporium maydis* race T by a root cap cell bioassay for HMT-toxin. Physiol. Plant Pathol. 24, 163–168.

Hinch J M and Clarke A E 1980 Adhesion of fungal zoospores to root surfaces is mediated by carbohydrate determinants of the root slime. Physiol. Plant Pathol. 16, 303–307.

Howie W J, Cook R J and Weller D M 1987 Effects of soil matric potential and cell motility on wheat root colonization by fluorescent pxseudomonads suppressive to take-all. Phytopathology 77, 286–292.

Knudson L 1919 Viability of detached root cap cells. Am. J. Bot. 6, 309–310.

Longman D and Callow J A 1987 Specific saccharide residues are involved in the recognition of plant root surfaces by zoospores of *Pythium aphanidermatum*. Physiol. Mol. Plant Pathol. 30, 139–150.

Lucas W J 1987 Functional aspects of cells in root apices. *In* Root Development and Function. Eds. P J Gregory, J V Lake and D A Rose pp 26–52. Cambridge University Press, Cambridge.

McCully M E 1987 Selected aspects of the structure and development of field grown roots with special reference to maize. *In* Root Development and Function. Eds. P J Gregroy, J V Lake and D A Rose. pp 52–70. Cambridge University Press, Cambridge.

Merckx R, Van Ginkel J H, Sinnaeve J and Cremers A 1986 Plant-induced changes in the rhizosphere of maize and wheat. Plant and Soil 96, 85–93.

Miki N K, Clarke K J and McCully M E 1980 A histological and histochemical comparison of the mucilages on the root tips of several grasses. Can J Bot 58, 2581–2593.

Moore R and Evans M L 1986 How roots perceive and respond to gravity. Am. J. Bot. 73, 574–587.

Moore R, McClelen C E, Fondren W M and Wang C L 1987 Influence of microgravity on root cap regeneration and the structure of columella cells in *Zea mays*. Am. J. Bot. 74, 218–223.

Moore R and McClelan C E 1983 Ultrastructural apsects of cellular differentiation in the root cap of *Zea mays*. Can. J. Bot. 61, 1566–1572.

Morre D J, Jones D D and Mollenhauer H H 1967 Golgi apparatus mediated polysaccharide secretion by outer root cap cells of *Zea mays*. Planta 74, 286–301.

Oades J M 1978 Mucilages at the root surface. J. Soil Sci. 29, 1–16.

Parke D, Ornston L N and Nester E W 1987 Chemotaxis to plant phenolic inducers of virulence genes is constitutively expressed in the absence of the Ti plasmid in *Agrobacterium tumefaciens*. J. Bacteriol. 169, 5336–5338.

Paull R E and Jones R L 1976 Studies on the secretion of maize root cap slime. V. The cel wall as a barrier to secretion. Z. Pflanzenphysiol. 79, 154–164.

Peters N K, Frost, J W and Long S R 1986 A plant flavone, luteolin, induces expression of *Rhizobium meliloti* nodulation genes. Science 233, 977–980.

Prikryl Z and Vancura V 1980 Root exudates of plants. Plant and Soil 57, 69–83.

Redmond J W, Batley M, Djordjevic M A, Innes R W, Kuempel P L and Rolfe B G 1986 Flavones induce expression of nodulation genes in *Rhizobium*. Nature 323, 623–635.

Rost T L and Jones T J 1988 Pea root regeneration after tip excisions at different levels: polarity of new growth. Ann. Bot. 61, 513–523.

Rovira A D 1969 Plant root exudates. Bot. Rev. 35, 35–37.

Rovira A D 1985 Manipulation of the rhizosphere microflora to increase plant production. *In* Reviews of Rural Science 6: Biotechnology and Recombinant DNA Technology in the Animal Production Industries. Eds. R A Leng, J S F Barker, D B Adams and K J Hutchinson. pp 185–197. University New England Press.

Scher F M, Ziegle J S and Kloepper J W 1984 A method for assessing the root colonizing capacity of bacteria on maize. Can. J. Microbiol. 30, 151–157.

Scher F M, Kloepper J W, Singleton C, Zaleska I and Laliberte M 1988 Colonization of soybean root by *Pseudomonas* and *Serratia* species: relationship to bacterial motility, chemotaxis, and generation time. Phytopathology 78, 1055–1059.

Shearman C A, Rossen L, Johnston A W B and Downie J A 1986 The *Rhizobium leguminosarum* nodulation gene nodF encodes a popypeptide similar to acyl carrier protein and is regulated by nodD plus a factor in pea root exudate. EMBO J.5, 647–652.

Sherwood R T 1987 Papilla formation in corn root cap cells and leaves inoculated with *Colletotrichum graminicola*. Phytopathology 77, 930–934.

Spencer P A and Towers G H N 1988 Specificity of signal compounds detected by *Agrobacterium tumefaciens*. Phytochemistry 27, 2781–2785.

Stachel S E, Messens E, Van Montagu M and Zambryski P 1985 Identification of the signal molecules produced by wounded plant cells that activate T-DNA transfer. Nature 318, 624–629.

Van Brussel A, Zaat S A, Cremers H C, Wijffelman C A, Pees E, Tak Teun and Lugtenberg B J J 1986 Role of plant root exudate and sym plasmid-localised nodulation genes in the synthesis by *Rhizobium leguminosarum* of Tsr factor, which causes thick and short roots of common vetch. J. Bacteriol. 165, 517–522.

Vermeer J and McCully M E 1982 The rhizosphere in Zea: new insight into its structure and evelopment. Planta 156, 45–61.

Vesper S J, Malik N S and Bauer W D 1987 Transposon mutants of *Bradyrhizobium japonicum* altered in attachment to host roots. Appl. Environ. Microbiol. 53, 1959–1961.

Voeller B R, Ledbetter M C, Porter K R 1964 The plant cell: aspects of its form and function. *In* The Cell, Vol. 6. Eds. J Bracket and E E Minsky. pp 245–312. Academic Press, New York and London.

Weller D M 1988 Biological control of soilborne plant pathogens in the rhizosphere with bacteria. Annu. Rev. Phytopathol. 26, 379–407.

Whipps J M and Lynch J M 1985 Energy losses by the plant in rhizodeposition. Ann. Proc. Phytochem. Soc. Eur. 26, 59–71.

Zaat S A J, Van Brussel A A N, Tak T, Pees E and Lugtenberg B J J 1987 Flavonoids induce *Rhizobium leguminosarum* to produce nodDABC gene related factors that cause thick, short roots and root hair responses on common vetch. J. Bacteriol. 169, 3388–3391.

D. L. Keister and P. B. Cregan (Eds.), The rhizosphere and plant growth, 61–71.

PLSO BARC07

Root growth and development*

RICHARD W. ZOBEL
USDA-ARS-USPSNL and Agronomy and Plant Breeding Departments, Cornell University, Ithaca, NY 14853, USA

Key words: environmental variation, genotypic variation, GxE interaction, root functionality, root terminology

Abstract

Much detailed basic research on root growth and development was undertaken during the early portion of this century. Recent research has explored some of the underlying anatomical, physiological and genetic mechanisms driving root growth. There have been three major constraints to plant rhizosphere research: 1) inconsistent use of an inexact terminology, 2) inadequate information about different root types (morpho-types) and their functional relationships, and 3) extreme variability. Information is accumulating which suggests that although roots all have apparently very similar anatomies, their detailed anatomy may differ dramatically, and their functions may be very different. The concept of continual growth and die-back of small roots impacts on the understanding of the bases for soil floral and faunal biodynamics. That some roots may have no active role in nutrient or water uptake forces a re-evaluation of even the most tightly held concepts. This paper suggests that genotype by environment interaction (a necessary evolutionary and adaptive condition for plant rooting) is responsible for the observed extreme variability. This variability has forced the conclusion that rooting patterns are under multi- or polygenic control. However, single gene root mutants discovered in tomato and statistical analyses designed to handle genotype by environment interaction suggest that this perceived genetic complexity is conditioned by a simple genetic system with extensive genotype by environment interaction.

Introduction

It can be suggested that: man knows more about the vast expanses of outer space than he does about the workings of the 24 inches of soil just below the earth's surface. This symposium addresses a portion of these 'workings', but only a portion. There are some similarities between the US Space program and rhizosphere research – their shuttles and our rhizosphere organisms have significant amounts of redundancy to guard against catastrophe. On the other hand, the two are dissimilar. The space program can be accurately called a shot into the blue and rhizosphere research is more often than not a shot in the dark. Not only do we not know exactly what we are going to find, but we work in an inherently opaque medium.

The nature of the task of adequately studying and describing roots began to be apparent during routine attempts at describing plant root systems in the late 1800's and earlier. The opacity of the soil media has continuously frustrated researchers attempting to develop a concept of root growth patterns, microorganism population distributions, and *in-situ* rhizoplane soil chemical and physical characteristics. Rhizosphere researchers must remain continually cognizant of the fact that they are looking at the results of a complex evolutionary phenomenon. The development of an integrated sometimes balanced, sometimes not, series of interactions between

* This paper is based, in part, on research undertaken as part of the program of the USDA/Cornell Center for Root-Soil Research.

fungi, microorganisms, soil flora and fauna, plant roots, soil physical and chemical characteristics and the weather is an evolutionary and ecological process of extreme proportions. As a result, what we see in the rhizosphere is evolutionarilly sensible and based on solid developmental principals. It is the task of the rhizosphere scientist to discover and elucidate those principals. Using relatively crude methods (a shovel and other small tools, for example) many good descriptions of natural plant root systems have been developed (see Weaver, 1919; 1920; 1926; Cannon, 1911, 1949; Kuschera, 1960). Because of the inherently tedious nature of available techniques, little replication was accomplished, and very few attempts at screening for genetic versus environmental variability attempted.

General pattern of growth and development

With one major exception, the general pattern of root growth and development is essentially the same for all species of plant. The resulting overall root system morphology is, however, highly diverse with species, cultivar and even family specific characteristics. The exception to the underlying similarity is the development of secondary vascular tissues in dicotyledonous but not monocotyledonous plants. Root growth begins with the first stages of activity in the embryo, and continues until well after the death of the shoot. Radicle elongation frequently signals outward signs of the beginning of germination. In species lacking a radicle, a seminal root usually acts as a replacement radicle. As this first root (referred to as the tap root) elongates, it develops root hairs and in two to three days the first lateral branches begin to develop at its base and in an acropetal direction (relative to the tap root tip). Depending on species or cultivar characteristics, seminal roots begin to emerge from the embryo/seedling with in the first day or two in an acropetal (relative to the shoot) direction (Weinhold, 1967). As these seminal roots elongate, they also develop root hairs and eventually lateral roots. Under appropriate environmental conditions the interactions between plant roots and the microorganisms of the rhizosphere can begin as the radicle elongates. Bradyrhizobia, for instance, can infect the first soybean root hairs developing on the radicle, producing effective nodules in plants less than two weeks old.

In dicotyledonous plants these first or primary roots, the tap root and seminal roots, have secondary vascular development. As a result, they enlarge at their base, establishing many late metaxylem elements, and thereby increasing their capacity to transport water and included minerals and other substances to the shoot. As these roots enlarge, they form the characteristic structure of a mature root system (Fig. 1); producing in mature trees very dramatic root systems. Monocotyledonous plants, on the other hand, have roots that do not have secondary vascular development. This restricts water movement capacity to that of the vascular system which the root initially developed. In the grasses we are most familiar with, (*i.e.* corn, wheat, barley, rice) the plant initiates adventitious roots from the upper nodes of the plant. The seminal roots develop from what has been termed the coleoptilar node or hypocotyl in dicotyledons, while the adventitious roots develop from stem nodes or other non root tissues. Adventitious roots in these grasses are initiated in nodal sequence with larger diameter roots developing on the upper root forming nodes. On corn, these adventitious roots can frequently be up to 1 cm in diameter. The pattern of larger and larger adventitious roots as you go up the plant and the pattern of internode elongation characterizes mature root system pattern of these plants (Fig. 2) If the temporal characteristics of a site are known (*i.e.* seasonal temperature, rainfall, and irrigation patterns) this developmental pattern can serve as a historical description of the conditions the plant perceived. The combination of tap, seminal and adventitious roots constitutes what has been called the axile root system. The lateral roots developing from the axile roots constitute the secondary root system, more on terminology later.

Lateral root diameters, in monocotyledonous plants, are determined by the size of the parent root. The larger lateral roots have branches of their own, sometimes resulting in branching patterns with fourth and fifth order, or more, laterals. Lateral roots come in many different sizes

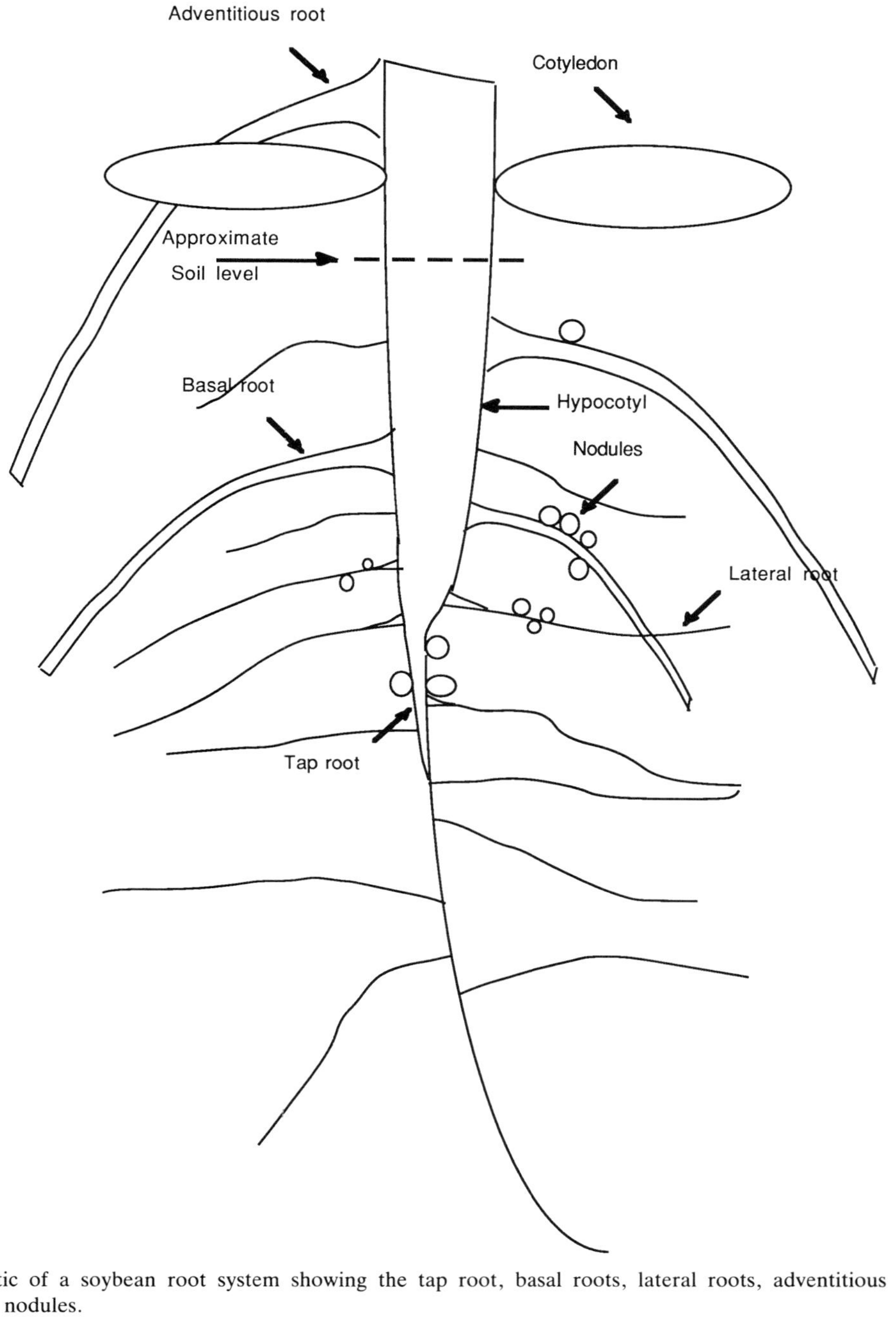

Fig. 1. Schematic of a soybean root system showing the tap root, basal roots, lateral roots, adventitious roots, hypocotyl, cotyledons, and nodules.

and shapes. Research has shown that lateral roots can have a very reduced anatomy with a cortex of only several cell layers and a single metaxylem element (McCully, 1987). In terms of water and nutrient uptake, these roots should be only slightly less efficient than root hairs. Research by Hackett and Rose (1972) clearly suggests that lateral roots of barley are determinant in growth. They describe two populations of lateral roots, one, the first order laterals, which attains an average maximum length of 2.5 cm and a second population, the second order later-

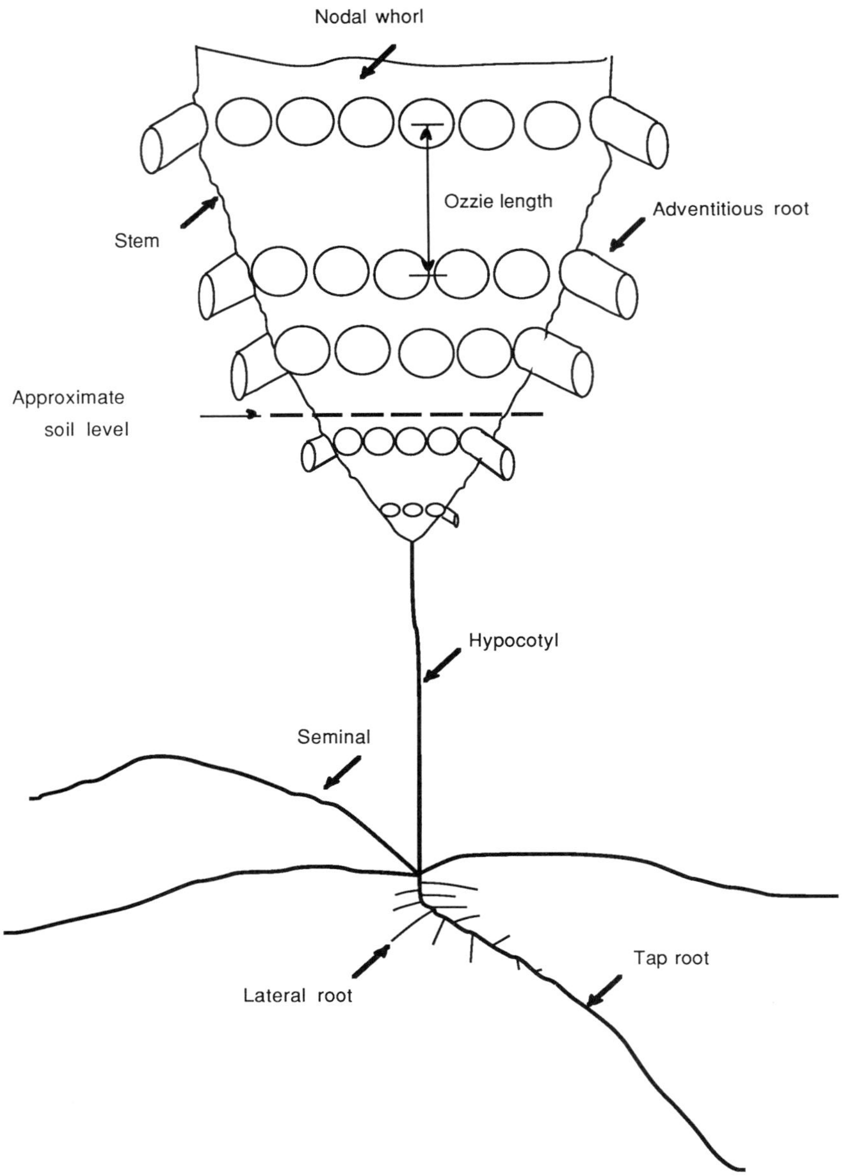

Fig. 2. Schematic of a corn root system showing the tap root, seminal roots, hypocotyl (coleoptilar node), lateral roots, adventitious roots and crown internodes (Ozzie length).

als, which attain an average maximum length of 0.7 cm. Using corn to confirm this finding, Cahn (1988) has shown a similar situation. In the corn study, the lateral roots could be further classified by diameter: roots less than 0.6 mm had a length of 2.5 cm or less and grew for an average of 2.5 days or less. On the other hand roots larger than 0.6 mm diameter grew indefinitely. Fusseder (1987) shows that roots of this type may loose their root caps, and maturate the apical cells, a response in keeping with the results of Cahn (1988) and DeFreitas (1988). Zobel (1989) has suggested that these small lateral roots are very short lived (approx. 2 weeks or less), and may provide a partial basis for the observed high rates of carbon dioxide production immediately under the plant (Buyanovsky and Wagner, 1983), by feeding the microbial population. Because of their lack of secondary thickening, the taproot, seminal roots and early adventitious roots effectively become vestigial in mature plants.

In dicotyledonous plants, the basal roots and tap roots continue to grow in diameter and therefore in water transport capacity. The early lateral roots also show significant secondary development as well. This type of development reduces the need for adventitious roots in most dicotyledonous species, yielding a mature root system which can be characterized as an expanded primary root system. On the other hand, the monocotyledonous root system is predominantly a secondary system.

The final form of a root system depends as much on the rhizosphere environment as it does on root types. Temperature, gases, soil structure, soil chemistry, and the microorganisms which modify these all control the rate and pattern of root development. Zobel (1989) suggests that CO_2 concentrations in excess of 0.2-0.5% at the plough layer are responsible for stimulating roots to grow outwards, leaving a 'vacant' region immediately underneath the plant. This vacant area can be observed in many field grown species like corn and soybean from a number of different locations. Irrigation patterns or rainfall intensity during the early part of the growing season will determine depth of penetration of roots into the soil profile. These patterns can be, and are, severely modified by soil physical conditions (*e.g.* compaction, soil pans, glacial till, *etc.*) and chemistry (*e.g.* acid subsoils). These environmental characters vary across regions and even across fields or small plots, resulting in variations in rooting patterns which are not entirely dictated by the genetic characteristics of the plant.

Constraints

Several constraints to further enhancement of knowledge of root growth and development have inhibited progress in root research: 1) Because we do not have an unequivocal basis for naming roots by morphology, developmental pattern, localization, or genetic control, a rather diverse set of names has developed. Because the naming of root types has been effectively random and based upon the particular interest of the researcher involved, a given type of root on a given plant may be labeled with different names by different individuals. The end result is that it is frequently difficult to read a research paper on roots and be certain which roots are being studied. 2) It is commonly assumed that all roots have the same, or very similar, functions. Part of this is based on their relatively uniform anatomy, part on the lack of evidence to the contrary. Increasing evidence shows that this may well be an erroneous assumption which leads to incorrect conclusions about roots and their functional relationship with the rhizosphere. 3) Because of environmentally induced changes in overall root system morphology in plants from even adjoining fields and plots, and a lack of clear genetic controls over rooting patterns, it has been assumed that the genetic control of root initiation and growth and function is at best complexly inherited. It also has not been possible to efficiently dissect the interaction patterns between genotype and environment. Counting the number of roots intersecting one line of a matrix of lines, does not discriminate between roots based on form, function, or site of origin, but it does provide the quantitative data needed for statistical analyses. Mounting evidence suggests that rooting characteristics may be simply inherited and their genotype by environment interactions predictable and describable.

These three constraints are closely interrelated, and the assumptions and conclusions predicated by one have a large impact on the others. This paper will attempt to establish a basis for alleviating these constraints, stimulating further research on plant roots.

Terminology

Esau (1965) has provided basic definitions for root classification. The first definition in order of ontogeny is: the embryonic radicle becomes the tap root when it elongates to become the primary root for the seedling plant. In this paraphrase of her definition, three different names for the same root are found. The first name, radicle, is a reference to a specific embryonic structure, and is relatively unambiguous. Tap root refers to the tendency of the first root to emerge from the germinating seedling to grow directly downward, or to 'tap' the soil. The term primary refers to the fact that this root is the first and the main root in supporting seedling growth and development. All very logical, but still three names for the same root. Frequently additional roots develop in the embryo and emerge shortly after the radicle begins elongation (Weinhold, 1967). These roots are said to be part of the primary root system. They are commonly called seminal roots in grasses, basal roots in dicotyledonous plants (Zobel, 1975), or grenzwürzeln (a term dating back to the mid 1800's and translated as: boundary roots; see Weinhold, 1966) in all species. Zobel (1986) equates grenzwürzeln to basal roots. Since in the dicotyledonous plants, these roots continue to be initiated up the hypocotyl during later stages of growth, use of the more restrictive term seminal would be inappropriate. On the other hand, these same roots are often referred to as secondary roots since their emergence is secondary to the emergence of the radicle. Other names are: lateral, adventitious, primary lateral, and primary adventitious roots. Eight names for this ontogenetically second order of roots, and this is not an attempt at an exhaustive survey of names. Unfortunately, most of our knowledge of root physiology has been derived from either or both of these first two classes of root.

Ontogentically the next class of root are those roots which branch from any other root: lateral roots. This name is clearly related to the site or organ from which the root arises, like radicle and seminal. Other names for these roots refer to their ontogeny of development: first, second, third, etc. order branches or laterals, or alternatively: secondary, tertiary, quaternary, etc., branches, laterals or roots. The specific name is dependent on the ontogeny of the root from which they arise. Additional names for lateral roots are: fibrous roots, feeder roots, stress roots, transient roots, collateral roots, etc. based on function, life cycle, appearance, or other classifying characteristics. It is assumed that the ectomycorrhizal roots are basically of this type.

The term secondary is sometimes used to refer to roots branching off of seminal roots, which are themselves frequently considered secondary roots. At other times secondary refers to the secondary root system, a classification which includes an ontogenetically fourth type of root: adventitious roots. Adventitious roots are defined by the site of origin and are restricted to roots arising from non root tissues. This differentiates them from lateral roots. Adventitious roots arise from stem, leaf, petiole, etc. tissues either as derivatives from a cambial layer, or as denovo meristematic development from parenchymatous cells. It is therefore commonly thought that the roots developing from tissue cultures should be classified as adventitious. Recent research (Zobel and Su, 1989) suggests that the tissue culture, or more specifically callus derived, roots of tomato are genetically lateral roots rather than adventitious roots, even though their origin appears to be essentially adventitious. Adventitious roots have been termed nodal roots (for the site of origin), stem born roots, prop roots (for their perceived function), aerial (aereal) roots, secondary roots (as opposed to the tap and seminal roots which are then termed primary roots), etc. Branches off of adventitious roots then are by definition lateral roots and use the same terminology as above, except that a few authors prefer to differentiate between laterals from adventitious roots and laterals from the tap or seminal roots.

Table 1 presents a collection of some of these terminologies and a suggested compromise terminology which is, in general, compatible with the historical naming conventions. I will use this terminology throughout the remainder of this paper. Where a descriptive name for a type of root is more appropriate, I will define it in terms of the suggested terminology. Basal roots of both monocotyledons and dicotyledons have been referred to as adventitious by some authors and as lateral roots by others. Aung (1982) reviewed the evidence from tomato, and concluded, based on the genetic evidence of Zobel (1975) and his developmental evidence, that the hypocotyl born roots of tomato were distinctly different from lateral and adventitious roots. The unintentional misclassification of one class of root into another class, naturally leads to difficulty in characterization of the genetic, developmental, and physio-

Table 1. Recommended, developmental and other names for different root types

Recommended name	Developmental name	Other names in use
Taproot	radicle	primary root, axile root
Basal	basal, seminal, grenzwürzeln	primary lateral, secondary, primary adventitious, lateral, adventitious, axile, . . .
Lateral first order second order *etc.*	lateral	secondary, fibrous, tertiary, feeder, first order laterals, stress, axile branches, transient, collateral, . . .
Adventitious	adventitious	axile, nodal, stem born, prop roots, aerial, . . .

logical characteristics of that later class of root. For economics and convenience much of the research on root physiology has been done on basal roots because, after the emergence of the radicle, they are the first roots to develop in seedlings. The resulting data may or may not have any basis in reality when applied to lateral or adventitious roots.

Root type/function

The extended radicle (or its replacement basal root in many monocotyledons) is called the tap root because of its usual growth pattern and primary function. The tap root provides the first inputs of water and minerals and the sensing of the root environment which is so critical to the functioning of the whole plant. Whether this root is genetically or physiologically different from basal roots is an open question. It certainly is not where it derives from a seminal replacement of the radicle. Because of the small size of the seedlings, this root is not required to be very efficient at nutrient and water uptake. The very similar basal roots emerge within hours or days of radicle elongation. The number of basals is genetically fixed in a given cultivar, but can vary widely due to environmental interaction. As mentioned earlier, the basal roots and tap root are the most thoroughly studied group of roots because of their early appearance in the seedling.

Adventitious roots when arising as nodal roots in the grasses or stem bound roots in dicotyledons can form a significant portion of the root system. In the grasses, including corn and wheat, these roots are the predominant structural component of the root system of mature plants from anthesis on. Historically, anatomical evidence has suggested that these roots are no different than the basal roots which have been so thoroughly investigated. Recent evidence from McCully's lab suggests that adventitious roots of corn are relatively nonfunctional in active uptake of water and nutrients for transport to the shoot (McCully, 1987; McCully and Canny, 1989). St. Aubin *et al.* (1986) describe field grown corn adventitious roots as having a mucilaginous sheath which can cover the apical meter of a two meter long root. When the anatomy of the late metaxylem in these roots was investigated they found that within that portion of the root covered by the sheath, the late metaxylem had intact cell walls and live nuclei, with no casparian strip development in the endodermis. Immediately distal to the sheath, they found that the root had matured, sloughing the epidermis, a mature hypodermis had formed, a casparian strip had developed in the endodermis, and the late metaxylem had lost their nuclei and appeared to have lost their end walls. Under these conditions, the nodal roots of corn provide open pipes to the shoot for the water and nutrients taken up by their lateral branches.

In other research they have shown that late metaxylem of soybean tap and basal roots also mature late in greenhouse soil grown plants (Kevekordes *et al.*, 1988). These similar results with basal and tap roots of a dicotyledonous plant suggests that this is a general phenomenon, and that large diameter roots grown *in situ* have reduced water flow during the early stages of their growth, relying on lateral roots to provide the large amounts of water and nutrients required for rapid plant growth and development. In the dicotyledonous plants with their secondary vascular development in the roots, the older tissues have a mixture of immature and mature late metaxylem. McCully's group provides no evidence to suggest similar late development of the late metaxylem in lateral roots. Indeed, some of McCully's (1987) anatomical studies and the diameter/growth rate studies of Cahn (1988) coupled with Fusseder's (1987) data suggest that the smaller lateral roots may be the equivalent of multicellular root hairs. These developmental differences and the genetic differences pointed out by Zobel (1986) indicate that there may be significant functional differences between different types of root.

Variability

The soil environment is relatively stable over periods of hours to days, but highly variable spatially. This results in an evolutionary/adaptive need for plants to be able to modify the root system to allow optimum growth in a given environment. Logically there are two, probably independent, mechanisms to do this: physiological adjustment or morphological adjustment. O'Toole and Bland (1987) have referred to the plants ability to adapt the morphology of the root system to a given environment as phenotypic plasticity. The opposite of this could be termed phenotypic rigidity or more simply morpho-stasis. Following this to its conclusion then, root systems can be morpho-plastic or morpho-static in response to differing soil environments. In terms of an evolutionary or ecological perspective, it is highly likely that a given plant has many more roots than the minimum necessary for optimum growth. This redundancy is a frequently observed phenomenon in plants that accounts for flower, seed, pod and nodule abortion or sloughage as the plant matures. Obviously this can lead to extensive variation in morphological characteristics across cultivars, species, soil types and years. Genetic variation, adds to this observed variation.

O'Toole and Bland (1987) in their review suggest that complex inheritance is responsible for our inability to dissect the genetics of root growth and development in the field. Zobel (1975, 1980, 1986, 1989) describes a series of single gene tomato mutants which condition variations in each of the different root types discussed above. This suggests a simple inheritance at least in tomato. Zobel additionally suggests that the discovery of these mutants in tomato but not other species is not fortuitous but rather due to inherent genic duplication in other species (Zobel, 1983; 1989) masking the presence of single gene characteristics. Lack of consistent terminology, poor separation of different types of root, and morpho-plasticity, suggest complex inheritance while the genetic data from single gene mutants suggest a simple underlying genetic pattern.

Morpho-plasticity is conditioned by the ability of a given genotype to interact with a given environment. In a standard analysis of variance, the interaction variance is pooled in the residual or G × E interaction term. Although many different approaches to statistically exploring this interaction term have been tried, from regression analysis to transformation, little progress has been made. Recently, Zobel *et al.* (1988) and Gauch (1988) reintroduced a statistical procedure (AMMI – Additive Main effects and Multiplicative Interaction) which is capable of detecting agronomically (horticulturally, ecologically)

Table 2. Abbreviated ANOVA tables of Sums of Squares for several root and shoot characteristics of seventeen soybean genotypes planted at 5 locations in 1985

	Root			Shoot	
	Basal number	Collateral number	Total nodule number	Height cm	Biomass gm
Location	1434	3487	744	42108	158069
Cultivar	765	2140	228	14677	31166
Loc. × cult.	1389	6413	232	16086	83000

significant patterns within the G × E interaction term of an ANOVA analysis. As Zobel (1989) observed in several of his data sets, shoot characteristics which have received breeding pressure have reduced amounts of variance in the G × E interaction term compared to root characters which have not had significant breeding pressure. Indeed, 50% or more of the treatment variance for a root characteristic may appear in the G × E term, with as much as 90% of that being characterized in the first interaction axis from their analysis (Table 2).

AMMI analyses of genetic by environment interaction effects on root numbers on field grown soybean plants demonstrate two different responses by morphologically different root types (Figs. 3 and 4). Basal root numbers on field grown soybeans (those roots arising from the hypocotyl with secondary thickening and over 1 mm in diameter – basal roots) have strong additive site/year and cultivar components (site/year and cultivar SS in Table 3; 36% and 9% of total treatment SS respectively) and a strong interaction (55% of the total treatment SS) between the cultivar and annual weather patterns (1984) and (1987) data across all sites in Fig. 3 have opposite interaction patterns). Fine lateral roots (no secondary thickening) on the same plants have a strong additive site/year component (30% of the total treatment SS), but little additive cultivar component (only 5% of the total treatment SS). The strong interaction component (65% of the total treatment SS) is characterized by site specific conditions (sites 1 and 6 across all years have opposite signs for their interaction scores in Figure 4). The genetic control of basal roots interacts with seasonal differences, while the genetic control for lateral roots interacts with site specific differences which are not weather related. If the data of these two types of root were pooled, no pattern would be apparent, thus suggesting lower heritability than is apparent when they are properly classified.

Table 3. Abbreviated ANOVA tabel for Sums of Squares for Basal root number and lateral root number of 9 soybean cultivars grown at 5 sites for 4 years

	Basal number	Lateral number
Site/Year	1646	11014
Cultivar	394	1944
SY × Cult.	2480	24164

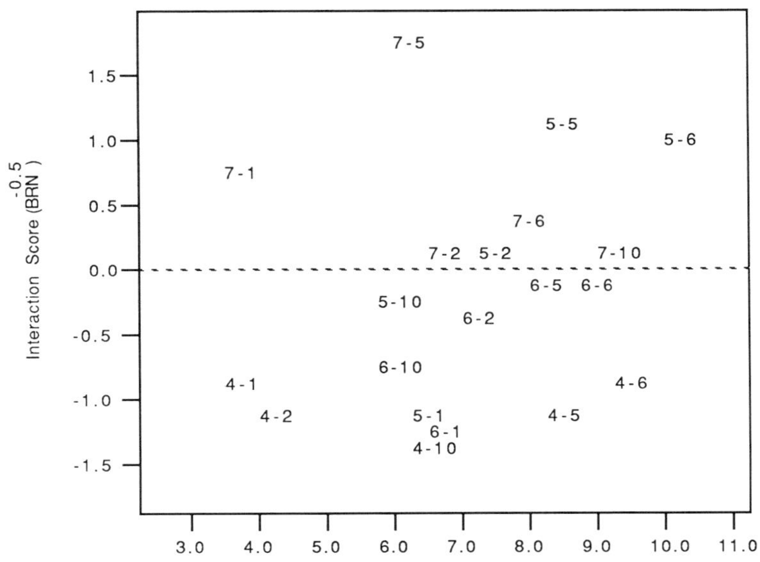

Fig. 3. Interaction plot for site/years for Basal root number. x axis is mean root number, y axis is the interaction score (value is in square root of basal root number). First digit of data points is year, second is site: 1984 at site 5 is 4–5. Data is for four years (1984, 1985, 1986, 1987) and 5 sites (1, 2, 5, 6, 10).

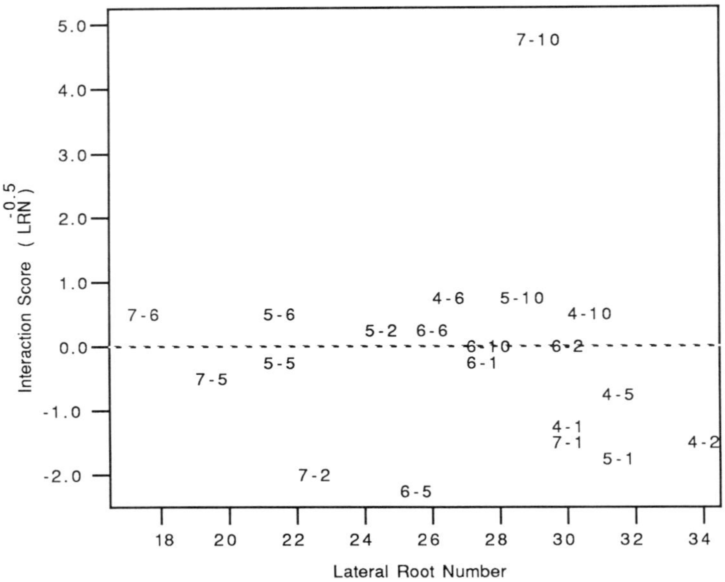

Fig. 4. Interaction plot for site years for lateral root number. Interaction score is in terms of square root of lateral root number. Other aspects are as in Figure 3.

These results reinforce the suggestion of a simple genetic basis complicated by a large G × E interaction.

Conclusions

Root system patterns are made up of a series of different roots and sequence of root development. Through carefully defining root differences and relating them to a common terminology clear genetic and environmental pictures of plant root development in the field can be developed. This will lead to a clearer understanding of the role of the root system in rhizosphere interactions. Physiological studies with basal roots grown under artificial conditions may have applicability to lateral roots in the field, but not to basal or adventitious roots grown under field conditions. The results from further studies of root anatomy and its changes during development of the root and in response to the environment will lead to a greater understanding of the functions of different root types. Finally the ability to dissect agronomic patterns out of the G × E interaction variance of an ANOVA should provide a powerful tool for field rhizosphere research.

References

Aung L H 1982 Root initiation in tomato seedlings. J. Am. Soc. Hortic. Soc. 107, 1015–1018.

Buyanovsky G A and Wagner G H 1983 Annual cycles of carbon dioxide level in soil air. Soil Sci. Soc. Am. J. 47, 1139–1145.

Cahn M D 1988 Growth and Morphology of the Maize Root System. M.S. Thesis, Cornell University. May, 1988. p. 120.

Cannon W A 1911 Root habits of desert plants. Carnegie Inst. Washington, Pub. 131.

Cannon W A 1949 A tentative classification of root systems. Ecology 30, 542–548.

Cooper A J 1973 Root Temperature and Plant Growth. Research Review No. 4, Commonwealth Bureau of Horticulture and Plantation Crops, Slough, England. p. 73.

Esau K 1965 Plant Anatomy. John Wiley and Sons, INC., New York. pp 481–538.

Freitas P L de 1988 Effects of Soil Structure on Root Growth and Function. PhD Thesis, Cornell University, January 1988, p. 213.

Fusseder A 1987 The longevity and activity of the primary root of maize. Plant and Soil 101, 257–265.

Gauch H G 1988 Model selection and validation for yield trials with interaction. Biometrics 88, 705–715.

Hackett C and Rose D A 1972 A model of the extension and branching of a seminal root of barley, and its use in studying relations between root dimensions, Part II. Results and inferences from manipulation of the model. Aust. J. Biol. Sci. 25, 681–690.

Kevekordes K G, McCully M E and Canny M J 1988 Late maturation of large metaxylem vessels in soybean roots:

significance for water and nutrient supply to the shoot. Ann. Bot. 62, 105–117.
Kutschera L 1960 Wurzelatlas mitteleuropäiser Ackerunkräuter und Kulturpflanzen. DLG-Verlag, Frankfurt am Main.
McCully M E 1987 Selected aspects of the structure and development of field-grown roots with special reference to maize. In Root Development and Function. Eds. P J Gregory, J V Lake and D A Rose. pp 53–70. Cambridge Univ. Press, London.
McCully M E and Canny M J 1989 Pathways and processes of water and nutrient movement in roots. In Structural and Functional Aspects of Transport in Roots. Eds. B C Loughman *et al.* pp 3–14. Kluwer Academic Publishers, Dordrecht.
St. Aubin G, Canny M J and McCully M E 1986 Living vessel elements in the late metaxylem of sheathed maize roots. Ann. Bot. 58, 577–588.
O'Toole J C and Bland W L 1987 Genotypic variation in crop plant root systems. Adv. Agron. 41, 91–145.
Weaver J E 1919 The ecological relations of roots. Carnegie Inst. Washington, Pub. 286, p. 128.
Weaver J E 1920 Root development in the grassland formation. Carnegie Inst. Washington, Pub. 292, p. 151.
Weaver J E 1926 Root Development of Field Crops. McGraw-Hill Inc., New York. p. 291.
Weinhold L 1967 Histogenische Studien zum Grenzwurzelproblem. Beitr. Biol. Pflanzen. 43, 367–454.
Zobel R W 1975 The genetics of root development, In The development and function of roots. Eds. J G Torrey and D F Clarkson, Academic Press, London. pp 261–275.
Zobel R W 1983 Genic duplication: A significant constraint to molecular and cellular genetic manipulation in plants. Comments Mol. Cell. Biophys. 1, 355–364.
Zobel R W 1986 Rhizogenetics (root genetics) of vegetable crops. Hortscience 21, 956–959.
Zobel R W 1989 Steady-state control and investigation of root system morphology. *In* Steady-state and Continuous Control of Root Growth. Eds. J G Torrey and L Winship. pp 165–182. Kluwer Academic Publishers, Dordrecht.
Zobel R W, Gauch H G and Wright M W 1988 Statistical analysis of a yield trial. Agron. J. 80, 388–393.
Zobel R W and Su Wenpei 1989 Lateral root initiation from callus culture. Tomato Genetics Coop. Reports. 39, 34–35.

D. L. Keister and P. B. Cregan (Eds.), The rhizosphere and plant growth, 73–80.

PLSO BARC08

The development of a conceptual model of root growth

J.H. BOWERS[1,3], G.N. SMERAGE[2] and D.J. MITCHELL[1]
[1]Department of Plant Pathology and [2]Department of Agricultural Engineering, University of Florida, Gainesville, FL 32611, USA. [3]Present address: Department of Plant Pathology, Ohio State University, OARDC, Wooster, OH 44691, USA

Key words: model, pepper, root growth

Abstract

A process-oriented model is presented for the description of root growth in dicotyledonous seedlings with taproots. The model is being developed to describe the relationships between root growth and colonization of the root by various microorganisms and infection by soilborne pathogens. The model is based on processes depicting the initiation and development of root segments. A root segment is defined as that portion of a root between successive branches or from the most terminal branch to the root tip. Each segment is identified by its root order, according to the morphometric root analysis system, and root type. Three root types are recognized: taproot, primary and secondary lateral roots, and basal roots. The model begins with planting, proceeds through seed germination and taproot formation, primary and secondary lateral branching, and basal root initiation and branching. The time delays between successive stages are major determinants of the dynamics of the model. The model is adaptable to fibrous root systems and monocotyledonous seedlings.

Introduction

Huisman (1982) argues that disease progress involving many soilborne pathogens should be described as a function of root growth and elongation. The argument is based on the fact that the ecology of many soilborne propagules is completely different from that of airborne spores, upon which most of the disease progress literature is based. With foliar diseases the host is normally immobile relative to the pathogen, whereas in root diseases the pathogen is often relatively immobile compared to the growing root system, and in particular, the movement of the root tips. Thus, the dynamics of disease caused by soilborne pathogens might be largely determined by the dynamics of the host as well as that of the pathogen. Ashworth *et al.* (1979), Tomimatsu and Griffin (1982), Kotcon *et al.* (1984), and English and Mitchell (1988) have incorporated root growth into data of infection by soilborne pathogens, but such studies are rare. Campbell (1986) and Jeger (1986) lament the general lack of studies and models of root infection and the relationship of infection to root dynamics. Jeger (1987), however, presented theoretical considerations on the influence of soilborne inoculum and root growth on the dynamics of root disease.

To understand the soilborne propagule-root interaction, one must know the dynamics of root growth in the soil and the effect of infection and disease development on root growth. This understanding is basic to any interpretation of soilborne pathogen-host dynamics when various treatments, such as resistance or biological control with rhizosphere organisms are applied, and is essential when attempting to model root disease. Bloomberg (1979a,b; 1988) modeled infection of Douglas-fir seedlings by *Fusarium oxysporum* and Douglas-fir stands by *Phellinus weirii* by describing root growth parameters, inoculum distribution, and root infection. Similarly, Smith and Walker (1981) and Walker and

Smith (1984) presented models of infection of roots of *Trifolium subterraneum* by vesicular-arbuscular mycorrhizal fungi. Gutierrez *et al.* (1983) modeled Verticillium wilt of cotton in relation to the growth and development of leaves, stems, and roots.

Before attempting to model the interaction between a soilborne pathogen and host roots, a clear understanding of healthy root growth must be accomplished. Root growth models in the literature have involved simulation studies with differential equations formulated via empirical data or theoretical analyses (*e.g.* Dexter, 1978; Hackett and Rose, 1972a,b; Hewitt and Dexter, 1979; Lungley, 1973; McCoy and Boersma, 1986; Page and Gerwitz, 1974; Porter *et al.*, 1986; Rose, 1983; Simmons and Pope, 1988; Stone *et al.*, 1983; Stone and Taylor, 1983) and have often been linked to carbon substrate and physiological processes (Brugge and Thornley, 1985) or to water uptake (Hayhoe, 1981; Hillel and Talpaz, 1976; Protopapas and Bras, 1987). In this study, a process-oriented, mechanistic model of root growth is presented that is compatible with the theory of system analysis. The model differs from those presented in the literature in that a process-oriented approach focuses on causal mechanisms and system structure as the basis for system behavior. The model describes root growth in dicotyledonous seedlings with taproots. The model, however, is adaptable to fibrous root systems and monocotyledons. The pepper (*Capsicum annuum* L.) root system was used to develop and define the model.

The model is based on root segments (Fitter, 1987). A root segment is defined as that portion of a root between successive branches or from the most terminal branch to the root tip. Each segment is identified by its root order (first, second, third, *etc.*) and type (the physical portion of the root system; taproot, lateral root, *etc.*). The morphometric root analysis system defined by Fitter (1982) will be used to assign root order (Fig. 1). The morphometric approach reverses the root order classification system used in the traditional developmental model and is based on geomorphological techniques developed for river systems (Fitter, 1982). In the morphometric root analysis system, first-order root segments terminate in a meristem. A second-order root segment is formed at the junction of two first-order root segments as laterals develop from first-order roots. A third-order root segment is formed at the junction of two second-order root segments, and so on for higher order root segments. First-order roots are physiologically the youngest segments. Physiological age increases with root order.

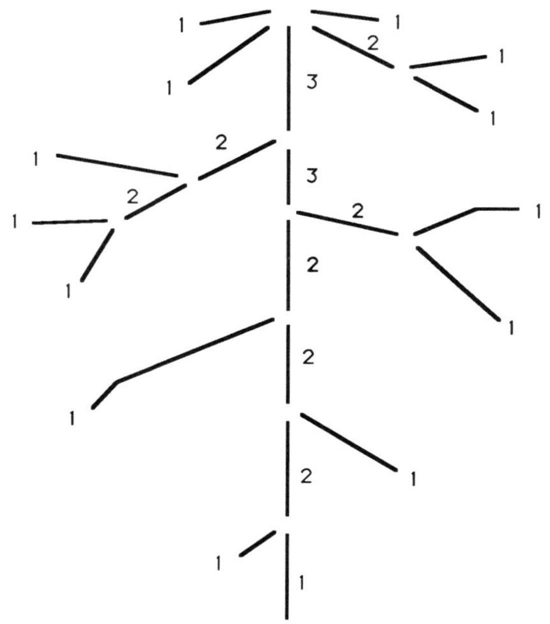

Fig. 1. Schematic representation of a pepper root system broken into root segments and described by the classification scheme of the morphometric root analysis system; 1, 2, 3 = root orders.

Four root types are recognized (Fig. 2) (Esau, 1960; Zobel, 1975). The radicle, which becomes the taproot, is of embryonic origin. Lateral roots arise from the pericycle of pre-existing roots. This includes primary laterals from the taproot and secondary laterals from the primary laterals. Adventitious roots arise from non-root or non-pericyclic tissues. Basal roots arise from the basal region of the hypocotyl. Basal roots differ from adventitious roots due to their anatomical region of origin (pericycle) and from lateral roots due to their location on the hypocotyl (Zobel, 1975). Adventitious roots do not impact on overall root growth during the seedling stage. Therefore, adventitious roots are excluded from the model for the present time.

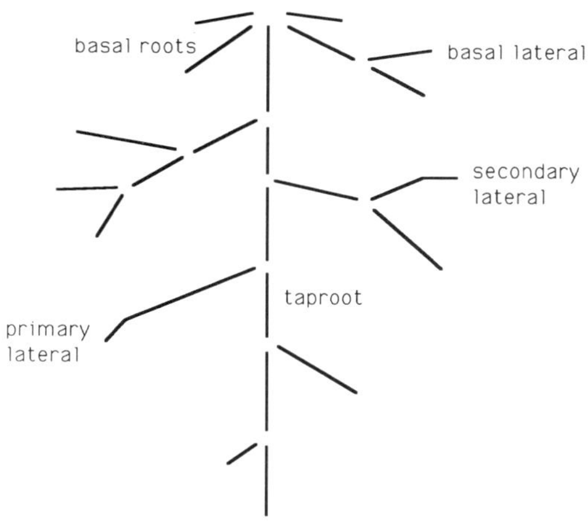

Fig. 2. Schematic representation of a pepper root system broken into root segments and described by root type.

Conceptual model

The modeling process consists of developing a conceptual model of the system in the form of a network of elementary process components, and then developing a mathematical model utilizing the components and structure of the conceptual model. The conceptual model is a qualitative statement of the components of a system, their generic descriptions, and their manner of interconnection to compose the whole system. It is based on the perceptions of components and their interconnections in the real system by the modeler. Conceptual models are expressed as diagrams that indicate component processes, their manners of interconnection, and important system variables and are supported by verbal descriptions.

A diagrammatic representation of root growth is presented in Figure 3. Movement of root segments through the model is in the direction of the arrows. The model begins at planting with an initial input of seeds into the system (flow source Γ_s). The seeds traverse delay network τ_0 and exit as germinated seeds. The delay network provides for dispersion; it accounts for variance of the germination interval because not all seeds germinate at the same time. Transformation χ_0, representing seed germination, establishes the initial population of plants and initiates a first-order radicle on each live plant that begins to grow. The radical develops into a taproot in delay network τ_1. A sink (triangular symbol) accumulates seeds that do not survive or germinate due to genetic variation, environmental conditions, or disease caused by a pathogen. Transformation χ_0 also directs hypocotyl tissue to delay network τ_5, which determines the time interval until initiation of basal roots. This portion of the model will be discussed later. In a similar manner, a model for plant top growth could be connected to χ_0.

The time interval between germination and the initiation of the first lateral root is accounted for by delay network τ_1. Branching transformation χ_1 represents the initiation of a population of first-order laterals (which then grow and develop in delay network τ_3), a population of second-order segments on the taproots from the branch point to the hypocotyl (store C_1), and retains the population of first-order taproots from the branch point to the root tip (which continue to grow and develop in delay network τ_2). Store C_1 accumulates second-order root segments on the tap root. A sink (triangle) representing mortality of first-order taproots is attached to χ_1.

The population of first-order taproots continue to elongate, and, after traversing delay network τ_2, the second lateral branches are initiated. Branching transformation χ_2 represents the initiation on each taproot of successive first-order laterals (which grow and develop in delay network τ_3), second-order segments on the taproot from the new branch point to the previous branch (C_1), and returns the first-order taproots from the new branch point to the root tip to the input of τ_2 for the next cycle of branching. As before, store C_1 accumulates second-order root segments. The third to k^{th} lateral branch on each taproot in the population is initiated via branching transformation χ_2 after time delay τ_2. To simplify the model it is assumed that the delay between successive lateral branchings is described by τ_2 and that the sequence of branch initiation is from the crown (the point where the hypocotyl and taproot join) towards the root tip (Fahn, 1974). The model, essentially, folds back on itself, but could be diagrammed in a linear fashion for each lateral branch initiated. A sink (triangle) representing mortality is associated with χ_2.

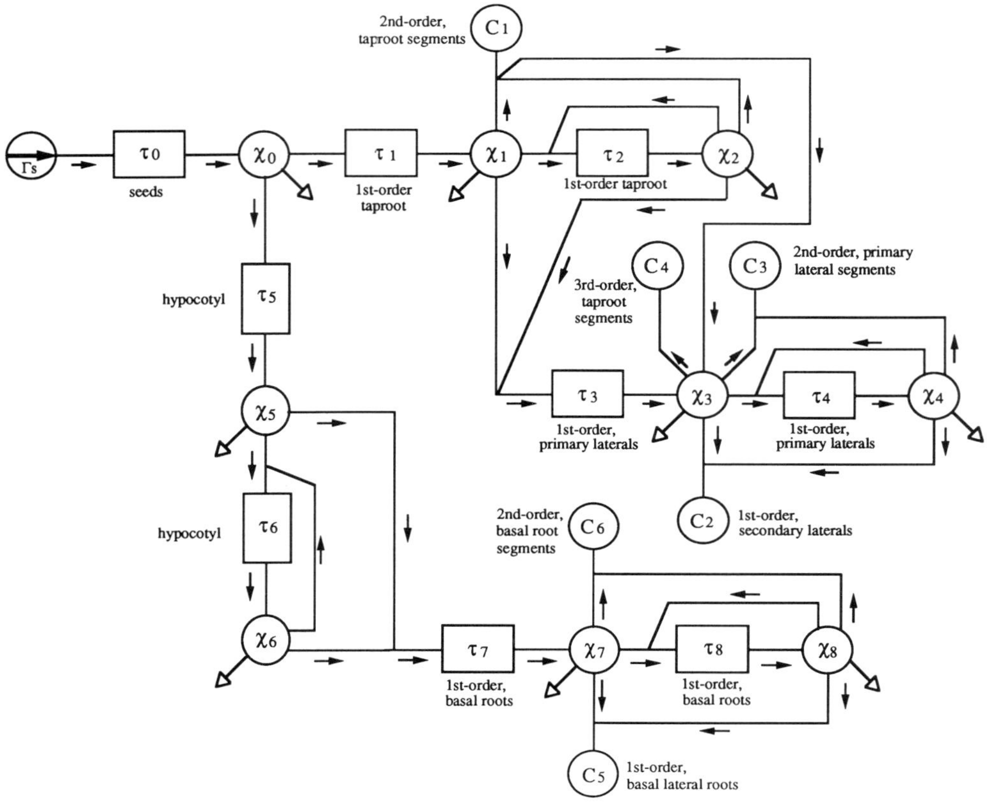

Fig. 3. Conceptual model of root growth in pepper seedlings; C_i are stores, χ_i are branching transformations, τ_i are Bessel delay networks, triangles represent sinks of dead or non-functional root segments, and flow is in the direction of the arrows.

Lateral roots, meanwhile, begin to grow and develop by traversing delay network τ_3. Branching transformation χ_3 represents the initiation on each primary lateral, a first-order, secondary lateral (C_2), a second-order segment of the primary lateral from the branch point to the taproot (C_3), and retains the first-order portion of the primary lateral (which continues to grow and develop in delay network τ_4). Store C_2 accumulates first-order, secondary, lateral roots and store C_3 accumulates second-order segments on primary lateral roots. Further development of secondary laterals is beyond the scope of the present model, but it would be treated in a manner similar to primary laterals. The second-order segment of each primary lateral now joins a second-order segment on the taproot. Thus, all second-order segments on the taproot above the branched primary lateral become third-order segments. To account for this reclassification, branching transformation χ_3 transfers segments from store C_1 to store C_4, where segments are accumulated. Once a segment becomes third-order, it remains so regardless of secondary lateral formation on the same primary lateral or on primary laterals below it. Only the formation of tertiary lateral branches will change a third-order segment to fourth-order. This is also beyond the scope of the present model.

The assumption has been made that the first primary lateral formed will be the first lateral to produce a secondary lateral and the second primary lateral formed will produce the next secondary lateral, and so on. This has been accounted for by branching transformation χ_3 taking one second-order segment from store C_1 and placing it in store C_4 when each primary lateral first branches. In practice, however, the first-formed primary lateral does not always produce the first secondary lateral. The *m*th-formed primary lateral may produce the first secondary lateral and primary laterals may not produce secondary laterals in successive order. In this instance, branching transformation χ_3 should transfer *m* second-order taproot segments to third-order based on a random process or an expected (average) value if the population is sufficiently large. A sink (triangle) representing mortality is associated with χ_3.

The population of primary laterals continues to grow and develop and traverse delay network τ_4, which corresponds to the time period between the initiation of the first secondary lateral and the next secondary lateral on each primary lateral. Branching transformation χ_4 represents the initiation for each branched, primary lateral, successive first-order, secondary laterals (C_2), second-order segments of the primary laterals from the new branch point to the previous branch point (C_3), and returns the first-order portion of the primary laterals to the input of τ_4 for the next cycle of branching. Store C_2 accumulates first-order, secondary, lateral roots and store C_3 accumulates second-order segments on primary lateral roots. As before, it is assumed that the delay between successive secondary laterals on each primary lateral is described by τ_4 and that the sequence of branching is from the point where the primary lateral joins the taproot to the root tip. Only the first-formed, secondary lateral on any primary lateral affects the root order of taproot segments. Successive secondary branching on any one primary lateral does not affect taproot root order. A sink (triangle) for mortality is associated with χ_4.

Transformation χ_0 initiates a population of taproots with one taproot per plant in the population. The portions of the model diagrammed to the right of this transformation deal with the lateral branching of taproots, as described above. These portions may be excluded when modeling root growth for a monocotyledon such as onion. The model can be modified to include other root types, such as seminal roots. The portions diagrammed below transformation x_0 deal with the development of basal roots, which may be thought of as secondary taproots (Zobel, 1975). Often, more than one basal root is produced per plant.

As with taproots, the initiation of basal roots are referenced to germination. Delay network τ_5 accounts for the time interval between seed germination and the initiation of the first basal root on each plant. Basal root tissue is considered dormant in hypocotyl tissue in delay networks τ_5 and τ_6. Branching transformation χ_5 represents the initiation of the first, first-order basal roots (which then grow and develop in delay network τ_7). A sink (triangle) for mortality is associated with χ_5. Delay network τ_6 corresponds to the time period between the initiation of the first basal root and successive basal roots on each plant. Branching transformation χ_6 represents the initiation of successive basal roots on each plant (which also develop in delay network τ_7), and returns the hypocotyl to the input of τ_6 for further branching. As before, the delay between successive basal roots on each plant is assumed to be described by τ_6. A sink (triangle) for mortality is associated with χ_6.

The population of basal roots, once initiated, begin to grow and develop. Initial first-order segments of basal roots traverse delay network τ_7, which corresponds to the time interval between the initiation of basal roots and the initiation of its first lateral root. Branching transformation χ_7 represents the initiation of a population of first-order, lateral roots (C_5), a population of second-order segments on the basal root from the branch point to the hypocotyl (C_6), and retains the population of first-order basal roots from the branch point to the root tip (which continues to develop in delay network τ_8). Store C_5 accumulates first-order, basal, lateral roots and store C_6 accumulates second-order segments on the basal roots. Further development of basal, lateral roots is not considered at this time, but would proceed in the manner described above for taproot lateral branching. A sink (triangle) for mortality is associated with χ_7.

The population of basal roots continue to grow over time and traverse delay network τ_8, which corresponds to the time period between the initiation of the first basal, lateral root and successive basal, lateral roots. Branching transformation χ_8 represents for each basal root the initiation of successive first-order, lateral roots (C_5), second-order segments on the basal root from the new branch point to the previous branch point (C_6), and returns the first-order basal roots from the new branch point to the root tip to the input of τ_8 for further branching. Store C_5 accumulates first-order, basal, lateral roots and store C_6 accumulates second-order segments on the basal roots. It is assumed, as before, that the delay between successive laterals on each basal root is described by τ_8, and that the sequence of branching is, once again, from the hypocotyl to the root tip. Secondary lateral branching of basal roots is not observed within the scope of the model. A sink (triangle) for mortality is associated with χ_8. Although the structure of lateral root development in the model is identical for both taproot laterals and basal root laterals, the two pathways are presented separately for clarity and because it is not yet known if delay networks τ_2 and τ_8 behave similarly. Also, it may be important to distinguish second-order, taproot segments (store C_1) from second-order, basal root segments (store C_6) when a pathogen is incorporated into the model.

Mathematical model

The conceptual model above is merely a qualitative statement of our perceptions of the processes of root growth. Model-based quantitative analyses of root growth dynamics for predictive purposes and additional understanding require development of the conceptual model into a corresponding mathematical model. Limited space prohibits presentation of the mathematical model that has been developed for this conceptual model of root growth. Instead, a brief overview of the basis of the mathematical model is given.

A mathematical model is derived from a conceptual model and relevant experimental data in three steps. First, mathematical descriptions of all processes identified in the conceptual model must be obtained. In this model, some of the processes are stores containing root segments; others are transformations initiating new segments and transferring existing segments between classes; the remaining processes pertain to delay networks accounting for time intervals of development between significant events of branching and changing classification. Generic mathematical identities of these processes have been presented (Smerage, 1979; 1985).

Experimentation to define the parameters of the process descriptions for the model have not been completed yet. Experiments to develop the parameters entail determination of the time for a cohort to complete a certain stage of development, dependency of that time on environmental variables, and mortality in the stage. This experimentation, however, requires repeated observation of individuals as they develop, and it is not possible to observe whole root systems of plants growing in soil over time. Thus, indirect techniques, such as destructive sampling, rhizotrons, or root boxes with observation windows, must be used. Plants growing in aeroponic or hydroponic systems may be more compatible with the experimentation necessary to develop and validate the model. General and theoretical discussions of the process descriptions and their determination have been presented by Smerage (1979, 1985). Currently, seed planting is the only input to the model. Environmental input variables, such as temperature and soil moisture, should be incorporated in the process descriptions for improved representation.

Second, the structure of the system is described by a set of algebraic equations involving the same variables used in describing the processes. These equations are readily derived from the conceptual model (Smerage, 1979; 1985).

In the third step, the complete sets of process and structural descriptions are combined by one of several, standard methods of formulation to obtain a mathematical model. The preferred method is state variables formulation. Once obtained, a mathematical model is available for behavior analyses of the dynamics of root development. The generic mathematical model consists of a set of first-order, differential equations in the state variables format. The state variables are the populations in all categories of root segments and their number. The size of the model, therefore, depends on the accuracy specified for predictive analyses, the root system studied, and the environmental conditions incorporated. Detailed discussions on formulation of the mathematical model can be found elsewhere (Bowers, 1989; Smerage, 1985).

The final aspect in model development is validation. A combination of experimentation and mathematical behavior analyses must be conducted on the whole system. We have not yet attempted to validate this model.

References

Ashworth L J Jr, Huisman O C, Harper D M and Stromberg L K 1979 Verticillium wilt disease of tomato: Influence of inoculum density and root extension upon disease severity. Phytopathology 69, 490–492.

Bloomberg W J 1979a A model of damping-off and root rot of Douglas-fir seedlings caused by *Fusarium oxysporum*. Phytopathology 69, 74–81.

Bloomberg W J 1979b Model simulations of infection of Douglas-fir seedlings by *Fusarium oxysporum*. Phytopathology 69, 1072–1077.

Bloomberg W J 1988 Modeling control strategies for laminated root rot in managed Douglas-fir stands: Model development. Phytopathology 78, 403–409.

Bowers J H 1989 The effects of inoculum density and soil-water matric potential on infection of pepper by *Phytophthora capsici* and the epidemiology of Phytophthora blight in the field. Ph.D. thesis. University of Florida, Gainesville, FL, 162 p.

Brugge R and Thornley J H M 1985 A growth model of root mass and vertical distribution, dependent on carbon substrate from photosynthesis and with non-limiting soil conditions. Ann. Bot. 55, 563–577.

Campbell C L 1986 Interpretation and uses of disease progress curves for root diseases. *In* Plant Disease Epidemiology: Population Dynamics and Management. Vol. 1. Eds. K J Leonard and W E Fry. pp. 38–54. Macmillan Publishing Company, New York, 372 p.

Dexter A R 1978 A stochastic model for the growth of roots in tilled soil. J. Soil Sci. 29, 102–116.

English J T and Mitchell D J 1988 Relationships between the development of root systems of tobacco and infection by *Phytophthora parasitica* var. *nicotianae*. Phytopathoogy 78, 1478–1483.

Esau K 1960 Anatomy of Seed Plants. John Wiley and Sons, New York, 376 p.

Fahn A 1974 Plant Anatomy. 2nd ed. Pergamon Press, Oxford, 611 p.

Fitter A H 1982 Morphometric analysis of root systems: Application of the technique and influence of soil fertility on root system development in two herbaceous species. Plant, Cell Environ. 5, 313–322.

Fitter A H 1987 An architectural approach to the comparative ecology of plant root systems. New Phytol. 106, 61–77.

Gutierrez A P, DeVay J E, Pullman G S and Friebertshauser G E 1983 A model of Verticillium wilt in relation to cotton growth and development. Phytopathology 73, 89–95.

Hackett C and Rose D A 1972a A model of the extension and branching of a seminal root of barley, and its use in studying relations between root dimensions. I. The model. Aust. J. Biol. Sci. 25, 669–679.

Hackett C and Rose D A 1972b A model of the extension and branching of a seminal root of barley, and its use in studying relations between root dimensions. II. Results and inferenes from manipulation of the model. Aust. J. Biol. Sci. 25, 681–690.

Hayhoe H 1981 Analysis of a diffusion model for plant root growth and an application to plant soil-water uptake. Soil Sci. 131, 334–343.

Hewitt J S and Dexter A R 1979 An improved model of root growth in structured soil. Plant and Soil 52, 325–343.

Hillel D and Talpaz H 1976 Simulation of root growth and its effect on the pattern of soil water uptake by a nonuniform root system. Soil Sci. 121, 307–312.

Huisman O C 1982 Interrelations of root growth dynamics to epidemiology of root-invading fungi. Annu. Rev. Phytopathol. 20, 303–327.

Jeger M J 1986 The potential of analytic compared with simulation approaches to modeling in plant disease epidemiology. *In* Plant Disease Epidemiology: Population Dynamics and Management. Vol. 1. Eds. K J Leonard and W E Fry. pp 255–281. Macmillan Publishing Co., New York, 372 p.

Jeger M J 1987 The influence of root growth and inoculum density on the dynamics of root disease epidemics: Theoretical analysis. New Phytol. 107, 459–478.

Kotcon J B, Rouse D I and Mitchell J E 1984 Dynamics of root growth in potato fields affected by the early dying syndrome. Phytopathology 74, 462–467.

Lungley D R 1973 The growth of root systems – a numerical computer simulation model. Plant and Soil 38, 145–159.

McCoy E L and Boersma L 1986 Dynamic equations of axial root growth from the empirical solution of a mechanistic model. Bot. Gaz. 147, 383–391.

Page E R and Gerwitz A 1974 Mathematical models, based on diffusion to describe root systems of isolated plants, row crops and swards. Plant and Soil 41, 243–254.

Porter J R, Klepper B and Belford R K 1986 A model (WHTROOT) which synchronizes root growth and development with shoot development for winter wheat. Plant and Soil 92, 133–145.

Protopapas A L and Bras R L 1987 A model for water uptake and development of root systems. Soil Sci. 144, 352–366.

Rose D A 1983 The description of the growth of root systems. Plant and Soil 75, 405–415.

Simmons G L and Pope P E 1988 Development of a root growth model for yellow popular and sweetgum seedlings grown in compacted soil. Can. J. For. Res. 18, 728–732.

Smerage G H 1979 Modeling theory for physiological systems. Trans. ASAE 22, 1488–1493.

Smerage G H 1985 Bessel delay networks for population models. Trans. ASAE 28, 1269–1278.

Smith S E and Walker N A 1981 A quantitative study of mycorrhizal infection in *Trifolium*: Separate determination of rates of infection and of mycelial growth. New Phytol. 89, 225–240.

Stone J A, Kasper T C and Taylor H M 1983a Predicting soybean rooting depth as a function of soil temperature. Agron. J. 75, 1050–1054.

Stone J A and Taylor H M 1983 Temperature and the development of the taproot and lateral roots of four indeterminate soybean cutivars. Agron. J. 75, 613–618.

Tomimatsu G S and Griffin G J 1982 Inoculum potential of *Cylindrocladium crotalariae*: Infection rates and microsclerotial density-root infection relationships on peanut. Phytopathology 72, 511–517.

Walker N A and Smith S E 1984 The quantitative study of mycorrhizal infection. II. The relation of the rate of infection and the speed of fungal growth to propagule density, the mean length of the infection unit and the limiting value of the fraction of the root infected. New Phytol. 96, 55–69.

Zobel R W 1975 The genetics of root development. *In* The Development and Function of Roots. Eds. J G Torrey and D T Clarkson. pp. 261–275. Academic Press, London, 618 p.

D. L. Keister and P. B. Cregan (Eds.), The rhizosphere and plant growth, 81–89.

PLSO BARC09

Methods for microbial autecology in the soil rhizosphere

EDWIN L. SCHMIDT
Department of Soil Science, University of Minnesota, St. Paul, MN 55108, USA

Key words: autecology, ecological molecular probes, immunofluorescence, microbial ecology, rhizosphere methods

Abstract

Developments in biotechnology have become a force to alter agriculture, and now place sharp focus on the biology of the rhizosphere as the site where modified plants and modified microorganisms will interact. Rhizosphere research, traditionally general and qualitative for the most part, must now address rhizosphere events in autecological terms to examine the specific microorganism in its interactions with its microenvironment and with other specific microbes therein. Extreme complexities of the rhizosphere call for innovative methodology. Molecular probes appear to offer the best approach to deal with the rhizosphere microhabitat with adequate specificity and sensitivity. Immunofluorescence, the first method to apply a molecular probe, has been used for autecological study of complex natural environments, including rhizospheres, for more than twenty years. Further developments in immunofluorescence, and its use together with refined complementary techniques will contribute increasingly to rhizosphere research. New molecular genetic probes are now emerging and, when successfully developed for application to soil systems, will provide a new generation of methodologies for autecological investigations of the rhizosphere.

Introduction

Most plants of direct agricultural and industrial concern are terrestrial, with complex root systems anchored in that most complex of all natural materials, the soil. This poses immense complications with respect to study of the rhizosphere. Still further complications are imposed by the microscopic size and extreme diversity of a rhizosphere biota that develops in close physical intimacy with the soil matrix and the root surfaces that comprise the rhizosphere. The temptation is to simplify the system to make it amenable to existing methodology. The likely result of this is that something other than the rhizosphere is investigated. Models derived from "artificial rhizospheres" must at best remain speculative until tested in nature.

The need to examine rhizospheres as they actually occur in nature has suddenly become obvious with the emergence of the early products of the new molecular-based biotechnology. Genetically engineered plants offer the prospect of improved productivity and greater disease resistance at lowered risk to the environment. Genetically engineered microorganisms offer enhanced plant nutrition, biological disease control, and reduced reliance on commercial fertilizers and other agricultural chemicals. Field tests already are under way for both genetically modified plants and microorganisms. Few can be unaware of the concerns that have been voiced worldwide as to the safety of the deliberate release of genetically modified organisms into the environment. Not only must the questions of safety be resolved at the level of the rhizosphere, but so also must the details and implications of rhizosphere events be addressed rigorously.

Rhizosphere interactions are currently understood only in general terms. The requirement now is for specific, autoecological investigation of the rhizosphere. How does the introduced organism of interest enter the rhizosphere, localize and reproduce in the rhizosphere, interact with other residents of the habitat, and how does it behave when the rhizosphere habitat disappears? How also does the modified plant modify the microenvironment of its roots and possibly the micro-community normal to the unmodified plant cultivar? Methods for autecological study of the rhizosphere are few at the present time, but considerable effort is directed to the development of new and innovative methodolgy. Whatever the method, it must accommodate the biological, physical, and chemical complexities of the natural rhizosphere, and the operator of that method must understand its limitations in the context of the rhizosphere's complexities.

Visualization of the rhizosphere

Some of the major advances in rhizosphere research are embodied in the micrographs derived from transmission and scanning electron microscopy of the natural rhizosphere. Such micrographs are invaluable guides to the structure of the rhizosphere and the sources of rhizosphere complexity. Reviews by Foster (1981), Foster *et al.* (1983), and Rovira and Campbell (1974) are particularly informative with respect to numerous problems in the application of any experimental approach to rhizosphere investigations.

Boundaries of the rhizosphere

The various parts of the rhizosphere are commonly represented by the terms: outer rhizosphere, inner rhizosphere, rhizoplane, and endorhizosphere. Such compartmentation is easy and useful conceptually but difficult practically when rhizospheres are prepared to study. Separation of the outer rhizosphere is particularly troublesome as it blends into the bulk soil on one side and the inner rhizosphere on the other. The outer rhizosphere can extend up to 5 mm from the root surface (Foster *et al.*, 1983). The inner rhizosphere may be difficult to distinguish from the surface of the epidermis (rhizoplane) because of dense mucilages variably secreted by the root (Foster, 1981; Jenny and Grossenbacher, 1963). The pragmatic differentiation of the rhizosphere for purposes of sampling and analysis usually follows procedures such as those reported by Campbell and Rovira (1973) and Old and Nicolson (1975).

Microorganisms in the rhizosphere

The rhizosphere represents an energy-rich microsite in the chronically energy-poor surrounding bulk soil. Concentrations of microorganisms in the rhizosphere can reach 10^{10}–10^{12} per gram of rhizosphere soil (Foster, 1988). Electron micrographs, however, show that only 7–15% of the actual root surface is occupied by microbial cells (Foster *et al.*, 1983). Bacteria dominate the rhizosphere, but appear to occur in small colonies embedded in polysaccharide of either bacterial or root origin (Rovira and Campbell, 1974). This view of the rhizosphere bacteria scattered in small colonies associated with a small portion of the root surface is a useful reminder that large populations do not equate with dense colonization in rhizospheres fashioned in soil.

Electron microscopy has shown that bacteria associated with roots grown in soil commonly penetrate intercellular spaces to constitute the endorhizosphere (Foster *et al.*, 1983). Old and Nicolson (1974) found evidence of rhizosphere bacteria penetrating several layers of root cells. The endorhizosphere zone may prove more accessible to microbial analysis than other zones since roots could be washed extensively prior to disruption and examination.

Electron microscopy, in combination with specialized histochemical techniques, has been very effective in describing the structure, development and complexity of the rhizosphere. It is especially intriguing to observe the distribution of microorganisms and their *in situ* localization on the root surface and in the rhizosphere soil associated with the root. Unfortunately, electron microscopy as presently developed is not a tool for autecological study since the organisms observed cannot be identified with certainty.

Selective culture techniques for autecology

Any culture medium is inherently selective. Early rhizosphere research frequently aimed at quantifying rhizosphere organisms relative to adjacent non-rhizosphere soil organisms (the "rhizosphere effect"). Culture media as non-selective as possible were used, but always with the realization that perhaps as few as 0.1% of microorganisms present in soil were cultivable by any method. Much attention has also been given to enhancement of culture medium selectivity to isolate specific microorganisms from complex mixed culture environments. Relatively few cultural techniques are sufficiently selective for autecological study of unmarked organisms, but little effort has been directed toward that objective. The use of marked organisms and complementary selective media has received more attention.

Selective culture of specific unmarked bacteria

Several media have been developed to study *Pseudomonas fluorescens* based on the intrinsic antibiotic resistance and detergent resistant properties of the fluorescent pseudomonads (Gould *et al.*, 1985). Two media were devised by Gould *et al.* (1985) that provided a high degree of selectivity for detection of fluorescent pseudomonads from soils with divergent properties. These media could possibly be used to study a specific pseudomonad in rhizospheres formed in soils that lack a fluorescent pseudomonad component.

Interest in strains of Azospirillum as agents for associative nitrogen fixation in the rhizosphere has generated numerous media selective for that genus. Several such media were reported by Levanony *et al.* (1987) to be inadequate for autecological application since they allowed the development of many Azospirillum strains. A semi-selective medium for *A. brasilense* proposed by Bashan and Levanony (1985) was used in combination with ELISA for rhizosphere studies involving detection of the specific strain, *A. brasilense* Cd (Bashan and Levanony, 1987; Bashan *et al.*, 1987; Levanony *et al.*, 1987).

Selective culture of antibiotic-marked bacteria

Spontaneous antibiotic resistance is a simple genetic marker that may be useful to monitor a specifical bacterial strain in nature. A recent example of the selection and evaluation of a spontaneous antibiotic resistant mutant that must precede its application in complex ecosystems was provided by Compeau *et al.* (1988). They isolated rifampicin-resistant mutants of *P. fluorescens* and *P. putida*, and were assured of their physiological similarity to the corresponding wild types before introducing the Rif^r mutants into a nonsterile soil. Recovery of the mutants from a nonsterile soil was successful with detection limits of 10^2 or 10^3 colony forming units/g of soil. Since stability of an antibiotic resistance trait in soil and the probable occurrence of other organisms with resistance to the test antibiotic are important considerations, more than one marker may be necessary. Graham and Istock (1979) used 3 antibiotic resistance markers in their study of *Bacillus subtilis* in soil.

Spontaneous antibiotic-resistance markers were avoided by van Elzas *et al.* (1986) in favor of transposon- and plasmid-generated mutants. These authors pointed out that strains containing transposon markers that do not interfere with genes needed for cellular development may be selected or may involve no chromosomal alteration as in the case of plasmid-borne markers. A strain of *P. fluorescens*, transposon Tn5 marked for kanamycin and streptomycin resistance was studied for survival in bulk and rhizospheres of two different soils. *P. fluorescens* showed a steady decline in both soils and no significant differences between rhizosphere and nonrhizosphere soils were detected.

Selective culture based on novel metabolic markers

The work of Drahos *et al.* (1986) provides an interesting example of how an organism of interest may be modified by genetic techniques to render it detectable on a simple selective medium. A recombinant strain of *P. fluorescens* expressing the delta endotoxin pesticide of *Bacil-*

lus thuringiensis was further modified by the insertion of *E. coli lac* Y and *lac* Z genes for beta-galactosidase production. When cultured on medium containing the galactose derivative, 5-chloro-4-bromo-3 indoyl-b-D-galactopyanoisde (Xgal), the *E. coli*-derived genes break down the Xgal to impart a bright blue-green color to colonies of the altered *P. fluorescens*. Rhizosphere applications of the technique have not been reported but other lactose fermenters are unlikely to occur in soil. Introduction of the *lux* genes for bioluminescence via transposon mutagenesis was recently described for the strains of the plant associated genera Pseudomonas, Agrobacterium and Rhizobium (Boivin *et al.*, 1988). Monitoring the occurrence of a specific organism in natural rhizosphere samples by chemiluminescence, however, would require prior application of a suitable selective or semi-selective medium.

Molecular probes–Immunofluorescence (IF)

The first molecular probe to be developed for the detection of a specific microorganism in the complex soil ecosystem was in the form of a fluorescent antibody to the fungus *Aspergillus flavus* (Schmidt and Bankole, 1962). Development of IF as a tool for autecological study of soil included its adaptation to soil bacteria, the blocking of nonspecific adsorption of FA to soil colloids and the creation of successful protocols to quantify specific bacteria in soil. The IF approach to microbial autecology, its inherent advantages and limitations, and the scope of its application as of 1980 has been reviewed (Bohlool and Schmidt, 1980). There have been scores of IF studies worldwide, a large number of which may provide insights relevant to the emerging focus on rhizosphere events.

IF study of bacteria recovered from rhizospheres

Most autecological studies of this rhizosphere by IF are made by collecting that soil adhering loosely to the roots and removed by gentle shaking (outer rhizosphere) and/or soil still adhering to the root after shaking but subject to removal by washing (inner rhizosphere). Rhizosphere soil material is then extracted to release bacteria and the released bacteria are concentrated on a membrane filter for IF analysis. The first application of IF to a rhizosphere ecosystem was that of Rennie *et al.* (1977) in a study relating to the controversy of whether or not wheat rhizospheres are inhibitory to the chemoautotrophic nitrite oxidizer *Nitrobacter agilis*.

Although much attention must be given to the monitoring of a specific microorganism introduced into the rhizosphere, it must be recognized that other microbes indigenous to the rhizosphere need be studied as well. Indigenous organisms may be of controlling importance to the plant as in the case of rhizobia and nitrogen fixation, or as for any number of plant disease agents. Even with the focus on the engineered strain, it is easily imagined that knowledge of its interaction with specific members of theindigenous community will be needed. Several studies indicate the capability of IF to deal with indigenous rhizosphere bacteria at the autecological level.

The dominance of one group of serologically related soil rhizobia in the nodulation of soybeans throughout the major production area of the United States is well documented. Members of this group, now referred to as serocluster 123 (Schmidt *et al.*, 1986) have been studied simultaneously with two other co-existing serotypes of *Bradyrhizobium japonicum* in the same rhizospheres. The aim was to examine the population

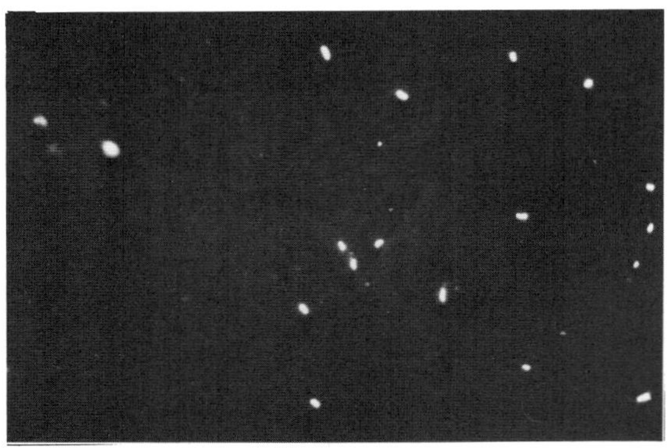

Fig. 1. Indigenous *B. japonicum* serocluster 123 cells as viewed for IF enumeration of an inner rhizosphere sample from 18-day field-grown soybean plant. Material collected on the surface of a polycarbonate membrane filter from a known volume of rhizosphere supernatant was stained with *B. japonicum* FA. Micrograph shows 123 cells in a single microscopic field at 800 × magnification.

dynamics of the competing indigenous strains for an explanation of the success of 123. The ability of IF to enumerate one indigenous strain among other competing rhizobia and among the vastly more abundant general rhizosphere community of bacteria is illustrated in Figures 1 and 2. Both micrographs represent the same volume of the same extract prepared from an 18 day-old field-grown soybean rhizosphere. Serocluster 123 is shown in Figure 1 trapped on the surface of a polycarbonate membrane, and visualized by IF. The companion membrane (Fig. 2) was stained with the nonspecific fluorochrome acridine orange to view 'total' bacteria associated with 123 in the rhizosphere.

Results of IF membrane filter counts during several such studies failed to explain the competitiveness of 123, but do illustrate the ability to deal concurrently with several indigenous target bacteria. Data compiled for Figure 3 show the population changes encountered by Moawad *et al.* (1984) for three *B. japonicum* groups during the growth of soybeans. All three indigenous groups were present in the one week rhizosphere at about the same level and all exhibited about the same modest rhizosphere response during five weeks of plant development. At week 9, the time of flowering, rhizosphere activity was greatly elevated as reflected in the high populations of all three groups even though nodule occupancy for 138, 110, and 123 was determined to be respectively 12, 4 and 84%. Robert and Schmidt

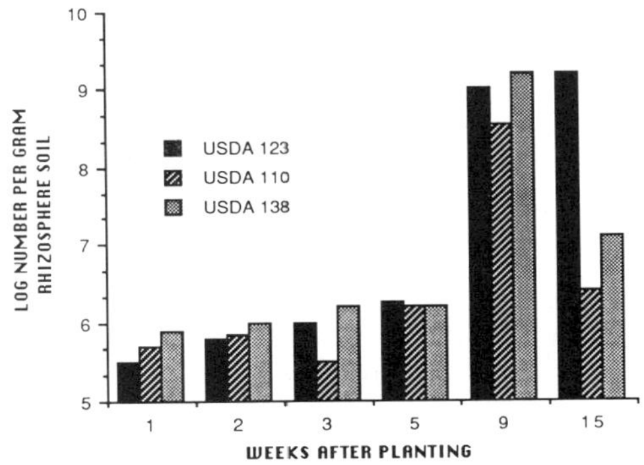

Fig. 3. Population dynamics of three groups of soil-indigenous *B. japonicum* in the inner rhizosphere of field-grown soybeans during a growing season. Flowering was at 9 weeks, late pod fill at 15 weeks. Unpublished data from study of Moawad *et al.* (1984).

Fig. 2. Total rhizosphere bacteria from same volume of same inner rhizosphere preparation as in Figure 1, but membrane was stained with the nonspecific fluorochrome acridine orange. Micrograph shows a single microscopic field at 800 × magnification.

(1985) extended the study to examine very early, pre-emergent rhizospheres of soybeans on almost a daily basis. By collecting as many as 50 pre-emergent seedlings per sampling it was possible to obtain sufficient inner rhizosphere soil to demonstrate that even at the earliest stages of nodule initiation, serocluster 123 had no population advantage over competitors.

IF for autoecology of bacteria in situ *in the rhizosphere*

Although IF has the capability to examine the localization of specific bacteria in the rhizosphere, that capability is not well developed. Direct examination of the rhizosphere of soil-grown plants is tedious, involving careful washing of roots, control of root autofluorescence, and usually the excision of root tissues for microscopic examination. The report by Reyes and Schmidt (1981) described a study of colonization of the rhizoplane of soybeans by indigenous *B. japonicum* serocluster 123 and introduced serogroup 138. Autofluorescence of the roots was controlled either by soaking in gelatin-rhodamine isothiocyanate (Bohlool and Schmidt, 1968) alone for several hours or in 0.5 M NaOH for 30–45 minutes prior to gel-rhodamine treatment. Figures 4 and 5 are photomicrographs taken during the Reyes and Schmidt (1981) study. Figure 4 illustrates a group of *B. japonicum* 138 cells localized about a soybean

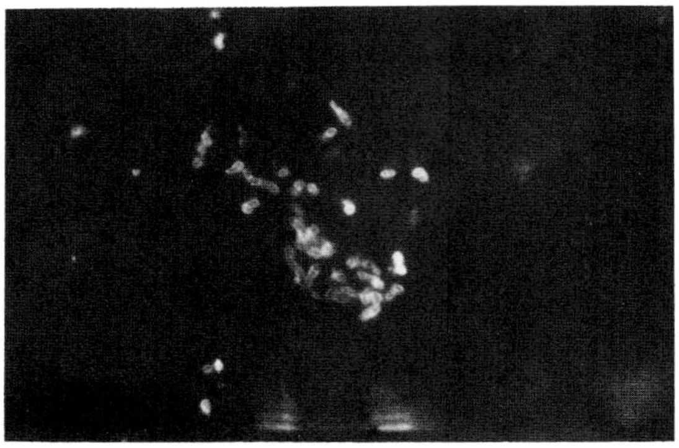

Fig. 4. In situ examination of the rhizoplane by IF. A group of *B. japonicum* serogroup 138 on the surface of a root hair of soybean. Magnification 1200 X. Unpublished data from study of Reyes and Schmidt (1981).

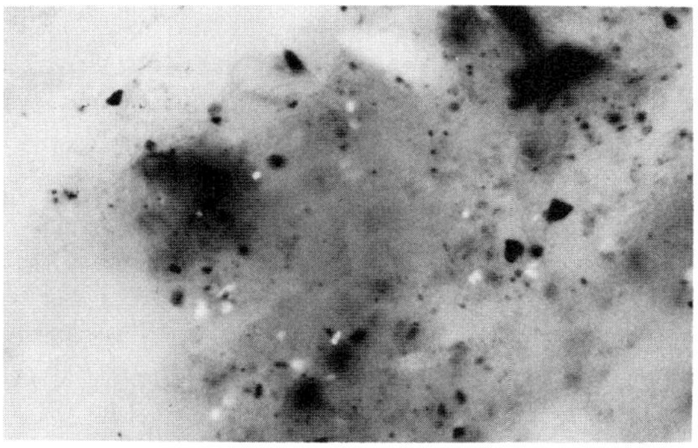

Fig. 5. In situ IF examination of the unwashed epidermis of a field-grown soybean root. Indigenous *B. japonicum* 123 cells seen among soil debris in rhizoplane. Magnification 630 X. Unpublished data from study of Reyes and Schmidt (1981).

root hair, and Figure 5 is from an unwashed epidermal root section. The bacteria seen distributed sparsely in the microscopic field are indigenous *B. japonicum* 123. In this study of the soybean rhizoplane the *B. japonicum* strains were always widely scattered as single, double or triple cells but seldom as microcolonies.

There is much opportunity for improvement in the application of IF to problems that demand the *in situ* examination of target organisms. Improved procedures to control root autofluorescence and enhance background contrast are available. Singleton and Bohlool (1984) described a pre-treatment of solution-grown roots with the dye Eriochrome Black. This combined with gelatin-rhodamine treatment may be equally effective for soil-grown roots. Similarly a single post-FA counterstaining of roots with crystal violet (Hockenhull, 1979) gives excellent results (B.B. Bohlool, personal communication)

IF and complementary techniques

The capability of IF can be expanded readily when used in conjunction with other approaches to autecology. An obvious application is in large-scale rhizosphere monitoring exercises when the primary autoecological approach is selective or semi-selective culture. IF is one of several techniques available to establish the efficiency and reliability of a given cultural procedure. Several other possibilities can be considered for exploiting or extending IF based on complementation with other special techniques.

Monoclonal antibodies

The polyclonal antibodies used thus far in autecological studies with IF are characteristically specific, often too specific for a given objective (Bohlool and Schmidt, 1980). Cross-reactivity at the generic rather than the strain level would be welcome to study the autecology of chemoautotrophic ammonia oxidizing populations, but nitrifier FAs are notably strain-specific (Belser and Schmidt, 1978). Where cross-reactivity among related forms is encountered and improved discrimination is sought in ecological applications it is often possible to carry out simple cross-adsorption procedures to gain the necessary specificity. As an example, cross-reactive members of *B. japonicum* serocluster 123 were distinguishable by adsorption techniques (Schmidt *et al.*, 1986).

IF with monoclonal FAs offers the ability to separate cross-reacting bacteria if cross-adsorption treatments of a polyclonal antibody are inadequate. One possible drawback to dependence in rhizosphere systems on a single antigenic epitope recognized by a monoclonal antibody is that the epitope might change or be lost. A monoclonal antibody capable of distinguishing *B. japonicum* serogroup 123 was used to examine the stability of the homologous epitope in the rhizosphere of soybeans (Kinkle and

Schmidt, unpublished). There was no loss in IF staining of strain 123 during residence in the rhizosphere for a period equivalent to 4 cropping seasons.

Autoradiography and other respiratory markers

A combination of IF and autoradiography was described by Fliermans and Schmidt (1975) to both recognize a particular bacterium and assess its metabolic activity. Metabolic activity was based on the incorporation of $^{14}CO_2$ by the autotroph *Nitrobacter agilis* on slides subsequently coated with photographic emulsion to detect the beta radiation from active cells. Finally the specific FA was applied to the developed emulsion to visualize the individual Nitrobacter cells. As part of the study active and inactive indigenous Nitrobacter cells were observed in a soil material.

Fliermans *et al.* (1981) subsequently combined uptake of the metabolic indicator INT [2-(p-iodophenyl)-3-(p-nitrophenyl)-5-phenyltetrazolium chloride] and IF for detection of *Legionella pneumophila* in a natural environment. Many bacteria reduce this dye during active metabolism with the formation of red intracellular granules of INT formazan, visible under the microscope. The procedure is much easier technically and more versatile than autoradiography.

Split-root growth system

Singleton (1983) devised a solution culture assembly which allowed the roots of soybeans to be split so that one half of the root system could be isolated from the other half during growth of the plant. The approach of providing two rhizospheres available for experimental manipulation during growth of the same plant is appealing and should be exploited for basic rhizosphere research. The technique was modified by Kosslak and Bohlool (1984) to grow soybeans with half root systems in vermiculite. They discovered that inoculation of one side of the split-root system with rhizobia suppressed nodule formation on the second side when that side was inoculated 4 or more days later. Apparently, host response initiated in one part of the root system led to host controls expressed in the other part of the root system.

Further development of the split-root technique to accommodate half-root systems growing in nonsterile soil should be feasible. Combined with IF in one obvious example, and with each root side growing in a different soil, the effect of soil properties on selected indicator bacteria in the rhizosphere could be determined.

Flow cytometry

Automated single-cell analysis can be accomplished with a cytofluorograph in the process known as flow cytometry. Cells are distinguished by size and fluorescence as they pass single file through a flow channel. FA-labeled cells can be detected by amplification of the fluorescent light emitted after excitation by an appropriate laser beam. IF and flow cytometry have been combined to detect pathogenic amoebae present in several natural enviroments (Muldrow *et al.*, 1982), however the amoebae were first enriched and separated from the environment by culture techniques. Obstacles to the direct examination of particulate environments such as rhizosphere soil by IF-flow cytometry appear to be severe. This coupled with the present high cost of cytofluorograph is likely to delay the desirable marriage of IF and flow cytometry for some time.

Molecular probes other than IF

Antibodies, both polyclonal and monclonal are widely used as probes to detect and identify microorganisms by many techniques other than IF. These methods are not useful in dealing with the complexities encountered in soil and rhizosphere samples. An apparent exception to this is found in the report of Levanony *et al.* (1987) who adapted the enzyme-linked immunosorbent assay (ELISA) to the enumeration of *Azospirillum brasilense* Cd in the root tissues (endorhizosphere) of cereal plants grown in soil. The adaptation of ELISA to other rhizosphere zones is probably not feasible because of interference from soil particles.

Exciting developments appear to be on the immediate horizon for the application of labeled genetic probes to the identification and autecological study of target microorganisms in soil and soil rhizospheres. The various types of probes and their potential for use in complex environments were reviewed by Ford and Olson (1988). The most direct attempt to date to adapt DNA probe methodology for use in soils is that of Holben *et al.* (1988). Details of this approach and its further development as directed to rhizosphere research are included elsewhere in this volume.

References

Bashan Y and Levanony H 1985 An improved selection technique and medium for the isolation and enumeration of *Azospirillum brasilense*. Can. J. Microbiol. 31, 947–952.

Bashan Y and Levanony H 1987 Horizontal and vertical movement of *Azospillum brasilense* Cd in the soil and along the rhizosphere of wheat and weeds in controlled and field environments. J. Gen. Microbiol. 133, 3473–3480.

Bashan Y, Levanony H and Ziv-Vecht O 1987 The fate of field-inoculated *Azospirillum brasilense* Cd in wheat rhizosphere during the growing season. Can. J. Microbiol. 33, 1074–1079.

Belser L W and Schmidt E L 1978 Serological diversity within an ammonia-oxidizing population. Appl. Environ. Microbiol. 36, 589–593.

Bohlool B B and Schmidt E L 1980 The immunofluorescence approach in microbial ecology. Adv. Microb. Ecol. 4, 203–241.

Bohlool B B and Schmidt E L 1968 Nonspecific staining: its control in immunofluorescence examination of soil. Science 162, 1012–1014.

Boivin R, Chalifour F P and Dion P 1988 Construction of a Tn5 derivative encoding bioluminescence and its introduction in *Pseudomonas*, *Agrobacterium* and *Rhizobium*. Mol. Gen. Genet. 213, 50–55.

Campbell R and Rovira A D 1973 The study of the rhizosphere by scanning electron microscopy. Soil Biol. Biochem. 5, 747–752.

Compeau G, Al-Achi B, Platsouka E and Levy S B 1988 Survival of rifampicin-resistant mutants of *Pseudomonas fluorescens* and *Pseudomonas putida* in soil systems. Appl. Environ. Microbiol. 54, 2432–2438.

Drahos D, Hemming B and McPherson S 1986 Tracking recombinant organisms in the environment: beta-galactosidase as a selectable non-antibiotic marker for fluorescent pseudomonads. Biotechnology 4, 439–444.

Fliermans C B and Schmidt E L 1975 Autoradiography and immunoflurescence combined for autecological study of single cell activity with *Nitrobacter* as a model system. Appl. Microbiol. 30, 676–684.

Fliermans C B, Soracco R J and Pope D H 1981 Measurement of *Legionella pneumophila* activity *in situ*. Current Microbiol. 6, 89–94.

Ford S and Olson B H 1988 Methods for detecting genetically engineered microorganism in the environment. Adv. Microbiol Ecol. 10, 45–79.

Foster R C 1981 The ultrastructure and histochemistry of the rhizosphere. New Phytol. 89, 263–273.

Foster R C 1988 Microenvironments of soil microorganism. Biol. Fertil. Soil. 6, 189–203.

Foster R C, Rovira A D and Cock T W 1983 Ultrastructure of the root-soil interface. American Phytopathological Society. St. Paul, MN, 157 p.

Gould W D, Hagedorn C, Bardinelli T R, and Zablotowicz R M 1985 New selective media for enumeration and recovery of fluorescent pseudomonads from various habitats. Appl. Environ. Microbiol. 49, 28–32.

Graham J B and Istock C A 1979 Gene exchange and natural selection cause *Bacillus subtilis* to evolve in soil culture. Science 204, 637–638.

Hockenhull J 1979 *In situ* detection of *Erwinia amylovora* antigen in symptomless petiole and stem tissue by means of the fluorescent antibody technique. Kgl. Vet. Landohojsk. Arsskr. 1979, 1–14.

Holben W E, Jansson J K, Chelm B K and Tiedje J M 1988 DNA probe method for the detection of specific microorganisms in the soil bacterial community. Appl. Environ. Microbiol 54, 703–711.

Jenny H and Grossenbacher K A 1963 Root-soil boundary zones as seen in the electron microscope. Soil Sci. Soc. Am. Proc. 27, 273–277.

Kosslak R M and Bohlool B B 1984 Suppression of nodule development on one side of a split-root system of soybeans caused by prior inoculation of the other side. Plant Physiol. 75, 125–130.

Levanony H, Bashan Y and Kahana Z E 1987 Enzyme-linked immunosorbenty assay for specific identification and enumeration of *Azospirillum brasilense* Cd in cereal roots. Appl. Environ. Microbiol. 53, 358–364.

Moawad H, Ellis W R and Schmidt E L 1984 Rhizosphere response as a factor in competition among three serogroups of indigenous *Rhizobium japonicum* for nodulation of field-grown soybeans. Appl. Environ. Microbiol. 47, 607–612.

Muldrow L L, Tyndall R L and Fliermans C B 1982 Application of flow cytometry to studies of pathogenic free-living amoebae. Appl. Environ. Microbiol. 44, 1258–1269.

Old K M and Nicolson T H 1975 Electron microscopical studies of the microflora of roots of sand dune grasses. New Phytol. 74, 51–58.

Rennie R J, Reyes V G and Schmidt E L 1977 Immunofluorescence detection of the effects of wheat and soybean roots on *Nitrobacter* in soil. Soil Sci. 124, 10–15.

Reyes V G and Schmidt E L 1981 Populations of *Rhizobium japonicum* associated with surfaces of soil-grown roots. Plant and Soil 61, 71–80.

Robert R M and Schmidt E L 1985 Response of three indigenous serogroups of *Rhizobium japonicum* to the rhizosphere of pre-emergent seedlings of soybean. Soil Biol. Biochem. 17, 579–580.

Rovira A D and Campbell R 1974 Scanning electron microscopy of microorganism on the roots of wheat. Microbiol. Ecol. 1, 15–23.

Schmidt E L and Bankole R O 1962 Detection of *Aspergillus flavus* in soil by immunofluorescent staining. Science 136, 776–777.

Schmidt E L, Zidwick M J and Abebe H M 1986 *Bradyrhizobium japnicum* serocluster 123 and diversity among member isolates. Appl. Environ. Microbiol. 51, 1212–1215.

Singleton P W 1983 A split-root growth system for evaluating the effect of salinity on components of the soybean *Rhizobium japonicum* symbiosis. Crop Sci. 23, 259–262.

Singleton P W and Bohlool B B 1984 Effect of salinity on nodule formation by soybean. Plant Physiol. 74, 72–76.

Van Elzas J D, Dijkstra A F, Govaert J M and Van Veen J A 1986 Survival of *Pseudomonas fluorescens* and *Bacillus subtilis* introduced into two soils of different texture in field microplots. FEMS Microbiol. Ecol. 38, 151–160.

D. L. Keister and P. B. Cregan (Eds.), The rhizosphere and plant growth, 91–97.

PLSO BARC10

Applications and limitations of rhizotrons and minirhizotrons for root studies

H.M. TAYLOR[1], D.R. UPCHURCH[2] and B.L. McMICHAEL[2]
[1]*Department of Agronomy, Horticulture and Entomology, Texas Tech University, Lubbock, TX 79409, USA and* [2]*Cropping System Research Laboratory, USDA, ARS, Route 3, Box 215, Lubbock, TX 79401, USA*

Key words: rhizosphere dynamics, rhizotrons, root studies

Abstract

This paper discusses two glass-wall techniques-rhizotrons and minirhizotrons. Rhizotrons are covered underground walkways with clear windows on one or both sides. Their design varies somewhat with the type of research to be conducted in them. A minirhizotron consists of a clear tube inserted into the soil then some type of viewing arrangement is lowered into the tube. In both techniques, roots and their rhizosphere organisms can be observed at the soil-wall boundary.

Rhizotrons usually are expensive to construct and operate. Their cost effectiveness depends on the research questions being asked and the number of scientists using the facility. Minirhizotrons vary considerably depending on the type of viewing instrument being used but require a substantial number of replications to characterize root dynamics in a field plot.

Rhizotrons and minirhizotrons both are transparent wall techniques that allow the researcher to observe the same plant roots and their rhizospheres and rhizoplanes on a serial basis while the roots are growing in soil.

Rhizotrons are one of the earliest nondestructive techniques for observing root growth in soil. Rhizotrons are covered underground cellars or walkways that have clear windows on one or both sides. The windows may be in contact with a natural soil profile that surrounds the passageway or may be one wall of a soil-filled box. A typical soil-filled box type of rhizotron is shown in Figure 1. Plants are grown in the soil adjacent to the window. Some of the roots intersect the window and can be observed. Rhizotron designs have been published by Fordham (1972), Freeman and Smart (1976), Glover (1967), Hilton *et al.* (1969), Huxley and Turk (1967), Rogers (1969), Soileau *et al.* (1974), Taylor (1969), and Taylor and Willatt (1981). Rhizotron construction techniques, their operation and types of experiments that have been conducted in them have been reviewed by Huck and Taylor (1982) and Taylor and Willatt (1981).

Rhizotrons have several advantages over most of the other root study methods. Successive measurements are easily made on the same individual root or the visible part of the root system. It is easy to find a specific portion of the root system when one wishes to make serial observations. Estimates of increases in root length can be made rapidly. Sensors can be installed easily in rhizotrons to measure soil conditions such as temperature, oxygen or carbon dioxide concentration, soil water potential or content, soil pH, or ionic concentration in the soil solution. There also is enough space in the central passageway to mount a camera for time-lapse photography. Rhizotrons can be shared by a number of scientists, each working within their specific subject matter field.

Rhizotrons also have several disadvantages, with cost being a primary one. The rhizotron at

Fig. 1. The Auburn rhizotron soon after it was built. **A.** Above ground portion, which was soon modified to reduce effects of the concrete roof of the central walkway. **B.** Underground cellar with transparent walls on either side.

Auburn Alabama cost about $40,000 when constructed in 1969 and the one at Ames, Iowa cost about $20,000 in 1973. Obviously, current costs would be substantially greater.

Although it is possible to enclose a natural soil profile in a box and transport it to the rhizotron site, the usual procedure is to have a disturbed soil in the soil-filled box. Natural soil also has been used by placing a window against a natural soil profile at the site. In this case, you have the advantage of natural soil structure but have no firm knowledge about the volume of soil being explored by the plant whose root system is being observed.

The very nature of a rhizotron facility results in some environmental modifications. In some rhizotrons, the concrete-covered walkway extends above ground level, thus modifying air flow. Even if a concrete-covered walkway does not project above ground level, the concrete will radiate more sensible heat during daytime than an equal area of moist soil or vegetation. The best option involves planting vegetation over the walkway but the shallow rooting zone of those plants may cause reduced shoot growth there. Rhizotrons require continual maintenance but are available for many experiments. They are most cost effective when located where several scientists interested in rhizosphere dynamics are stationed nearby.

Minirhizotrons were originally proposed by Bates (1937). This technique involves installing a clear tube in the soil and lowering a device into the tube to observe roots at the soil-tube interface. The tubes commonly are installed a few days after a crop is planted. The tubes are installed at an angle thirty to forty-five degrees from the vertical to prevent roots from following the tube for long distances. Soil physical factors may differ drastically between the soil-tube interface and bulk soil. These differences cause substantial distortion of the results. With the thirty to forty-five degree angles, roots usually will not follow the upper surface of the tube but may follow the underside if intimate contact is not established between the tube and soil. Care must be taken while creating the hole for inserting the tube in order to prevent soil compaction and soil shear, which will affect root growth at the interface (Box *et al.*, 1989).

Bates (1937) originally used a mirror and battery-operated lamp mounted on the end of a stick to see roots intersecting a glass tube. The mirror and stick method of observation limited the usefulness of this technique because roots at depths below about one meter could not be seen without the aid of a telescope. With the application of new technologies to the observation procedure, results improved. Waddington (1971) replaced the mirror and stick arrangement with a coherent fiber-optic and right-angle viewing attachment, which improved the quality of the root image. Sanders and Brown (1978) included a 35 mm camera with a fiber-optic duodenoscope to further improve images of the intersecting roots. Dyer and Brown (1980) replaced the 35 mm camera with a black and white video camera. Upchurch and Ritchie (1984) described a system involving a color video camera with a right-angle viewing head. While the camera was lowered into a tube, a picture of the intersecting roots was recorded on video tape. This battery-operated system was mounted on a back-pack which could be carried in the field (Fig. 2). This technology has improved the quality of the image such that any root intersecting the tube can be seen.

Various techniques are available for quantifying the intersections of roots with minirhizotron tubes. Bates (1937) drew pictures of the roots that intersected the tube. Later, Gregory (1979) measured the length of roots in contact with the tube. This length was used for relative comparisons between depths or treatments or converted to an estimate of root length density. The conversion to root length density involves one of several assumptions concerning the volume of soil to associate with the measured root length.

Upchurch and Ritchie (1983) proposed that the number of roots visible at the soil-tube interface, rather than their length, was the appropriate parameter. Upchurch (1985) developed a theoretical relationship between the number of intersecting roots and the root length density in the soil surrounding the minirhizotron. The results from this relationship adequately predicted the measured root length density.

One of the greatest limitations of the minirhizotron system is the number of tubes required to accurately estimate rooting. Upchurch and

Fig. 2. A backpack video camera-recorder-viewer system being used in a minirhizotron.

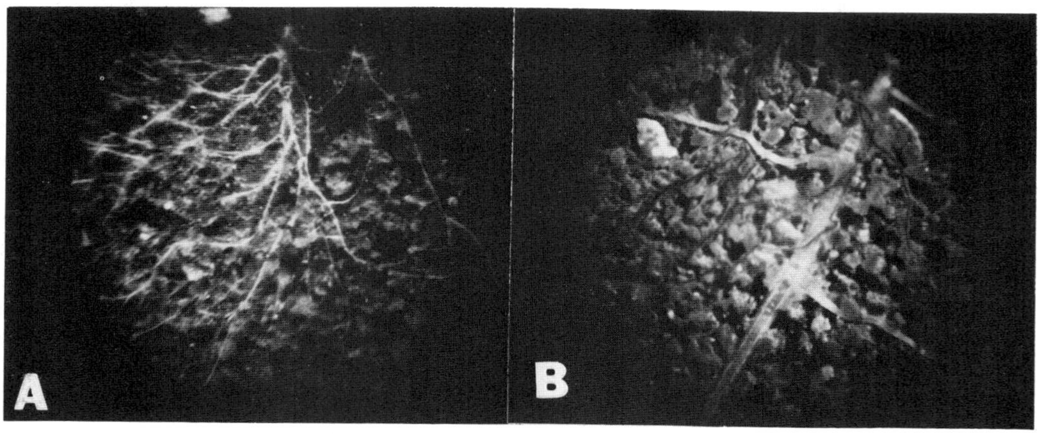

Fig. 3. Phymatotrichum omnivorum observed using a minirhizotron. **A.** Strands growing through the soil toward a cotton root. **B.** Strands on a cotton root. Photographs are from study of Rush *et al.*, 1984.

Ritchie (1983, 1984) and Upchurch (1985) reported observations that suggest a minimum of eight tubes may be required for the mean value to correlate with root length density of a single plot. This number is not substantially different, however, from the number of soil cores required for the same estimate. Although no explanation has been provided, near-surface (upper 0.2 to 0.4 m) estimates of rooting from this sytem are frequently low. Accuracy is also low at or near the maximum rooting depth, where rooting density is low. Data from soil cores exhibits similar variability at depth. However, data in Box *et al.* (1989) suggest that this problem may be a soil and/or installation procedure related phenomenon.

The minirhizotron system provides the opportunity for repeated observations of the sample location, eliminating one source of error. Although it requires thirty to forty-five minutes to install each minirhizotron tube, routine observation of roots in each tube requires about ten to fifteen minutes by two experienced individuals. At present, the equipment needed to observe and record color video pictures of roots is commercially available at a cost of about $15,000 (Circon Corporation, Santa Barbara, California). Magnification lens systems are also available (Bartz Technology Co., Santa Barbara, CA).

Several extensions to the use of minirhizotrons have been reported or are under development. The minirhizotron system has been used to study root pathogens (Rush *et al.*, 1984). Upchurch, McMichael and Taylor (1988) described a method using the ratio of top to bottom root intersections to determine the dominant orientation of the root system. Procedures are also being developed to use this system in various

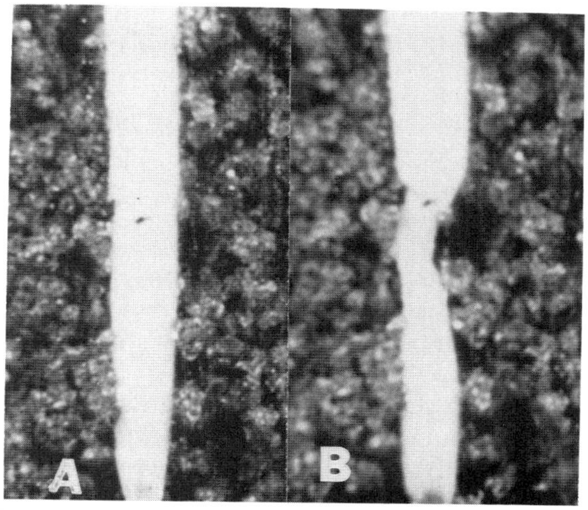

Fig. 4. Death of a cotton root caused by lack of oxygen supply for 180 minutes. **A.** Before oxygen deprivation. **B.** About 120 minutes after oxygen supply had been returned to the roots. Photos courtesy of M.G. Huck.

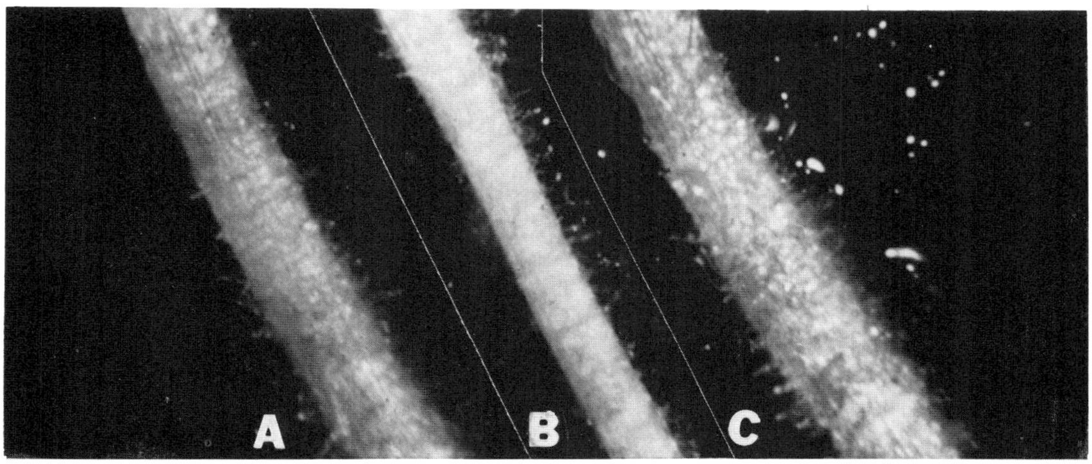

Fig. 5. Cotton root located in the Auburn rhizotron about. **A.** midnight, **B.** 1700, **C.** midnight the following day. Photos courtesy M.G. Huck.

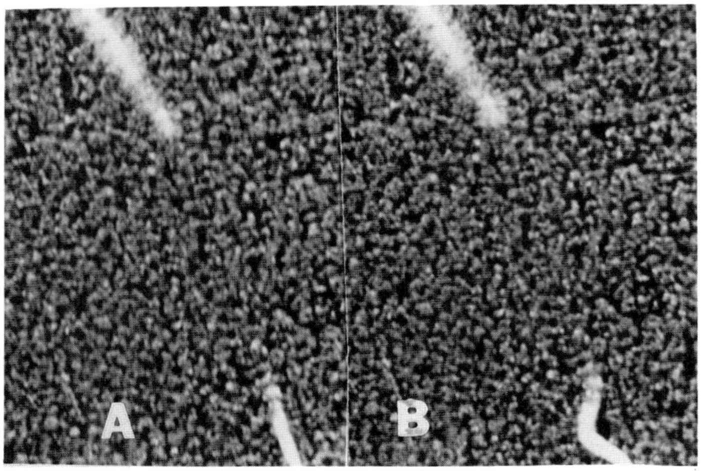

Fig. 6. Cotton root. **A.** As it grows through loose soil approaching a compact, high-strength layer and **B.** after the tip has encountered the layer.

studies of rhizosphere flora and fauna, especially in ecological studies. A complete review of this technique is available in a symposium publication (Taylor, 1987).

It is interesting scientifically to observe the spread of a fungal hyphae (Fig. 3), the death of an individual root as its experiences an oxygen deficit (Fig. 4), the swelling and shrinkage of a root as the soil water potential changes around it (Fig. 5), or the bulging of a root as it encounters an impeding object (Fig. 6). All of these, and many more, observations can be made using time-lapse imagery in either the rhizotron or minirhizotron. There is a broad window of opportunity for studying rhizosphere dynamics by using the rhizotron and minirhizotron techniques.

References

Bates G H 1937 A device for the observation of root growth in the soil. Nature 139, 966–967.

Box J E, Jr., Smucker A J M and Ritchie J T 1989 Minirhizotron installation techniques for investigating root responses to drought and oxygen stresses. Soil Sci. Soc. Am. J. 53, 115–118.

Dyer D and Brown D A 1980 *In situ* root observation using fiber optic/video and fluorescence. *In* Agron. Abstracts p. 80, Madison, WI.

Fordham R 1972 Observations on the growth of roots and shoots of tea (*Camellia sinensis* L.) in Southern Malawi. J. Hortic. Sci. 47, 221–229.

Freeman B M and Smart B E 1976 A root observation laboratory for studies with grapevines. Am. J. Enol. Viticulture. 27, 36–39.

Glover J 1967 The simultaneous growth of sugarcane roots and tops in relation to soil and climate. Proc. South African Sugar Tech. Assn. 41, 143–159.

Gregory P J 1979 A periscope method for observing root growth and distribution in the field. J. Expt. Bot. 30, 205–214.

Hilton R J, Bhar D S and Mason G F 1969 A rhizotron for in situ root growth studies. Can. J. Plant Sci. 49, 101-104.

Huck M G and Taylor H M 1982 The rhizotron as a tool for root research. Adv. Agron. 35, 1–35.

Huxley P A and Turk A 1967 A new root observation laboratory for coffee. Kenya Coffee 156–159.

Rogers W S 1969 The East Malling root-observation laboratories. *In* Root Growth. Ed. W J Whittington. pp 361–376. Butterworth, London.

Rush C M, Upchurch D R and Gerik T J 1984 *In situ* observation of *Phymatotrichum omnivorum* with a borescope minirhizotron system. Phytopathology 74, 104–105.

Sanders J L and Brown D A 1978 A new fiber optic technique for measuring root growth of soybeans under field conditions. Agron. J. 70, 1073–1076.

Soileau J M, Mays D A, Khasawneh F E and Kilmer V J 1974 The rhizotron-lysimeter research facility at TVA, Muscle Shoals, Alabama. Agron. J. 66, 828–832.

Taylor H M 1969 The rhizotron at Auburn Alabama . . . a plant root observation laboratory. Auburn U. Agr. Expt. Sta. Circ. 197, 9 p.

Taylor H M 1987 Minirhizotron Observation Tubes: Methods and Applications for Measuring Rhizosphere Dynamics. ASA Special Publication 50, American Society of Agronomy, Madison, WI.

Taylor H M and Willatt S T 1981 Utilization of rhizotrons in root research. *In* The Soil/Root System in Relation to Brazilian Agriculture. Eds. R S Russell, K Igue and Y R

Mehta. pp 319–337. Instituto Agronomica do Parana, Londrina, Brazil: IAPAR.
Upchurch D R 1985 Relationship between observations in minirhizotrons and true root length density. Ph.D. Diss. Texas Tech University Press. Lubbock.
Upchurch D R, McMichael B L and Taylor H M 1988 Use of minirhizotrons to characterize root system orientation. Soil Sci. Soc. Am. J. 53, 319–323.
Upchurch D R and Richie J T 1983 Root observations using a video recording system in minirhizotrons. Agron. J. 75, 1009–1015.
Upchurch D R and Ritchie J T 1984 Battery-operated color video camera for root observations in minirhizotrons. Agron. J. 76, 1015–1017.
Waddington J 1971 Observations of plant roots *in situ*. Can. J. Bot. 49, 1850–1852.

D. L. Keister and P. B. Cregan (Eds.), The rhizosphere and plant growth, 99–102.

Conversion factors relating thymidine uptake to growth rate with rhizosphere bacteria

H. CHRISTENSEN
Institute of Population Biology, Copenhagen University, Universitetsparken 15, DK-2100 Copenhagen Ø, Denmark

Key words: bacteria, DNA, empirical, growth rate, Pseudomonas

Abstract

Growth rates of some rhizosphere bacteria can be measured by the use of ^{3}H-thymidine (^{3}H-TdR) labelling. However fluorescent pseudomonads of soil origin were found to incorporate little or no ^{3}H-TdR into DNA. The conversion factor relating cellular growth rate to ^{3}H-TdR incorporation for exponentially growing cultures of non-pseudomonads was about 2×10^{18} cells mole TdR^{-1}. This value is comparable to constants used in aquatic environments.

Introduction

The present ^{3}H-thymidine labelling methodology is being developed with the purpose of using bacterial growth rate as a quantitative *in situ* bioassay of rhizosphere activity. The efficiency with which bacteria incorporate the label and the derivation of constants for quantification are prerequisites for use of the method.

Bowen and Rovira (1973) proposed ^{3}H-TdR labelling for rhizosphere studies, to distinguish between growing and dormant cells by microautoradiography. Few attempts have been made to use the method in terrestrial environments (Christensen *et al.*, 1989). A filter method originally proposed for *Bacillus cereus* growth has become the favorite method for study of bacterial growth in aquatic environments (Moriarty, 1986). Although studied in detail for *Escherichia coli* and *Salmonella typhimurium* (Neuhard and Nygaard, 1987), information of the efficiency with which different bacteria are labelled with thymidine under a range of environmental conditions is very limited.

At present, bacterial growth can be compared in relative terms between different locations within the root-system (Christensen, unpublished). More useful is the absolute amount of bacterial biomass produced. An empirically derived conversion factor has provided a constant relating incorporation rate of labelled thymidine to production of cells (Kirchman *et al.*, 1982). This differs from conversion factors based on DNA content per cell and mole % T in DNA, representing the maximal efficiency with which TdR can be incorporated.

Materials and methods

The following bacterial isolates were investigated: *Pseudomonas chlororaphis* ATCC43928 (I14) and *P. putida* (I19) both soil denitrifiers. The fluorescent pseudomonads D38, D69 and D102 were lettuce rhizosphere bacteria possibly all *P. putida* based on negative tests for gelatine hydrolysis. In addition two non-identified soil isolates, H5 and H9, and Mix, a culture of at least 5 different bacteria without fluorescent pseudomonads were included.

Batch cultures were grown in nutrient broth (Difco) aerated with sterile filtered air (flow rate 5–7.5 vol.$^{-1}$ culture min^{-1}). Exponential growth was monitored by $OD_{450\ nm}$ for sufficient time to allow growth rates to be calculated. Subsamples were taken for plate counts on nutrient agar (Difco), and for ^{3}H-TdR labelling to be performed in sterile polyethylene beakers continuously aerated with sterile filtered air. A stock solution of ^{3}H-TdR (89 or 77 Ci/mmole, methyl-^{3}H-thymidine, Amersham, UK), diluted with unlabelled TdR (Sigma Co., St. Louis, USA) was added at 0.5–3.0 Ci mmole^{-1} to 1.00 μM final TdR conc. The labelling time was 2 min at 28°C and 10 min at 10°C (in time course experiments 2–10 min at 28°C or 5–15 min at 10°C). Incorporation was stopped by adding formalin to 10% final conc.

Extraction, filtration and liquid scintillation counting followed Riemann and Søndergaard (1984). In short, 2 mL subsamples of labelled culture in 3 replicates were treated with equal vol. of ice cold 10% trichloroacetic acid (TCA) (15 min, 0°C), or 0.5 *N* NaOH final conc. (60°C, 1 h, cooled and acidified) or boiled in 20% TCA final conc. (30 min, and chilled to 0°C). The extracts were filtered on 0.45 μm cellulose nitrate filters (Sartorius, Göttingen, W. Germany) to collect the precipitated labelled macromolecules. RNA, DNA and protein were collected by cold TCA treatment, DNA and protein were extracted with warm alkali and only protein was extracted when samples were boiled in acid. Activity in RNA, DNA, and protein was calculated by subtraction and converted into mole TdR min^{-1} ml^{-1}. Data were corrected for recovery of added ^{14}C-DNA (*E. coli* extract, Amer-

Table 1. Comparison of conversion factors. Empirically derived factors were obtained by relating increase in optical density to ^{3}H-TdR incorporation and represent actual incorporation efficiency, the efficiency is low if conversion factors are high. An absolute expression of efficiency is gained by dividing maximal by empirical values

Sample	Temp. °C	Conversion factors (10^{18} cells mole Tdr^{-1})			
		Empirical		Maximal	
		RNA + DNA + prot. fraction	DNA fraction	Measured[a]	Literature[b]
P. chlororaphis					
I14	28	43.3	160	0.146	0.12
	10	17.2	ud[c]		
P. putida					
I19	28	33.3	62.8		0.16
	10	68	ud		
fl. Pseudomonas					
D38	28	6.33			
	10	1.60	ud		
D69	28	0.479–30.9	26–ud		
	10	46.5	31.0		
D102	28	1.27–33.5	6–ud	0.10	0.16
	10	0.513–1.91	0.513–12.7		
soil-isolate					
H5	28	3.01	3.01		
H9	28	2.91		0.537	
E. coli					
K12	28	0.323	0.323	0.092	
Mix	28	0.960	1.55		0.15
	10	1.73	2.08		0.15

[a] Based on flow cytometric measurements of DNA (Christensen H., Bakken L.R., Olsen R.A. and Steen H., unpublished) T mole % as in b) was used.

[b] Genome sizes and T mole % given in Herdman (1985), 2 genome equivalents per cell assumed. For mix the mean value of G- bacteria and 25 mole % T was used.

[c] ud: no significant activity found in DNA fraction conversion factor increases infinitely.

sham) and were 67% in the cold TCA treatment and 59% for the warm base.

Results and discussion

Labeling of Pseudomonas

Highly variable and very high empirical conversion factors were obtained with all Pseudomonas tested (Table 1). The incorporation of ^{3}H-TdR was not coupled to cellular growth because incorporation did not increase with labelling time. In contrast bacteria other than Pseudomonas incorporated ^{3}H-TdR at linear rates. *E. coli* up to 15 min with only 200 n*M* ^{3}H-TdR in the medium (data not shown). With Pseudomonas, RNA was either the only compound labelled significantly or contained the majority of label (Table 2). The major labelled macromolecule of the other bacteria was always DNA (Table 2).

Three ways were tried to improve labelling of DNA. Different labelling times did not change the relative composition of labelled macromolecules, neither did the addition of 1.00 μM adenosine (data not shown). Adenosine is known to inhibit the enzyme thymidine phosphorylase, degrading thymidine to thymine, thereby improving incorporation of ^{3}H-TdR (Boyse and Setlow, 1962; Neuhard and Nygaard, 1987). Thirdly, incorporation was not increased at 10°C (Table 1), as would be the case if the bacteria were not able to fulfil their thymidine requirement from exogenous sources because of too rapid growth (Moriarty, 1986).

The most likely reason why ^{3}H-TdR was not incorporated in DNA is the lack of thymidine kinase (TK), not found or produced during any phase of growth in *P. aeruginosa*, *P. cepacia*, *P. putida* and *P. pseudomallei* whereas *P. fluorescens* and *P. maltophilia* showed a significant activity of 9 and 51% respectively of *E. coli* K12 (Saito *et al.*, 1985). The lack of TK in fungi probably also accounts for non-incorporation of ^{3}H-TdR into DNA (Grivell and Jackson, 1968). *S. cerevisiae* GR558 and GR624 included in the present study also was not labelled in the DNA fraction (data now shown).

Mean conversion factors

Even with bacteria other than Pseudomonas, incorporation of ^{3}H-TdR into DNA is far from maximal (Table 1), with the highest efficiency around 30% with *E. coli*, and the lowest with

Table 2. Relative distribution of ^{3}H-labelled macromolecules following incubation of exponential growing cultures with ^{3}H-thymidine

Sample	Temp. °C	Percent of activity		
		RNA	DNA	protein
fl. *Pseudomonas*				
D69	28	93	7.0	0.0
	10	100	0.0	0.0
P. chlororaphis				
I14	28	73	27	0.0
P. putida				
I19	28	47	53	0.0
soil-isolate				
H5	28	0.0	100	0.0
H9	28	35	65	0.0
Mix	28	38	62	0.0
	10	17	83	0.0

mixed cultures around 10%. Based on the present very fragmentary results, the empirical conversion factor was about 2×10^{18} cells mole TdR^{-1} for cultured soil bacteria. The same value was proposed as a round figure for aquatic environments by Moriarty (1986).

Conclusion

The relative comparison of growth of the pseudomonad D102 in a sterile soil-plant system between rhizosphere and non-rhizosphere soil done by Christensen *et al.* (1989) is still valid. However the growth rates given were apparently much too low since a conversion factor 2.2×10^{17} cell mole TdR^{-1} based on DNA content was used for quantification. Empirical conversion factors of the present study cannot be used for recalculation because of their highly variable nature.

Rovira and Sands (1971) found the highest proportion of fl. pseudomonads to total bacteria to be 16% on slightly decomposed wheat straw and 5.6% on wheat roots, else less. Underestimation of growth rates will be a serious problem only if pseudomonads are growing faster *in situ* than the rest of the population.

In situ derived conversion factors are in the moment measured by time course root colonization experiments where loss of bacteria caused by lysis and grazers is eliminated and direct relations between bacterial biomass increase and ^{3}H-TdR incorporation are measured.

References

Bowen G D and Rovira A D 1973 Are modelling approaches useful in rhizosphere biology? Bull. Ecol. Res. Comm. (Stockholm) 17, 443–450.

Boyce R P and Setlow R B 1962 A simple method of increasing the incorporation of thymidine into deoxyribonucleic acid of *Escherichia coli*. Biochim. Biophys. Acta 61, 618–620.

Christensen H, Funck-Jensen D and Kjøller A 1989 Growth rate of rhizosphere bacteria measured directly by the tritiated thymidine incorporation technique. Soil Biol. Biochem. 21, 113–117.

Grivell A R and Jackson J F 1968 Thymidine kinase: evidence for its absence from *Neurospora crassa* and some other micro-organisms, and the relevance of this to the specific labelling of deoxyribonucleic acid. J. Gen. Microbiol. 54, 307–317.

Herdman M 1985 The evolution of bacterial genomes. *In* The Evolution of Genome Size. Ed. T Cavalier-Smith. pp. 37–68. John Wiley and Sons, Ltd., New York.

Kirchman D, Ducklow H and Mitchell R 1982 Estimates of bacterial growth from changes in uptake rates and biomass. Appl. Environ. Microbiol. 44, 1296–1307.

Moriarty D J W 1986 Measurement of bacterial growth rates in aquatic systems from rates of nucleic acid synthesis. Adv. Microbiol. Ecol. 9, 245–292.

Neuhard J and Nygaard P 1987 Purines and pyrimidines. *In Escherichia coli* and *Salmonella typhimurium*. Ed. F C Neidhardt. Vol. 1. pp 445–473. American Society for Microbiology, Washington.

Riemann B and Søndergaard M 1984 Measurements of diel rates of bacterial secondary production in aquatic environments. Appl. Environ. Microbiol. 47, 632–638.

Rovira A D and Sands D C 1971 Fluorescent pseudomonads – a residual component in the soil microflora? J. Appl. Bacteriol. 34, 253–259.

Saito H, Tomioka H and Ohkido S 1985 Further studies of thymidine kinase: Distribution pattern of the enzyme in bacteria. J. Gen. Microbiol. 131, 3091–3098.

Short papers

D. L. Keister and P. B. Cregan (Eds.), The rhizosphere and plant growth, 105.

PLSO BARC36

Growth rate of rhizosphere bacteria as measured by the thymidine method

H. CHRISTENSEN
Institute of Population Biology, Copenhagen University, Universitetsparken 15, DK-2100 Copenhagen Ø, Denmark

Key words: cucumber, DNA, humic acid

Bacterial growth rates in soil in different proximity to roots was determined by adding ^{3}H-thymidine and measuring incorporation of the compound into DNA. The method is presented in short together with preliminary results. The soil-plant system under study was separated into root and soil fractions and labeled immediately. Samples equivalent to 50–100 mg dry weight (DW) of root segments with adhering soil or soil without roots, was taken to microtubes and labeled as a slurry with ^{3}H-thymidine (200 μl, 200 nM, 30 Ci mmole^{-1}) for 15 min. Soil samples added formalin (10% final conc.) before labeling were included as blanks. Extraction was performed with 0.5 M NaOH (1 h, 60°C) followed by centrifugation (10,000 xg, 10 min). Supernatants were diluted 10 times with distilled water, acidified to pH 2 at 0°C and filtered through 0.45 μm cellulose nitrate filters. The amount of activity collected on the filter was measured by liquid scintillation.

Effect of sample size on recovery and incorporation of ^{3}H-thymidine

Recovery of ^{14}C-DNA internal standards was 30–70% depending on type of soil and amount of sample. Expressed on a soil DW basis, DNA recovery increased slightly but ^{3}H-thymidine incorporation decreased at increased amount of soil. The effect on DNA recovery is probably because humid acid aids in DNA precipitation. The opposite effect on ^{3}H-thymidine incorporation was either because of higher incorporation in more diluted samples (even though slurries could be considered saturated with thymidine), or related to the inclusion of larger low bacterial activity minerals in samples of higher weight. The lower limit of sample size to be measured, is set by the availability of ^{3}H-thymidine with a sufficient high specific activity, the upper by the humus content of soil.

Spatial variability of bacterial growth within a cucumber root system

Bacterial growth in pots with cucumber, grown for 3 weeks in a sandy loam soil mixed with coarse sand, was measured with the method. When 36 samples were compared based on DW of soil in individual samples, 8% of bacterial growth was in soil without roots, 32% on root tips and 60% on older root parts. The coefficient of variation between 4 replicate samples of the same origin was around 50%. The variation between top, middle and bottom sections of the pot was very high, with low to insignificant incorporation in the top-soil, independent of sample type, probably because of daily cycles of drying and rewetting.

D. L. Keister and P. B. Cregan (Eds.), The rhizosphere and plant growth, 106.

Chromosome and symbiotic plasmid diversity within a naturally-occurring population of clover rhizobia

DAVID H. DEMEZAS[1], TERRY B. REARDON[2], JOHN M. WATSON[1] and ALAN H. GIBSON[1]
[1]*CSIRO, Division of Plant Industry, Canberra, A.C.T., 2601, Australia, and* [2] *Evolutionary Biology Unit, South Australian Museum, Gilles Plains, 5086, South Australia, Australia*

Key words: genetic diversity, rhizobia, Sym plasmid

Naturally-occurring populations of rhizobia have been shown by serological and protein electrophoresis techniques to be diverse. However, the genetic structure of such populations is poorly understood. We have used multilocus enzyme electrophoresis and restriction fragment length polymorphism (RFLP) analysis around chromosomal and Sym plasmid-located genetic regions to characterize the genetic diversity of chromosomal and Sym plasmid types within a naturally-occurring population of *Rhizobium leguminosarum* bv. *trifolii*.

Southern blots of genomic DNAs from 121 root-nodule isolates of *R. leguminosarum* bv. *trifolii* with a specific Sym plasmid-located DNA repeat sequence, RtRS, were undertaken to characterize the diversity of Sym plasmid types. Each Sym plasmid type was characterized by a unique RtRS-hybridization pattern of between two and eight strongly-hybridizing restriction fragments. There were 24 uniquely different Sym plasmid types.

The same Southern blots as above were probed with two chromosomally-derived DNA sequences to establish the chromosomal RFLP type of each isolate. There were 13 chromosomal RFLP types. A sub-population of 68 isolates, representative of each chromosomal RFLP/Sym plasmid type combination, was characterized further by multilocus enzyme electrophoresis at 18 different enzyme loci. Multilocus enzyme electrophoretic analysis complemented the RFLP analysis. Enzyme electrophoretic analysis revealed 17 different electrophoretic types (ETs). The genetic diversity between ETs ranged from 6% to 94% of loci examined. The genetic diversity amongst several ETs (8) was less than 20% of loci examined. The RFLP patterns of these same ETs were also very similar.

There was a non-random association between the chromosomal-types and Sym plasmid types, such that similar Sym plasmid types were confined generally to genetically related chromosomal types. However, similar Sym plasmid types were occasionally observed in genetically quite distinct chromosomal types. Furthermore, very different Sym Plasmid types were harbored by the same chromosomal type. These observations imply that Sym plasmids may be transferred between strains of a naturally-occurring population of rhizobia. These results are relevant to future studies on the fate of genetically-engineered strains of rhizobia, which may be released into agricultural soils already harboring strains of homologous rhizobia.

D. L. Keister and P. B. Cregan (Eds.), The rhizosphere and plant growth, 107.

PLSO BARC38

Stability of antibiotic-resistance markers in *Bacillus cereus* UW85

L.J. HALVERSON and J. HANDELSMAN
Department of Plant Pathology, University of Wisconsin-Madison, Madison, WI 53706, USA

Key words: antibiotic resistance, *Bacillus cereus*, biocontrol

Bacillus cereus UW85 is a potential biocontrol agent for pre- and post-emergence damping-off caused by oomycete fungi. Establishment of UW85 in the rhizosphere may be required for it to be an effective biocontrol agent. Understanding the rhizosphere population dynamics of UW85 may help predict its effectiveness in biocontrol. We used spontaneous antibiotic resistance mutations in *B. cereus* UW85 as markers to distinguish it from the indigeneous *B. cereus* population. In this summary we present evidence suggesting that antibiotic resistance was rapidly lost from the *B. cereus* population in the field and that this loss could not be predicted from laboratory experiments.

Neomycin- and streptomycin- resistant colonies were present in UW85 cultures at a frequency of 10^{-8}. Strain UW11 (Str^r) produced orange colonies (Ora^+) which were distinguishable from the wild-type cream-colored colonies. UW2 (Neo^r) and UW11 revert to the antibiotic-sensitive phenotype at similar frequencies. Doubling times of UW2 and UW11 varied depending on the media in which they were grown. Stability of antibiotic-resistance in the population of UW2 and UW11 was assessed in growth chamber experiments. After 46 days the antibiotic-resistant population equalled the total *B. cereus* population obtained from soybean root segments.

Plants in plots inoculated with UW2 had significantly greater populations of *B. cereus* than those in the uninoculated plots, suggesting that inoculation increased the total *B. cereus* population. However, in the inoculated plot the Neo^r population was significantly smaller than the total *B. cereus* population. The total *B. cereus* population in plots inoculated with UW11 was always greater than the total *B. cereus* population in the uninoculated plots. Approximately 90% of the colonies of the total *B. cereus* population in the plots inoculated with UW11 were Ora^+ whereas the Ora^+ phenotype was not detected in the uninoculated plots. However, no streptomycin-resistant population could be recovered 21 days after planting from soybean roots inoculated with UW11.

Although inoculation with *B. cereus* increased the total *B. cereus* population on soybean roots, the antibiotic resistance in the inoculated strains appears to be strongly selected against in the field. The antibiotic-resistant population appears to be replaced by antibiotic-sensitive revertants.

D. L. Keister and P. B. Cregan (Eds.), The rhizosphere and plant growth, 108.

PLSO BARC39

Bacterial characteristics important to rhizosphere competence

E. HOZORE and M. ALEXANDER
Department of Agronomy, Cornell University, Ithaca, NY 14853, USA

Key words: adherence, agglutination, chemotaxis, mobility

Inoculation of bacteria onto seeds and roots has attracted growing interest for the purpose of biological control, nitrogen fixation, and plant growth promotion. A major impediment to fully exploiting the potential of microorganisms in these areas has been a lack of knowledge about factors determining rhizosphere competence. Recently, studies have been conducted which evaluate the contribution of specific characteristics to the root-colonizing ability of individual bacteria. The purpose of the present study was to compare bacteria isolated from roots and from soil with respect to several characteristics that have been suggested as important to successful root colonization.

Ten bacteria were isolated from soil, and ten from soybean roots. For each organism, six traits were measured: rate and extent of growth on root exudates, chemotaxis to root exudates, and tolerance of low osmotic potentials.

Root-isolated bacteria differed significantly from soil isolates in their ability to move along roots. Root isolates were able to colonize segments farther from the point of inoculation in greater numbers. In other characteristics measured there were not significant differences between the two groups.

The results indicate that mobility along the root is important to successful rhizosphere colonization. Identifying traits which are most useful in predicting rhizosphere competence across a range of organisms would contribute to making inoculation technologies practical and understanding the forces at work in bacterial colonization of roots.

D. L. Keister and P. B. Cregan (Eds.), The rhizosphere and plant growth, 109.

PLSO BARC40

Growth of genetically altered *Pseudomonas solanacearum* in soil and rhizosphere

J.W. WILLIAMSON[1], P.G. HARTEL[1] and M.A. SCHELL[2]
Departments of Agronomy[1] and Microbiology[2], University of Georgia, Athens, GA 30602, USA

Key words: endoglucanase, plasmid, tomato, transposon

A bacterium can grow in soil and rhizosphere because the bacterium possesses genes for this growth. Therefore, it may be possible to genetically alter these genes in order to better understand bacterial growth in the soil and in the rhizosphere. The objective of our experiments was to determine the effect of the genetic alteration of the *egl* gene encoding for β-1, 4-endoglucanase on the growth of *Pseudomonas solanacearum* in soil and in the rhizosphere of a solanaceous host, tomato (*Lycopersicon esculentum*); a nonsolanaceous host, common purslane (*Portulaca oleracea*); and a nonhost, pearl millet (*Pennisetum glaucum*). Endoglucanase production by *P. solanacearum* was either eliminated by Tn*5* insertion mutagenesis or enhanced two-fold by increasing the *egl* copy number using a recombinant plasmid (pHE 3). To measure the growth of the bacterial strains in the rhizosphere and nonrhizosphere, seedlings of individual plant species were grown in Styrofoam cup-membrane assemblies (Hartel *et al.*, 1989). Rhizosphere and nonrhizosphere pads were inoculated with bacterial suspensions, and initial counts compared with counts following 48 h of exposure to rhizosphere or nonrhizosphere conditions. After soil inoculation of the Styrofoam cup-membrane assemblies with the individual strains, the plants were observed for wilting for 14 days. In soil solutions extracted from Hayesville sandy loam and Eel silt loam, all the genetically altered strains had increased generation times compared to the wild-type strain. A highly significant correlation ($r = -0.97$) was observed between the generation time and the maximum amount of growth of each strain in freshly moistened Hayesville soil. Under conditions of soil inoculation, the genetically altered strains did not wilt tomato seedlings significantly faster than the wild-type strain. Counts of all strains, including the wild-type did not increase in the nonrhizospheres and rhizospheres of the three plant species. The data suggest that genetic alterations have an effect on the growth of *P. solanacearum* in soil. However, the ability to produce endoglucanase does not significantly affect growth in the rhizosphere. It may be that endoglucanase is important in speeding disease development within the plant, but is not absolutely required in root infections. Since the bacterial counts did not increase in the rhizosphere, there may be other factors which obscured enhancement.

Reference

Hartel P G *et al.* 1989 Appl. Environ. Microbiol. 55, 1291–1294.

D. L. Keister and P. B. Cregan (Eds.), The rhizosphere and plant growth, 110.

PLSO BARC41

Growth of genetically-engineered *Pseudomonas aeruginosa* and *Pseudomonas putida* in soil and rhizosphere

K.-H. A. YEUNG[1], M.A. SCHELL[1] and P.G. HARTEL[2]
Departments of Microbiology[1] and Agronomy[2], University of Georgia, Athens, GA 30602, USA

Key words: plasmid, polygalacturonase, *Pseudomonas solanacearum*, tomato

The ability of a bacterium to grow in soil and rhizosphere exists because the bacterium possesses genes for this growth. Using recombinant DNA techniques, it may be possible to add a specific gene to a bacterium and determine whether this gene is favorable or deleterious to growth of this bacterium. In this manner, the ecological significance of the gene can be assessed. The objective of our experiment was to determine the effect of the addition of a recombinant plasmid containing a *pgl* A gene encoding for an α-1,4-endopolygalacturonase from *Pseudomonas solanacearum* on the growth of *Pseudomonas aeruginosa* and *Pseudomonas putida* in soil and rhizosphere. The recombinant plasmid pJE8 containing the *pgl* A gene on a 30-kb *P. solanacearum* DNA fragment in the 20-kb cosmid vector pLAFR3 and the vector alone were conjugally transferred into both *P. aeruginosa* and *P. putida*. The generation times, growth, and plasmid stability of the *P. aeruginosa* and *P. putida* strains were determined in nonsterile potting soil. To determine bacterial growth in the rhizosphere, bacteria were placed in the rhizosphere of tomato plants (*Lycopersicon esculentum*) in a Styrofoam cup-membrane assembly (Hartel *et al.*, 1989). With the presence of plasmid pJE8 in *P. putida* and *P. aeruginosa*, both strains produced almost 5- and 21-fold more extracellular polygalacturonase specific activity compared to that produced by the wild-type *P. solanacearum* AW (0.11 μmol of galacturonic acid reducing equivalents min^{-1} 10^8 $cells^{-1}$), respectively. In the absence of pJE8, none of bacteria had any detectable level of polygalacturonase. Under growth conditions in freshly-moistened potting soil, the cell densities of all strains increased at least 1000-fold during the 4-day incubation period. Both *Pseudomonas* species reached their maximum cell density within 2 days. The generation times and final cell densities of the two genetically-engineered strains with either pLAFR3 or pJE8 were not significantly different from the wild-type strains. At a density of 10^5 colony-forming units g^{-1} of soil, all strains either remained at the initial inoculation density or declined in density in the rhizosphere of tomato. Nearly all strains declined more in the rhizosphere than in the non-rhizosphere. No differences in plasmid stability were observed under survival conditions $\leq$12 days. Thus, despite high level of polygalacturonase production by genetically-engineered *P. putida* 12633 (pJE8) and *P. aeruginosa* PAO1 (pJE8), the results suggest that polygalacturonase production had little effect on the bacterial growth of these strains in soil and rhizosphere.

Reference

Hartel PG *et al.* 1989 Appl. Environ. Microbiol. 55, 1291–1294.

D. L. Keister and P. B. Cregan (Eds.), The rhizosphere and plant growth, 111.

Soil invertebrates are concentrated on roots

J. LUSSENHOP[1] and R. FOGEL[2]
[1]*Biological Sciences Department, University of Illinois at Chicago, Box 4348, Chicago, IL 60680, USA and* [2]*University of Michigan Herbarium, Ann Arbor, MI 48109, USA*

Key words: invertebrates, rhizotron, roots

The density of invertebrates on roots and in soil was determined. This is the first time such observations have been reported; they were made by direct observation through windows of the Soil Biotron at the University of Michigan Biological Station, Pellston, MI. The Soil Biotron is located in a Bigtooth Aspen-mixed hardwood stand on a sandy podzol. Horizontal transects in the top eight cm of soil were censused through a 50 power dissecting microscope mounted on a trolley. Light from the microscope did not affect behavior of the invertebrates. The roots were principally Bigtooth Aspen (*Populus grandidentata*) and Bracken (*Pteridium aqualinum*). Total area of soil censused in July was 262 cm^2, 3037 cm^2 in August, and 446 cm^2 in October, 1988.

Invertebrate density was at least two orders of magnitude greater on roots than in the surrounding soil (Fig. 1). Invertebrate groups with highest density on roots were bacterivores (Enchytraeidae), microbivores (Tydeidae, Oribitida, Collembola), and fungivores (Protura). Predatory Mesostigmata were least concentrated on roots. The food of Pauropoda is unknown, but they consistently occurred at root tips.

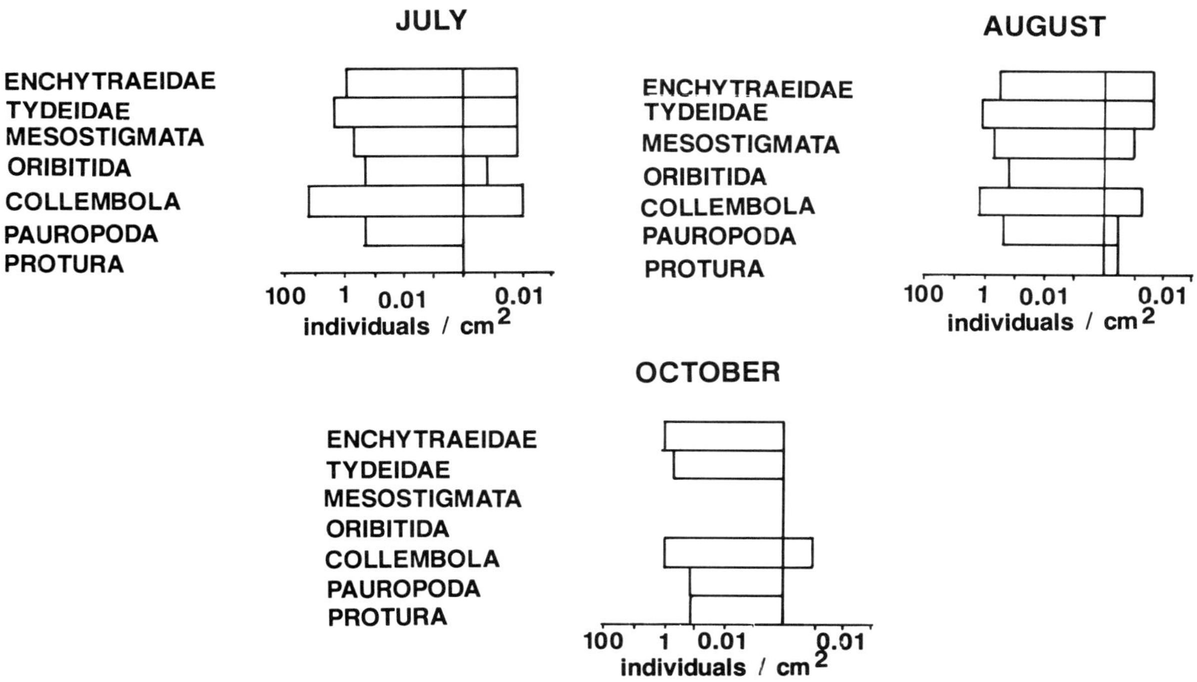

Fig. 1. Invertebrate density on roots (left) and in soil (right) of a Bigtooth Aspen-mixed hardwood stand. Direct counts of invertebrates and measurement of root area were made through windows in the top eight cm of soil.

D. L. Keister and P. B. Cregan (Eds.), The rhizosphere and plant growth, 112.

Distribution of assimilated carbon within the plant and rhizosphere of *Lolium perenne*: Comparison of field and laboratory grown plants

A.A. MEHARG and K. KILLHAM
Department of Soil Science, Meston Building, The University of Aberdeen, Aberdeen AB9 2UE, Scotland, UK

Key words: carbon flow, field grown plants, rhizosphere,

A number of environmental factors were found to affect carbon distribution within the plant and rhizosphere under laboratory conditions. These factors included pH, anaerobiosis, water potentials, temperature and microorganisms. The stage of plant development also influenced carbon distribution within the plant and rhizosphere. In addition, the carbon economy of the plant may differ significantly between rye grass plants grown under laboratory and field conditions.

Carbon distribution within rye grass was determined in laboratory and field grown plants at different stages of plant development by pulse labelling shoots with $^{14}C-CO_2$. Microcosms were designed which enabled plants to be pulse-labelled after growth either in the laboratory or in the field. The microcosms and pulse labelling techniques are described by Meharg and Killham (1988).

As both field and laboratory grown plants developed, less carbon was translocated below ground as a percentage of net assimilated pulse label. Loss of label from roots and the ability of the plants to assimilate label also decreased as plants developed. There was a greater percentage of label invested below ground and lost from plant roots in the field grown plants, suggesting that laboratory studies may underestimate rhizosphere carbon-flow under natural conditions. Carbon balances for the pulse labelled plants were also more variable under field conditions, and this was attributed to the soil water potential regime.

Reference

Meharg A A and Killham K 1988 Plant and Soil 112, 225–231.

D. L. Keister and P. B. Cregan (Eds.), The rhizosphere and plant growth, 113.

PLSO BARC44

Rhizosphere effect on soil organic matter decomposition

WEIXIN CHENG and DAVID C. COLEMAN
Institute of Ecology, University of Georgia, Athens, GA 30602, USA

Key words: carbon-14, decomposition, rhizosphere

In our study, ^{14}C-labelled rye straw was incubated in fertilized or unfertilized soil with or without plants for 49 days under semi-controlled conditions. The soil was a well-drained sandy clay loam (66% sand, 13% silt, 21% clay). Carbon-14 labelled rye straw was produced from pulse labelling in plastic chambers. The respired $^{14}CO_2$ was trapped in 1 *M* NaOH solution and counted by a liquid scintillation counter. Total and ^{14}C microbial biomass-C in the soil were measured using the fumigation-incubation method of Jenkinson and Powlson (1976) taking $K_c = 0.4$. Radioactivity of shoots, roots and soils was determined by the dry combustion method using an OX-300 Biological Oxidizer. Our objective was to study the rhizosphere effect on soil organic matter mineralization under different soil mineral nutrient conditions.

The planted treatment had a higher $^{14}CO_2$ loss and a higher efficiency of ^{14}C-labelled material utilization by the microorganisms. Fertilization decreased $^{14}CO_2$ loss. Percent ^{14}C in microbial biomass was positively correlated with percent ^{14}C respired ($P < 0.001$). Total microbial biomass, ^{14}C remaining, and ^{14}C-labelled microbial biomass of the rhizosphere soil were higher compared to the bulk soil. The rhizosphere soil in the fertilized treatment had a higher percent labelled microbial biomass and a lower total microbial biomass than that in the unfertilized treatment. This suggests that fertilization reduces the rhizosphere effect.

The presence of living roots significantly affected soil organic matter decomposition. Both higher microbial biomass and higher soil organic matter decomposition rate occurred in the planted treatments. This suggested a stimulating effect of living roots on soil organic matter decomposition due to the higher microbial activity induced by the roots.

Very little of the ^{14}C was taken up by plants (<0.4% of ^{14}C respired) in our experiment. This result is in conflict to those of Reid and Goss (1982) and Sparling *et al.* (1982), but in accordance with Sallih and Bottner (1988).

Application of N–P–K fertilizer reduced the effect of living roots on soil organic matter decomposition in the rhizosphere. More attention should be paid to the mechanisms and other processes involved in the rhizosphere in order to solve the controversy about the effect of living roots on soil organic matter decomposition.

References

Jenkinson D S and Powlson D S 1976 Soil Biol. Biochem. 8, 167–177.
Reid J B and Goss M J 1982 J. Soil Sci. 33, 387–395.
Sallih Z and Bottner P 1988 Biol. Fertil. Soils 7, 67–70.
Sparling G P *et al.* 1982 J. Soil Sci. 33, 89–100.

D. L. Keister and P. B. Cregan (Eds.), The rhizosphere and plant growth, 114.

PLSO BARC45

Influence of nitrogen fertilization on photosynthate distribution and utilization by rhizosphere micro-organisms in wheat

E. LILJEROTH, J.A. VAN VEEN and H.J. MILLER
Research Institute Ital, P.O. Box 48, 6700 AA Wageningen, The Netherlands

Key words: bacteria, microbial biomass, nitrogen fertilization, rhizosphere

Rhizosphere interactions between plants and microbes are to a large degree determined by the supply of carbon from root exudates but the availability of mineral nutrients to microbes and plants is also of importance. Root exudates from maize were found to be utilized more efficiently when a higher level of a mixed mineral fertilizer was given (Merckx *et al.*, 1987).

In order to investigate the effect of different levels of nitrogen on the microbial utilization of root exudates wheat plants were grown in a ^{14}C-labelled atmosphere and fertilized with two different levels of nitrogen, 50 mg N (low) and 150 mg N (high) per kg soil, respectively. During plant growth the soil-root respiration was monitored and at two times during the experiment (after 33 and 54 days) the distribution of ^{14}C over different plant and soil compartments was determined. Significantly more ^{14}C was released by the roots of plants grown at high nitrogen level as compared to roots of plants grown at the low nitrogen level. However, the root biomass did not significantly differ between the two nitrogen levels. The amount of ^{14}C-labelled microbial biomass in the soil, as well as the bacterial counts (colony forming units), was significantly higher at high nitrogen level. It appeared that part of the increase in microbial biomass resulted from an increased exudation rate at the high nitrogen level. However, the differences in microbial biomass were larger than the differences in the release of carbon from the roots. This suggests that the root exudates were more efficiently utilized at high nitrogen level. While high nitrogen treatment stimulated the decomposition and microbial utilization of root-derived materials it seemed to have an opposite effect on native soil organic matter since the rate of respiration of unlabelled carbon from the soil decreased. This is in agreement with results from soil microcosm experiments (van Veen *et al.*, 1989) where it was found that high levels of mineral nutrients stimulated the turnover of added ^{14}C-glucose but decreased the decomposition rate of added ^{14}C-labelled crop residue.

The results show that a changed nitrogen status of the soil can significantly influence the turnover of both root-derived organic material in the rhizosphere and the decomposition rate of native soil organic matter. The different reactions of the soil nutrient levels on the decomposition of simple sugars or root exudates versus complex soil organic matter and crop residues requires further research.

References

Merckx *et al.* 1987 Biol. Fertil. Soils 5, 126–132.
Van Veen J A *et al.* 1989 Plant and Soil 115, 179–188.

D. L. Keister and P. B. Cregan (Eds.), The rhizosphere and plant growth, 115.

PLSO BARC46

In vitro studies on the interactions of *Agrobacterium* spp. and *Pseudomonas* spp. isolated from opine environments

C.R. BELL[1], L.W. MOORE[2] and M.L. CANFIELD[2]
[1]*Department of Biology, Acadia University, Nova Scotia, Canada BOP 1X0 and* [2]*Department of Botany and Plant Pathology, Oregon State University, Corvallis, OR 97331, USA*

Key words: chemostats, competition models, octopine catabolism

Introduction

The opine concept is central to the understanding of the ecology of *Agrobacterium tumefaciens* in galled environments (Tempé and Petit, 1982). This study quantifies the growth of *A. tumefaciens* strains B6 and 15955 on octopine under defined conditions in chemostats. Comparisons are made to octopine catabolizing *Pseudomonas fluorescens* strains B99A and E175D and competitions between the four strains are examined.

Materials and methods

The pseudomonads were both isolated from apple crown galls. The μmax, Ks and yield were determined by standard methods in media containing octopine or glutamate either as limiting nitrogen or limiting carbon. The competition models were based on those of Powell (1958) for single substrates and on those of Yoon and Blanch (1977) for double substrates.

Results and discussion

The maximum specific growth rates of the pseudomonads and agrobacteria on octopine as substrate were very comparable. The substrate affinities (Ks) indicated that octopine was utilized much less efficiently as a substrate than glutamate. Yield coefficients revealed that the octopine could be utilized much more efficiently when supplied as the source of limiting nitrogen than as the sole source of limiting carbon. Monod curves generated with these parameters produced models of the growth and competition of these stains. *A. tumefaciens* strain B6 was notable for its poor competitive potential when growing on octopine. Dual culture chemostat runs of pseudomonads + agrobacteria were always dominated by the pseudomonad sometimes in contradiction of the models. Although both pseudomonads are fluorescent, siderophores are not considered the prime antagonistic agent because inhibition occurred when fluorescence was absent.

These physiological data suggest that in the rhizosphere and the tumorosphere there could be a succession of opine catabolizing bacteria when opines are the limiting substrate.

References

Powell E O 1958 J. Gen. Microbiol. 18, 259–268.
Tempé J and Petit A 1982 *In* Molecular Biology of Plant Tumors. Eds. G Kahl and J S Schell. pp 451–459. Academic Press, New York.
Yoon H and Blanch H W 1977 J. Appl. Chem. Biotechnol. 27, 260–268.

D. L. Keister and P. B. Cregan (Eds.), The rhizosphere and plant growth, 116.

PLSO BARC47

Colonization of wheat roots by *Pseudomonas fluorescens*: Scanning electron microscopy and biochemical analysis

R. DE MOT[1], H. JOOS[2], A. VAN GOOL[1] and J. VANDERLEYDEN[1]
[1]*F.A. Janssens Memorial Laboratory of Genetics, Catholic University of Leuven, B-3030 Heverlee, Belgium and* [2] *Plant Genetic Systems, B-9000 Gent, Belgium*

Key words: fluorescent Pseudomonas, wheat root colonization, SEM, 2D analysis

The interaction of selected fluorescent Pseudomonads (*P. fluorescens* and *P. putida*) with the roots of wheat seedlings was studied using a gnotobiotic system with sand. The strains were obtained as predominant isolates from wheat and maize rhizospheres. Fingerprint characterization at the strain level by two-dimensional (2D) analysis of total bacterial proteins aided in the inventory of our Pseudomonas collection.

Scanning electron microscopy (SEM) of colonized root surfaces revealed the occurrence of rod-like structures apparently connecting bacterial cells with the root (Fig. 1), clearly different from fimbriae or flagella. Our initial observations with *P. fluorescens* OE 28.3 were extended to several others, including strain NRRL B-15132 (2-79).

2D electrophoresis of bacterial proteins indicated the presence of a major constitutively produced 33 kDa protein by most of the strains generating the structures at the root surface. A heterogeneity with regard to isoelectric point was observed among different strains. In the outer membrane preparation of strain OE 28.3, this protein is by far the most abundant one.

Comparative 2D analysis of root extracts from OE 28.3-inoculated and axenic plants (wheat, barley, maize) demonstrated a single additional acidic protein on washed roots from inoculated plants, which was subsequently shown to be the bacterial 33 kDa protein. It remained attacted to the roots only when possessing a specific pI value, as evidenced by analysis of roots inoculated with several other Pseudomonas strains. *In vitro* experiments with bacterial protein extracts (whole cells or outer membrane) and isolated wheat roots also indicated a selective adsorption of the 33 kDa protein. In order to characterize this protein, 2D PAGE-purified material is being processed for microsequencing.

Future experiments, including genetic analysis, will be directed at elucidating the possible involvement of the 33 kDa protein and the rod-like appendages in root colonization.

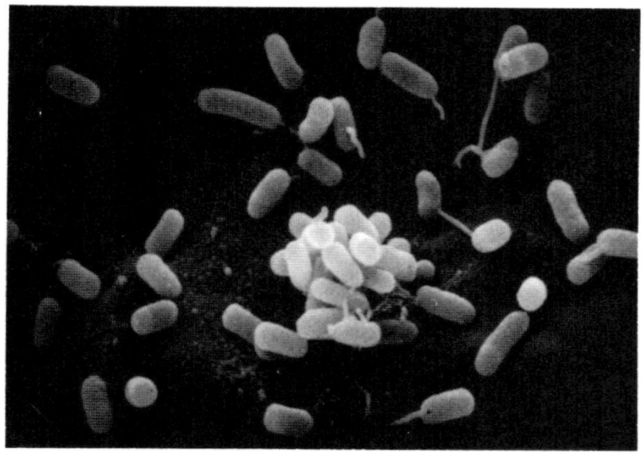

Fig. 1. SEM of *P. fluorescens* OE 28.3 colonizing the surface of wheat seedling roots.

D. L. Keister and P. B. Cregan (Eds.), The rhizosphere and plant growth, 117.

PLSO BARC48

Pectic enzymes of *Azospirillum brasilense*

C. SEKAR and N.N. PRASAD
Department of Agricultural Microbiology, Annamalai University, Annamalainagar-608002, India

Key words: *Azospirillum brasilense*, pectic enzymes, rice

The production and activity of pectic enzymes elaborated by *Azospirillum brasilense* (strain C.A. 10), a diazotroph isolated from the rhizosphere of rice (*Oryza sativa* L.), was studied as this organism is known to enter through the middle lamella to colonize the histosphere of rice roots.

Azospirillum brasilense produced adaptively exo-polygalacturonase (Exo-PG), pectin *trans*-eliminase (PTE) and pectic acid *trans*eliminase (PATE) *in vitro* with either pectin or pectic acid as sole sources of carbon. The production of endo-polygalacturonase (Endo-PG) was found in trace amounts and that of pectin methyl esterase (PME) was nil.

The regulatory role of different levels of pH (5.5 to 8.5) *per se* on the production and activity of pectic enzymes was studied. This diazotroph was metabolically inactive at pH values 7.5 to 8.5. Growth of the diazotroph positively correlated with Exo-PG production. The production and activity of Exo-PG were found to decrease with increase in pH levels from 5.5 to 7.0. Further, the production of *trans*eliminases was found to be at pH 6.5 to 7.0. Since the optimum activity of the *trans*eliminases is at pH 8.7, it is concluded that the hydrolases play a relatively more significant role than the *trans*eliminases in the entry of this diazotroph into the histosphere of rice roots.

D. L. Keister and P. B. Cregan (Eds.), The rhizosphere and plant growth, 118.

PLSO BARC49

Water uptake patterns in loblolly pines as seen with magnetic resonance microscopy

J.S. MacFALL[1] and G.A. JOHNSON[2]
[1]*School of Forestry and Environmental Studies, and* [2]*Department of Radiology, Duke University, Durham, NC 27713, USA*

Key words: magnetic resonance imaging, roots, uptake, water

Proton (^{1}H) magnetic resonance microscopy (MRM) has been used to describe water localization in plant stems (Johnson *et al.*, 1987). The objective of this study was to determine if magnetic resonance microscopy could be used to describe water uptake from sand surrounding plant roots. In the present study, MRM was used to observe water removal from sand by loblolly pine (*Pinus taeda*) seedlings. Sand surrounding seedling roots imaged as a region of very low signal intensity (dark regions), while sand in other portions of the potting containers appeared brighter. It was proposed that the dark regions of sand surrounding the roots were regions of water depletion while the brighter regions were areas of higher water content.

Proton MRM will primarily detect water as the proton source in plant studies. The relationship between water content and signal intensity can be described by the general equation:

$$S(t) = N(^1H)\, e^{-TE/T2}(1 - e^{-TR/T1})$$

(S(t) = signal intensity, N(^{1}H) = proton content, TE = echo time, TR = pulse sequence repetition time, T2 = spin-spin relaxation time, T1 = spin-lattice relaxation time) (Herfkens *et al.*, 1981).

It was necessary to evaluate the effect of water content on signal intensity in order to interpret the images. A sand phantom was constructed containing tubes filled with sand mixed with varied amounts of distilled water (0–25%). Cross-sectional images of the tubes were made on a General Electric 2T CSI imager (General Electric Medical Systems, Milwaukee, WI 53029). Signal intensities and T1 values were measured from the images.

There was a linear positive relationship between signal intensity and percentage water ($R^2 > 0.97$). Therefore, observations of dark regions in the images of sand surrounding plant roots can be interpreted to be areas of water depletion and bright regions as areas of greater water content.

The relationship was further complicated, however, by the observation that the calculated T1 values increased with increasing water content. This increase in value with increased water content is likely reflecting more free water. Further work is in progress to characterize this relationship with MRM.

Conclusions which can be drawn from the results reported here are 1) dark regions surrounding roots show areas of low water content and 2) that the technique of magnetic resonance microscopy offers the potential of quantification of water contents in sand surrounding plant roots.

References

Herfkens R *et al.* 1981 Radiology 141, 211–218.
Johnson G A *et al.* 1987 Proc. Natl. Acad. Sci. 84, 2752–2755.

D. L. Keister and P. B. Cregan (Eds.), The rhizosphere and plant growth, 119.

PLSO BARC50

The occurrence of cluster roots in actinorhizal plants

I. LOUIS, S. RACETTE and J.G. TORREY
Harvard Forest, Harvard University, Petersham, MA 01366, USA

Key words: actinorhizal plants, cluster (proteoid) roots, phosphorus

Determinate lateral root proliferation clustered at intervals along a root axis, as exhibited by many species of the Proteaceae, was first referred to as proteoid roots (Purnell, 1960). Subsequently, analogous structures were reported also in the legumes. Recognizing their more general occurrence beyond the Proteaceae, Lamont (1982) used the term 'cluster roots' to describe these specialized root branching patterns. This study reports their occurrence in some species of the nitrogen-fixing actinorhizal plants.

Cluster root formation in the legumes has been suggested to occur in response to low phosphate levels (Gardner *et al.*, 1982; Walker and Pate, 1986). In our experiments, we grew several actinorhizal plants in water culture, using 1/4-strength Hoagland's solution with (8 ppm P as KH_2PO_4) and without P for 3 weeks. Cluster roots were regularly observed in some of these actinorhizal plants when grown in water culture lacking P (Table 1).

Cluster roots were also observed on *Alnus* species and *Casuarina cunninghamiana* and *C. glauca* but not in response to the elimination of P from the nutrient solution. This study extends the observation on occurrence of cluster roots to several actinorhizal species and shows the influence of phosphorus on their formation. Research on their function and mechanism of formation is in progress.

References

Gardner W K *et al.* 1982 Plant and Soil 68, 33–41.
Lamont B 1982 Bot. Rev. 48, 597–689.
Purnell H M 1960 Aust. J. Bot. 8, 38–50.
Walker B A and Pate J S 1986 Aust. J. Plant Physiol. 13, 305–319.

Table 1. Cluster root formation in actinorhizal plant species grown in nutrient solution lacking P

Cluster root formation (−P)	
Myricaceae:	*Comptonia peregrina, Myrica cerifera, Myrica gale, Myrica pensylvanica*
Casuarinaceae:	*Gymnostoma papuanum*
No cluster root formation (−P)	
Casuarinaceae:	*Allocasuarina campestris*
Elaeagnaceae:	*Elaeagnus umbellata*
Rhamnaceae:	*Ceanothus americanus*

Part 2

The Rhizobium-legume symbiosis

D. L. Keister and P. B. Cregan (Eds.), The rhizosphere and plant growth, 123–134.

PLSO BARC12

Legume inoculation: Successes and failures

R.H. MILLER[1] and S. MAY
Department of Soil Science, North Carolina State University, Raleigh, NC 27650, USA. [1]*Present address: University of Rhode Island, Kingston, RI 02881, USA*

Key words: Bradyrhizobium, legumes, legume inoculation, Rhizobium

Abstract

The contribution of leguminous plants including food, forage and tree legumes to the nitrogen needs of agricultural systems is well recognized. At the same time the efficacy of inoculation with specific rhizobia to improve biological nitrogen fixation remains less certain. In a historical perspective it is interesting to note that attempts at improving N-fixation through inoculation of soil or seed began almost simultaneously with the successful demonstration that symbiotic N-fixation was associated with legume root nodules. At the same time the problems which plagued these early investigators, *i.e.* differences in response to method of inoculation, problems in assuring adequate inoculant quality, excessive claims of efficiacy by some inoculation producers and the inability of inoculant rhizobia to compete against indigenous rhizobia, remain with us today. This is true even though our journals contain a voluminous literature covering inoculation trials, the intricacies of the symbiosis, methods of inoculation, rhizobium ecology, the genetics of rhizobia, *etc.* The challenge to successfully and selectively introduce selected or genetically engineered highly effective rhizobium strains remains a high priority. Methods such as massive inoculation, use of highly competitive strains, improved inoculation technology, co-inoculation, chemical modification of the rhizosphere, and altered legume-rhizobium specificity show promise in meeting this challenge.

Introduction

The major part of the nitrogen found in natural wooded and savanna ecosystems of the tropics and some temperate grasslands is derived from biological N_2 fixation by legumes (Döbereiner and Campelo, 1977; Mulder *et al.*, 1977). This is particularly true in tropical forested areas which almost always contain a variety of leguminous tree species (Döbereiner and Campelo, 1977).

Legumes also play an important role in farming systems around the world and have for many millenia. Grain legumes represent an important source of protein for human nutrition and in tropical third world countries often compete successfully with animal proteins as the major source of protein in diets. Effective nodulation and N_2 fixation by these legumes assures an important protein resource for many of the world's poor. In more developed countries grain legumes, *e.g.* soybeans, serve as a source of both protein and oil for a variety of non-food products and uses. In both temperate and tropical countries N_2 fixing forage legumes provides a high protein diet for livestock. Many of these same legumes if incorporated as green manure crops can markedly reduce the nitrogen fertilizer needed for the production of non-leguminous crops such as maize, sorghum, wheat, rice, *etc.* The widespread interest in a more sustainable agriculture has enhanced the opportunities for a research on farming systems using a variety of legumes for green manure.

While the importance of N_2 fixing leguminous plants to the N economy of this planet remains unequivocal, the importance of the practice of legume inoculation with specific rhizobia to improve yields of economically important legumes remains an extensively studied but scientifically equivocal practice. It thus seems appropriate 103 years after Hellriegel (1886) first reported on the nitrogen nutrition of leguminous plants to examine the scientific literature in an attempt to determine what factors influence the success and failure of legume inoculation. In the second part of the paper we will attempt to summarize some methods and practices which offer the opportunity for improving the success of inoculation. The potential for enhanced N_2 fixation and yields by introducing highly effective rhizobia must remain an attractive goal.

Early history of legume inoculation (1877–1932)

The classical reference, "Root Nodule Bacteria and Leguminous Plants," (Fred *et al.*, 1932) effectively documents the early attempts at legume inoculation and should be considered "must reading" for everyone interested in leguminous biological nitrogen fixation (BNF) and early attempts at inoculation.

Soil transfer was the first method of legume inoculation. Within a year of Hellreigel's report (1886) on the N nutrition of plants, a German scientist by the name of Salfield in 1888 reported on soil transfer inoculation studies. In these studies Salfield (1888) transferred and spread by hand 3600 lbs per acre of soil from a field where horse beans (*Vicia faba* L.) had been grown, to an experimental plot in a newly drained swamp. Treatments included a control, Chilian saltpeter ($NaNO_3$), soil alone, and $NaNO_3$ plus soil with three legumes, grey Prussian peas (*Pisum sativum* L.), field peas (*Pisum sativum* L), and garden peas (*Pisum sativum* L.). Salfield found a pronounced yield response from the use of soil inoculation. This success stimulated him to expand his studies to other legumes and different kinds and rates of soils as a inoculum. He consistently obtained positive results from inoculation with cultivated soils but not from soils from an uncultivated area. Some effects of rate of inoculation were found with the greatest response at the 3600 lbs per acre application. Salfield continued his research and published many papers on the subject between 1888 and 1898 (Fred *et al.*, 1932).

The interest in Salfield's research stimulated research in many areas of the world. Included are van Feilitzen in Sweden; Fleischer in Germany; Atwater, Wood and Duggar in the U.S.; Laws and Gilbert and Warington in Great Britain; Nobbe, Hiltner, and Wilfarth of Germany, and Deherain and Demoussy in France. In general, positive results occurred in soils recently brought into production, sandy or burned-over soils, with little response in old cultivated or loamy or clay soils. The amount of soil transferred for inoculation remained high although Fleischer in 1893 obtained positive results with as little as 90 lbs per acre.

Dugger of Alabama in 1897 experimented with seed inoculation of hairy vetch (*Vicia villosa*) using a soil-water mixture with the soil derived from a field where common vetch (*Vicia sativa*) had been grown. Remarkable yield increases of 10 × the control were obtained. Studies on the necessary rates of broadcast soil inoculation, "soil paste or muddy water" inoculation, or soil paste inoculation after the seed is coated with glue or sugar solution were common in the US until the 1930s. As might be expected the success of these soil inoculation methods varied depending on the source of the soil used as an inoculum and the rate.

The amount of labor involved in soil inoculation led very quickly to experimentation using pure cultures of isolated rhizobia for an inoculant. Historically noteworthy were the early studies of Nobbe and Hiltner published in 1896. These Germans investigated direct seed inoculation as well as broadcast application of soil to which a pure culture of rhizobia were added. The latter was considered most effective. In 1885, Nobbe and Hiltner applied for patents for this process in both the U.S. and Great Britain. The patented cultures of Nobbe and Hiltner was marketed and sold under the same "Nitragin" in Germany beginning in 1896. Nobbe and Hiltner believed that each leguminous plant required a specific rhizobia and 17 kinds of "Nitragin" were prepared and sold.

Nitragin at first was prepared in medium of sugar, asparagin, water extract of a leguminous plant and gelatin contained in 8–10 oz glass bottles (1 acre size) inoculated with the specific Rhizobium culture. As might be expected these pure culture products were extensively investigated with mixed results. When compared with the soil transfer method of inoculation, "Nitragin" was most often clearly inferior. In response to criticism about their product, Nobbe and Hiltner tried many other formulations including milk, peptone, and agar solid formulations or pure liquid cultures. Hiltner also commented that the claims of the commercial company marketing "Nitragin" were often greatly exaggerated.

In the early twentieth century other commercial products for legume inoculation appeared in the marketplace. For example, Coates (USA) patented a process in which sterilized legume meals became carriers for specific pure cultures of rhizobia. Others included the Hartleb culture using artificially formed "bacteroids," the Nitro-culture of Moore (USDA) where the bacterium was grown in "N-free" medium, the Azotogen culture which involved moistened sterilized soil plus lime inoculated with pure cultures, solid cultures adsorbed on ground raw bone, silica, calcium carbonate, calcium phosphate, peat, carbon black, *etc.*, and products in many kinds of containers such as tubes, bottles and tin boxes. Many of these product approaches were patented and much research was stimulated within the US. Inoculation response studies and inoculant methodology development then, as now, were active researchable topics. Many small beneficial responses and even dramatic yield increases were obtained and dutifully reported.

What is apparent from this interesting cursory review of the early history of attemptes at legume inoculation is that the problems which plagued these early investigators (see Table 1), *i.e.* inconsistent results, problems in assuring inoculant product quality, concern with methods of inoculation and sometimes excessive zeal by inoculant producers in marketing products remain with us today.

Table 1. Legume inoculation research (1888–1932)

Accomplishments
Legume inoculation begun using soil
First evaluation of pure culture inoculants
Recognition that all soils did not need inouclation for legume nodulation
First use of carrier materials and stickers
Experiments with differing culture media for rhizobium inoculant production
Problems
Inconsistent results
Commercial product quality
Methods of inoculation
Excessive zeal in marketing

Legume inoculation (1932–1959)

The intense period of discovery, experimentation and general excitement about inoculation successes which was so evident in the early part of the 20th century was very much less evident in the period from 1932–59. No attempt was made for an exhaustive search of the literature but three journals which now extensively publish inoculation and BNF research were gleaned for articles relevant to inoculation responses. Plant and Soil which published Vol. 1 in 1948, had no articles on inoculation up to 1959. The Soil Science Society of America Proceedings (now the Journal) which began in 1937 had only eight articles on inoculation, six in Vol 8, 1943 which published the proceedings of a special program on legume inoculation at the previous annual meeting. The Journal of the American Society of Agronomy (since 1949 the Agronomy Journal) published only ten articles on inoculation trials and BNF during this 27 year period. Two other special publications, (Allen and Allen, 1950; Erdman, 1959) are worthy of perusal by the interested reader. The latter article has 415 reference citations of all phases of BNF prior to 1950.

The publications from this period provided estimates on the quantity of N_2 fixed by legumes, cross-inoculation and nodulation studies on less well known legumes and inoculation response studies for introductions of grain and forage legumes into new areas and farming systems of the U.S (Table 2). For instance, Duggar (1935) and Albrecht (1943) found major yield increases when peanuts were introduced into non-traditional growing areas of Alabama; Erdman (1943) reported frequent positive inoculation responses of grain soybeans in Wisconsin and Alabama; Moodie and Vandecaveye (1943) reported yield responses of 37–74% for chick pea (*Cicer arietinum*) when first introduced into the Palouse of Washington, Norman (1943) demonstrated

Table 2. Legume inoculation research (1933–1959)

Accomplishments
Inoculation studies with new legume introductions, generally positive
First estimates of N_2 fixed as legumes
Cross-inoculation and nodulation studies on less common legumes
Problems
Seed fungicide treatments detrimental to inoculants

that inoculated soybeans in new soybean growing areas of Iowa yielded one-third better than uninoculated controls, and Collins (1943) evaluated the response of nine forage and grain legumes to commercial inoculants on reclaimed land in Georgia. In this study all the treatments increased legume nodulation and vigor and only soybean and Crimson clover showed some nodule formation in the check plots. One other paper of this period, that by Lynch and Sears (1952) is interesting in that inoculation studies on eight field locations in Illinois where soybeans had been raised before did not show a yield increase.

It was also during the early 1940s that copper or mercurial seed treatments, *e.g.* Cerasan and Spergon, were introduced and as might be expected nodulation problems ensued. A number of studies were undertaken to find alternatives to seed inoculation of legumes. Albrecht (1943) achieved success by mixing a commercial inoculum into 300 lbs of finally ground stable manure and distributing this amount of manure inoculum over each acre before planting the specific legume. Baur (1943) prepared a soil applied preparation by mixing 4, bu cans of commercial pea inoculum with 100 lbs of either a sandy loam soil, muck, steamed bone meal, lime or bone black. This preparation was then distributed in the seed farrow at the rate of 100 lbs per acre and proved effective in increasing pea nodulation and yield. Erdman (1943) commented that J. Burton was successful with a preparation of "Nitragin" with wheat middlings predrilled into the soil. This research was the precursor for the soil implant inoculation strategies of the 1970s.

The expansion of U.S. Agriculture and legumes, particularly soybeans, stimulated what Burton (1967) called the greatest historic expansion of the legume industry during the period 1929–44, production increased from 1.5 to 11 million bu units of inoculum. This increase occurred during a largely unregulated environment so problems with inoculant quality were probably present but not specifically noted.

Legume inoculation (1960–present)

Inoculation studies and field plot research during this time period have been extensive for a large number of grain and forage legumes in the temperate region. During the decades of the 70s and 80s increasing interest in tropical grain, forage, and tree legumes has also resulted in an extensive literature. Any attempt to review and comment on all successes or failures would be difficult and probably not very instructive. For this reason only selected references will be emphasized to develop significant concepts.

Soybeans (*Glycine max* (L). Merr) is the dominant legume grown in the US and the majority of inoculation studies in the US have focused on this legume. When soybeans were first introduced to this country from China in the early 20th century soil inoculation was used to establish an effective symbiosis. Later as the soybean growing region was expanded some commercial inoculants were used and it is presumably this population which has established itself in soils in the USA. Inoculation response studies in Iowa (Damirgi *et al.*, 1967; Ham *et al.*, 1971); Minnesota (Ellis *et al.*, 1984; Ham *et al.*, 1971; Kvein *et al.*, 1981); Illinois (Kapusta and Rouwenhorst, 1973); Indiana (Nelson *et al.*, 1978) and Ohio Miller, unpublished data) have shown that seed inoculation at normal rates will not effectively displace the dominant indigenous *Bradyrhizobium japonicum* serogroup 123 from soybean root systems. The percent occupancy of the inoculant strain is generally less than 20% and often 5% of less. Studies with soybean inoculation have been done in Maryland by Vest *et al.* (1971) with a similar result although the serogroups found were different (C1 (3-24-44), C2 (46-71a-121), C3 (31-40-120), 76, 94, 110, and 122).

Soybean inoculation has not been successful in the establishment but comparatively new soybean growing area of the Southeast as well (Giddens *et al.*, 1982), and field results have shown no increase in seed yield or nodule occupancy because of inoculation.

One extensive inoculation study in North Carolina deserves special mention. Dumphy *et al.* (1981) conducted 254 inoculation comparison studies between 1975 and 1980 in farmers' fields. The 23 on-farm tests were on soils with no history of soybeans being grown in the previous eight years or more and 16 of the fields had never grown soybeans. The 110 fields inoculated with dry peat-based inoculum on the average had no yield response while liquid seed inoculation (45 fields) or soil inoculation (28 fields) average about a 0.5 bu yield increase. Seventy granular soil implant studies averaged 1.25 bu yield increases because of inoculation.

There is less extensive data on dominant indigenous serogroups in the southeast compared to that of the midwest. Keyser *et al.* (1984) found serogroups 31 and 76 dominant in five southeastern states. One recent study by Mpofu (1988) at five locations in North Carolina found serogroups 76, M1 (31, 94), 94, 24 and 122 to represent 50–84 percent of the total isolates. Interestingly, an evaluation of strain effectiveness on the cultivar Centennial found that 58 percent of 506 isolates were as effective as USDA strain 110. This provides a reason to be optimistic that yields could be increased if the 42 percent of less effective strains could be replaced.

Although limited data are available for other regions of the US the data of Abel and Erdman (1964) indicated that commercial inoculation did not result in adequate nodulation when soybeans were first introduced into the Southwestern US in 1953. Actual studies reported in this paper did show inoculation responses and strain differences in field plots where soybeans were never grown but no responses where previous inoculation had been practiced.

Soybean is a relatively new crop in most of the tropical and sub-tropical areas of the world and few indigenous rhizobium are present. Positive yield responses to inoculation are the norm although some failures have occurred. Graham (1984) summarized the data from 25 field studies in 16 tropical or sub-tropical countries and reported soybean yield increases from 22–446% with 18 of these studies having yield increases greater than 50%.

In summary, we conclude that the inoculation of soybeans cannot be recommended except where soybeans have never grown. It is perhaps fortunate that the indigenous population seems to collectively have reasonable effectiveness since yield responses to N fertilization seldom exceeds that of the symbiotic system. The recent study by Mpofu (1988) might be one argument that yield improvement could occur with the introduction of more effective strains.

Observation and inoculation response studies with forage legumes (*Trifolium subterraneum* and *Medicago sativa*) in Northwestern US and Australia demonstrate differences in effectiveness of indigenous rhizobia as well as in the ability to alter nodule occupancy by inoculation with competitive and effective strains.

Weber and Leggett (1966) described a condition of alfalfa in eastern Washington and northern Idaho called alfalfa sickness which was charaterized by unthrifty, spindly, chlorotic plants. Subsequent studies indicated that effective indigenous *Rhizobium meliloti* were scarce although most plants were nodulated. Subsequent greenhouse pot-studies demonstrated that various effective strains of *R. meliloti* could successfully nodulate alfalfa roots and produce highly vigorous high yielding plants. Inoculation rates of 3.5×10^4 cells/g of soil were necessary for optimum yields although benefits were demonstrated at lower rates.

Hagedorn (1978) found that subterranean clover (*Trifolium subterraneum* L.) pastures in southwest Oregon were often of poor quality with poor forage yields. The indigenous rhizobia were found to be largely or poorly effective on subclover. It was shown that inoculation with a peat inoculum of two strains (TA1 and C6) of *R. trifolii* and lime pelleting was successful in establishing highly effective symbiosis at five field locations. The range of nodule occupancy by inoculant strains was 45–100% against background indigenous *R. trifolii* of 58 to 1700/g soil. Fertilization with P, S and Mo significantly increased the overall nodulation of plants.

Ineffective indigenous populations of *R. trifolii* have also been reported by Holland (1970) in range pastures of subclover in Northern California, by Gibson *et al.* (1975) in southeastern Australia, and for Scottish hill pastures with *Trifolium repens* by Holding and King (1963). The problem situations were corrected in both Australia (Roughley *et al.*, 1976) and northern California (Holland, 1970) by seed inoculation with effective strains of *R. trifolii*.

The response of forage legumes (alfalfa, birdsfoot trefoil, and crimson, arrow leaf and white clovers) to inoculation in Southeastern US has generally been positive if various kinds of adhesives and stickers are utilized to increase the number of rhizobium retained by the seed (Giddens et al., 1982). Few of the studies included in this review reported on the background population of indigenous rhizobia but those that have indicated that nodulation was generally poor or slight without inoculation.

This review would not be complete without a brief coverage of inoculation trials with tropical legumes. Two multi-country, cooperative inoculation trials centered in the tropics will be emphasized. The early results of International Bean Inoculation Trial (IBIT) in Central and South America was summarized by Graham *et al.* (1982). The response to inoculation of *Phaseolus vulgaris* at the 12 locations in seven countries was highly variable with yield increases varying from 0 to 61.1%. Interestingly a significant response to inoculation was obtained in five trials including two with a relatively high indigenous population of *Rhizobium leguminosarum* bivar *phaseoli* (*i.e.* 10^4 rhizobia/g.).

The second study is part of the International Network of Legume Inoculation Trials (INLIT) designed and supervised by NifTAL in Hawaii. (Bohlool, 1988). This inoculation response study involved 13 different legumes, and 196 different experiments in Asia, Africa, and Latin America. A preliminary summary of the data for experiments considered to be statistically valid is shown in Table 3. For the eight legumes evaluated 37% of the experiments resulted in positive yield increases.

Other inoculation studies involving tropical legumes are reported in Sections 5 of Graham and Harris (1982). Reports from eastern Africa, India and the Sudan involved *Phaseolus vulgaris*, *Arachis hypogaea*; *Vigna* spp., *Dolichos* spp., *Cajanus cajan*, *Vigna radiata* and *Vigna unguiculata*. As with the regional multi-country networks both successful and unsuccessful responses to inoculation were reported.

Table 3. Response of 8 legume species to rhizobial inoculation in the field. Bohlool (1988)

Host Legume	*No. Expts*[a]	
	A	+
Arachis hypogaea (peanut)	8	2
Cajanus cajan (pigeon pea)	5	0
Cicer arietinum (chickpea)	14	7
Glycine max (soybean)	16	8
Lens culinaris (lentil)	9	4
Phaseolus vulgaris (common bean)	6	2
Vigna mungo (urd bean)	8	5
V. radiata (mungbean)	15	3
V. unguiculata (cowpea)	5	1
TOTAL	86	32

[a] Experiments: No. conducted that were statistically acceptable (A), no. resulting in significant yield increase due to inoculation (+).

The most instructive, even humbling, conclusion to be drawn from a review of inoculation studies is that we have not progressed all that far since the early twentieth century. The literature is still resplendent with failed attempts to introduce the inoculant strain on legume root systems at a significant level. The major successes then, and now, occur when a legume is introduced into an agricultural region for the first time or where soil constraints to an effective symbiosis have been corrected.

Our purpose in pointing out the apparent lack of progress is not to suggest that we as scientists have been only reinventing the wheel. On the contrary, the research dealing with inoculation responses, rhizobia ecology, and host-rhizobia responses during the past three decades has been both exciting and fruitful (see Table 4). It would appear that we can be optimistic that our ability to introduce highly effective strains on to legume root systems whether obtained by selection or through genetic engineering will be markedly enhanced in the future. The next section will briefly address some possible ways this might be achieved.

Approaches to improve inoculation success

Research must continue on improving our understanding of those physiological and genetic fac-

Table 4. Legume inoculation research (1960–present)

Accomplishments
Recognition of rhizobia serogroups
Methodologies for ecological studies on fate of inoculant strains
Extensive studies on biotic and abiotic factors influencing "saprophytic competence"
Recognition of plant genetic control of nodulation
Studies on strain efficacy and competitiveness
Extensive studies on tropical legume response to inoculation
Problems
Commercial inoculant quality
Inoculant strain competition with indigenous strains

tors of both the host plant and rhizobium which enhance the competitiveness of rhizobium strains for nodule occupancy in the presence of an indigenous population. It seems apparent from the literature that a reasonable chance of success with current inoculation methodology occurs only when the population of indigenous rhizobia of the same legume species is 10^2/g or less (Bohlool, 1988). Our future ability to utilize highly effective strains of specific rhizobia to enhance N_2 fixation depends upon two things 1) an improved understanding of the physiological and genetic basis of competition and 2) strategies to enhance the opportunity of inoculation success. Both of these topics will be covered in part as part of this symposium. It seems instructive, however, to expand slightly on topic 2 as a conclusion to this paper.

Table 5 summarizes some of the approaches which have demonstrated or have promise as methods to enhance inoculation success in the presence of a high population of indigenous rhizobia.

Table 5. Approaches to improve inoculation success

Procedures
Massive inoculation
Repeated inoculation with competitive strains
Use of improved inoculation technology
Selected, mutagenized or genetically engineered indigenous competitive strains
Co-inoculation with rhizosphere isolates or other microorganisms
Use of soil applied chemicals
Legume host-rhizobium specificity

Massive inoculation

There have been some successes in overcoming the indigenous population of rhizobia by using seed or soil inoculation at rates very much higher than those possible with commercial inocula. Bohlool and Schmidt (1973) were able to demonstrate the nature of this response in greenhouse studies with soybeans in modified Leonard jars with sterilized sand-vermiculite growth medium. Increasing populations of *R. japonicum*, strain 110 from 0 to 500,000 cells/g were mixed with soil used as an inoculum. The background population of indigenous population was 30–200 cells/g of soil. The resulting nodule occupancy data was plotted as the logarithm of numbers of added 110 cells versus the percentage of 110 nodules recovered. A sigmoidal curve (resembling a buffer curve) was obtained which showed a small increase in nodule occupancy until a critical level of 2×10^3 cells/g. At this point the inoculant strain reached approximately 50 percent occupancy. From this point on a rapid increase in occupancy occurred with each small incremental change in inoculum size.

Field plot studies with soybeans have also demonstrated that massive inoculation at rates very much higher than those commonly employed can have some success in increasing nodule occupancy by the inoculant strain. Johnson *et al.* (1965) in Iowa and Maryland soils were able to increase the proportion of nodules from 5 percent with commercial inoculum to an average of 25 percent with a rate of 25×. Increasing inoculum rates to 400× or 800× continued to increase nodule occupancy but differences were not statistically significant. Kapusta and Rouwenhorst (1973) in Illinois and Weaver and Frederick (1974) in Iowa found that an inoculum rate of 1000× that of the indigenous population provided by soil or seed inoculation was necessary for 50 percent nodule occupancy by the inoculant strain over a range of indigenous populations. Although this approach required inoculation rates as high as 3.3×10^8 cells/cm of row the possibility of successfully altering nodule occupancy with highly effective strains might make this approach economically viable. As an interesting aside Smith *et al.* (1981) found that populations of log 2.59 and 3.59 cells/cm row were not successful in establishing nodules on soybean in a Puerto Rican soil without an indigenous population of *R. japonicum*.

Repeated inoculation with competitive strains

Most research on altering rhizobium occupancy of legume nodules is interested in immediate shifts in strain distribution. As an ideal, immediate change is most desirable because a farmer would obviously be intersted in immediate yield responses. A recent paper by Dunigan *et al.* (1984) in Louisiana suggests the possibility that repeated inoculation with a competitive strain can eventually change nodule occupancy. In this 7 year study, *R. japonicum* strain USDA 110 was used as a soil inoculum at the rate of 10^4 or 10^8 cells/cm of row for three years (indigenous population of 10^5 g/soil). Initially the recovery of 110 in soybean nodules was low, approximately 4, 6, and 7 percent for the first three years. Recovery reached 17 percent the fourth year and 54 percent by year 7. During the entire 7 years there was relatively little difference between the 10^4 and 10^8 rates in percent recovery. There is no information provided in this paper which would provide a hypothesis to explain this response.

Use of improved inoculation technology

Inoculant technology involving choice of carriers or method of inoculant delivery have challenged researchers since the latter part of the 19th century and continues unabated. Solid carriers for legume seed inoculant include peat, soil, coal dust, charcoal, stable manure, filtered mud, compost, pulverized lignite, ground plant materials, sugar cane bagasse, corn cobs, wheat and rice straw, vermiculite, bentonite, kaolinite, and recently polyacrylamide or alginate gels (Jawson *et al.*, 1989; Jung *et al.*, 1982). Liquid seed inoculation has utilized both cultures and more recently vegetable oils (Kremer and Peterson, 1983). Numbers of rhizobia introduced by seed inoculation using liquid or solid carriers have been enhanced by use of natural or artificial stickers or glues. The environment around the seed has been altered by pelleting seeds using lime, rock phosphate, or even bauxite (Brockwell, 1977). Various pre-inoculation strategies have also been used including pelleting, clay mineral coatings, or pressure or vacuum seed impregnation. During the 1970s, the use of concentrated liquid or solid granular materials sprayed or directly drilled into the soil have been used to effectively increase rhizobium numbers or reduce the impact of insecticide or fungicide seed treatments. Inoculation through irrigiation has also been attempted (Ciafardini and Barbieri, 1987). All of the above materials and inoculation techniques have been effective under some circumstances. None have been effective in supplying the numbers of rhizobia to supplant the indigenous population except direct soil inoculation strategies (Bezdicek, 1978) and potentially the vegetable oil-lyophilized cell inoculant of Kramer and Peterson (1983b). It is certain that inoculant technology studies will continue in the future, hopefully, with a more successful outcome.

Selected, mutagenized, or genetically engineered competitive strains

The mechanisms that confer a competitive advantage to a strain of Rhizobium unfortunately remains poorly understood. Yet it has been possible through selection to find highly competitive strains with the ability to successfully compete for nodule sites with indigenous strains (Herridge and Roughley, 1975; May and Bohlool, 1983). Recently Klubek *et al.* (1988) evaluated the success of granular soil implant inocula for 19 strains of *B. japonicum* at nine field locations in four mid-western states. Although none of the strains was as competitive as desired, a maximum of 15.7 percent occupancy in 1984 and 26.6 percent in 1985, three strains clearly were more competitive than the others. The authors considered these strains to be potential recipients for the introduction of new genetic information that could enhance symbiotic efficiency.

A very recent paper by Paau (1989) reports on an exciting approach. Dominant indigenous *B. japonicum* strains from the Mississippi delta were mutagenized to achieve greater N_2 fixing effectiveness. Inoculation trials with 5 mutants in a vermiculite-based inoculant (delivered $> 10^5$ bacteria/seed) were conducted at 7 locations in three Mississippi delta states. Soybeans yield increases averaging 169 kg/ha (range 39–392) and 6.7 percent greater than the control plots were achieved. Interestingly the highest yields were obtained in Louisiana where the isolates were initially obtained. The authors consider this strategy to be potentially useful in other geographical regions requiring superior inoculant strains.

The opportunity to design highly competitive and effective strains by the tools of genetic engineering will depend on an improved understanding of those factors which determine competitiveness. More information concerning possible factors will be addressed in other papers of this symposium.

Co-inoculation with other microorganisms

There are a number of studies in the literature which show that the use of inocula which contain both the specific rhizobium and another microorganism which complements the Rhizobium can enhance rhizobium colonization and nodulation.

Some success has been achieved by Li and Alexander (1988) using co-inoculants of an antibiotic producing *Pseudomonas* sp. and an antibiotic resistant *R. meliloti* strain and an antibiotic producing *Bacillus sp.* with an antibiotic resistant *B. japonicum*. Fuhrmann and Wollum (1989) found that four *Pseudomonas* sp. and one *Serratia* sp. when used as co-inoculant strains with different *B. japonicum* strains in combination, could cause significant changes in nodule occupancy on roots of soybean plants grown under low iron conditions in sterile vermiculite culture. The pseudomonads were fluorescent, siderophore producing strains. Some of the pseudomonads were effective even when ferric-hydroxide was added to the rooting media. Hodgson *et al.* (1985) have utilized bacteriocin-resistant strains of *R. trifolii* in co-inocula with bacteriocin producing strains of *R. trifolii* to alter the normal nodulation pattern of *Trifolium subterraneum*.

Other organisms tried as co-inoculant strains include non-symbiotic N-fixing Azospirillum, Azotobacter and Beijerinckia with various strains of Rhizobium (Apte and Iswaran, 1974; Plazinski and Rolfe, 1985; Singh and Subba Rao, 1979.) Both positive and negative effects on nodulation and the quantity of nitrogen fixed were reported. These co-inoculant organisms are all considered to produce plant growth hormones and responses are considered to be due to an alteration in available sites for nodule initiation. VA-mycorrhizal co-inoculation with rhizobium is known to enhance N_2 fixation in legumes, particularly with tropical legumes growing in high P fixing soils (Ganry *et al.*, 1985; Mosse, 1976). Less well known and described are any changes in nodule occupancy by specific rhizobium strains in the inoculum.

Use of soil applied chemicals

The previous section discussed the use of a second "co-inoculant" microorganism with a rhizobium strain to overcome competition from indigenous soil and rhizosphere microorganisms. This idea is predicated on the co-inoculant microorganism producing a chemical which would adversely affect the indigenous strains and competing rhizosphere microorganisms but not the desired strain which has resistance.

Coincidentally with these studies, researchers have investigated the direct addition of a chemical, often a fungicide or antibiotic, to the soil or seed along with of a rhizobium strain or strains resistant to the chemical. Lennox and Alexander (1981) found a greater percentage of the nodules on *Phaseolus vulgaris* were formed by a thiram resistant strain of *R. phaseoli* if the seeds were treated with thiram. In a very similar study, Jones and Giddens (1983) inoculated thiram treated soybean seed with a mutant of strain 110 resistant to thiram and were successful in increasing the proportion of thiram resistant isolates in the soybean root nodules. Li and Alexander (1986) found that the addition of streptomycin and a streptomycin resistant strain of *R. meliloti* to non-sterile soil suppressed the numbers of bacterial cells in the rhizosphere of alfalfa (*Medicago sativa* L) and increased the number of nodules of the alfalfa roots. It should be noted that none of these studies were done under field conditions where an evaluation could be made of the efficacy of this approach in changing strain occupancy of legume nodules. Such field studies are badly needed to test the hypothesis raised by

Lennox and Alexander (1981) and Li and Alexander (1986) that a primary reason for a lack of success by inoculant strains in inhibition by rhizosphere microorganisms early in the infection process.

Legume host-rhizobium specificity

The ability of some legume genotypes to exclude one, two or all strains of its specific rhizobia from nodules provides the potential opportunity for excluding the often less desirable indigenous rhizobia and allow for nodulation only by the inoculant strain(s). The inability of certain soybean varieties to form nodules was recognized as early as 1915 and this characteristic, rj_1, rj_1, has been utilized effectively to develop non-nodulating isolines for comparative research on nitrogen fixation (Vest *et al.*, 1973). Later Caldwell and Vest (1968) and Kvien *et al.* (1981) found that some soybean genotypes excluded one or more serogroups from nodules. Exclusion results in soybeans with a totally different nodule occupancy even though growing in the same field. Hobbs (1985) has indicated that some nodulation specificity occurs with *Pisum sativum* and *R. leguminosarum*.

In recent years Cregan and Keyser (1986) have developed soybean PI lines selected for exclusion of *B. japonicum* serogroup 123, the dominant indigenous serogroup in midwestern soils. Although the original selections have totally restricted nodulation by strain USDA 123 they were only partially effective against 20 recently isolated field isolates of *B. japonicum* serogroup 123 (Keyser and Cregan, 1987). Four of these isolates when evaluated against USDA strain 110 on the same PI lines were not competitive for nodule occupancy. It thus appears that through the process of breeding and selection it may be possible to develop soybean cultivars which would allow the introduction of highly effective strains of *B. japonicum* in the presence of a high population of a less desirable indigenous strains. A similar strategy should be effective for other important legumes as well.

A totally different approach, that of attempting to obtain legumes cultivars which develop a highly effective symbiosis with most of the native soil rhizobia (promiscuous nodulation) has been attempted by Nagju (1980) and Pulver *et al.* (1985). Both authors propose this approach for developing countries growing soybeans but lacking the facilities to produce and distribute high quality inoculants. Since most soils of the tropics have large indigenous populations of the cowpea-miscellany this approach has some potential merit. Dreyfus *et al.* (1988) questions the wisdom of this approach because of the risk of nodulation by ineffective or marginally effective strains.

References

Abel G H and Erdman L W 1964 Response of Lee soybeans to different strains of *Rhizobium japonicum*. Agron. J. 56, 423–424.

Albrecht H R 1943 A method for inoculating small lots of legume seed. J. Am. Soc. Agron. 35, 914–917.

Allen E K and Allen O N 1950 Biochemical and symbiotic properties of the rhizobia. Bacteriol. Rev. 14, 273–330.

Apte R and Iswaran V 1974 Cultures of rhizobium inoculants with those of Beijerinckia and Azotobacter. Proc. Ind. Natl. Sci. Acad., Pt. B., Biol. Sci. 40, 482–485.

Baur K 1943 The use of bulk inoculum for peas in western Washington. Soil Sci. Soc. Am. Proc. 8, 223–225.

Bezdicek D F, Evans D W, Abede B and Witters R E 1978 Evaluation of peat and granular inoculum for soybean yield and N fixation under irrigation. Agron. J. 70, 865–868.

Bohlool B B and Schmidt E L 1973 Persistence and competition aspects of *Rhizobium japonicum* observed in soil by immunofluorescence microscopy. Soil Sci. Soc. Am. Proc. 37, 561–564.

Bohlool B 1988 Rhizobium technology: applications for international agricultural developments in the tropics. *In* Nitrogen Fixation: 100 years Later. Ed. H Bothe. Gustav Fischer, Stuttgart, FRG.

Brockwell J 1977 Application of legume seed innoculants. *In* A Treatise on Dinitrogen Fixation. Eds. R W F Hardy and A H Gibson. Section 4, pp 277–309. John Wiley and Sons, New York.

Burton J C 1967 Rhizobium culture and use. *In* Microbial Technology. Ed. H J Holmes. pp 1–33. Van Nostrand, Reinhold, NY.

Caldwell B E and Vest G 1968 Nodulation-interactions between soybean genotypes and serogroups of *Rhizobium japonicum*. Crop Sci. 8, 680–682.

Ciafardini G and Barbieri C 1987 Effects of cover inoculation of soybeans on nodulation, nitrogen fixation and yield. Agron. J. 79, 645–648.

Collins W O 1943 Preliminary report on legume inoculation studies. Soil Sci. Soc. Am. Proc. 8, 221–222.

Cregan P B and Keyser H H 1986 Host restriction of nodulation by *Bradyrhizobium japonicum* strain USDA 123 in soybean. Crop Sci. 26, 911–916.

Domergi S M, Frederick L R and Anderson I C 1967 Serogroups of *Rhizobium japonicum* in soybean nodules as affected by soil types. Agron. J. 59, 10–12.

Devine T E and Weber D F 1977 Genetic specificity of nodulation. Euphytica 26, 527–535.

Döbereiner J and Campelo A B 1977 Importance of legumes and their contribution to tropical agriculture. *In* A Treatise on Dinitrogen Fixation. Section IV. Agronomy and Ecology. Eds. R W F Hardy and A H Gibson. pp 191–220. John Wiley and Sons, New York.

Duggar J F 1935 The effects of inoculation and fertilization of Spanish peanuts on root nodule number. J. Am. Soc. Agron. 27, 128–133.

Dunigan E P, Bollich P K, Hutchinson R L, Hicks P M, Zaunbrecker F C, Scott S G and Mowers R P 1984 Introduction and survival of an inoculant strain of *Rhizobium japonicum* in soil. Agron. J. 76, 463–466.

Dreyfus B L, Diem H G and Dommergues Y R 1988 Future directions for biological nitrogen fixation research. Plant and Soil 108, 191–199.

Erdman L W 1943 New developments in legume inoculation. Soil Sci. Soc. Am. Proc. 8, 213–216.

Erdman L W 1959 Legume inoculation: What it is–What it does. US Dept. Agr. Farmers Bul. 2003, 16 pp.

Ellis W R, Ham G E and Schmidt E L 1984 Persistence and recovery of *Rhizobium japonicum* inoculum in a field soil. Agron. J. 76, 573–576.

Fred E B, Baldwin I L and McCoy E 1932 Root Nodule Bacteria and Leguminous Plants. pp 229–256, University of Wisconsin Studies in Science, No. 5, Madison, WI.

Fuhrmann J and Wollum A G 1989 Nodulation competition among *Bradyrhizobium japonicum* strains as influenced by rhizosphere bacteria and iron availability. Biol. Fertil. Soils 7, 108–112.

Ganry F, Diem H G, Wey J and Dommerques Y R 1985 Inoculation with *Glomus mosseae* improves N_2 fixation by field grown soybeans. Biol. Fert. Sols 1, 15–23.

Gibson A H, Curnow B C, Bergerson F J, Brockwell J and Robinson A C 1975 Studies of field populations of Rhizobium: effectiveness of strains of *Rhizobium trifolii* associated with *Trifolium subterraneum* L. pastures in southeastern Australia. Soil Biol. Biochem. 7, 95–102.

Giddens J E, Dunigan E P and Weaver R W 1982 Legume Inoculation in the Southeastern USA. Southern Cooperative Series Bulletin 283. The University of Georgia, College of Agriculture Experiment Stations.

Graham P H and Harris S C 1982 (Eds.) Biological Nitrogen Fixation Technology for Tropical Agriculture. 768 p. Centro Internacional de Agricultura Tropical, Cali, Columbia.

Graham P H, Apolitano C, Ferrera-Cerrato R, Halliday J, Lepiz R, Menendez O, Rios R, Saito S M T and Viteri S 1982 The international bean inoculation trial (IBIT) *In* Biological Nitrogen Fixation Technology for Tropical Agriculture. pp 223–229. Eds. P H Graham and S C Harris. Centro Internacional de Agricultura Tropical, Cali, Columbia.

Graham P H 1984 Problems of soybean inoculation in the tropics. *In* World Soybean Research Conference III: Proceedings. pp 951–959. Ed. R Shibles. Westview Press, Boulder, CO.

Hagedorn C 1978 Effectiveness of *Rhizobium trifolii* populations associated with *Trifolium subterraneum* L. in southwest Oregon soils. Soil Sci. Soc. Am. J. 42, 447–451.

Hagedorn C 1979 Nodulation of *Trifolium subterraneum* L. by introduced rhizobia in southwest Oregon soils. Soil Sci. Soc. Am. J. 43, 515–519.

Halliday J 1982 Present constraints to, and a future strategy for, fuller utilization of legume-fixed nitrogen for crop production in tropics. Food Fert. Tech. Center, Technical Bulletin No. 60.

Ham G E, Frederick L R and Anderson I C 1971 Serogroups of *Rhizobium japonicum* in soybean nodules. Agron. J. 57, 179–185.

Ham G E, Cardwell V B and Johnson H W 1971 Evaluation of *Rhizobium japonicum* inoculants in soils containing naturalized populations of rhizobia. Agron. J. 63, 301–303.

Hellreigel H 1886 Welche Stickstoffe stehen der Pflanze zu Gebote? Ztschr. des Rübenzucker Industrie Deutschen Reich 36, 863–877.

Herridge D F and Roughley R J 1975 Nodulation of and nitrogen fixation by *Lablab purpureus* under field conditions. Aust. J. Exp. Agric. Anim. Husb. 15, 264–269.

Hobbs S L A 1985 Nodulation specificity in *Pisum sativum*. *In* Nitrogen Fixation Progress, Ed. H J Evans, H J Bottomley and W E Newton. pp 33. Martinus Nijhoff Publishers, Dordrecht.

Hodgson A L M, Roberts W P, Ward J S 1985 Regulated nodulation of *Trifolium subterraneum* inoculated with bacteriocin-producing strains of *Rhizobium trifolii*. Soil Biol. Biochem. 17, 475–478.

Holding A J and King J 1963 The effectiveness of indigenous populations of *Rhizobium trifolii* in relation to soil factors. Plant and Soil 18, 191–198.

Holland A A 1970 Competition between soil- and seed-borne *Rhizobium trifolii* in nodulation of introduced *Trifolium subterraneum*. Plant and Soil 32, 293–302.

Jawson M D, Franzluebbers A J and Berg R K 1989 *Bradyrhizobium japonicum* survival in and soybean inoculation with fluid gels. Appl. Environ. Microbiol. 55, 617–622.

Johnson H W, Means V M and Weber C R 1965 Competition for nodule sites between strains of *Rhizobium japonicum* applied as inoculum and strains in the soil. Agron. J. 57, 179–185.

Jones R and Giddens 1983 Introduction of effective N_2-fixing rhizobial strains into the soybean plant by use of fungicide resistance. Agron. J. 76, 599–602.

Jung G, Mugnier J, Diem H G and Dommergues Y R 1982 Polymer-entrapped rhizobium as an inoculant for legumes. Plant and Soil 65, 219–231.

Kapusta G and Rouwenhort D L 1973 Influence on inoculum size on *Rhizobium japonicum* serogroup distribution frequency in soybean nodules. Agron. J. 65, 916–919.

Keyser H H, Weber D F and Uratsu S L 1984 *Rhizobium japonicum* serogroup and hydrogenase phenotype distribution in 12 states. Appl. Environ. Microbiol. 47, 613–615.

Keyser H H and Cregan P B 1987 Nodulation and competition for nodulation of selected soybean genotypes among *Bradyrhizobium japonicum* serogroup 123 isolates. Appl. Environ. Microbiol. 53, 2631–2635.

Klubek B P, Hendrikson L L, Zablotowicz R M, Skwaro J E,

Varsa E C, Smith S, Isleib T G, Maya J and Valdes M 1988 Competitiveness of selected *Bradyrhizobium japonicum* strains in midwestern USA soils. Soil Sci. Soc. Am. J. 52, 662–666.

Kremer R J and Peterson H L 1983a Effect of carrier and temperature on survival of *Rhizobium* spp. in legume inocula: Development of an improved type of inoculant. Appl. Environ. Microbiol. 45, 1790–1794.

Kremer R J and Peterson H L 1983b Field evaluation of selected rhizobium in an improved legume inoculant. Agron. J. 75, 139–143.

Kvien C S, Ham G E and Lambert J W 1981 Recovery of introduced *Rhizobium japonicum* strains by soybean genotypes. Agron. J. 73, 900–905.

Lennox L B and Alexander M 1981 Fungicide enhancement of nitrogen fixation and colonization of *Phaseolus vulgaris* by *Rhizobium phaseoli* Appl. Environ. Microbiol. 41, 404–411.

Li D M and Alexander M 1986 Bacterial growth rates and competition affect nodulation and root colonization by *Rhizobium meliloti*. Appl. Environ. Microbiol. 52, 807–811.

Li D M and Alexander M 1988 Co-inoculation with antibiotic-producing bacteria to increase colonization and nodulation by rhizobia. Plant and Soil 108, 211–219.

Lynch D L and Sears O H 1952 The effect of inoculation upon yields of soybean on treated and untreated soils. Soil Sci. Soc. Am. Proc. 16, 214–216.

May S N and Bohlool B B 1983 Competition among *Rhizobium leguminosarum* strains for nodulation of lentils (*Lens esculenta*). Appl. Environ. Microbiol. 45, 960–965.

Moodie C D and Vandecaveye S C 1943 Yield and nitrogen content of chickpea, *Cicer arietinum* as affected by seed inoculation. Soil Sci. Soc. Amn. Proc. 8, 229–233.

Mosse B 1976 Plant growth and responses to vesicular-arbuscular mycorrhiza. IX. Interactions between VA mycorrhiza, rock phosphate and symbiotic fixation. New Phytol. 76, 331–342.

Mpofu S M 1988 Characterization of indigenous *Bradyrhizobium japonicum* in North Carolina soils: Serotype and symbiotic potential. M. S. Thesis, Dept. of Soil Science, North Carolina State University.

Mulder E G, Lie T A and Houwers A 1977 The importance of legumes under temperate conditions. *In* A Treatise on Dinitrogen Fixation. Section IV. Agronomy and Ecology Eds. R W F Hardy and A H Gibson. pp 221–242. John Wiley and Sons, New York.

Nangju D 1980 Soybean response to indigenous rhizobia as influenced by cultivar origin. Agron. J. 72, 403–406.

Nelson D W, Swearengin M L and Beckham L S 1978 Response of soybeans to commercial soil-applied inoculants. Agron. J. 70, 517–518.

Nobbe F and Hiltner L 1886 Bodenimpfung für Anbau von Leguminosen. Sächs. Landw. Ztschr. 44, 90–92.

Norman A G 1943 The nitrogen nitrogen of soybeans: I. Effect of inoculation and nitrogen fertilizer on the yield and composition of beans on Marshall silt loam. Soil Sci. Soc. Am. Proc. 8, 226–228.

Paau A S 1989 Improvement of rhizobium inoculants. Appl. Environ. Microbiol. 55, 862–865.

Plazinski J and Rolfe B G 1985 Influence of *Azospirillum* strains on the nodulation of clovers by *Rhizobium* strains. Appl. Environ. Microbiol. 49, 984–989.

Pulver E L, Kueneman E A and Ranga-Rao V 1985 Indentification of promiscuous nodulating soybean efficient in N_2 fixation. Crop Sci. 25, 660–663.

Roughley R J, Blowes W M and Herridge D F 1976 Nodulation of *Trifolium subterraneum* by introduced rhizobia in competition with naturalized strains. Soil Biol. Biochem. 8, 403–407.

Salfield A 1888 Eine. Verwertung der Hellriegelschen Versuche mit Leguminosen im landwirtschaftlichen Betriebe. Dt. Landw. Presse. 15, 630–631.

Singh C S and Subba Rao N S 1979 Associative effect of *Azospirillum brasilense* with *Rhizobium japonicum* on nodulation and yield of soybean. Plant and Soil 53, 387–392.

Smith R S, Ellis M A and Smith R E 1981 Effect of *Rhizobium japonicum* inoculant rates on soybean nodulation in a tropical soil. Agron. J. 73, 505–508.

Vest G, Weber D F and Sloger C 1971 Nodulation and nitrogen fixation. Ch. 10. *In* Soybeans: Improvement, Production and Uses. Agronomy Monograph #16. Ed. B E Caldwell. American Society of Agronomy, Madison, WI.

Vest G, Weber D F and Sloger C 1973 Nodulation and nitrogen fixation. *In* Soybeans: Improvement Production and Uses. Ed. B E Caldwell, pp 353–390. Agronomy Monograph No. 16, American Society of Agronomy, Madison, WI.

Weaver R W and Frederick L R 1974 Effect of inoculum rate on competitive nodulation on *Glycine max* L. Merr. II. Field Studies. Agron. J. 66, 233–236.

Weber D F and Leggett G E 1966 Relation of rhizobia to alfalfa sickenss in eastern Washington. ARS. 41–117. USDA.

D. L. Keister and P. B. Cregan (Eds.), The rhizosphere and plant growth, 135–140.

PLSO BARC13

Importance of saprophytic competence for introduced rhizobia*

PETER J. BOTTOMLEY[1,2], SHARON P. MAGGARD[1], KAM LEUNG[1] and MATT D. BUSSE[2]
[1]*Department of Microbiology and* [2]*Department of Soil Science, Oregon State University, Corvallis, OR 97331-3804, USA*

Key words: Bradyrhizobium, composition, density, responsiveness to environmental stimuli, Rhizobium, soil populations, viability

Abstract

The ambitious goal of expanding or enhancing legume production on marginal agricultural land led to the development of a concept that rhizobia in possession of special symbiotic and saprophytic attributes would be required. Although 20 years have passed since the term saprophytic competence was applied to describe soil colonization, survival, and persistence by rhizobia, there is no direct evidence for the possession of specific phenotypes being critical to the success or failure of rhizobia in soil. In this paper, selected data are summarized on the density and dynamics of soil populations of rhizobia. We reaffirm old observations that Bradyrhizobium and Rhizobium might be rather distinct in their saprophytic behavior. Progress in the field of competitive nodulation continues to be slow due not only to our poor understanding of the composition of soil populations, but due to inadequate attention given to the behavior of rhizobia in the soil and rhizosphere phases. Some of our own data are presented which suggest that saprophytic competence may be different between the members of a naturalized population of *Rhizobium leguminosarum* bv. *trifolii*. Such a conclusion is derived from differences measured in: i) population densities of serotypes, ii) percentage of viable cells within serotypes, and iii) kinetics of substrate responsiveness of serotypes. Rhizosphere biologists must begin to integrate the large data base being generated on different 'specialist' organisms if a comprehensive understanding of microbial activity, and population dynamics at the root-soil interface is to be achieved.

Introduction

Bacteria of the genera Rhizobium and Bradyrhizobium are widely distributed in soils throughout the world as a result of both the natural distribution of the Leguminosae, and through the migratory and agricultural activities of man (Brockwell, 1981). Although failure of legume crops directly attributable to the absence of appropriate rhizobia in soil has become increasingly rare, the question is often raised whether soil rhizobia are capable of realizing the N assimilatory potential of legumes growing under field conditions. Numerous reports have shown that legumes nodulated with representatives of soil rhizobia populations are incapable of fixing dinitrogen in quantities comparable with the N which can be assimilated from mineral forms (Bergersen, 1970; Caldwell and Vest, 1970; Jansen van Rensberg and Strijdom, 1985). Furthermore, information has been available for more than 20 years which shows that *Bradyrhizobium japonicum* strains superior in symbiotic dinitrogen fixation are equally unexceptional in competition with soil rhizobia for nodule occupancy (Johnson *et al.*, 1965). In the case of Rhizobium, questions have been raised not only in regard to nodulating competitiveness of superior dinitrogen fixing strains, but also with regard to their persistence in soil, and the stability of their dinitrogen fixing phenotype under field soil conditions (Chatel *et al.*, 1968; Dudman and Brock-

* Technical Paper No. 8888 of the Oregon Agricultural Experiment Station.

well, 1968; Jansen van Rensberg and Strijdom, 1985).

Given these concerns, considerable effort has been spent in attempting to identify phenotypes which are important to competitiveness and persistence. The term 'saprophytic competence' was introduced to the ecology of rhizobia twenty years ago (Chatel *et al.*, 1968) and was defined to include: 1) *colonization* of soil; 2) *survival* through non-growth periods of environmental stress; and 3) *persistence* during periods conducive for microbial growth. The objective of this paper is to present an overview of recent attempts to expand our understanding of the soil-borne phase of rhizobial existence.

Density of soil rhizobial populations

Through use of the technically simple plant infection-soil dilution procedure, the magnitude of soil rhizobial populations has been shown to vary widely (Parker *et al.*, 1977). A relationship between the magnitude of soil rhizobial populations and a variety of biological and abiological parameters has been reported on numerous occasions. However, it has been extremely rare for researchers to monitor simultaneously the density of different rhizobia in the same soil (Chatel and Parker, 1973; Woomer *et al.*, 1988), or to relate the density of rhizobia to that of total soil bacteria (Bottomley and Dughri, 1989), or to monitor population dynamics over agriculturally meaningful time periods. Mahler and Wollum (1981) showed seasonal variation in the soil populations of *R. leguminosarum* bv. *trifolii* under pastures of unspecified clover species. The largest population densities were measured during the transition between winter dormancy and the active period of pasture growth in spring. In contrast, Australian studies revealed population dynamics to be more complex, with increases in a soil population of *R. leguminosarum* bv. *trifolii* under pasture occurring both during the period of active plant growth, and also at the time of plant death and nodule senescence (Richardson and Simpson, 1988). In the case of *B. japonicum*, soil populations have been observed to increase at the end of the soybean growing season (Mahler and Wollum, 1982; Moawad *et al.*, 1984). Since this phenomenon is attributed to nodule senescence, it is appropriate to consider the possibility that dynamics in *B. japonicum* soil populations might be partially a consequence of superior symbiotic attributes rather than through soil colonizing ability *per se*. In this regard, soybean bacteroids retain competence at cell division (Tsien *et al.*, 1977; Zhou *et al.*, 1985) whereas clover bacteroids do not (Zhou *et al.*, 1985). Caution should be exercised when attempts are made to compare data on soil populations of Bradyrhizobium with those concerning Rhizobium.

Spatial distribution is an issue of potential significance to the interpretation of population density values. Wollum and Cassel (1984) observed the population density of *B. japonicum* to vary widely over short distances (0.3 m) in a field, and to be influenced in a complex manner by soybean row orientation and soil properties. Chatel and Greenwood (1973a) estimated that the nodule fraction of senesced subclover plants could contribute between 70 to 98% of the viable noduating population of *R. leguminosarum* bv. *trifolii* under undisturbed subclover pasture. Richardson and Simpson (1988) calculated that 80 to 94% of the total *R. leguminosarum* bv. *trifolii* population found in the 0–25 cm depth increment of an acidic subclover pasture soil were located in the top 0 to 2 cm. Bottomley and Dughri (1989) found the contribution of non-viable, non-nodulating cells to an *R. leguminosarum* bv. *trifolii* soil population increased abruptly from 14 to 70% between the 10- to 15-cm and 15- to 20-cm soil depth increments. Obviously, microsite and depth effects on population size can be easily overlooked during sampling.

Composition of soil rhizobial populations

The results of numerous nodule occupancy experiments show that certain rhizobial strains can occupy a fraction of nodules disproportionate to their relative contribution to inoculant strain mixtures or rhizosphere populations (Date and Brockwell, 1978). Although the outcome of competitive nodulation has often been directly correlated with the relative abundance of the different strains in inocula or rhizospheres (Amarger and Lobreau, 1982), data generated under nonsterile soil conditions have been limited because of our lack of understanding of the composition of indigenous soil populations of rhizobia.

For many years a comprehensive picture of soil population composition was only available for *B. japonicum*. Between six and nine different serogroups of *B. japonicum* were determined in soybean nodules at each of several locations throughout the U.S.A. (Johnson and Means, 1963). Although soil populations of *Rhizobium* species can also be complex, the extent of diversity is influenced by the particular method of delineation used by different researchers. There is a need to ascertain which of these methods correlate with each other. For example, Valdivia *et al.* (1988) identified eight serotypes of *R. leguminosarum* bv. *trifolii* in one soil and Young (1985) identified nine enzyme polymorphism patterns within a soil population of *R. leguminosarum* bv. *trifolii*. Other delineation methods, which included phage typing, plasmid and protein profiles, and restriction fragment length polymorphisms within the symbiotic plasmid have revealed an even more complex picture of population composition (Bromfield *et al.*, 1986; Brockman and Bezdicek, 1989; Young and Wexler, 1988). Although many of these methods can identify both the dominant and incompetent nodule occupying types, only immunofluorescence can be used to accurately study the different naturalized types in their normal, nonsterile soil phase. At this time there is a need to ascertain which identification methods provide information of most relevance to identifying phenotypes important to soil and rhizosphere colonization and to competitive nodulation.

Determination of population densities within the subpopulations of soil rhizobia

Despite numerous studies having revealed extensive phenotypic diversity within Rhizobium and Bradyrhizobium, little direct evidence has accumulated from nonsterile field soil studies for the existence of sufficient interstrain variation in saprophytic competence to account for nodulation success or failure (Chatel *et al.*, 1968; Chatel and Parker, 1973). The pioneering work in immunofluorescence by E. L. Schmidt at the University of Minnesota has made it possible to detect, and make estimates of the size of subpopulations of rhizobia in soil (Schmidt *et al.*, 1968). The impetus behind many of these studies was to gain a better understanding of the soil and rhizosphere phases of the highly competitive *B. japonicum* serogroup 123. The non-sterile soil studies revealed many significant findings. Despite the overwhelming dominance of 123 in nodule occupancy, neither bulk soil, nor host or non-host rhizosphere populations of *B. japonicum* serogroups 123, 110, and 138 were significantly different from each other (Moawad *et al.*, 1984). Although members of serogroup 123 are released in large numbers from nodules at the end of the growing season, no persistent numerical advantage of serogroup 123 was carried over into the next cropping season (Moawad *et al.*, 1984).

Immunofluorescence studies from our laboratory, indicate that populations of *Rhizobium leguminosarum* bv. *trifolii* can be maintained at significantly greater densities in undisturbed pasture soils than the densities of *B. japonicum* in cultivated soils. Furthermore, statistically significant differences have been measured between the densities of serotype subpopulations within the *R. leguminosarum* bv. *trifolii* population (Almendras and Bottomley, 1987; Bottomley and Dughri, 1989; Valdivia *et al.*, 1988). Earlier work had also revealed differences between different strains of *R. leguminosarum* bv. *trifolii*, and between different *Rhizobium* species in their ability to colonize host and non-host rhizospheres (Chatel and Greenwood, 1973b; Robinson, 1967). It is unclear whether such findings are related to fundamental differences between the two genera, or, are related to differences between the population ecology of rhizobia in undisturbed, high organic matter-containing pasture soils, and in cultivated, intensively managed soils in crop production.

Viability of soil rhizobia enumerated by immunofluorescence

Since fluorescent antibodies are usually targeted

at heat stable somatic antigens of rhizobia (Schmidt *et al.*, 1968), it is reasonable to question what proportion of viable cells exist within a soil population detected by immunofluorescence. There is circumstantial evidence for the existence of nonviable rhizobia in soil. First, an extremely low percentage of the total soil population of bacteria is competent at cell division (Bottomley and Dughri, 1989). Second, immunofluorescence studies have documented the existence of morphologically atypical cells of *B. japonicum* and *R. leguminosarum* bv. *trifolii* in naturalized soil populations (Bohlool and Schmidt, 1970; Bottomley and Dughri, 1989). In the latter study, cells sufficiently small (1.0 by 0.3 μm) to pass through 0.4 μm pore size filters accounted for substantial portions of the direct immunofluorescence counts of certain indigenous serotypes. No evidence was obtained for these cells possessing vital signs or nodulating capability (Bottomley and Dughri, 1989). Thirdly, data presented in Table 1 show that within the Abiqua soil at least eight serotypes of *R. leguminosarum* bv. *trifolii* exist, and which represent a total direct count population of approximately 9×10^6 per gram of soil. Serotype 36, which is numerically dominant (ca. 50% of the bv. *trifolii* population), is rarely recovered in either field nodules, or in nodules on plants of the terminal positive soil dilution in plant infection bioassays. From these data it is not possible, however, to distinguish between the possibilities that: i) either a large portion of the population of serotype 36 lacks viability, or ii) the members are viable but are noncompetitive at nodulating, or iii) the members are viable with no nodulating capability.

Table 1. Composition of the *Rhizobium trifolii* population in Abiqua soil

Serogroup	Population density g of soil^{-1}, 10^6	Fraction of *R. trifolii* population (%)
36	4.60	46.6
6	3.20	32.4
23	0.55	5.6
21	0.46	4.7
27	0.37	3.7
16	0.31	3.1
17	0.22	2.2
G4	0.16	1.6
Σ individual Σ serogroups	9.87	
cocktails[a]	9.70	

[a] Individual fluorescein labeled conjugates were mixed together in combination cocktails and the density of the total *R. trifolii* population determined.

To address this concern we have resorted to a technique developed by aquatic microbiologists. The principle of the substrate responsive assay used by Kogure *et al.* (1979) involves the action of the DNA gyrase inhibitor, nalidixic acid. The latter antibiotic can prevent DNA replication and cell division without inhibiting RNA and protein synthesis. As a result of incubation in the presence of appropriate substrates and nalidixic acid there is significant elongation of those cells which are viable and biosynthetically competent. We have extended this approach to soil and confirmed the findings made in aquatic systems. Data in Table 2 show that $\leqslant$1.1% of the total bacterial cells recovered from a soil population can be cultured. In contrast, between 65 and 74% of the microscopically observed cells undergo significant cell elongation ($\geqslant$7.0 μm) after 16 to 20 h of exposure to yeast extract (200 mg L^{-1}) (Table 3).

The substrate responsive assay has been combined with immunofluorescence to study the viability of indigenous serotypes of *R. leguminosarum* bv. *trifolii*. Under pure culture conditions, yeast extract supplemented with mineral salts will support short-term exponential growth of various indigenous serotype representatives of *R. leguminosarum* bv. *trifolii* at rates (3 h gen. time) equivalent to complete medium. Regardless of serotype affiliation the maximum percentage of elongated cells can be observed within 6 to 9 h of exposure to yeast extract and nalidixic acid. Data in Fig. 1 show the kinetics of substrate responsiveness of different indigenous serotypes of bv. *trifolii* exposed to assay conditions within 1 h of separation from the soil matrix. Serotypes 6 and 21, which represent highly competitive nodule forming groups (Bottomley and Dughri, 1989), respond to the presence of substrate rapidly with 50% of the population significantly elongated by 8 to 12 h. In contrast, members of the noncompetitive serotypes 17 and 36 require significantly longer to respond despite the final percentage of elon-

Table 2. Comparison of the density of culturable bacteria with the total direct count of bacteria in Abiqua soil

Experiment	Population density g of soil^{-1}, 10^7			
	YE[b]	BH	YP	TS
1	0.5 (0.3)[a]	0.6 (0.4)[a]	0.6 (0.4)[a]	1.0 (0.6)[a]
2	0.2 (0.1)	0.2 (0.1)	0.4 (0.3)	0.7 (0.4)
3	0.8 (0.5)	N.D.	N.D.	1.2 (0.8)
4	0.4 (0.4)	1.0 (1.1)	1.0 (1.2)	0.7 (0.8)

[a] Values in parentheses represent the percentages that the plate counts constitute of the total epifluorescence direct count of soil bacteria. The mean of the latter was $1.7 \pm 0.3 \times 10^9$ per gram oven dry soil.

[b] The abbreviations are: YE, yeast extract (0.2 g per liter); BH, brain heart infusion (0.4 g per liter); YP, yeast extract and bactopeptone (0.1 g per liter of each); TS, tryptic soy (0.3 g per liter).

Table 3. Comparison of the density of substrate responsive elongated bacteria with the total direct count of bacteria

Experiment	Epifluorescence count of total bacteria g of dry soil^{-1}, 10^9	Substrate responsive elongated cells (% of total)[a]
1	1.8	66.0
2	1.0	73.9
3	1.7	65.0

[a] Values were obtained after 16 to 24 h of incubation.

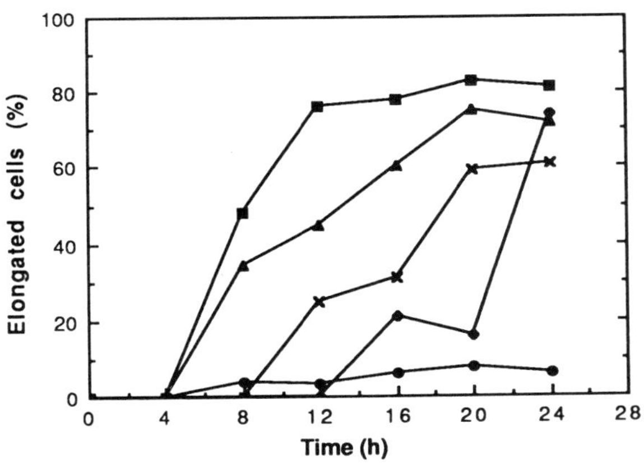

Fig. 1. Kinetics of cell elongation in different indigenous serogroups of *R. trifolii* in Abiqua soil. ■, serogroup 21; ▲, serogroup 6; X, serogroup 36; ◆, serogroup 17; ●, serogroup 23.

gating cells being similar to serotype 6. Less than 10% of the cells of serotype 23 show measurable cell elongation over the time course of the assay. The differences which we observed between serotypes in terms of: (i) population densities, (ii) proportions of viable, substrate responsive cells, and (iii) kinetics of the cell elongation response to substrate presence, suggest that saprophytic competence does vary between members of the same indigenous soil population of Rhizobium. The substrate responsive assay has the potential to extend the power of immunofluorescence to evaluate indigenous serotype populations both in terms of percent viability *per se*, and to compare the kinetics of recovery of physiological activity after exposure to stress conditions, or in response to different nutrient sources.

Acknowledgements

The research from our own laboratory described in this paper was supported by USDA competitive research grant no. 87-CRCR-1-2471 and the Oregon Agricultural Experiment Station. We express appreciation to C. Pelroy for processing the manuscript.

References

Almendras A S and Bottomley P J 1987 Influence of lime and phosphate on nodulation of soil-grown *Trifolium subterraneum* L. by indigenous *Rhizobium trifolii*. Appl. Environ. Microbiol. 53, 2090–2097.

Amarger N and Lobreau J P 1984 Quantitative study of nodulation competitiveness in *Rhizobium* strains. Appl. Environ. Microbiol. 44, 583–588

Bergersen F J 1970 Some Australian studies relating to the long-term effects of the inoculation of legume seeds. Plant and Soil 32, 727–736.

Bohlool B B and Schmidt E L 1970 Immunofluorescent detection of *Rhizobium japonicum* in soils. Soil Sci. 110, 229–236.

Bottomley P J and Dughri M H 1989 Population size and distribution of *R. leguminosarum* bv. *trifolii* in relation to total soil bacteria and soil depth. Appl. Environ. Microbiol. 55, 959–964.

Brockman F J and Bezdicek D F 1989 Diversity within

serogroups of *R. leguminosarum* bv. *viciae* in the Palouse region of eastern Washington as indicated by plasmid profiles, intrinsic antibiotic resistance, and topography. Appl. Environ. Microbiol. 55, 109–115.

Brockwell J 1981 A strategy for legume nodulation research in developing regions of the old world. Plant and Soil 58, 367–382.

Bromfield E S P, Sinha I B and Wolynetz M S 1986 Influence of location, host cultivar, and inoculation on the composition of naturalized populations of *Rhizobium meliloti* in *Medicago sativa* nodules. Appl. Environ. Microbiol. 51, 1077–1084.

Caldwell B E and Vest G 1970 Effects of *Rhizobium japonicum* strains on soybean yields. Crop Sci. 10, 19–21.

Chatel D L and Greenwood R M 1973a The location and distribution in soil of rhizobia under senesced annual legume pastures. Soil Biol. Biochem. 5, 799–808.

Chatel D L and Greenwood R M 1973b Differences between strains of *Rhizobium trifolii* in ability to colonize soil and plant roots in the absence of their specific host plants. Soil Biol. Biochem. 5, 809–813.

Chatel D L, Greenwood R M and Parker C A 1968 Saprophytic competence as an important character in the selection of *Rhizobium* for inoculation. Proc. Ninth Intern. Congr. Soil Sci. 2, 65–73.

Chatel D L and Parker C A 1973 The colonization of host-root and soil by rhizobia-1. Species and strain differences in the field. Soil Biol. Biochem. 5, 425–432.

Date R A and Brockwell J 1978 *Rhizobium* strain competition and host interaction for nodulation. *In* Plant Relations in Pastures. Ed. J R Wilson, pp 206–216. CSIRO East Melbourne, Australia.

Dudman W F and Brockwell J 1968 Ecological studies of root-nodule bacteria introduced into field environments. Aust. J. Agric. Res. 9, 739–747.

Jansen van Rensburg H and Strijdom B W 1985 Effectiveness of *Rhizobium* strains used in inoculants after their introduction into soil. Appl. Environ. Microbiol. 49, 127–131.

Johnson H W and Means U M 1963 Serological groups of *Rhizobium japonicum* recovered from nodules of soybeans (*Glycine max*) in field soils. Agron. J. 55, 269–271.

Johnson H W, Means U M and Weber C R 1965 Competition for nodule sites between strains of *Rhizobium japonicum* Agron. J. 57, 179–185.

Kogure K, Simidu U and Taga N 1979 A tentative direct microscopic method for counting living marine bacteria. Can. J. Microbiol. 24, 415–420.

Mahler R L and Wollum A G 1981 Seasonal variation of *Rhizobium trifolii* in clover pastures and cultivated fields in North Carolina. Soil Sci. 132, 240–246.

Mahler R L and Wollum A G 1982 Seasonal fluctuation of *Rhizobium japonicum* under a variety of field conditions in North Carolina. Soil Sci. 134, 317–324.

Moawad H A, Ellis W R and Schmidt E L 1984 Rhizosphere response as a factor in competition among three serogroups of indigenous *Rhizobium japonicum* for nodulation of field-grown soybeans. Appl. Environ. Microbiol. 47, 607–612.

Parker C A, Trinick M J and Chatel D L 1977 Rhizobia as soil and rhizosphere inhabitants. *In* A Treatise on Dinitrogen Fixation. Section IV: Agronomy and ecology. Eds. R W F Hardy and A H Gibson, pp 311–352. John Wiley and Sons, New York.

Richardson A E and Simpson R J 1988 Enumeration and distribution of *Rhizobium trifolii* under a subterranean clover-based pasture growing in an acid soil. Soil Biol. Biochem. 20, 431–438.

Robinson A C 1967 The influence of host on soil and rhizosphere populations of clover and lucerne root nodule bacteria in the field. J. Aust. Inst. Agric. Sci. 33, 207–209.

Schmidt E L, Bankole R and Bohlool B B 1968 Fluorescent antibody approach to study rhizobia in soil. J. Bacteriol. 95, 1987–1992.

Tsien H C, Cain P S and Schmidt E L 1977 Viability of *Rhizobium* bacteroids. Appl. Environ. Microbiol. 34, 854–856.

Valdivia B, Dughri M H and Bottomley P J 1988 Antigenic and symbiotic characterization of indigenous *Rhizobium leguminosarum* bv. *trifolii* recovered from root nodules of *Trifolium pratense* L. sown into subterranean clover pasture soils. Soil Biol. Biochem. 20, 267–274.

Wollum A G and Cassel D K 1984 Spatial variability of *Rhizobium japonicum* in two North Carolina soils. Soil Sci. Soc. Am. J. 48, 1082–1086.

Woomer P, Singleton P W and Bohlool B B 1988 Ecological indicators of native rhizobia in tropical soils. Appl. Environ. Microbiol. 54, 1112–1116.

Young J P W and Wexler M 1988 Sym plasmid and chromosomal genotypes are correlated in field populations of *Rhizobium leguminosarum*. J. Gen. Microbiol. 134, 2731–2739.

Zhou J C, Tchan Y T and Vincent J M 1985 Reproductive capacity of bacteroids in nodules of *Trifolium repens* L. and *Glycine max* (L.) Merr. Planta 163, 473–482.

D. L. Keister and P. B. Cregan (Eds.), *The rhizosphere and plant growth*, 141–148.

PLSO BARC14

Genetic stability in rhizobia in the field

A.H. GIBSON, D.H. DEMEZAS, R.R. GAULT, T.V. BHUVANESWARI and J. BROCKWELL
CSIRO Division of Plant Industry, Canberra, A.C.T., 2601, Canberra, Australia

Key words: antigenic characters, *Bradyrhizobium japonicum*, genetic variability, RFLP analysis, *Rhizobium leguminosarum* bv. *trifolii*, symbiotic effectiveness

Abstract

Genetic instability within strains of rhizobia maintained on laboratory media is well recognized, although rarely has the mutation been characterized. Variability within a strain introduced into the field is very difficult to recognise due to poor understanding of naturally-occurring populations of rhizobia. We have examined populations of *Rhizobium leguminosarum* bv. *trifolii* from both laboratory cultures and field populations and found significant variation in symbiotic effectiveness within both. In Australia, the only significant introduction of *Bradyrhizobium japonicum* has been strain CB1809 (=USDA136b). Symbiotic tests on field reisolates obtained by plant entrapment indicate little or no change in symbiotic effectiveness up to nine years after introduction. The RFLP pattern, using the RSα probe (Hahn and Hennecke, 1987a) was unchanged but marked differences in serological characters were observed.

Introduction

Genetic variability within a culture or strain has been a distraction for many Rhizobium workers, not least those providing commercial inoculants. Colony morphology changes are relatively common but they are not necessarily associated with changes in symbiotic characters. Mutations in genes generally occur at a frequency of 10^{-6} to 10^{-8}, but unless such mutations have a competitive growth advantage, they will not assume dominance in a culture. The most notable case with rhizobia involves *Bradyrhizobium japonicum* strain USDA110 which produced ineffective mutants that were able to utilise mannitol whereas the parental strain made very poor use of this carbon substrate (Kuykendall and Elkan, 1976). As most laboratory media used to culture rhizobia have been based on mannitol, it was not surprising that in our laboratory in 1977, three cultures of this strain, derived from different sources, were found to form ineffective associations with soybeans. Further studies have revealed that mutations leading to mannitol utilisation and ineffectiveness are not related (Mathis *et al.*, 1986) but obviously a mutation affecting effectiveness in a mannitol-utilising strain could give it a competitive advantage. It is possible that a similar set of circumstances involved the build-up of ineffective mutants within the *Rhizobium leguminosarum* bv. *trifolii* strain WU290 (Labandera and Vincent, 1975) although the nutritional aspects were not examined. In this laboratory, a culture of this strain declined in symbiotic effectiveness in standard tests over three years (Gibson *et al.*, 1976); eventually this culture failed to form nodules and a year later it was dead. A four-year field trial to examine persistence of strains (Gibson *et al.*, 1976) produced a number of ineffective reisolates of strain WU290 but it could not be determined whether these were descendents of ineffective mutants in the inoculant or resulted from mutation in the field; their proportion appeared to increase with time but numbers tested were low. In the same trial, low numbers of less effective reisolates of strains WA67 and WU95, but not CC242 or CC2480a, were identified. Examination of the ineffective WU290 culture isolates failed to show any changes in the plasmid profile (C. Boucher and A. Gibson, unpublished) or the hybridization profile using the RtRS DNA probe of Watson and Schofield (1985).

In a brief review of variability in rhizobia, Roughley (1976) suggested that *R. leguminosarum* bv. *trifolii* strains were more prone to producing ineffective mutants than *R. meliloti*, and pointed to apparent dangers of high levels of particular amino acids, as may be found in some yeast extract preparations (see also Skinner *et al.*, 1977). Some embarrassment occurred in Australia in 1963 when a nodule isolate from a plant inoculated with a fully-effective culture of *Rhizobium leguminosarum* bv. *trifolii* TA1 was used directly to provide stock cultures for the inoculant manufacturers. This culture was only 70% effective (Gibson, 1968) and was replaced rapidly. Perhaps the greatest variability recorded has been within the *Bradyrhizobium* sp. strain CB756 (Herridge and Roughley, 1975; Weaver and Wright, 1987), both for colony type and effectiveness. Different cultures of this strain used to provide stock cultures for inoculant manufacturers, showed wide variability in symbiotic effectiveness. Small colony types were more effective than large colony types, and in this context it is interesting that small colony isolates were more active in ex-planta nitrogenase activity than larger colony isolates (A. Gibson, unpublished). Colony dimorphism is a common occurrence among strains of rhizobia, and movement from large to small colony types, and *vice versa*, should not be mistaken for changes in effectiveness (Sylvester-Bradley *et al.*, 1988).

Pankhurst (1977) and others have warned of the dangers of effectiveness loss in natural and induced mutants resistant to various antibiotics.

Field variability

Although detection of variability in laboratory cultures can be difficult, it is comparatively easy relative to the examination of a field population which is usually dependent on plant entrapment, possible unconscious bias in nodule selection and the difficulty of identifying reisolates as the inoculant strain when there is a complex naturally-occurring population of rhizobia. Changes in the level of effectiveness, in contrast to complete loss of effectiveness, were detected in *R. leguminosarum* bv. *trifolii* populations as part of a study of persistence (Gibson *et al.*, 1976) and this was investigated in greater depth. On the other hand, with only one *B. japonicum* strain introduced into Australian soils in recent years, opportunity was taken to examine variability in this strain where it was established in the field; this topic has relevance to recent observations of apparent variability in this species in the U.S.

Rhizobium leguminosarum *bv.* trifolii

In 1974, eight strains of *Rhizobium leguminosarum* bv. *trifolii* were introduced into five field sites in south-eastern Australia in order to determine their competitiveness against naturally-occurring strains (Brockwell *et al.*, 1982). Three years later, nodule reisolates giving a reaction of immunological identity with the inoculant strain were examined for symbiotic effectiveness. For each isolate, 10 plants of *Trifolium subterraneum* cv. Bacchus Marsh (Gibson and Brockwell, 1968) were grown for 28 days with the roots inside test-tubes (Gibson, 1980) (22° root temperature, 22/15° shoot temperature, 14 h daily light period with photon flux density of 300 μmol $m^{-2} s^{-1}$). As standards, 14–30 single colony isolates of lyophilised cultures, from the same batch used prepare the initial inoculants, were also tested for symbiotic effectiveness. All isolates, from field and ampoules, formed effective associations with the host, but, within a strain, significant differences in shoot dry weight were observed between the parental culture isolates (Table 1) and between field isolates (Table 3). Re-examination of isolates from both ends of the effectiveness spectrum confirmed that the observed differences were not a statistical aberration in that, for four of the seven strains examined, the correlation between performance in the two tests was significant at $P < 0.05$ and for two strains, P was 0.10–0.15.

To examine further this variability in two strains, (i) a single colony isolate from the freshly-opened lyophilised culture was used to provide 18 further single colony isolates and (ii) from one nodule on a plant inoculated with

Table 1. Symbiotic effectiveness of single colony isolates derived from lyophilised cultures tested with *Trifolium subterraneum* over 28 days. Correlation coefficient (r) between growth in two tests for (n) high and low effectiveness isolates. (N = 10)

Strain	No. Isolates	Shoot dry weight (mg)		l.s.d.	Comparison		
		Range	$\bar{x}$		n	r	P
WA67	29	43.6–76.1	68.4	13.2	13	0.68	<0.01
WU290	14	37.8–73.9	58.7	14.4	14	0.63	<0.01
WU95	26	41.1–63.4	53.8	12.5	11	0.45	0.10–0.15
CC2480a	30	41.5–70.5	55.0	12.1	8	0.42	0.10–0.15
CC242	27	75.2–105.9	91.7	16.0	11	0.58	<0.05
TA1	17	47.3–70.8	60.3	12.5	9	−0.12	n.s.
UNZ29	19	56.5–83.5	71.2	13.4	4	0.77	0.05
NA30	17	59.1–83.0	73.7	14.9	–	–	–

growth from the single colony isolate, 18 single colony isolations were made. For both strains, there were significant differences between the symbiotic effectiveness of the single colony isolates from both sources (Table 2). For strain CC2480a, but not for strain WA67, the two population means differed significantly, although none was different from that of a mass-culture derived from the heavy growth when the ampoules were first streaked on YMA. Similar results have been reported for *Bradyrhizobium* sp. strains (Weaver and Wright, 1987). In neither case was the cause of lower, or higher, effectiveness determined either genetically or physiologically. However the results indicate that variability for symbiotic characteristics is probably higher than that associated with physiological functions dependent on one or two genes for their expression. They also indicate the need for extreme caution in handling cultures, especially where culture transfer is dependent on single colony isolates.

Table 2. Symbiotic effectiveness (mg shoot dry weight) of 18 single colony isolates derived directly from a single colony isolate or after plant passage, and compared with the original single colony isolate. The host was *Trifolium subterraneum*. (N = 10)

Source		Strain CC2480a	Strain WA67
Original single colony	$\bar{x}$	70.7	81.7
-ex original	$\bar{x}$	77.7	69.3
single colony	Range	(59.2–95.3)**	(44.0–82.5)**
-ex nodule	$\bar{x}$	60.6	73.6
	Range	(48.9–79.5)**	(47.4–88.1)**
Significant $P = 0.05$		60.6 < 77.7	n.s.

**P < 0.01 within group of isolates.

With respect to the field reisolates of *R. leguminosarum* bv. *trifolii*, variable numbers were obtained for each strain at each site, indicating site effects on the competitiveness and persistence of the strains. The recovery of three strains (TA1, NA30, UNZ29) was consistently poor and only strain WA67 was recovered at high frequency from all sites (data not shown). There were significant differences in symbiotic effectiveness between the isolates in 11 of the 19 strain × site treatments with three or more reisolates (Table 3), with strains WA67, WU290 and WU95 showing a higher degree of variability than the others. For seven of the site × strain comparisons, the population mean exceeded significantly the mean of the parental culture isolates. Although this supports the commonly-held thesis that strains of higher effectiveness are likely to be more competitive in nodulation, it should be noted that this thesis does not have

Table 3. Examination of differences within field populations of nodule isolates from up to 5 sites, and the relationship between the population means of field isolates and parental culture (P.C.) single colony isolates

Strain	Differences within field populations	Field population mean compared with parental culture mean
WA67	4/5	1 > P.C.
WU290	2/2	2 > P.C.
WU95	2/3	3 > P.C.
CC2480a	2/4	0
CC242	1/3	1 > P.C.
UNZ29	0/1	0
TA1	0/1	1 < P.C.

universal acceptance. Furthermore, no physiological or genetic basis has been established to explain the greater competitiveness of more effective strains. At the only site where the inoculant strain TA1 was recovered, all four isolates were inferior to the parental culture. This observation, in conjunction with a similar previous observation (Gibson *et al.*, 1975) suggests either that there is a high degree of variability in this strain or that mutation to lower effectiveness is associated with either greater persistence or enhanced competitiveness.

Bradyrhizobium japonicum

Australian commercial inoculants contain only one strain of rhizobia, and since 1966, the commercial soybean inoculant strain has been CB1809, a derivative of USDA136b. Prior to 1966, very limited areas were sown to soybeans inoculated with other strains (mainly in Queensland), and natural nodulation of uninoculated soybeans by rhizobia nodulating native *Glycine* species is rare. This situation is probably unique among soybean-growing countries and provides an excellent opportunity to study variability within introduced field populations and to establish a genetic baseline before other strains are introduced. In 1987, soil was sampled from six sites where CB1809-inoculated soybeans had been sown since 1979. No recoveries were possible where seed was sown prior to 1983. Isolations were made from nodules formed on *G. max* growing in Leonard jars, or on *G. soja* (*G. ussuriensis*) growing in vermiculite in test tubes (Gibson, 1987), following the addition of soil or soil suspensions. From three sites, soil dilutions were plated in agar medium containing bacterial antibiotics, anti-fungal agents and zinc (Schwinghamer and Gault, unpublished) and putative colonies of rhizobia were selected. Symbiotic effectiveness tests involved *G. max* cv. Lincoln growing in pots of perlite and vermiculite, carefully maintained to exclude contamination, under controlled conditions (27/22°, 14 h natural light period) in the Canberra phytotron (Gibson, 1987). At harvest (5-6 weeks) two plants of each treatment were assayed for H_2 evolution (Turner and Gibson, 1980) and shoot dry weight ($n = 9$) determined. Gel immuno-diffusion assays (Schwinghamer and Dudman, 1980) were made using CB1809 antiserum. Numerous isolates and standard laboratory strains of *Bradyrhizobium* sp., *B. japonicum* and *B. lupini* were subjected to restriction fragment length polymorphism (RFLP) analysis, using HindIII restriction endonuclease digested total DNA probed with the RSα DNA probe (Hahn and Hennecke, 1987a).

Colleambally

Fifty eight nodule isolates were made from an acid soil sown to CB1809-inoculated soybeans for three years (1982–4) and immediately following two crops of irrigated rice (1985, 1986). When 28 isolates were inoculated onto *Macroptilium atropurpureum* (Siratro) grown on agar slopes in foil-capped test-tubes (Gibson, 1987) all formed nodules promptly (9-12 days) and with the same low level of symbiotic effectiveness. The same result was observed for 7 cultures of CB1809 derived from lyophilised cultures put down at various intervals from 1971 to 1984 by R.J. Roughley of the Australian Inoculant Research and Control Service (Gosford, N.S.W.) these were representative of the stock cultures provided annually to the Australian inoculant manufacturers. In addition, two antiobiotic resistance mutants of CB1809, kindly provided by Dr Roughley, were included (NA626 Str^rSpc^r, ex-Niftal, and NA627 Str^r, ex-AIRCS) as well as a culture of CB1809 from the CSIRO Canberra Rhizobium Collection. With the exception of NA626, which nodulated slowly and fixed little or no nitrogen, these laboratory cultures behaved similarly to the others. Symbiotic effectiveness tests on all these cultures, using *G. max* cv. Lincoln, showed the 7 AIRCS stock cultures to be similar (Fig. 1) and the range of the means was significantly less than that of field isolates classified as serologically-identical to CB1809, or as serological cross-reactive (major lipopolysaccharide band missing or not identical with that of CB1809). The means for the three populations were not different statistically ($P = 0.05$) and in a subsequent re-examination of the 'poorer' and 'better' isolates, they did not differ from the CB1809 stock cultures ($r = 0.01$ over the two trials). All laboratory cultures and field isolates were Hup^+ in symbiosis.

RFLP analysis of 16 standard *B. japonicum*

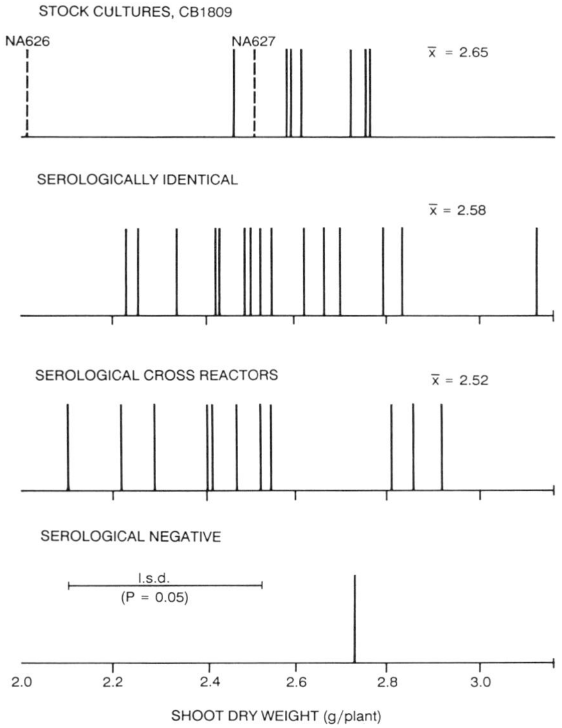

Fig. 1. Distribution of shoot dry weight in groups of laboratory cultures of *Bradyrhizobium japonicum* CB1809, field reisolates serologically-identical to CB1809 and field reisolates cross-reaction, or failing to react, with CB1809 antiserum. (N = 9).

strains revealed 14 unique patterns, with USDA136b (isolated Maryland, 1961), its derivative CB1809 (taken in 1963) and USDA122 (isolated Mississippi, 1960) providing identical patterns. Three *Bradyrhizobium* sp. strains (CB751, CB756, 8A9) and *B. lupini* strain WU425 failed to hybridise with RSα, although *B. lupini* strain 606B gave one faint band. These results confirm the observations of Hahn and Hennecke (1987a) regarding the specificity of RSα, and the value of the probe for distinguishing between strains of *B. japonicum*. Twenty-two field isolates, including 10 that were serological cross-reactive and one serologically-negative isolate (Fig. 1), plus four stock cultures, the CSIRO Collection culture, and strain NA627, all had the same hybridization profile. Strain NA626, the slower-nodulating, less-effective antibiotic-resistant mutant, lacked one fast-moving band. The serological analysis of the 58 isolates revealed 26 giving a reaction of identity with CB1809, 30 cross-reacters and two, both with CB1809-RFLP profiles, were not reactive with the antiserum.

Southern New South Wales

Five soil samples from each of three fields sown to CB1809-inoculated soybeans once only between 1978 and 1984, were used to inoculate *G. max* growing in Leonard Jars, and used to make suspensions to inoculate *G. soja* growing in test-tubes, or YMA plates containing antibiotics. Reisolations were made (18–20) from the *G. max* nodules; all were examined against CB1809 antiserum, 11 or 12 were tested for symbiotic effectiveness, and isolates not serologically identical with CB1809 were subjected to RFLP analysis. The results for the three sites were similar and have been pooled (Table 4). Two of the isolates were less effective than the CB1809 controls, one marginally (4.1 *cf.* 5.1 g shoot dry weight) and the other markedly (1.9 g). This latter isolate did not react with CB1809 antiserum or with the RSα-probe, and was, probably, a strain nodulating native *Glycine* species occurring in the region. Two further reisolates in this category were found among the 20 isolates not tested for effectiveness but found to nodulate *G. max* poorly. All *G. soja* isolates tested were fully effective on *G. max*, and although two were not serologically-identical to CB1809, their RSα-hybridizations were identical. Of 12 putative *B. japonicum* isolates isolated directly from agar, four were identified as CB1809 by serology and DNA hybridization; the remainder were unidentified soil bacteria. Although this is a relatively low recovery rate, the technique does offer promise to select *B. japonicum* CB1809 directly from the soil population, and further improvement could be expected by using DNA colony hybridization (Hodgson and Roberts, 1983). The fact that two of these isolates, from different sites, were marginally lower in symbiotic effectiveness (4.0 and 4.1 g shoot weight) indicates that further examination of field populations by this approach is warranted.

Tamworth

Samples from alkaline soils in northern N.S.W. that had been sown with CB1809-inoculated soybeans 2–5 years previously were used to prepare

Table 4. Symbiotic effectiveness and both serological and RFLP identity to CB1809 for isolations from 3 sites isolated from *G. max*, *G. soja* or directly on YMA-amended medium. Data pooled for the 3 sites

Isolated from	Symbiotic effectiveness[a]			Serology (CB1809)		RFLP (CB1809)		Nodulation (*G. max*)
	ie	e	E	+	–	+	–	
G. max	1	1	33	32	3	2	1	+
	Not tested			18	2	0	2	+
G. soja	0	0	7	5	2	2	0	+
	Not tested			8	3	3	0	+
Soil	0	2	2	4	0	4	0	+
	Not tested			0	8	0	8	–

[a] E = not different from CB1809; e = effective but less than CB1809; ie = low level of nitrogen fixation.

suspensions to inoculate *G. soja* in an MPN plant dilution series. Isolates from the nodulating plants at the lowest dilution were examined against CB1809 antiserum (22/23 positive) and 12, including the serological cross-reacting isolate, were shown by RFLP analysis to be identical to CB1809.

Discussion

These observations, preliminary in that only a limited number of sites have been examined, indicate that there has been little loss of symbiotic effectiveness in field populations of CB1809 at least up to 8 years after introduction to the soil. There has been remarkable stability in the RFLP patterns, indicating that no obvious genomic rearrangement has occurred, at least in the regions associated with the RSα-sequences. However changes in antigenic properties were found, due primarily to alteration, or loss, of the major lipopolysaccharide antigen. These changes were greatest in a soil subjected to continuous irrigation for two periods of four months (rice crops), conditions which may be expected to increase mutation rates due to higher levels of mutagenic or toxic elements (Mn, Al). However these serological changes did not affect symbiotic effectiveness.

These findings appear to be in contrast to the situation in the U.S., especially regarding isolates reacting with antiserum to USDA123. Previously, 20 isolates collected from Iowa soils were classified into 3 groups based on agglutination reactions with USDA123, USDA127 and USDA129 antisera, into 11 groups based on gel immuno-diffusion assays against these antisera, and 18 groups when other characters were considered (Gibson *et al.*, 1971); isolates classified as serotype 129 were homogeneous.

Schmidt *et al.* (1986) using bacteriophage sensitivity, and Sadowsky *et al.* (1987) using various approaches, also showed wide diversity among serogroup 123 field isolates, although the latter authors were able to detect broad groupings in their collection of 20 isolates. In the absence of knowledge about the initial introduction of *B. japonicum* into U.S. soils, it must be assumed that many different serogroup 123 strains were introduced or that there has been a high rate of mutational change in a limited number of introductions. Credence for the latter suggestion is provided by recent preliminary results obtained by Dr Bhuvaneswari in Ohio. Twenty two serogroup 123 isolates (ELISA technique) from crown nodules on 5 *G. max* plants grown in Leonard Jars filled with soil from one field, gave 21 distinct RFLP patterns when total DNA, restricted by *Xho* I nuclease, were probed with RSα. It is very difficult to believe that 21 serologically-similar strains would have been introduced to this soil deliberately or accidentally. The change may only involve genomic rearrangement (symbiotic characters were not examined) but relative to the results obtained with CB1809, it does indicate that some strains are more likely to undergo variation than others. In this context, it is interesting that symbiotic performance with respect to different *G. max* introductions did not always follow the basic groupings (Sadowsky *et al.*, 1987).

RFLP analysis, using different restriction enzymes and different DNA probes has become a popular means to identify strains but caution is necessary. Flores *et al.* (1988) have demonstrated genomic instability within strains of *Rhizobium leguminosarum* bv. *phaseoli*, confirming earlier results from that laboratory (Soberon-Chavez *et al.*, 1986) which also showed that these rearrangements could modify symbiotic performance. Although the basis of genomic rearrangement is not known, Hahn and Hennecke (1987b) suggest that sites of repeated sequences could facilitate recombination, with subsequent rearrangement and possible deletions.

The changes in symbiotic effectiveness found in the populations of clover rhizobia are cause for concern, especially if the mutants become dominant. Symbiotic effectiveness is a broadly-based phenotype, poorly defined in physiological terms, that is dependent on at least 30–40 genes. Alterations within any one of these genes, or others that may not appear to be directly involved in symbiosis, could affect this character. This may explain why variability in effectiveness appears much greater than would be expected for a single gene character. The relationship between these changes and any effect on ecological adaptation is currently unknown but worthy of further study.

In considering genetic variability in field populations, consideration must be given to genetic exchange. While this can be achieved commonly in the laboratory, demonstration of such exchanges in the field is difficult. In our laboratory, we have been looking for isolates with genetic elements of two strains in the field population, using DNA probes specific for chromosomal DNA and symbiotic plasmid DNA. Our results (Schofield *et al.*, 1987; Demezas *et al.*, 1990) suggest that this exchange occurs at a low level but as yet we have been unable to define the conditions under which it occurs. The situation is further complicated by evidence suggesting genetic recombination between resident and transferred plasmids, indicating the need for different approaches to resolving this question.

Variability, be it due to mutation, genome rearrangement or genetic exchange is a fact of life. Further efforts should be directed towards understanding the basis of differences between strains in their propensity to vary or to exchange genetic information. Inoculant strains should be chosen for minimum variability, not only to achieve stable soil populations but to minimise potential disasters in inoculant production. Through understanding genetic exchange, and the factors involved, it may even be possible to use this knowledge to introduce new genetic material into adapted, ecologically-stable, populations, or to minimise genetic exchange. Without this knowledge, opponents to the introduction of microbes, especially those developed through genetic engineering, can be justifiably critical of efforts to utilise the potential of modern technology in a beneficial manner.

References

Brockwell J, Gault R R, Zorin M and Roberts M J 1982 Effects of environmental variables on the competition between inoculum strains and naturalised populations of *Rhizobium trifolii* for nodulation of *Trifolium subterraneum* L. and on rhizobia persistence in the soil. Aust. J. Agric. Res. 331, 803–815.

Demezas D H, Reardon T B, Watson J M and Gibson A H 1990 Chromosome and symbiotic plasmid diversity within a naturally-occurring population of clover rhizobia. *In* The Rhizosphere and Plant Growth. Eds. D L Leister and P B Cregan. p 106. Kluwer Academic Publishers, Dordrecht, The Netherlands.

Flores M, Gonzalez V, Pardo M A, Leija A, Martinez E, Romero D, Pinero D, Davila G and Palacios R 1988 Genomic instability in *Rhizobium phaseoli*. J. Bacteriol. 170, 1191–1196.

Gibson A H 1968 Nodulation failure of *Trifolium subterraneum* L. cv. Woogenellup (syn. Marrar.). Aust. J. Agric. Res. 19, 907–918.

Gibson A H 1980 Methods for legumes in glasshouses and controlled environment cabinets. *In* Methods for Evaluating Biological Nitrogen Fixation. Ed. F J Bergersen. pp 139–184. Wiley-Interscience, Chichester, UK.

Gibson A H 1987 Evaluation of nitrogen fixation by legumes in the greenhouse and growth chamber. *In* Symbiotic Nitrogen Fixation Technology. Ed. G H Elkan, pp 321–369. Marcel Dekker, New York, USA.

Gibson A H and Brockwell J 1968 Symbiotic characteristics of subspecies of *Trifolium subterraneum* L. Aust. J. Agric. Res. 19, 891–905.

Gibson A H, Curnow B C, Bergersen F J, Brockwell J and Robinson A C 1975 Studies of field populations of *Rhizobium*: effectiveness of strains of *Rhizobium trifolii* associated with *Trifolium subterraneum* L. pastures in south-eastern Australia. Soil Biol. Biochem. 7, 95–102.

Gibson A H, Date R A, Ireland J A and Brockwell J 1976 A comparison of competitiveness and persistence amongst five strains of *Rhizobium trifolii*. Soil Biol. Biochem. 8, 395–401.

Gibson A H, Dudman W F, Weaver R W, Horton J C and Anderson I C 1971 Variation within serogroup 123 of *Rhizobium japonicum*. *In* Biological Nitrogen Fixation in Natural and Ecological Habitats. Eds. E G Mulder and T A Lie. pp 33–37. Martinus Nijhoff, The Hague, Netherlands.

Hahn M and Hennecke H 1987a Conservation of a symbiotic DNA region in soybean root nodule bacteria. Appl. Environ. Microbiol. 53, 2252–2255.

Hahn M and Hennecke H 1987b Mapping of a *Bradyrhizobium japonicum* DNA region carrying genes for symbiosis and an asymmetric accumulation of reiterated sequences. Appl. Environ. Microbiol. 53, 2247–2252.

Herridge D F and Roughley R J 1975 Variation in colony characteristics and symbiotic effectiveness of *Rhizobium*. J. Appl. Bacteriol. 38, 19–27.

Hodgson A L M and Roberts W P 1983 DNA colony hybridization to identify *Rhizobium strains*. J. Gen. Microbiol. 129, 207–212.

Kuykendall L D and Elkan G H 1976 *Rhizobium japonicum* derivatives differing in nitrogen-fixing efficiency and carbohydrate utilisation. Appl. Environ. Microbiol. 32, 511–519.

Labandera C A and Vincent J M 1975 Loss of symbiotic capacity in commercially useful strains of *Rhizobium trifolii*. J. Appl. Bacteriol. 39, 209–211.

Mathis J N, Barbour W M, Miller T B, Israel D W and Elkan G H 1986 Characterisation of a mannitol-utilizing, nitrogen-fixing *Bradyrhizobium japonicum* USDA 110 derivative. Appl. Environ. Microbiol. 52, 81–85.

Pankhurst C E 1977 Symbiotic effectiveness of antibiotic resistant mutants of fast- and slow-growing strains of *Rhizobium* nodulating *Lotus* species. Can. J. Microbiol. 23, 1026–1033.

Roughley R J 1976 The production of high quality inoculants and their contribution to legume yield. *In* Symbiotic Nitrogen Fixation in Plants. Ed. P S Nutman. pp 125–136. Cambridge University Press, Cambridge, UK.

Sadowsky M J, Tully R E, Cregan P B and Keyser H H 1987 Genetic diversity in *Bradyrhizobium japonicum* serogroup 123 and its relation to genotype-specific nodulation of soybean. Appl. Environ. Microbiol. 53, 2624–2630.

Schmidt E L, Zidwick M J and Abebe H M 1986 *Bradyrhizobium japonicum* serocluster 123 and diversity among member isolates. Appl. Environ. Microbiol. 51, 1212–1215.

Schofield P R, Gibson A H, Dudman W F and Watson J M 1987 Evidence for genetic exchange and recombination of *Rhizobium* symbiotic plasmids in a soil population. Appl. Environ. Microbiol. 53, 2942–2947.

Schwinghamer E A and Dudman W F 1980 Methods of identifying strains of diazotrophs. *In* Methods for Evaluating Biological Nitrogen Fixation. Ed. F J Bergersen. pp 337–365. Wiley-Interscience, Chichester, UK.

Skinner F A, Roughley R J and Chandler M R 1977 Effect of yeast extract concentration on viability and cell distortion in *Rhizobium* spp. J. Appl. Bacteriol. 43, 287–297.

Soberon-Chavez G, Najera R, Olivera H and Segovia L 1986 Genetic rearrangements of a *Rhizobium phaseoli* symbiotic plasmid. J. Bacteriol. 167, 487–491.

Sylvester-Bradley R, Thornton P and Jones P 1988 Colony dimorphism in *Bradyrhizobium* strains. Appl. Environ. Microbiol. 54, 1033–1038.

Turner G L and Gibson A H 1980 Measurement of nitrogen fixation by indirect means. *In* Methods for Evaluating Biological Nitrogen Fixation. Ed. F J Bergersen. pp 111–138. Wiley-Interscience, Chichester, UK.

Watson J M and Schofield P R 1985 Species-specific, symbiotic plasmid-located repeated DNA sequences in *Rhizobium trifolii*. Mol. Gen. Genet. 199, 279–289.

Weaver R W and Wright S F 1987 Variability in effectiveness of rhizobia during culture and in nodules. Appl. Environ. Microbiol. 53, 2972–2974.

D. L. Keister and P. B. Cregan (Eds.), The rhizosphere and plant growth, 149–154.

PLSO BARC15

Rhizosphere flavonoids released by alfalfa

D.A. PHILLIPS[1], C.A. MAXWELL[1], U.A. HARTWIG[1], C.M. JOSEPH[1] and J. WERY[2]
[1]*Department of Agronomy and Range Science, University of California, Davis, CA 95616, USA.*
[2]*Laboratoire d'Agronomie, ENSA-INRA, Place Viala, F-34060 Montpellier Cedex, France*

Key words: alfalfa, dinitrogen fixation, exudates, flavonoids, *Medicago sativa* L., *nod* genes, root nodules

Abstract

Flavonoids released from seeds and roots of alfalfa (*Medicago sativa* L.) are important chemical components of the seedling rhizosphere. Some flavonoids induce transcription of nodulation (*nod*) genes in *Rhizobium meliloti* as a first step toward root nodule formation and symbiotic N_2 fixation, but many others with unknown functions also are released from imbibing seeds. Three flavonoid *nod* inducers exuded by alfalfa roots and others released from imbibing seeds have been identified. Interactions among those compounds indicate that the chemical domains created by seed and roots govern expression of rhizobial *nod* genes and, perhaps, other microbial activities in the soil. As modern tools increase our knowledge of rhizosphere chemistry, the specific signals that control interactions between plant roots and microbes may become management tools for achieving agricultural objectives.

Introduction

Plant roots release a diverse mixture of organic molecules into the rhizosphere (Rovira, 1969). Flavonoids were first reported in root exudates by Lundegardh and Stenlid (1944), but exudation rates of specific flavonoids were calculated much later (D'Arcy-Lameta, 1986). Functional roles of these rhizosphere compounds also have been defined only recently. Those effects include their capacity to induce transcription of nodulation (*nod*) genes in Rhizobium and Bradyrhizobium bacteria (see reviews: Rolfe, 1988; Long, 1989) and their positive chemotactic effects on rhizobia (Aguilar *et al.*, 1988; Armitage *et al.*, 1988; Caetano-Anolles *et al.*, 1988). Thus flavonoids are important plant signals to microbes that form root nodules and later reduce N_2.

Assessing the importance of flavonoids in the rhizosphere is difficult because most of these compounds have been studied in plant extracts rather than in root exudates or seed rinses. Thus it is not known if the 20 flavonoids identified in Medicago seed extracts (Gehring and Geiger, 1980) also are released into the soil during imbibition and germination. The importance of knowing which flavonoids are released from plant roots is emphasized by the observation that some nonleguminous plants contain, but do not exude, factors that induce *nod* genes in *R. meliloti* (Yelton *et al.*, 1987). Thus processes governing the release of flavonoids into the rhizosphere may be one control over root nodulation. The purpose of this report is to indicate how alfalfa seedlings create a specific chemical environment in the rhizosphere by releasing flavonoids from seeds and roots. Some of these molecules induce *nod* genes in *R. meliloti*, but other effects may remain to be discovered.

Flavonoids and *nod* induction in alfalfa

Background

Mulligan and Long (1985) found that a factor from alfalfa plants induced transcription of *nodABC* in *R. meliloti*, and Peters *et al.* (1986) identified one active compound as 3′,4′,5,7-tetrahydroxyflavone, a flavonoid with the trivial name luteolin. Those workers characterized luteolin by its capacity to induce β-galactosidase from an *Escherichia coli lacZ* gene fused to *nodABC* on plasmid pRmM57 in *R. meliloti* (Mulligan and Long, 1985). Parallel discoveries identified flavonoids active in other leguminous systems (Table 1). The luteolin-induced products of *nodABC* are required for root hair curling and cortical cell divisions which lead to root nodule formation, but 10^{-9} *M* luteolin also induces a positive chemotaxis in *R. meliloti* (Caetano-Anolles *et al.*, 1988). Thus flavonoids apparently function in the rhizosphere both as attractants and as signals which incite the actual rhizobial infection.

How flavonoids induce *nod* transcription is unknown, but several lines of evidence support the concept that the *nodD* gene product (NodD), which is required for flavonoid induction, is a DNA-binding transcriptional activator (Long, 1989). Whether the flavonoid binds directly to NodD or affects it through other factors is not known. All rhizobia thus far examined have *nodD*, and *R. meliloti* has three functional forms, *nodD1*, *nodD2* and *nodD3* (Honma and Ausubel, 1987), that interact differently with seed rinses from various host legumes (Györgypal *et al.*, 1988). Identifying flavonoids that interact with specific *nodD* alleles will help clarify the processes leading to nodulation.

Structures of flavonoid nod *inducers from alfalfa*

Peters *et al.* (1986) identified luteolin as a *nod* inducer in alfalfa seed extracts. Other undefined *nod* inducers also were detected in seed extracts by those workers, but root exudates were not examined. During the past year we have purified *nod* inducers released from 'Moapa 69' alfalfa seeds during the first 4 hr of imbibition and from root exudates of 72-h-old seedlings. Compounds were characterized by UV-visible absorbance, ^{1}H-nuclear magnetic resonance, and mass spectroscopy, and their *nod*-inducing capacities were assessed as β-galactosidase production from a *nodC-lacZ* fusion on plasmid pRmM57 in *R. meliloti*.

Large quantities of *nod*-inducing flavonoids are released from imbibing alfalfa seeds. The *nod*-inducing activity in unfractionated seed eluate released during the first 24 h of imbibition is about four-fold greater than that exuded from roots between days 1 and 8 (Fig. 1). Luteolin and various methoxylated derivatives are active components of the seed exudate, but large amounts of the more soluble luteolin-7-0-glucoside also are released (Hartwig, unpublished data). The hydrophobic luteolin aglycone probably moves poorly in soil, but the more

Table 1. Flavonoid *nod* inducers identified from legumes. Compounds were characterized by their capacity to induce transcription of *nod* genes in rhizobia that form root nodules on the indicated legumes

Source	Compound(s)	Reference
Alfalfa seed extract	luteolin (3′,4′,5,7-tetrahydroxyflavone)	Peter *et al.* (1986)
Alfalfa root exudate	4,4′-Dihydroxy-2′-methoxychalcone 4′,7-Dihydroxyflavanone 4′,7-Dihydroxyflavone	Maxwell *et. al.* (1989)
Alfalfa seed rinse	luteolin chrysoeriol (3′-methoxyluteolin)	Hartwig *et al.* (1990)
Pea seed rinse	apigenin-7-0-glucoside (4′,5,7-trihydroxyflavone-7-0-glucoside) eriodictyol (3′,4′,5,7-tetrahydroxyflavanone)	Firmin *et al.* (1986)
Vetch root exudate	3,3′,5,7-tetrahydroxy-4′-methoxyflavanone	Zaat *et al.* (1988)
White clover seedling extract	4′,7-dihydroxyflavone geraldone (4′,7-dihydroxy-3′-methoxyflavone) 4′-hydroxy-7-methoxyflavone	Redmond *et al.* (1986)
Soybean seedling extract	genistein (4′,5,7-trihydroxyisoflavone) daidzein (4′,7-dihydroxyisoflavone)	Kosslak *et al.* (1987)

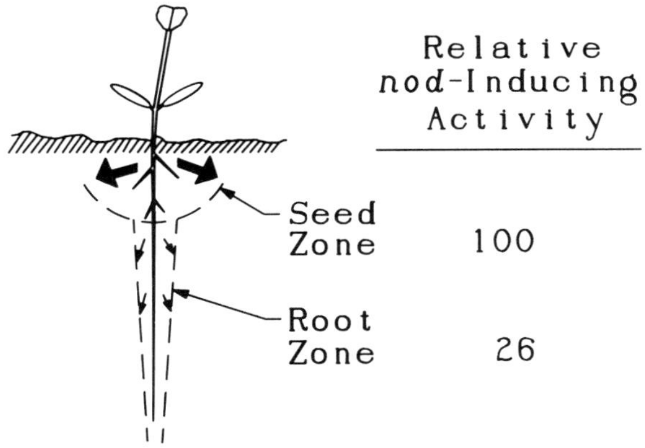

Fig. 1. Relative amounts of *nod*-inducing activity in crude, unpurified alfalfa seed rinse (0 to 4 hr of imbibition) and root exudates (1 to 8 days). Activities are based on β-galactosidase induction from a *nodC-lacZ* fusion on pRmM57 in *R. meliloti*.

water soluble luteolin-7-0-glucoside may enlarge the seed zone by diffusion. Most glucosides are degraded in soils, and this molecule probably releases luteolin quickly. Thus luteolin may influence rhizobia before a rhizosphere develops around an emerging radical.

Alfalfa roots do not release luteolin, but they do exude three other *nod*-inducing flavonoids (Maxwell *et al.*, 1989). The strongest *nod* inducer, 4,4′-dihydroxy-2′-methoxychalcone, produces a half-maximum *nod* induction at a much lower concentration than luteolin (Table 2). No chalcone has been identified previously as a natural inducer in a Rhizobium-legume symbiosis, although this molecule was isolated previously from pea (Carlson and Dolphin, 1982). Alfalfa roots also exude 4′,7-dihydroxyflavone and 4′,7-dihydroxyflavanone, which are *nod*-inducers in *R. leguminosarum* biov. *trifolii* (Djordjevic *et al.*, 1987).

Amounts and interactions of nod *inducers*

Assessing quantities of *nod* inducers released from alfalfa is complicated by the fact that unfractionated root exudate and seed rinses contain both weak and strong *nod* inducers. Djordjevic *et al.* (1987) showed that activity of the natural clover inducer 4′,7-dihydroxyflavone was decreased in the presence of the synthetic weak inducers chrysin (5,7-dihydroxyflavone) and kaempferol (3,4′,5,7-tetrahydroxyflavone). Tests

Table 2. Relative activties of *nod*-inducing flavonoids released from alfalfa seeds and roots. Flavonoids were assayed for their capacity to induce β-galactosidase activity from a *nodC-lacZ* fusion on pRmM57 in *R. meliloti* (Maxwell *et al.*, 1989)

Compound	Half maximum induction	
	Concentration (n*M*)	Relative activity (%)
Luteolin	15	83
4′,7-Dihydroxyflavanone	10	18
4′,7-Dihydroxyflavone	20	56
4,4′-Dihydroxy-2′-methoxychalcone	1	100

with natural *nod* inducers in the Rhizobium-alfalfa system show that activities of the strong inducers luteolin and 4,4′-dihydroxy-2′-methoxychalcone are decreased significantly by the weaker inducers 4′,7-dihydroxyflavone and 4′,7-dihydroxyflavanone (Hartwig *et al.*, 1989). This fact suggests that these two weaker inducers, which are significant components of alfalfa seedling root exudate but not seed rinses (Table 1), attenuate the strong nodulation signal produced in the soil by luteolin from imbibing seeds (Fig. 1).

As a first step toward understanding how alfalfa *nod* inducers interact in the rhizosphere, mixtures of 4,4′-dihydroxy-2′-methoxychalcone, 4′-7-dihydroxyflavone, and 4′,7-dihydroxyflavanone (1:2.5:1) were tested for *nod*-inducing activity. The ratio of inducers was based on quantities of these three compounds which were exuded into a hydroponic medium by 72-h-old alfalfa roots (Maxwell *et al.*, 1989). Although it is impossible to calculate the actual concentration of these compounds present in a normal rhizosphere or on the root surface, the ratio of the three compounds may remain constant. Different concentrations of these flavonoids in 1:2.5:1 proportions consistently induced a lower level of *nod* transcription than 4,4′-dihydroxy-2′-methoxychalcone did alone (Hartwig *et al.*, 1989). However, when this artificial ‘root exudate’ was combined with a non-inductive concentration of luteolin from the seed rinse, a syngergistic increase in *nod* induction occurred (Table 3). This observation suggests trace amounts of luteolin from the seed can have major effects on root nodulation in alfalfa. One hypothesis that may explain this

Table 3. Interactions among *nod*-inducing flavonoids released by alfalfa seeds and roots. Treatments B, C, and D were defined by the relative amounts of the specified flavonoids measured in alfalfa root exudate: 1:2.5:1, B:C:D. The *nod*-inducing capacity of all treatments was assessed as β-galactosidase activity produced form a *nodC-lacZ* fusion on pRmM57 in *R. meliloti* (Hartwig *et al.*, 1989)

Treatment	Flavonoid added	β-Galactosidase activity (U)
A.	None	6
B.	2.1 n*M* 4,4′-dihydroxy-2′-methoxychalcone	70
C.	5.25 n*M* 4′,7-dihydroxyflavone	13
D.	2.1 n*M* 4′,7-dihydroxyflavanone	4
E.	2.1 n*M* luteolin	4
F.	'Root exudate' (B + C + D)	26
G.	'Root exudate' + 2.1 n*M* luteolin (F + E)	63
	LSD 0.01	8

phenomenon is that NodD1, NodD2, and/or NodD3 function synergistically while interacting separately with luteolin and the *nod* inducers in root exudates.

Synthesis of alfalfa nod *inducers*

Flavonoid metabolism begins as products of phenylpropanoid metabolism combine with three molecules of malonyl-CoA (Beggs *et al.*, 1986; Hahlbrock and Grisebach, 1979). Organs where alfalfa *nod*-inducers are synthesized have not been identified, but flavonoids released from alfalfa seeds and roots probably are formed at different sites. Possible metabolic relationships among the *nod* inducers released from roots can be inferred from pathways known in other organisms (Fig. 2). If enzymes responsible for the reactions in Figure 2 are present in alfalfa roots, then alfalfa can form both strong and weak *nod* inducers from 2′,4,4′-trihydroxychalcone. Understanding those reactions may clarify how alfalfa controls release of *nod* inducers into the rhizosphere. Glycosylation of precursors, intermediates, or final products may offer other control points for the plant and could complicate future studies.

Importance of nod *inducers for nodulation and* N_2 *fixation*

Extrapolating from the molecular level, where flavonoid *nod* inducers function, to the organis-

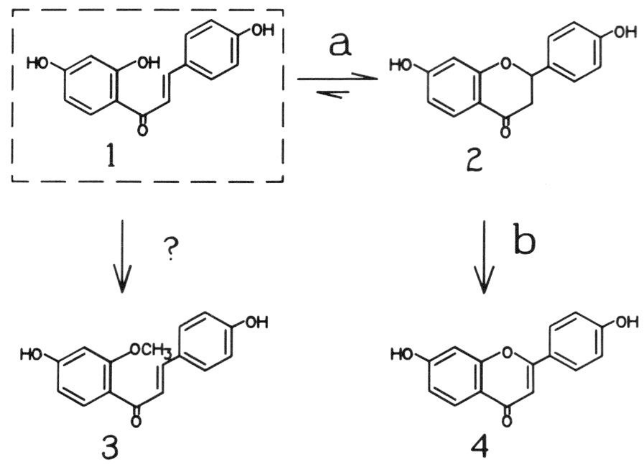

Fig. 2. Possible metabolic relationships among three *nod*-inducing flavonoids isolated from alfalfa root exudates. Compounds 2 (4′,7-dihydroxyflavanone), 3 (4,4′-dihydroxy-2′-methoxychalcone), and 4 (4′,7-dihydroxyflavone) were characterized as weak, strong, and moderate *nod* inducers, respectively. Compound 1 (2′,4,4′-trihydroxychalcone) was not detected in root exudate. Enzymes known in other systems, which may function in alfalfa include a, chalcone isomerase, and b, flavone synthase. No methyltransferase capable of converting compound 1 to 3 has been reported.

mic level, where root nodules fix N_2, requires care. At the molecular level *nod* inducers operate as on-off switches controlling gene transcription in each bacterial cell. If a seedling releases only enough *nod* inducers to prime a few rhizobial cells for nodulation, that quantity of *nod* inducers probably will not produce optimum nodulation and N_2 fixation. Evidence supporting this conclusion is found in numerous classical reports which indicate that increasing the number of effective rhizobial cells promotes root nodulation and N_2 fixation in N-deficient soils (Vincent, 1974). In other words, inducing the rhizobial cell is a molecular event, but forming an optimum number of nodules is a population phenomenon that requires a large number of induced bacteria. Thus it is not surprising Kapulnik *et al.* (1987) observed that adding luteolin to an alfalfa rhizosphere increased root nodule number and N_2 fixation. A different number of rhizobial cells in the rhizosphere or a cultivar that released more natural *nod* inducers could have produced a different result. Kapulnik *et al.* (1987) used that reasoning to suggest that additional production of flavonoid *nod*-inducers by a cultivar selected for enhanced N_2 fixation was partially responsible for increased nodulation and more N_2 fixation.

Other rhizosphere flavonoids

Not all flavonoids released from alfalfa are *nod* inducers. We find that spectral characteristics (230 to 400 nm) of 25 compounds released from imbibing Moapa 69 alfalfa seeds suggest that at least 20 are flavonoids, but only five show *nod*-inducing activity. A few compounds in seed exudates may be inactive, conjugated forms of *nod* inducers, but many others, including glycosides of quercetin (3,3′,4′,5,7-pentahydroxyflavone) and kaempferol, which do not induce *nod* genes (Peters and Long, 1988), also are present. Some of these compounds may protect seeds from insect or microbial predation by inhibiting various enzymes (Harborne, 1979), but no such effects have been defined as specifically as those of the *nod* inducers.

One challenge for rhizosphere biologists is to identify specific functional roles for seed flavonoids. Because many of the compounds released from seeds are glycosylated, they probably move short distances in moist soil and contribute to the biochemical diversity of seedling rhizospheres. The model of *nod*-inducing flavonoids, suggests by analogy that other similar molecules may be required for mycorrhizal associations. If rhizobia have adapted a general mechanism for transducing the *nod*-inducing signals, then that same system may be coupled to important processes in other soil microbes. In that case, plants may have evolved the capacity to release specific flavonoids which either promote or inhibit those microbial processes. Because so many plant-microbe interactions have both positive and negative effects on plant productivity, rhizosphere biologists can make major contributions by defining how procaryotes and eucaryotes interact in the soil. As modern analytical tools increase our ability to identify and understand chemical signals between different organisms, we may be able to modify those signals and use them as management tools for achieving agricultural objectives.

Acknowledgements

This work was supported in part by USDA CRGO grant 87-CRCR-1-2552 and BARD grant IS-1348-87. Awards from the Swiss National Research Foundation to UAH and from NATO to JW were appreciated.

References

Aguilar J M M, Ashby A M, Richards A J M, Loake G J, Watson M D and Shaw C H 1988 Chemotaxis of *Rhizobium leguminosarum* biovar *phaseoli* towards flavonoid inducers of the symbiotic nodulation genes. J. Gen. Microbiol. 134, 2741–2746.

Armitage J P, Gallagher A and Johnston A W B 1988 Comparison of the chemotactic behaviour of *Rhizobium leguminosarum* with and without the nodulation plasmid. Mol. Microbiol. 2, 743–748.

Beggs C J, Wellmann E and Griesebach H 1986 Photocontrol of flavonoid biosynthesis. *In* Photomorphogenesis in Plants. Eds. R E Kendrick and G H M Kronenberg. pp 467–499. Martinus Nijhoff Publishers, Dordrecht.

Caetono-Anolles G, Crist-Estes D K and Bauer, W D 1988 Chemotaxis of *Rhizobium meliloti* to the plant flavone luteolin requires functional nodulation genes. J. Bacteriol. 170, 3164–3169.

Carlson R E and Dolphin D H 1982 *Pisum sativum* stress metabolites: two cinnamylphenols and a 2′-methoxychalcone. Phytochemistry 21, 1733–1736.

D'Arcy-Lameta A 1986 Study of soybean and lentil root exudates II. Identification of some polyphenolic compounds, relation with plantlet physiology. Plant and Soil 92, 113–123.

Djordjevic M A, Redmond J W, Batley M and Rolfe B 1987 Clovers secrete specific phenolic compounds which either stimulate or repress *nod* gene expression in *Rhizobium trifolii*. EMBO J. 6, 1173–1179.

Firmin J L, Wilson K E, Rossen L and Johnston A W B 1986 Flavonoid activation of nodulation genes in *Rhizobium* reversed by other compounds present in plants. Nature 324, 90–92.

Gehring E and Geiger H 1980 Die Flavonoide der Samen von *Medicago* × *varia* Martyn c.v. Cardinal (Fabaceae). Z. Naturforsch. 35c, 380–383.

Györgypal Z, Iyer N and Kondorosi A 1988 Three regulatory *nodD* alleles of diverged flavonoid-specificity are involved in host-dependent nodulation by *Rhizobium meliloti*. Mol. Gen. Genet. 212, 85–92.

Hahlbrock K and Grisebach H 1979 Enzymic controls in the biosynthesis of lignin and flavonoids. Annu. Rev. Plant Physiol. 30, 105–130.

Harborne J B 1979 Flavonoid pigments. *In* Herbivores II. The Chemical Participants. Eds. G A Rosenthal and D H Janzen. pp 619–655. Academic Press, New York.

Hartwig U A, Maxwell C A, Joseph C M and Phillips D A

1989 Interactions among flavonoid *nod* inducers released from alfalfa seeds and roots. Plant Physiology 91, 1138–1142.

Hartwig U A, Maxwell C A, Joseph C M and Phillips D A 1990 Chrysoeriol and luteolin released from alfalfa seeds induce *nod* genes in *Rhizobium meliloti*. Plant Physiology 92, 116–122.

Honma M A and Ausubel F M 1987 *Rhizobium meliloti* has three functional copies of the *nodD* symbiotic regulatory gene. Proc. Natl. Acad. Sci., USA 84, 8558–8562.

Kapulnik Y, Joseph C M and Phillips D A 1987 Flavone limitations to root nodulation and symbiotic nitrogen fixation in alfalfa. Plant Physiol. 84, 1193–1196.

Kosslak R M, Bookland R, Barkei J, Paaren H E and Appelbaum E R 1987 Induction of *Bradyrhizobium japonicum* common *nod* genes by isoflavones isolated from *Glycine max*. Proc. Natl. Acad. Sci., USA 84, 7428–7432.

Long S R 1989 Rhizobium-legume nodulation: Life together in the underground. Cell 56, 203–214.

Lundegardh H and Stenlid G 1944 On the exudation of nucleotides and flavanone from living roots. Ark. Bot. 31A (10), 1–27.

Maxwell C A, Hartwig U A, Joseph C M and Phillips D A 1989 A chalcone and two related flavonoids released from alfalfa roots induce *nod* genes in *Rhizobium meliloti*. Plant Physiol. 91, 842–847.

Mulligan J T and Long S R 1985 Induction of *Rhizobium meliloti nodC* expression by plant exudate requires *nodD*. Proc. Natl. Acad. Sci. USA 82, 6609–6613.

Peters N K, Frost J W and Long S R 1986 A plant flavone, luteolin, induces expression of *Rhizobium meliloti* nodulation genes. Science 233, 977–980.

Peters N K and Long S R 1988 Alfalfa root exudates and compounds which promote or inhibit induction of *Rhizobium meliloti* nodulation genes. Plant Physiol. 88, 396–400.

Redmond J W, Batley M, Djordjevic M A, Innes R W, Kuempel P L and Rolfe B G 1986 Flavones induce expression of nodulation genes in *Rhizobium*. Nature 323, 632–635.

Rolfe B G 1988 Flavones and isoflavones as inducing substances of legume nodulation. BioFactors 1, 3–10.

Rovira A D 1969 Plant root exudates. Bot. Rev. 35, 35–57.

Vincent J M 1974 Root-nodule symbioses with *Rhizobium*. *In* The Biology of Nitrogen Fixation. Ed. A Quispel. pp. 265–341. North-Holland Publ. Co., Amsterdam.

Yelton M M, Mulligan J T and Long S R 1987 Expression of *Rhizobium meliloti nod* genes in *Rhizobium* and *Agrobacterium* backgrounds. J. Bacteriol. 169, 3094–3098.

Zaat S, Schripsema J, Wijffelman C, Tak T, van Brussel A and Lugtenberg B 1988 Analysis of *Rhizobium leguminosarum nod* gene inducers in root exudate, and comparison to inducers present in root extract of *Vicia sativa*. *In* Nitrogen Fixation: Hundred Years After. Eds. H Bothe and F J de Bruijn. Gustav Fischer, Stuttgart. 469 p.

D. L. Keister and P. B. Cregan (Eds.), The rhizosphere and plant growth, 155–162.

PLSO BARC16

Chemotaxis, induced gene expression and competitiveness in the rhizosphere

WOLFGANG D. BAUER[1] and GUSTAVO CAETANO-ANOLLÉS[2]
[1]*Agronomy Department, Ohio State University, Columbus, OH 43210, USA and* [2]*Plant Molecular Genetics, University of Tennessee, Knoxville, TN 37916, USA*

Key words: chemoattractants, luteolin, *nod* gene, nodulation, phenolics, Rhizobium, symbiosis

Abstract

Rhizobia are soil bacteria which symbiotically infect legume roots and generate nodules in which they fix atmospheric nitrogen for the plant in exchange for photosynthetically fixed carbon. A crucial aspect of signal exchange between these symbionts is the secretion of phenolic compounds by the host root which induce *nod*ulation gene expression in the bacteria. Stimulation of *nod* gene expression by host phenolics is required for nodule formation, is biochemically specific at 10^{-6} *M*, and is mediated by *nodD*. We and others have shown that rhizobia display chemotaxis to 10^{-9} *M* of the same phenolic compounds. Chemotaxis to inducer phenolics is selectively reduced or abolished by mutations in certain *nod* genes governing nodulation efficiency or host specificity. Conversely, mutations in rhizobia that affect general motility or chemotaxis have substantial effects on nodulation efficiency and competitiveness. These findings suggest that microbes entering the rhizosphere environment may utilize minor, non-nutrient components in root exudates as signals to guide their movement towards the root surface and elicit changes in gene expression appropriate to this environment.

Introduction

Chemically and physically, the rhizosphere is perhaps the most complex and changeable of all environments (Curl and Truelove, 1986). In this kind of environment, any microbe which is chemotactic, which can move to optimize its chemical inputs as things change, could have an important competitive edge. Unfortunately, there have been few studies to address the physiological ecology of microbial chemotaxis in the rhizosphere (*cf* Ames and Bergman, 1981; DeWeger *et al.*, 1987; El Haloui *et al.*, 1986; Griffin and Quail, 1968; Howie *et al.*, 1987; Hunter and Fahring, 1980; Madsen and Alexander, 1982; Scher *et al.*, 1988). A great deal has been learned in recent years about the cellular workings and machinery of chemotaxis in *E. coli* and Salmonella. Over 60 genes have been characterized, encoding signal detectors, transducers, wires, switches, motors, shafts, propellors, and system integrators (Koshland, 1981; McNab, 1987; Wolfe *et al.*, 1987). However, almost nothing is known about this machinery in any rhizobacterial species, and essentially nothing is known about the ecological role and functioning of this machinery in any bacteria, including *E. coli* (Chet and Mitchell, 1976; McNab, 1987). So, while chemotaxis is a common feature among soil bacteria (Bergey 1986), it demands a significant fraction of their total metabolic/genetic resources (McNab, 1987), and may be crucial to rhizosphere dynamics, this review describes an infant field, with all the dangers of error and oversimplification which that entails.

Presumably the adaptive value of chemotaxis involves movement towards or away from certain substances in the environment. But for bacteria that live in soil, what are the substances to seek and avoid? At what concentrations do they need to be detected? What niches are occupied by chemotactic bacteria that cannot be occupied successfully by non-motile species? How important is chemotaxis to successful colonization of the rhizosphere? What substances from the root leak through an established surface microflora in high enough concentrations to elicit taxis? How are chemotactic responses to root substances integrated with changes in gene expression appropriate to the rhizosphere environment? Is there appreciable movement of bacteria on the root surface within the mucigel layer? What controls exist to regulate flagellar synthesis and activity levels in different soil and root environments? In our view, answers to such questions are fundamental to understanding the dynamics of microbial competition in the soil and rhizosphere.

Chemotaxis and the lifestyle of rhizobia

This review is concerned with recent studies on the role of chemotaxis in the symbiotic lifestyle of rhizobia. Rhizobia thrive in three very different environments: bulk soil, rhizosphere and host issue. They survive in fallow soils for many years, where they compete against diverse chemoorganotrophic soil bacteria and maintain population densities between 10^2 and 10^5 cells/g soil (Bowen and Rovira, 1976; Demezas and Bottomley, 1986; Rovira, 1961; Schmidt, 1979). Rhizobia appear to be excellent scavengers of organic compounds, and may depend largely on materials that escape utilization by faster-growing bacteria. For example, some rhizobia grow faster on 10^{-7} *M* succinate than certain strains of Enterobacter and Pseudomonas, although not as rapidly as these strains at higher concentrations (Humbeck *et al.*, 1985). The scavenging ability of rhizobia is also indicated by the fact that they (and other soil bacteria) can survive for long periods, and multiply extensively in distilled water (Crist *et al.*, 1984; Humbeck *et al.*, 1985). While it is known that rhizobia can move chemotactically through samples of autoclaved soils (Soby and Bergman, 1983) and move to micromolar concentrations of various organic and aromatic acids (Parke *et al.*, 1985), the role of chemotaxis in long-term survival in the soil does not appear to have been investigated for rhizobia or any other soil bacterium.

Rhizobia also are successful competitors in the rhizosphere of both host and non-host plants, achieving population densities 5 to 100 times higher than in bulk soil (*e.g.* Mowad *et al.*, 1984; Schmidt, 1979). They have a slower growth rate than dominant species like Pseudomonas, but nonetheless can establish and maintain themselves at about 2% to 20% of the total number of bacteria in the rhizosphere of their host (Mowad *et al.*, 1984). Recent studies with Pseudomonas (DeWeger *et al.*, 1987; Howie *et al.*, 1987) indicate that chemotaxis can be either very important or unimportant in rhizosphere colonization, depending on circumstances. The importance of chemotaxis to rhizosphere colonization by rhizobia has never been examined.

Although very respectable soil saprophytes and rhizosphere colonizers, rhizobia have received most attention for their ability to establish intimate symbiotic associations with roots of legumes (Bauer, 1981; Djordjevic *et al.*, 1987; Rolfe and Gresshoff, 1988). During the infection process, the first visible events are movement of rhizobia to the root and their attachment to the surface (Gulash *et al.*, 1984; Pueppke, 1984). Firm attachment is frequently polar (Bohlool and Schmidt, 1976) and may involve pili (Vesper and Bauer, 1986). Shortly after inoculation, rhizobia on or near the root surface induce at least two crucial responses in the host: the curling of growing root hairs and the localized dedifferentiation of root cells in the root cortex. The elicitation of these host responses depends on the expression of certain *nod*ulation genes (*nodABC*) in the bacterium. The expression of these *nod* genes is elicited, in turn, by exposure of the bacteria to specific phenolic compounds secreted by the host root. Induction of the *nodABC* genes is known to involve a regulatory gene, *nodD* (Mulligan and Long, 1985; Rossen *et al.*, 1986; Spanik *et al.*, 1987).

In most of the host legumes studied, only young root hairs can be induced to curl sufficiently to entrap Rhizobium cells in a pocket of host cell walls. In soybean and alfalfa, such entrapment is complete within 8 to 12 h after contact and seems to be required for subsequent penetration and infection (Dudley *et al.*, 1987;

Turgeon and Bauer, 1985). Diffusible substances from rhizobia on (or near) the root surface rapidly induce the host-specific formation of many localized centers of cell division in the outer cortex. Further rounds of cell division lead to the creation of new meristems. The bacteria-induced meristems differentiate to form nodules in which the rhizobia, packaged densely within host cells, convert atmospheric N_2 into NH_3 and receive fixed carbon from the host plant.

Individual root cells are susceptible to symbiotic induction of root hair curling and cortical cell division for a period of only 6 to 12 h (Bhuvaneswari *et al.*, 1981; Calvert *et al.*, 1984). The acquisition and loss of susceptibility to rhizobia appears to be governed by acropetal root development, restricting new infections to the zone of emergent root hairs near the root tip. Nodule initiation is therefore a very dynamic process, one which a bacterial cell can elicit only if it gets to the right place on the surface, and if it does so in time to respond to host phenolics that induce *nodABC* expression and produce enough of the *nodABC*-related signal substance(s) to provoke the necessary developmental changes in nearby host cells. These constraints suggest that nodule initiation might depend significantly on rapid, directed movement of bacteria to, and on, the root surface.

The role of chemotaxis in nodule formation by rhizobia

Chemotaxis appears to have an important role in nodule initiation. Early work established that non-motile mutants of rhizobia were able to infect and nodulate their hosts (Ames *et al.*, 1980; Ames and Bergman, 1981; Hunter and Fahring, 1980; Napoli and Albersheim, 1980), demonstrating that chemotaxis is not required for nodulation. However, non-motile mutants have been shown to be at a serious competitive disadvantage (Ames and Bergman, 1981; Caetano-Anollés *et al.*, 1988b; El Haloui *et al.*, 1986; Hunter and Fahring, 1980). When alfalfa roots were inoculated with an equal mixture of wild-type and non-motile mutant cells, wild-type *R. meliloti* cells generated 2 to 6 times more nodules than the mutant. Similarly, wild-type *B. japonicum* 110 formed about 6 times more nodules on soybean than equal numbers of a mutant with reduced motility. Based on the log-linear dose-response behavior of strain 110, the inoculum mixture would have to contain approximately 100 times more cells of the mutant than the wild-type to obtain equal nodule occupancy.

Motility and chemotaxis in rhizobia have not been well characterized. The only genetic analysis to date is in *R. meliloti* (Ames *et al.*, 1980; Bergman *et al.*, 1988; Ziegler *et al.*, 1987). Some work has been done to characterize the components of flagellar complexes or the dynamics of movement during chemotaxis in rhizobia (*cf.* DeLey and Rassel, 1965; Gotz and Schmitt, 1987; Gotz *et al.*, 1982; Krupski *et al.*, 1985; Schleicher and Bergman, 1981; Schmitt *et al.*, 1974). Others have assayed chemotaxis of rhizobia towards root exudates, or towards sugar and amino acid components of root exudates (Burg *et al.*, 1982; Currier and Strobel, 1976, 1981; Gaworzewska and Carlisle, 1982; Gitte *et al.*, 1978; Kandsamy and Prasad, 1979). Videomicroscopy has revealed that chemotaxis to the host root might be rather more focused, specific and interesting than previously imagined: chemotactic swarms of *R. meliloti* cells were observed to form at certain highly localized sites in the infectible zone of alfalfa roots and new swarms developed at the same sites after rinsing (Gulash *et al.*, 1984).

Our initial studies (Caetano-Anollés *et al.*, 1988b) sought to determine when and how chemotaxis contributed to nodulation. It appears that chemotaxis makes important contributions to several stages of the infection process, not just a single stage. Even after flooding roots with mutant/parent mixtures, the wild-type occupied 3 to 10 times more of the first-formed nodules than non-chemotactic mutants. These results demonstrated for the first time that chemotaxis enhances steps in nodule initiation during the first few hours after contact. Dose-response studies showed that 10 to 50 times more cells of the non-chemotactic mutants would have to be added to initiate the same number of nodules as the wild-type. The wild-type strain attached to the root surface in greater numbers than the mutants, but these differences accounted for less than half of its higher nodulation efficiency

(Caetano-Anollés and Bauer, 1988) suggesting that chemotaxis must confer other symbiotic advantages. Infections by non-motile mutants were somewhat slower to develop after initiation, so that the emergence of nodules formed by non-chemotactic mutants was delayed about one day relative to wild-type nodules.

In unpublished studies, we examined the ability of *R. meliloti* cells to move and multiply on the root surface by placing a 10 nanoliter droplet containing just 1 to 10 bacteria onto the surface of alfalfa roots maintained in sterile growth pouches. Both nodules and bacteria appeared in younger regions of the root, several cm below the point of inoculation. The bacteria multiplied extensively on the root surface in more mature regions of the root. In general, however, the rate of bacterial spread and multiplication were not sufficient to maintain a substantial population density of *R. meliloti* in the zone of emerging root hairs, the only zone susceptible to infection in alfalfa (Bhuvaneswari *et al.*, 1981). In agriculture, it may be the failure of inoculum strains to maintain an adequate presence in the infectible regions of the root, near the growing tips, that most limits their effectiveness in the field.

Using spot-inoculation techniques, we found that the wild-type bacteria formed 1.5 to 5 times more nodules in the initially susceptible region of the root than non-motile mutants. Since the bacteria were placed directly onto the root and remained there, enhanced nodulation by the wild-type must have been due to movement *on* the surface rather than movement to the root. This suggests that chemotaxis may enable rhizobia to reach microsites more suitable for infection after initial contact with the root. Such movement and its regulation could have a major bearing on symbiotic competition between strains of rhizobia.

In unpublished work, we have enriched non-mutagenized cultures of *Bradyrhizobium japonicum* I-110 through 7 cycles for cells from the outer edge of swarm colonies. At least three independent isolates were obtained which formed swarm colonies that increased in diameter 2–3 times faster than the wild-type on soft agar swarm plates. The basis for the enhanced chemotaxis has not yet been determined. Since only *some* of the isolates we obtained through this serial enrichment showed enhanced chemotaxis, it appears that the more active isolates are in fact spontaneous mutants. One of these 'hypermotile' mutants was recently tested for nodulation ability on soybean and was found to be significantly more efficient in nodule initiation than the parent. Dose-response studies indicate that 5 to 10 times as many wild-type bacteria would have to be added to achieve the same level of nodulation as this mutant. Enhanced nodulation by the hypermotile mutant is evidence that chemotaxis contributes to the efficiency of nodule initiation, as is reduced nodulation by the non-motile mutant.

Chemotaxis to phenolic compounds from the host root

Recent studies in Nester's and Shaw's labs revealed that *A. tumefaciens* was chemically attracted to several of the same host root phenolics that are required for induction of the *vir*ulence genes in this bacterium (Ashby *et al.*, 1987, 1988; Parke *et al.*, 1987; Shaw *et al.*, 1988a,b). Acetosyringone (2,4,6-trimethoxy phenol) is selectively produced in wounded tissue, and wounds are the site of *A. tumefaciens* infection. Acetosyringone was found to elicit maximum chemotaxis at concentrations of about 10^{-7} *M*, which is about 100-fold lower than the concentration required for maximal induction of the *vir* genes. From these results, it appears that the pathogen is able to detect very low concentrations of this wound-specific compound, swim towards it by chemotaxis, and then turn on its *vir*ulence genes when it gets close to the wound site.

Just as *A. tumefaciens* responds chemotactically towards phenolic inducers of *vir* gene expression, we have shown that *R. meliloti* responds chemotactically to host flavonoids that act as inducers of *nod* gene expression. An earlier study by Peters *et al.* (1986) identified the flavone luteolin from alfalfa seeds as a potent inducer of *nod* genes in *R. meliloti*, active at concentrations of about 10^{-6} to 10^{-7} M. We found that luteolin elicits half-maximal chemotactic responses in *R. meliloti* at concentrations of 10^{-9} M or less (Caetano-Anollés *et al.*, 1988a). The addition of naringenin, a close structural relative of luteolin, blocked chemotaxis of *R. meliloti* to luteolin, just as it specifically blocked *nodD1*-mediated induction of the *nodABC* genes by luteolin (Peters and Long, 1988). This suggests that *nod* gene induction and chemotaxis elicited by luteolin may share a common receptor or other components of their signal transduction pathways. The addition of apigenin, another close structural relative of luteolin, also had parallel effects on both induction of *nod* genes and chemotaxis. At the genetic level, we observed that *nodD1*::Tn5 mutants of *R. meliloti* failed to respond to luteolin as a chemoattractant, whereas these same mutants responded normally to amino acids and other common attractants, indicating that the *nodD1* gene may have some role in chemotactic responses to luteolin (Caetano-Anollés *et al.*, 1988a). Similarly, Shaw *et al.* (1988a) reported that the *virA* and *virG* genes were required for chemotaxis of *A. tumefaciens* to acetosyringone. Thus, there is both biochemical and genetic evidence to indicate that gene induction by host phenolics and chemotaxis towards these compounds are highly specific and interrelated processes. Similar evidence of chemotaxis towards *nod* gene-inducing host phenolics has been obtained with other rhizobia, including *R. leguminosarum* bv. *viciea* (Armitage *et al.*, 1988) and *R. leguminosarum* bv. *phaseoli* (Aguilar *et al.*, 1988).

Chemotaxis towards host phenolics is not always linked to induction of new gene expression. In *R. leguminosarum* and *A. tumefaciens*, it seems that some host phenolics serve only as attractants or only as inducers of new gene expression, but not both. And different strains of a symbiont or pathogen may prove to differ in their responses to a particular phenolic compound, reflecting different coevolutionary histories and regulatory strategies. Nonetheless, it seems likely from present evidence that many microorganisms, pathogenic, symbiotic and saprophytic, may rely on specific members of their host's array of secreted phenolics to trigger both taxis and new gene expression appropriate to the rhizosphere environment.

Two points of caution deserve mention in this regard. First, it is important to recognize that the flavonoids or other phenolics that a host plant secretes into the rhizosphere may be quite different from those sequested in the seed or root tissues. Furthermore, the spectrum of phenolics in root exudates may change dramatically with plant age, genotype, growth environment and exposure to microorganisms. The recent studies of Maxwell *et al.* (1989) and Hartwig *et al.* (1989) provide particularly noteworthy examples of analysis, but there is a great deal still to be learned about the secretion of host phenolics before their biological role in plant-microbe interactions can be properly assessed and usefully manipulated.

Second, it is our experience, shared by others, that the chemotactic responses of bacteria to host phenolics are weak and variable, at least under laboratory culture and assay conditions. Chemotactic responses to acetosyringone and the various *nod* gene inducing flavonoids typically range between 2 to 5 times higher than background, in contrast to the 10- to 100-fold responses elicited by organic and amino acids. The weaker chemotaxis of *Rhizobiaceae* towards host phenolics may be due to the much lower concentrations of phenolics which elicit maximal responses. Peak responses to the phenolics usually occur at concentrations about 10^4-fold lower than those eliciting maximal responses to amino acids or other common nutrients. It may be that a gradient of attractant ranging between 10^{-8} M and 10^{-9} M simply cannot provide enough of a differential in attractant molecules bound to bacterial transducer proteins to generate a strongly biased random walk, just as it's hard to locate where smoke is coming from if it is so dilute that you can barely detect it. Alternatively, it may be that chemotactic responses to the phenolics are suppressed in bacteria cultured under laboratory conditions. In this regard, Peters and Long (1988) observed that *R. meliloti* cultured on a rich medium was about 20-fold less responsive to luteolin induction of *nod* genes than when cultured on a minimal medium. With regard to variability in chemotaxis towards host phenolics, we note that wild-type *R. meliloti* RCR2011 has twice spontaneously lost its responsiveness to luteolin in the past three years, forcing us to use isolates from other labs. And while our earlier studies indicated that chemotactic responses to luteolin were abolished by mutations in the *nodA* and *nodC* genes as well as by mutations in the *nodD1* gene (Caetano-Anollés *et al.*, 1988a), the chemotactic responses of these *nod* mutants to luteolin have been inconsistent in more recent experiments, usually negative but occasionally positive. Strains carrying sequences that complement the mutations have shown similarly inconsistent chemotaxis towards luteolin. These results suggest that chemotaxis to luteolin is an unstable trait in *R. meliloti*, at least in strain RCR2011, and may not be directly dependent on the expression of the *nod* genes.

Bergman *et al.* (1988) recently described a mutant of *R. meliloti* which is defective in general chemotaxis towards common nutrients of soft agar plates, but is perfectly capable of forming chemotactic swarms at specific sites on the host root surface. These results strongly imply that *R. meliloti* has two relatively independent pathways of chemotactic response: one set of responses to common nutrients and another set of responses to substances from the host root, perhaps non-metabolized compounds like luteolin. This concept of a dual chemotaxis pathway, as proposed by Bergman *et al.*, suggests a more general possibility: that rhizosphere bacteria may rely on certain poorly metabolized host substances as important sources of information about the proximity and genotype of a nearby root or seed. If this idea is correct, as the available evidence indicates, then the molecular details of signal transduction and response to such compounds should be of considerable importance to the competitiveness of bacteria in colonizing the rhizosphere.

Conclusions and directions

In summary, there are now several good reasons to believe that chemotaxis is important to rhizosphere interactions and symbiotic infection by rhizobia: non-motile mutants are less competitive in nodule initiation and occupancy; they are also slower in infection development; hypermotile mutants have improved ability to initiate nodules; rhizobia form chemotactic swarms at local sites on host roots; they respond chemotactically to host substances that induce *nod* gene expression; and expression of several *nod* genes is required for chemotaxis to these compounds.

These results suggest that further studies on the role of chemotaxis in rhizosphere colonization and host interactions are warranted. Studies are needed to establish when, how, and how much, chemotaxis contributes to the growth and survival of bacteria in soil or rhizosphere environments. These questions would seem to be of basic importance to soil microbiology, particularly in regards to the dynamics of competition for energy sources. Initial studies are needed to provide at least some measure of the overall costs and benefits associated with chemotaxis in soil rhizosphere environments. The isolation and characterization of defined mutants with altered chemotaxis can provide effective tools for assessing such costs and benefits. Information from such experiments may be useful in devising ways to enhance the effectiveness of genetically engineered inoculum bacteria or to limit their spread and long term survival in the soil.

References

Aguilar J M M, Ashby A M, Richards A J M, Loake G J, Watson M D and Shaw C H 1988 Chemotaxis of *Rhizobium leguminosarum* towards flavonoid inducers of the symbiotic nodulation genes. J. Gen. Microbiol. 134, 2741–2746.

Ames P and Bergman K 1981 Competitive advantage provided by bacterial motility in the formation of nodules by *Rhizobium meliloti*. J. Bacteriol. 148, 728–729.

Ames P, Schluederberg S A and Bergman K 1980 Behavioral mutants of *Rhizobium meliloti*. J. Bacteriol. 141, 722–727.

Armitage J P, Gallagher A and Johnston A W B 1988 Comparison of the chemotactic behaviour of *Rhizobium leguminosarum* with and without the nodulation plasmid. Mol. Microbiol. 2, 743–748.

Ashby A M, Watson M D and Shaw C H 1987 A Ti-plasmid determined function is responsible for chemotaxis of *Agrobacterium tumefaciens* towards the plant wound product acetosyringone. FEMS Microbiol. Lett. 41, 189–192.

Ashby A M, Watson M D, Loake G J and Shaw C H 1988 Ti plasmid-specific chemotaxis of *Agrobacterium tumefaciens* C58C toward *vir*-inducing phenolic compounds and soluble factors from monocotyledonous and dicotyledonous plants. J. Bacteriol. 170, 4181–4187.

Bauer W D 1981 Infection of legumes by rhizobia. Annu. Rev. Plant Physiol. 32, 407–449.

Bergey's Manual of Systematic Bacteriology 1984. Williams and Wilkins, Baltimore.

Bergman K, Gulash-Hoffee M, Hovestadt R E, Larosiliere R

C, Ronco P G and Su L 1988 Physiology of behavioral mutants of *Rhizobium meliloti*: evidence for a dual chemotaxis pathway. J. Bacteriol. 170, 3249–3254.

Bhuvaneswari T V, Bhagwat A A and Bauer W D 1981 Transient susceptibility of root cells in four common legumes to nodulation by rhizobia. Plant Physiol. 68, 1144–1149.

Bohlool B B and Schmidt E L 1976 Immunofluorescent polar tips of *Rhizobium japonicum*: possible site of attachment or lectin binding. J. Bacteriol. 125, 1188–1194.

Bowen G D and Rovira A D 1976 Microbial colonization of plant roots. Annu. Rev. Phytopath. 14, 121–144.

Burg D, Guillaume J and Tailliez R 1982 Chemotaxis by *Rhizobium meliloti*. Arch Microbiol. 133, 162–163.

Caetano-Anolles G, Crist-Estes D K and Bauer W D 1988a Chemotaxis of *Rhizobium meliloti* to the plant flavone luteolin requires functional nodulation genes. J. Bacteriol. 170, 3164–3169.

Caetano-Anolles G, Wall L G, DeMicheli A T, Macchi E M, Bauer W D and Favelukes G 1988b Motility and chemotaxis affect nodulating efficiency of *Rhizobium meliloti*. Plant Physiol. 86, 1228–1235.

Caetano-Anolles G and Bauer W D 1988 Enhanced nodule initiation on alfalfa by wild-type *Rhizobium meliloti* co-inoculated with *nod* gene mutants and other bacteria. Planta 174, 385–395.

Calvert H E, Pence M K, Pierce M, Malik N S A and Bauer W D 1984 Anatomical analysis of the development and distribution of *Rhizobium* infections in soybean roots. Can. J. Bot. 62, 2375–2384.

Chet I and Mitchell R 1976 Ecological aspects of microbial chemotactic behavior. Annu. Rev. Microbiol. 30, 221–239.

Crist D K, Wyza R E, Mills K K, Bauer W D and Evans W R 1984 Preservation of *Rhizobium* viability and symbiotic infectivity by suspension in water. Appl. Environ. Microbiol. 47, 895–900.

Curl E A and Truelove B 1986 The Rhizosphere. Springer-Verlag, Berlin.

Currier W W and Strobel G A 1976 Chemotaxis of *Rhizobium* species to plant root exudates. Plant Physiol. 57, 820–823.

Currier W W and Strobel G A 1981 Characterization and biological activity of trefoil chemotactin. Plant Sci. Lett. 21, 159–165.

De Ley J and Rassel A 1965 DNA base composition, flagellation and taxonomy of the genus *Rhizobium*. J. Gen. Microbiol. 41, 85–91.

Demezas D H and Bottomley P J 1986 Autecology in rhizospheres and nodulating behavior of indigenous *Rhizobium trifolii*. Appl. Environ. Microbiol. 52, 1014–1019.

DeWeger L A, Vlugt van der C I M, Wijfjes A H M, Bakker P A H M, Schippers B and Lugtenberg B 1987 Flagella of a plant-growth stimulating *Pseudomonas fluorescens* strain are required for colonization of potato roots. J. Bacteriol. 169, 2769–2773.

Djordevic M A, Gabriel D W and Rolfe B G 1987 *Rhizobium* – the refined parasite of legumes. Annu. Rev. Phytopathol. 25, 145–168.

Dudley M E, Jacobs T W and Long S R 1987 Microscopic studies of cell divisions induced in alfalfa roots by *Rhizobium meliloti*. Planta 171, 289–301.

El Haloui N E, Ochin D, and Tailliez R 1986 Infection competitivity between *Rhizobium meliloti* strains: role of motility. Plant and Soil 95, 337–344.

Gaworzewska E T and Carlile M J 1982 Positive chemotaxis of *Rhizobium leguminosarum* and other bacteria towards exudates from legumes and other plants. J. Gen. Microbiol. 128, 1179–1188.

Gitte R R, Vittal Rai P and Patil R B 1978 Chemotaxis of *Rhizobium* spp towards root exudate of *Cicer arietinum* L. Plant and Soil 50, 533–566.

Gotz R and Schmitt R 1987 *Rhizobium meliloti* swims by unidirectional, intermittent roation of right-handed flagellar helices. J. Bacteriol. 169, 3146–3150.

Gotz R, Limmer N, Ober K and Schmitt R 1982 Motility and chemotaxis in two strains of *Rhizobium* with complex flagella. J. Gen. Microbiol. 128, 789–798.

Griffin D M and Quail G 1968 Movemet of bacteria in moist, particulate systems. Aust. J. Biol Sci. 21, 579–582.

Gulash M, Ames P, LaRosiliere R C and Bergman K 1984 Rhizobia are attracted to localized sites on legume roots. Appl. Environ. Microbiol. 48, 149–152.

Hartwig U A, Maxwell C A, Joseph C M and Phillips D A 1989 Interactions among flavonoid *nod* gene inducers released from alfalfa seeds and roots. Plant Physiol. 91, 1138–1142.

Harwood C S, Rivelli M and Ornston L N 1984 Aromatic acids are chemoattractants for *Pseudomonas putida*. J. Bacteriol. 160, 622–628.

Howie W J, Cook R J and Weller D M 1987 Effects of soil matric potential and cell motility on wheat root colonization by fluorescent pseudomonads suppressive to take-all. Phytopathology 77, 286–292.

Humbeck C, Thierfelder H, Gresshoff P M and Werner D 1985 Competitive growth of slow growing *Rhizobium japonicum* against fast growing *Enterobacter* and *Pseudomonas* species at low concentrations of succinate and other substrates in dialysis culture. Arch Microbiol. 142, 223–228.

Hunter W J, and Fahring C J 1980 Movement of *Rhizobium* and nodulation of legumes. Soil Biol. Biochem. 12, 537–542.

Kandsamy D and Prasad N N 1979 Colonization by rhizobia of the seed and roots of legumes in relation to exudation of phenolics. Soil Biol. Biochem. 11, 73–75.

Koshland D E 1981 Biochemistry of sensing and adaptation in a simple bacterial system. Annu. Rev. Biochem. 50, 765–782.

Krupski G, Gotz R, Ober K, Pleier E and Schmidt R 1985 Structure of complex flagellar filaments in *Rhizobium meliloti*. J. Bacteriol. 162, 361–366.

Madsen E L and Alexander M 1982 Transport of *Rhizobium* and *Pseudomonas* through soil. Soil. Sci. Soc. Am. J. 46, 557–560.

Maxwell C A, Hartwig U A, Joseph C M and Phillips D A 1989 A chalcone and two related flavonoids released from alfalfa roots induce *nod* genes of *Rhizobium meliloti*. Plant Physiol. 91, 842–847.

McNab R 1987 Motility and chemotaxis. pp. 732–759. *In Escherichia coli* and *Salmonella typhimurium*. Cellular and Molecular Biology. Eds. F C Neidhart *et al.* Eds. Am Soc. Microb., Washington, DC.

Moawad H A, Ellis W R and Schmidt E L 1984 Rhizosphere response as a factor in competition among three serogroups of indigenous *Rhizobium japonicum* for nodulation of field-grown soybeans. Appl. Environ. Microbiol. 47, 607–612.

Mulligan J T and Long S R 1985 Induction of *Rhizobium meliloti nodC* expression by plant exudate requires *nodD*. Proc. Nat. Acad. Sci. USA 82, 6609–6613.

Napoli C A and Albersheim P 1980 Infection and nodulation of clover by non-motile *Rhizobium trifolii*. J. Bacteriol. 141, 979–980.

Parke D, Rivelli M and Ornston L N 1985 Chemotaxis to aromatic and hydroaromatic acids: comparison of *Bradyrhizobium japonicum* and *Rhizobium trifolii*. J. Bacteriol. 163, 417–422.

Parke D, Ornston L N and Nester E W 1987 Chemotaxis to plant phenolic inducers of virulence genes is constitutively expressed in the absence of the Ti plasmid in *Agrobacterium tumefaciens*. J. Bacteriol. 169, 5336–5338.

Peters N K, Frost J W and Long S R 1986 A plant flavone, luteolin, induces expression of *Rhizobium meliloti* nodulation genes. Science 233, 977–980.

Peters K N and Long S R 1988 Alfalfa root exudates and compounds which promote or inhibit induction of *Rhizobium meliloti* nodulation genes. Plant Physiol. 88, 396–400.

Pueppke S G 1984 Adsorption of slow- and fast-growing rhizobia to soybean and cowpea roots. Plant Physiol. 75, 924–928.

Rolfe B G and Gresshoff P M 1988 Genetic analysis of legume nodule initiation. Annu. Rev. Plant Physiol. 39, 297–319.

Rossen L, Schearman C A, Johnston A W B and Downie J A 1986 The *nodD* gene of *Rhizobium leguminosarum* is autoregulatory and in the presence of plant exudate induces the *nodABC* genes. EMBO J. 43, 3369–3373.

Rovira A D 1961 *Rhizobium* numbers in the rhizosphere of red clover and paspalum in relation to soil treatment and numbers of bacteria and fungi. Aust. J. Agric. Res. 12, 77–83.

Scher F M, Kloepper J W, Singleton C, Zaleska I and Laliberte M 1988 Colonization of soybean roots by *Pseudomonas* and *Serratia* species: relationship to bacterial motility, chemotaxis, and germination time. Phytopathol. 78, 1055–1059.

Schleicher L R and Bergman K 1981 The amino acid composition of *Rhizobium meliloti* flagellin: occurrence of a single tryptophan residue. Biochem. Biophys. Res. Commun. 100, 1634–1641.

Schmidt E L 1979 Initiation of plant root-microbe interactions. Annu. Rev. Microbiol. 33, 355–76.

Schmitt R, Bamberger I, Acker G and Mayer F 1974 Feinstrukturanalyse der komplexen Geisseln von *Rhizobium lupini*. Arch. Microbiol. 100, 145–162.

Shaw C H, Ashby A M, Brown A, Royal C and Loake G J 1988 *virA* and *virG* are the Ti-plasmid functions required for chemotaxis of *Agrobacterium tumefaciens* toward acetosyringone. Mol. Microbiol. 2, 413–418.

Shaw C, Ashby A, Loake G, Brown A and Watson M 1988 Chemotaxis of *Agrobacterium tumefaciens* is the initial step in crown gall formation. Fourth Intl. Symp. on Molecular Genetics of Plant-Microbe Interactions. Acapulco, Mexico. Abstr. Ib-15.

Soby S and Bergman K 1983 Motility and chemotaxis of *Rhizobium meliloti* in soil. Appl. Environ. Microbiol. 46, 995–998.

Spaink H P, Wijffelman C A, Pees E, Okker R J H and Lugtenberg B J J 1987 *Rhizobium* nodulation gene *nodD* as a determinant of host specificity. Nature 328, 337–339.

Turgeon B G and Bauer W D 1985 Ultrastructure of infection-thread development during the infection of soybean by *Rhizobium japonicum*. Planta 163, 328–349.

Vesper S J and Bauer W D 1986 Role of pili (fimbriae) in attachment of *Bradyrhizobium japonicum* to soybean roots. Appl. Environ. Microbiol. 52, 134–141.

Wolfe A J, Conley M P, Kramer T J and Berg H C 1987 Reconstitution of signaling in bacterial chemotaxis. J. Bacteriol. 169, 1878–1885.

Ziegler R J, Pierce C and Bergman K 1987 Mapping and cloning of a *fla-che* region of the *Rhizobium meliloti* chromosome. J. Bacteriol. 168, 785–790.

D. L. Keister and P. B. Cregan (Eds.), The rhizosphere and plant growth, 163–171.

PLSO BARC17

Gene-for-gene interaction in the legume-Rhizobium symbiosis

P.B. CREGAN, M.J. SADOWSKY[1] and H.H. KEYSER[2]
Soybean and Alfalfa Research Laboratory, USDA-ARS, Bldg. 011, HH-19, BARC-West, Beltsville, MD 20705, USA. Present addresses: [1]Department of Soil Science and Department of Microbiology, University of Minnesota, St. Paul, MN 55108, USA and [2]The NifTAL Project, University of Hawaii, Paia, HI 96779, USA

Key words: *Bradyrhizobium japonicum*, nitrogen fixation, nodulation, plant-microbe interaction, soybean

Abstract

The gene-for-gene hypothesis of plant-pathogen interaction as proposed by H.H. Flor states that for each resistance gene in the host there is a specific, complementary gene conditioning pathogenicity in the parasite. Numerous examples of this type of plant-pathogen interaction have been elucidated in the past 40 years. This model of plant-microbe interaction also provides a simple framework in which to view symbiotic plant-microbe interactions. We have identified 10 instances in the literature in which interaction between plants in the family Leguminosae and bacterial symbionts in the genera Rhizobium or Bradyrhizobium appear to function in a gene-for-gene fashion. In addition, we describe two further instances of apparent gene-for-gene interaction in the soybean-*B. japonicum* symbiosis. These last examples were identified in our laboratory as a result of an intensive search for soybean germplasm that would restrict or resist nodulation by highly competitive, but relatively poor nitrogen fixing strains of *B. japonicum* that are indigenous to soybean production areas of the U.S.A. The practical goal of this work is firstly, the development of a soybean genotype with genetic factors conferring 'resistance' to a broad spectrum of ineffective indigenous bradyrhizobia and secondly, the construction of a highly effective inoculant strain possessing the complementary genes necessary to optimally nodulate such a soybean genotype. This dual approach to develop both partners in the legume-rhizobial symbiosis represents a new application of the gene-for-gene system.

Introduction

In contrast to saprophytes, plant pathogenic and symbiotic organisms possess at least three types of additional genetic factors. These include genes necessary for growth in plant tissue, genes that condition specificity for the genera or species of plants that can be colonized, and genes that define the degree of colonization on specific plant genotypes within a host species. It is this latter class of genetic factors and their relationship to host plant genes that are the focus of this review. In particular, we will focus on host genetic factors in the Leguminosae that appear to interact with specific genotypes in the bacterial genera Bradyrhizobium and Rhizobium.

The first report of the mode of inheritance of resistance to a plant pathogen was published by R.H. Biffin (1905). He suggested that resistance was controlled by a single gene that segregated in a Mendelian fashion. According to Ellingboe (1976), less than 10% of subsequent reports concerning the inheritance of disease resistance suggested that there were a large number of genes controlling resistance. Thus, in most instances disease resistance appeared to be controlled by one or a few Mendelian genes. One of the earliest reports of the genetic control of pathogenicity of a plant disease organism (Newton *et al.*, 1930) did not elucidate clearly the mode of inheritance of pathogenicity in wheat stem rust (*Puccinia graminis tritici*) but suggested that in one case two, and in a second case three genetic factors were involved. Numerous subsequent reports have indicated that pathogenicity on a given host plant genotype is frequently under the control of single genetic factors.

H. H. Flor and the Gene-for-Gene hypothesis

A unique early research effort was that of H. H. Flor who studied the inheritance of resistance in flax (*Linum usitatissimum*) to the flax rust fungus, *Melampsora lini*. He also examined the genetic control of pathogenicity in the rust fungus. By the mid-1940's Flor (1946) had determined that at least 19 pairs of host genetic factors were involved in conditioning the reaction of rust to 16 different flax varieties. Similarly, he reported (Flor, 1947) that in one race of *M. lini* at least 5 genetic factors, and in a second race, 16 or more genetic factors controlled pathogenicity. He concluded that in the flax-rust interaction, host resistance was always dominant to susceptibility and that in the fungus, virulence was recessive to avirulence. More importantly, Flor (1955) proposed the gene-for-gene hypothesis as the simplest explanation of the results obtained in the above reports. He proposed that on varieties of flax with one allele for resistance, F_2 cultures of the rust segregate into two classes for virulence and avirulence with the ratio of 1:3, virulent:avirulent. On varieties with 2, 3 or 4 alleles for resistance, the F_2 cultures segregate approximately into bi-, tri-, or tetrafactorial ratios, respectively. Thus, Flor concluded that 'for each resistance gene in the host there is a specific, complementary gene conditioning pathogenicity in the parasite.' When a flax genotype with a particular resistance gene is inoculated with a rust strain with the complementary avirulence allele, a resistant interaction occurs. If the same flax genotype is inoculated with the rust strain carrying the complementary virulence allele, a susceptible reaction results. If the flax plant does not carry the resistance gene a susceptible reaction occurs regardless of the genotype of the rust.

The data in Table 1 illustrate the operation of the gene-for-gene hypothesis in the flax/flax rust system (Flor and Comstock, 1971). The first rust genotype carries the virulence alleles L and M^3 and the avirulence allele N^1. Because each host genotype possesses the complementary N^1 allele for resistance, this rust genotype produces a 'resistant' reaction. In a similar manner, the second rust genotype would produce a resistance reaction with flax genotypes 3 and 4 because these genotypes carry the M^3 resistance gene complementary to the A_{M^3} avirulence allele. In contrast, the second rust strain produces a susceptible reaction with the first two flax genotypes. This would be predicted from the fact that neither of these two flax genotypes carry the M^3 resistance gene corresponding to the only avirulence allele (A_{M^3}) in the rust. In the data presented in Table 1 the reaction of each host genotype with a given rust genotype produces either a susceptible or resistant reaction. It is important to note that the degree of pathogenicity does not necessarily have to range from resistant to susceptible but rather that there is a distinguishable difference between pathogen genotypes with the 'virulent' or 'avirulent' allele when interacting with the 'resistant' host.

One explanation of the mode of operation of the gene-for-gene interaction of a plant and pathogen suggests that the product of the resist-

Table 1. Predictability of the gene-for-gene model of host plant-pathogen interaction as demonstrated by genes at three loci in flax and three corresponding loci in flax rust (From Flor and Comstock, 1971)

Flax rust genotype indicating the avirulence (A) and virulence (a) alleles present	Resistance alleles present in the flax plant			
	N^1	L, N^1	M^3, N^1	L, M^3, N^1
a_La_L, $a_{M^3}a_{M^3}$, $A_{N^1}A_{N^1}$	R	R	R	R
a_La_L, $A_{M^3}A_{M^3}$, $a_{N^1}a_{N^1}$	S	S	R	R
A_LA_L, $a_{M^3}a_{M^3}$, $a_{N^1}a_{N^1}$	S	R	S	R

R = Resistant, S = Susceptible

ance gene in the host functions in some manner to recognize the product of the avirulence allele in the pathogen. Thus, when the resistant genotype interacts with the avirulent pathogen, an 'incompatible' or resistant phenotype results. In contrast, if the avirulence gene product is not clearly recognized or is not detected at all, because of the lack of the resistance allele in the host, the pathogen is only partially recognized or not recognized at all, and susceptibility results. Likewise, if the avirulence allele is replaced by its virulent alternative in the pathogen, a nonrecognizable product is produced and the host is 'susceptible' whether the corresponding allele for resistance is present or not. One can readily understand that under these circumstances selection pressure (in respect to the host) would move the pathogen in the direction of virulence, *i.e.* the production of a product that does not trigger a response by the corresponding resistance locus in the host plant. Similarly, selection pressure would favor the plant genotype that possesses the allelic form that functions most efficiently in the recognition of the prevalent disease races and minimizes their detrimental effects upon plant reproduction, *i.e.* towards a superior pathogen detection system.

The gene-for-gene model and the legume-rhizobial symbiosis

The gene-for-gene model can be used to describe the genetic relationship controlling the outcome of interaction between a plant and a plant pathogen. Such interactions have been demonstrated in the interaction of numerous fungal pathogens and their hosts (reviewed by Day, 1974) and more recently in a number of instances in bacterial pathogens (Gabriel *et al.*, 1986; Staskawicz *et al.*, 1984). Of recent interest has been the attempt to apply the gene-for-gene concept to the interaction between plants in the family Leguminosae and bacterial symbionts in the genera Rhizobium and Bradyrhizobium (Djordjevic *et al.*, 1987). The report by Sadowsky *et al.* (1990) in this volume focuses upon genetic aspects of the microsymbiont, while we will briefly discuss genetic determinants in the Legume host that appear to act in a gene-forgene fashion.

Evidence of gene-for-gene interaction could include rhizobial strain specificity in 1.) complete or partial restriction of nodule formation or 2.) the lack of proper functioning of nodules. Our review of the literature of the genetics of Legume-rhizobial interactions indicates both types of strain specific interactions can be found.

In pea (*Pisum sativum* L.), a series of symbiotic (Sym) genes have been defined, at least three of which appear to act in a gene-for-gene manner (Table 2). Lie (1971) reported that the primitive pea cultivar Afghanistan was very resistant to nodulation and formed only a few nodules with some strains of *Rhizobium leguminosarum* bv. *viciae*. Holl (1975) demonstrated that the non-nodulation was controlled by a single recessive gene that was later referred to as sym_2 (Holl and LaRue, 1975). Shortly thereafter, Lie (1978) reported finding strains Tom+ +, F_{13}, and Him of *R. leguminosarum* isolated in Turkey, Israel, and Afghanistan, respectively, that abundantly nodulated Afghanistan pea. Thus, sym_2sym_2 genotypes specifically exclude from nodulation *R. leguminosarum* strains native to Europe and North America while allowing normal nodulation with at least some strains from the Middle East. A second gene that conditions strain specific reduced nodulation was found by Lie (1984) in the pea genotypes JI 261 and AIII*. These genotypes carry the dominant allele Sym_4 that confers resistance to nodulation by strain 310a. A third gene was found in the cultivar Afghanstan that prevents fully effective N_2 fixation with strain F13. Lie and Timmermans (1983) suggested that this phenotype was conditioned by the recessive allele sym_6.

In red clover (*Trifolium pratense*), three genes are reported to condition symbiotic defects with specific strains of *R. trifolii* (*R. leguminosarum* bv. *trifolii*) (Table 2). When inoculated with strain A121111 (subsequently called 0403), plants with the homozygous recessive genotype i_1i_1 form small and ineffective nodules (Nutman, 1954) that contain no leghemoglobin (Bergersen and Nutman, 1957). With other strains tested, nodulation of this red clover genotype was normal (Nutman, 1969). Plants homozygous for the recessive *n* allele form ineffective nodules of normal number and size with little bacterial infection when inoculated with certain strains

Table 2. Examples of host genetic factors in legume species that appear to function in a gene-for-gene manner with specific strains of their homologous rhizobial microsymbiont

Legume host and plant genotype	Description
Pea (*Pisum sativum*)	
sym_2sym_2	In cv. Afghanistan, restricts nodulation of many European *Rhozibium leguminosarum* bv. *viciae*
Sym_4	In pea genotypes JI 261 and AIII*, resists nodulation by *R. leguminosarum* bv. *viciae* strain 310a
sym_6sym_6	In cv. Afghanistan, prevents fully effective N_2 fixation with *R. leguminosarum* bv. *viciae* strain F13
Red clover (*Trifolium pratense*)	
i_1i_1	Plants with this genotype form small ineffective nodules with strain A121111 (subsequently referred to as strain 0403) of *R. leguminosarum* bv. *trifolii*
nn	Normal nodulation but no N_2 fixation with specific strains of *R. leguminosarum* bv. *trifolii*
dd	Many small ineffective nodules with specific strains or *R. leguminosarum* bv. *trifolii*
Soybean (*Glycine max*)	
Rj_2	Restricts nodulation of *B. japonicum* strains of serogroups 122 and cl
Rj_3	Restricts nodulation of *B. japonicum* strain USDA 33
Rj_4	Restricts nodulation of *B. japonicum* strain USDA 61 and a few related strains
Unnamed dominant allele in 'Kent' soybean	Restricts nodulation of *R. fredii* strain USDA 205

(Nutman, 1968). Nodulation and N_2 fixation with other strains are normal. Compared to normal genotypes, red clover with the *dd* genotype forms more nodules that are smaller, ineffective, and lack leghemoglobin. This phenotype was confined to four of eight strains tested (Nutman, 1968).

In the soybean (*Glycine max*), three examples of strain specific nodulation defects have been reported. The dominant allele Rj_2, first found in the cultivar Hardee, restricts the formation of nodules by strains of *B. japonicum* serogroup 122 and some additional strains (Caldwell, 1966). Hardee nodulates normally with all other strains with the exception of USDA 33 whose nodulation is also severely restricted by this unique soybean genotype. This latter phenomenon is controlled by a second dominant allele Rj_3 (Vest, 1970). A third strain specific nodulation defect in soybean is defined by the dominant allele Rj_4 (Vest and Caldwell, 1972). Soybean genotypes that carry the Rj_4 allele such as the cultivar Hill, restrict nodulation by strain USDA 61 and a few other similar strains. Keyser *et al.* (1982) reported the existence of a new species of rhizobia that nodulates soybean and is characterized by a 'fast-growing' phenotype *in vitro*. This group of strains was distinct from the traditional 'slow-growing' rhizobia (bradyrhizobia) that had long been assumed to be the only microsymbiont that could nodulate soybean. This species was named *Rhizobium fredii* (Scholla and Elkan, 1984). The initial *R. fredii* strains isolated by Keyser *et al.* (1982) formed few nodules with certain soybean cultivars such as Kent, while many other cultivars were nodulated almost normally. Dinitrogen fixation was generally ineffective with cultivated soybean. However, with the primitive soybean 'Peking' nodulation was normal and fixation was quite effective. Devine (1984) reported the presence of a single dominant allele in Kent soybean that conditioned restricted nodulation of *R. fredii* strain USDA 205 in the analysis of a cross between Kent and Peking soybean. One can assume that this gene functions to 'resist' nodulation by a whole genera of soybean rhizobia and would therefore be distinct from the Rj_2, Rj_3, and Rj_4 genes described above. This type of host genetic factor would be more analogous to the *hsn* genes in rhizobia that control nodulation of all members of given plant species (see Sadowsky *et al.*, this volume) and thus may represent a different level in the hierarchy of factors controlling legume-rhizobial specificity.

Evidence suggests that the legume host genes described above may act in a gene-for-gene fashion. However, the sym_2 and sym_6 genes in pea, and the three genes in red clover (i_1, n, and d) are recessive and therefore would not easily fit into the gene-for-gene model as described by Flor. Interestingly, a gene-for-gene system has been demonstrated in the case of sym_2 and the complementary rhizobial gene *nodX* (Davis *et al.*, 1988; Sadowsky *et al.*, 1990). This suggests that the details of the gene-for-gene model as defined by Flor may vary from system to system and may be quite different in a symbiotic versus a pathogenic system. Nonetheless it is interesting that in the soybean-rhizobial interaction, the host genes defined to date are all dominant (Table 2) in a manner similar to most alleles for resistance. The exclusively dominant nature of the soybean genes found to date may be a result of the allotetraploidy of the soybean, thus making the detection of recessive mutants unlikely. This system of dominant genes in soybean may exist because of some selective advantage for certain genotypes to eliminate the nodulation of specific strains or perhaps it is the remnant of a host-pathogen recognition system in which there is currently no selective advantage for the presence of host *Rj* genes or complementary virulence or avirulence factors in the *B. japonicum*. Whatever the explanation, it was the precedent of the Rj_2, Rj_3, and Rj_4 alleles that stimulated our research group to investigate the presence of additional such genes in soybean to accomplish a very specific research goal and in so doing to positively impact the N_2 fixation and productivity of soybean grown in U.S. soils.

Ineffective indigenous bradyrhizobia – a problem in soybean production

The Midwest is the major soybean production area of the U.S.A. In this region, and particularly in the northern Midwest, the indigenous strains belonging to serocluster 123 compete with other indigenous and applied *B. japonicum* strains to form the majority of nodules on soybeans. The use of the term serocluster 123 was used by Schmidt *et al.* (1986) to describe the serological relationship among strains in serogroups 123, 127, and 129. Previous reports that referred to strains of 'serogroup 123', have included strains in this category that would more accurately be classified as serogroup 127 or 129. Reports by a number of workers including Damirgi *et al.*, (1967); Ellis *et al.*, (1984); Ham *et al.*, (1971b); and Kapusta and Rouwenhorst, (1973) have indicated the prevalence of strains of 'serogroup 123' in the nodules of field-grown soybean in the northern Midwest. Moawad *et al.* (1984) noted that even when other strains were present in the soil in numbers similar to serogroup 123, most nodules were formed by strains of this serogroup. Thus, the prevalence of strains of serocluster 123 is not simply a function of numbers, but results from their inherent competitiveness for nodule formation.

In addition to their competitiveness, strains of serocluster 123 are purported to be less effective in N_2 fixation than many other strains. Three reports (Caldwell and Vest, 1970 Chamber *et al.*, 1983; Ham, 1980) indicated that yields from soybean grown in soybean-rhizobia-free soil and inoculated with the serotype strain USDA 123 were inferior to yields obtained with inoculant-quality strains such as USDA 110 or USDA 138. Similarly, Kvien *et al.* (1981) indicated that significant yield increases were obtained whenever 50% or more of the nodules of soybean were occupied by strains UDA 110 or USDA 138, rather than by strains of indigenous serogroup 123. These reports support the contention that the displacement of serocluster 123 in favor of more effective *B. japonicum* strains should result in increased soybean productivity.

A long-standing problem of those studying N_2 fixation by soybean has been to effect the displacement of undesired indigenous rhizobia with more effective strains. Even when a high rate of an inoculant strain is applied to soil with an indigenous population, the recovery of the inoculant strain in the nodules is generally less than 20% (Caldwell and Vest, 1970; Ham *et al.*, 1971a; Kuykendall and Weber, 1978). Weaver and Frederick (1974) indicated that to establish an introduced inoculum strain in 50% of the nodules of a soybean plant, the inoculation rate needed to be at least 1000 times greater (g^{-1} soil) than the indigenous population. Thus, with current inoculation technology, the displacement of indigenous strains with those carried in inoculum is difficult and expensive.

Displacement of indigenous soybean bradyrhizobia using gene-for-gene interaction

Given the precedence of the Rj_2, Rj_3, and Rj_4 genes in soybean, it seemed plausible that a gene or genes existed in soybean germplasm that would restrict the nodulation of serocluster 123 strains. If such genes existed and could be incorporated into agronomically superior soybean genotypes, then nodulation with the indigenous serocluster 123 strains could be eliminated or reduced in favor of a highly effective inoculant. Cregan and Keyser (1986) initiated a search of the USDA soybean germplasm collection and reported the existence of 13 soybean genotypes that restricted the nodulation of the serotype strain USDA 123. Two of these genotypes, PI 371607 and PI 377578 (PI genotypes), in addition to the standard production cultivar Williams, were tested more extensively. The PI genotypes nodulated quite well with the highly effective strain USDA 110 and also were found to reduce the competitiveness of strain USDA 123 (Table 3).

In a subsequent study, Keyser and Cregan (1987) inoculated Williams and the two PI genotypes with 20 serocluster 123 isolates collected from the U.S. and the People's Republic of China. All of the isolates abundantly nodulated Williams, but the USDA 123-restricting PI genotypes restricted nodulation by four of the 20 isolates in a manner similar to the serotype strain, USDA 123. It was concluded that the USDA 123-restricting PI genotypes reduced nodulation and competitiveness of only a portion of the 123 serocluster, and that additional soybean genotypes would have to be identified that would restrict nodulation by other serocluster 123 members.

Accessions from the USDA Soybean Germplasm Collection were inoculated with the serocluster 123 isolate MN1-1c that produces normal nodulation on PI 371607 and PI 377578. This screen identified PI 417566, that restricted the nodulation of isolate MN1-1c (Cregan *et al.*, 1989). When PI 417566, along with Williams and PI 377578, were inoculated with approximately equal numbers of cells of MN1-1c and the highly effective strain CB1809, at least 90% of the nodules formed by Williams and PI 377578 were occupied by MN1-1c (Table 4). However, in the case of PI 417566, only 23% of the nodules

Table 3. Nodulation, total N accumulation, and bradyrhizobial strains recovered from nodules of Williams and two PI genotypes grown in soybean-rhizobia-free soil inoculated as indicated (From Cregan and Keyser, 1986)

Soybean genotype	Inoculum applied					
	USDA 123		USDA 110		USDA 123 and USDA 110	
	Nodule dry weight	Total N	Nodule dry weight	Total N	Strain recovered from nodules: USDA 123	USDA 110
	mg/plant				%	
Williams	211 a[+]	66 a	138 a	98 a	79 a	45 b
PI 371607	22 b	29 b	156 a	107 a	6 b	99 a
PI 377578	37 b	30 b	180 a	106 a	9 b	97 a

[+] Means within a column not followed by the same letter differ significantly ($P = 0.05$).

Table 4. Competition for nodulation between strains of *B. japonicum* on three soybean genotypes grown in soybean-rhizobia-free soil (From Cregan *et al.*, 1989)

Soybean genotype	Competition treatment			
	MN1-1c versus CB 1809		MN1-1c versus USDA 123	
	Percent nodules occupied by:			
	MN1-1c	CB 1809	MN1-1c	USDA 123
Williams	92 a[+]	9 b	75 a	44 b
PI 377578	97 a	8 b	100 a	8 c
PI 417566	23 b	70 a	4 b	100 a

[+] Means within a column not followed by the same letter differ significantly ($P = 0.05$).

contained MN1-1c, demonstrating the ability of this genotype to alter nodule occupancy.

In our search to find soybean genotypes resistant to nodulation by MN1-1c, we also identified PI 283326 that appeared to nodulate poorly with a number of *B. japonicum* strains (Cregan *et al.*, 1989). We determined the breadth of nodulation restriction conditioned by PI 417566 and PI 283326. These genotypes were grown in rhizobia-free soil and inoculated with 20 isolates of serocluster 123. Of the 15 serocluster 123 isolates whose nodulation is not restricted by PI 377578, 12 were restricted by either PI 417566, PI 283326, or both. Thus, of the 20 serocluster 123 isolates tested, the nodulation of only three was not restricted by at least one of the three PI genotypes (Keyser *et al.*, 1988). This result suggests that we have identified a number of strain specific host factors that appear to function in a gene-for-gene fashion. We are close to having identified the host factors necessary to restrict the nodulation and reduce the competitiveness of the whole spectrum of indigenous serocluster 123 strains.

Host genetic control of restricted nodulation in PI 377578 and PI 417566

We investigated the inheritance of restricted nodulation of strain USDA 123 by PI 377578 in the greenhouse using F_1 and F_2 plants, as well as the parents, of a cross between Williams and PI 377578 grown in rhizobia-free soil and inoculated with strain USDA 123. The distribution of nodule number per plant of the two parents did not overlap. The F_1 nodule number was similar to that of PI 377578. The distribution of F_2 plants was bimodal, with plants falling in the ranges of the two parents. This preliminary result approximated a 3:1 ratio of restricted to non-restricted nodulation and suggested the likelihood that nodulation restriction of USDA 123 is controlled by a single dominant gene.

In the case of PI 417566, rather than determining the inheritance of nodulation restriction, we determined the inheritance of reduced competitiveness of MN1-1c in a cross of Williams × PI 417566. The results of these experiments indicated that the reduced competitiveness of isolate MN1-1c is also conditioned by a single dominant gene (Cregan *et al.*, 1988).

Conclusion

The gene-for-gene hypothesis of plant-pathogen interaction provides a simple framework in which to view symbiotic plant-microbe interactions. For many years plant breeders and pathologists have exploited the results of gene-for-gene interaction in the development of disease resistant plant cultivars. Our work has sought a similar goal of developing a soybean genotype that resists infection by undesired indigenous rhizobia. We have the additional goal of engineering a strain of soybean bradyrhizobia that will efficiently nodulate such a soybean genotype. Thus, the work reported by Sadowsky *et al.* (1990) in this volume has two goals. The first is to define the *B. japonicum* genes controlling genotype specific nodulation of soybean and to conclusively demonstrate gene-for-gene interaction in the soybean-*B. japonicum* symbiosis. The second, is to identify and isolate the genetic factors necessary for the construction of a *B.*

japonicum strain capable of nodulating a soybean genotype that excludes most or all indigenous bradyrhizobia. The engineered bradyrhizobia must also condition highly effective N_2 fixation. Our dual approach to develop both partners in the plant-microbe interaction represents a new application of the gene-for-gene system.

References

Bergersen F J and Nutman P S 1957 Symbiotic effectiveness in nodulated red clover. IV. The influence of the host factors i_1 and *ie* upon nodule structure and cytology. Heredity 11, 175–184.

Biffen R H 1905 Mendel's Laws of inheritance and wheat breeding. J. Agric. Sci. 1, 4–48.

Caldwell B E 1966 Inheritance of a strain-specific ineffective nodulation reaction in the soybean *Glycine max* L. Merrill. Crop Sci. 495–496.

Caldwell B E and Vest G 1970 Effect of *Rhizobium japonicum* strains on soybean yields. Crop Sci. 10, 19–21.

Chamber Perez M A, Orive Echevarrieta R, Temprano Vera F J and Lavandera Gonzalez G 1983 Seleccion de raza de *Rhizobium japonicum* par la produccion de inoculantes para soja. Eurosoya 1, 21–25.

Cregan P B and Keyser H H 1986 Host restriction of nodulation by *Bradyrhizobium japonicum* strain USDA 123 in soybean. Crop Sci. 26, 911–916.

Cregan P B, Keyser H H and Sadowsky M J 1988 A single dominant gene in PI 417566 controls competitiveness for nodulation of *B. japonicum* isolate MN1-1c. Am. Soc. Agron. Abstracts. Madison, WI, p 78.

Cregan P B, Keyser H H and Sadowsky M J 1989 Soybean genotype restricting nodulation of a previously unrestricted serocluster 123 bradyrhizobia. Crop Sci. 29, 307–312.

Damirgi S M, Frederick L R and Anderson I C 1967 Serogroups of *Rhizobium japonicum* in soybean nodules as affected by soil types. Agron. J. 59, 10–12.

Davis E O, Evans I J and Johnston A W B 1988 Identification of nodX, a gene that allows *Rhizobium leguminosarum* biovar *viceae* strain TOM to nodulate Afghanistan peas. Mol. Gen. Genet. 212, 531–535.

Day P R 1974 Genetics of Host-parasite Interaction. W H Freeman and Co., San Francisco, 238 p.

Devine T E 1984 Inheritance of soybean nodulation response with a fast-growing strain of *Rhizobium*. J. Hered. 75, 359–361.

Djordjevic M A, Gabriel D W and Rolfe B G 1987 *Rhizobium* – The refined parasite of legumes. Ann. Rev. Plant Path. 25, 145–168.

Ellingboe A H 1976 Genetics of host-parasite interactions. *In* Encyclopedia of Plant Pathology. Vol. 4, Physiological Plant Pathology. Eds. R Heitfuss and P H Williams. pp 761–778. Springer-Verlag, Heidelberg, FRG.

Ellis W R, Ham G E and Schmidt E L 1984 Persistence and recovery of *Rhizobium japonicum* inoculum in a field soil. Agron. J. 76, 573–576.

Flor H H 1946 Genetics of pathogenicity in *Melampsora lini*. J. Agric. Res. 73, 335–357.

Flor H H 1947 Inheritance of reaction to rust in flax. J. Agric. Res. 74, 241–262.

Flor H H 1955 Host-parasite interaction in flax rust – its genetics and other implications. Phytopathology 45, 680–685.

Flor H H and Comstock V E 1971 Flax cultivars with multiple rust-conditioning genes. Crop Sci. 11, 64–66.

Gabriel D W, Burges A and Lazo G R 1986 Gene-for-gene interactions of five cloned avirulence genes from *Xanthomonas compestris* pv. *malvacearum* with specific resistance genes in cotton. Proc. Natl. Acad. Sci. USA 83, 6415–6419.

Ham G E 1980 Interactions of *Glycine max* and *Rhizobium japonicum*. *In* Advances in Legume Science, Eds. R J Summerfield and A H Bunting. pp 289–296. Royal Botanical Gardens, Kew, UK.

Ham G E, Cardwell V B and Johnson H W 1971a Evaluation of *Rhizobium japonicum* inoculants in soils containing naturalized populations of rhizobia. Agron. J. 63, 301–303.

Ham G E, Frederick L R and Anderson I C 1971b Serogroups of *Rhizobium japonicum* in soybean nodules. Agron. J. 63, 69–72.

Holl F B 1975 Host plant control of the inheritance of dinitrogen fixation in the *Pisum-Rhizobium* symbiosis. Euphytica 24, 767–770.

Holl F B and LaRue T A 1975 Genetics of legume plant hosts. *In* Proc. 1st Int. Sym. Nitrogen Fixation. Eds. W E Newton and C J Nyman. pp 391–399. Washington St. Univ., Pullman, WA.

Kapusta G and Rouwenhorst D L 1973 Influence of inoculum size on *Rhizobium japonicum* serogroup distribution frequency in soybean nodules. Agron. J. 65, 916–919.

Keyser H H, Bohlool B B, Hu T S and Weber D F 1982 Fast-growing rhizobia isolated from the root nodules of soybean. Science 215, 1631–1632.

Keyser H H and Cregan P B 1987 Nodulation and competition for nodulation of selected soybean genotypes among *Bradyrhizobium japonicum* serogroup 123 isolates. Appl. Environ. Microbiol. 53, 2631–2635.

Keyser H H, Cregan P B and Sadowsky M J 1988 Soybean genotypes which restrict nodulation of strains of *Bradyrhizobium japonicum* serocluster 123. *In* Nitrogen Fixation: Hundred Years After. Eds. H Bothe, F J de Bruijn and W E Newton. pp 784. Gustav Fisher, Stuttgart, FRG.

Kuykendall L D and Weber D F 1978 Genetically marked *Rhizobium* identifiable as inoculum strain in the nodules of soybean plants grown in fields populated with *Rhizobium japonicum*. Appl. Environ. Microbiol. 36, 915–919.

Kvien C S, Ham G E and Lambert J W 1981 Recovery of introduced *Rhizobium japonicum* strains by soybean genotypes. Agron. J. 73, 900–905.

Lie T A 1971 Symbiotic nitrogen fixation under stress conditions. Plant and Soil, Special Vol., 117–127.

Lie T A 1978 Symbiotic specialization in pea plants: The

requirement of specific *Rhizobium* strains for peas from Afghanistan. Ann. Appl. Biol. 88, 462–465.

Lie T A 1984 Host geens in *Pisum sativum* L. conferring resistance to European *Rhizobium leguminosarum* strains. Plant and Soil 82, 415–425.

Lie T A and Timmermans P C J M 1983 Host-genetic control of nitrogen fixation in the legume-Rhizobium symbiosis: complication in the genetic analysis due to maternal effects. Plant and Soil 75, 49–453.

Moawad H A, Ellis W R and Schmidt E L 1984 Rhizosphere response as a factor in competition among three serogroups of indigenous *Rhizobium japonicum* for nodulation of field grown soybeans. Appl. Environ. Microbiol. 47, 607–612.

Newton M, Johnson T and Brown A M 1930 A study of the inheritance of spore colour and pathogenicity in crosses between physiologic form of *Puccinia graminis tritici*. Sci. Agric. 10, 775–798.

Nutman P S 1954 Symbiotic effectiveness in red clover. II. A major gene for ineffectiveness in the host. Heredity 8, 47–60.

Nutman P S 1968 Genetic effectiveness in nodulated red clover. V. The *n* and *d* factors for ineffectiveness. Heredity 22, 537–551.

Nutman P S 1969 Genetics of symbiosis and nitrogen fixation in legumes. Proc. Roy. Soc. B. 172, 417–437.

Sadowsky M J, Cregan P B, Rodriguez-Quinones F and Keyser H H 1990 Microbial influence on gene-for-gene interactions in legume-*Rhizobium* symbioses. Plant and Soil 129, 53–60.

Schmidt E L, Zidwick M J and Abede H H 1986 *Bradyrhizobium japonicum* serocluster 123 and diversity among member isolates. Appl. Environ. Microbiol. 51, 1212–1215.

Scholla M H and Elkan G H 1984 *Rhizobium fredii* sp. nov., a fast-growing species that effectively nodulates soybeans. Int. J. Syst. Bacteriol. 34, 484–486.

Staskawicz B J, Dahlbeck D and Keen N T 1984 Cloned avirulence gene of *Pseudomonas syringae* pv. *glycinea* determines race specific incompatibility on *Glycine max* (L.) Merr. Proc. Natl. Acad. Sci. USA 81, 6024–6028.

Vest G 1970 Rj_3 – A gene conditioning ineffective nodulation in soybean. Crop Sci. 10, 34–35.

Vest G and Caldwell B E 1972 Rj_4 – A gene conditioning ineffective nodulation in soybean. Crop Sci. 12, 692–693.

Weaver R W and Frederick L R 1974 Effect of inoculum rate on competitive nodulation of *Glycine max* (L.) Merr. II. Field studies. Agron. J. 66, 233–236.

D. L. Keister and P. B. Cregan (Eds.), The rhizosphere and plant growth, 173–180.

PLSO BARC18

Microbial influence on gene-for-gene interactions in legume-Rhizobium symbioses

MICHAEL J. SADOWSKY[1,2], PERRY B. CREGAN[2], FRANCISCO RODRIGUEZ-QUINONES[2] and HAROLD H. KEYSER[2,3]

[1]*Department of Soil Science and Department of Microbiology, University of Minnesota, St. Paul, MN 55108, USA,* [2]*Soybean and Alfalfa Research Laboratory, USDA-ARS, Beltsville, MD 20705, USA and* [3]*The NifTAl Project, Paia, HI 96779, USA*

Key words: avirulence loci, genotype-specific nodulation, host-bacterial interactions, molecular biology, nodulation restriction, rhizobia/bradyrhizobia

Abstract

Recent advances in our understanding of the molecular genetics of legume-Rhizobium symbioses have indicated that relatively few bacterial genes are required for nodulation. While some of these genes are functionally similar and shared among microsymbionts nodulating genetically diverse legumes, others appear to encode host-specific nodulation (*hsn*) functions which allow for nodulation of plants within a given legume genus. More recently, genotype-specific nodulation (GSN) determinants have been identified in *R. leguminosarum* bv. *viceae* strain TOM and in *B. japonicum* strain USDA 110. GSN determinants refer to those bacterial sequences that allow for nodulation of specific plant genotypes within a given legume species. In contrast to the *avr* loci of several plant pathogens, rhizobial host-range determinants (*hsn* and GSN) have been shown to positively affect nodulation. That is, the introduction of exogenous *hsn* and GSN loci extends host-range. Since GSN loci have been reported to interact with single host plant alleles, it suggests that gene-for-gene interactions occur in rhizobial-legume symbioses and contribute to nodulation specificity at the host genotype level.

Introduction

Certain plant pathogen races elicit a hypersensitive response on a particular host plant depending on the presence of single dominant alleles in the plant host and corresponding dominant alleles in the pathogen. In the host plant these alleles are termed avirulence genes (Flor, 1956). If either bacterial or plant allele is recessive, the plant does not recognize the pathogen and plant defense mechanisms are not activated. The end result of this interaction is that the plant becomes diseased (Fig. 1). On the other hand, if the pathogen is recognized, then the plant activates a series of disease defense mechanisms which culminate in what is termed a hypersensitive reaction (Keen, 1982; Staskawicz, 1987). From this, one can view the hypersensitive response itself as the end product of the interaction of the alleles of both partners. The analogy we would like to draw to this is that legume nodule initiation and formation is similarly the end result of the interaction of specific plant alleles and specific bacterial genes. Regardless of the system we are discussing, the overall interaction between pathogen or symbiont and its host can thus be viewed as the genetic interrelationship of the host-range determinants of both partners. Moreover, the similarity between both systems suggest that many of the same mechanisms that are operating to control specificity in plant pathogen-host interactions may be operating in legume-Rhizobium symbioses. The similarity between the pathogenic and symbiotic sys-

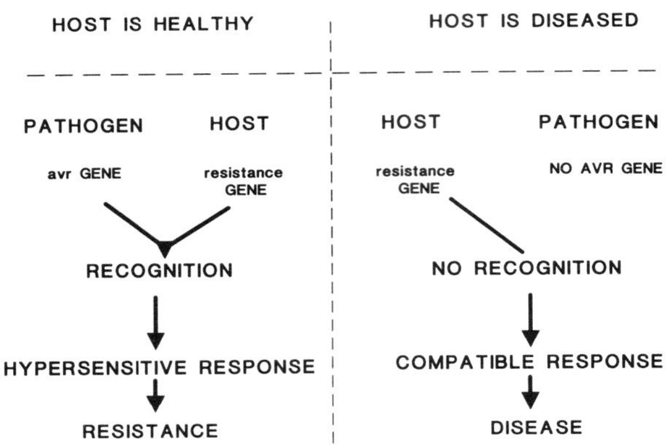

Fig. 1. Model describing the gene-for-gene interaction between a pathogen and its host leading to hypersensitive and compatible responses.

tems is further exemplified by the fact that results from several studies have indicated that gene-for-gene recognition in plant pathogen interaction is superimposed on an organism's ability to parasitize (Ellingboe, 1976; Gabriel, 1986; Stall, 1986). That is, an organism must have the basic genetic ability to be phytopathogenic before it can induce a hypersensitive response on a particular host plant. The same idea is apparently true for Rhizobium/legume interactions; the specificity of host-specific nodulation is superimposed on the ability to nodulate itself.

Models for host-microbe interactions

While the avirulence gene-resistance gene model (Flor, 1946) elegantly describes the genetic interaction between the dominant alleles of the flax plant and its fungal pathogen, flax rust, results from several studies indicate that there may be more than one way in which host and microbe genetic specificity determinants can interact. While some of the bacterial specificity determinants act negatively to control plant microbe interactions (Djordjevic *et al.*, 1985 and 1987; Leroux *et al.*, 1987), others act positively (Davis *et al.*, 1988; Horvath *et al.*, 1987; Keen and Staskawicz, 1988; Staskawicz *et al.*, 1984; Yanofsky and Nester, 1986). In addition, some plant host-specificity determinants act dominantly (Cregan *et al.*, 1989b; Crute, 1986; Flor, 1971; Mukherjee *et al.*, 1966; Vest and Caldwell, 1972), while others are recessive (Holl, 1975). These multiple mechanisms appear to be operating to control specificity in legume-Rhizobium symbioses and will be discussed in more detail below. But before beginning a discussion of specific rhizobial host-range determinants and how they interact with the host in a gene-for-gene manner, we will briefly review some of what is known about two types of rhizobial host-range genes which are involved in the early stages of legume-microbe interactions.

Genetics of Rhizobium and Bradyrhizobium

Recent advances in our understanding of the molecular genetics of nodulation have been achieved mostly with the fast-growing legume nodulating bacteria *Rhizobium meliloti* and *R. leguminosarum* biovars *viceae* and *trifolii* (see reviews by Denarie *et al.*, 1981; Kondorosi and Johnston, 1981 and Long, 1984 and 1989). This is mostly due to the fact that nodulation and nitrogen fixation genes are located on large indigenous symbiotic, or Sym, plasmids in these species (Long, 1989; Sadowsky and Bohlool, 1983). In *Rhizobium meliloti*, one of the better studied species, the essential nodulation genes, $nodD_1ABC$, are closely linked (Fig. 2) and located in a gene cluster on the Sym plasmid (Debelle *et al.*, 1986; Denarie *et al.*, 1981; Kondorosi *et al.*, 1984; Long *et al.*, 1989). Data from hybridization and transcription fusion studies indicated that the *nodABC* genes are transcribed in the same direction and that $nodD_1$ is located upstream of *nodA* and is divergently transcribed (Egelhoff *et al.*, 1985; Horvath *et al.*, 1986; Rossen *et al.*, 1984). Interspecies complementation studies have demonstrated that *nodABC* genes are functionally similar among rhizobial species and thus they have been termed common nodulation genes (Kondorosi *et al.*, 1984). In *R. leguminosarum* biovars *viceae* and *trifolii* and in *R. meliloti*, mutations in the common *nod* genes result in a root hair curling minus (Hac^-) phenotype (Djordjevic *et al.*, 1987; Egelhoff *et al.*, 1985; Torok *et al.*, 1984). Thus, these genes are thought to be involved in the very early stages of symbioses. In *R. meliloti*, another nodulation

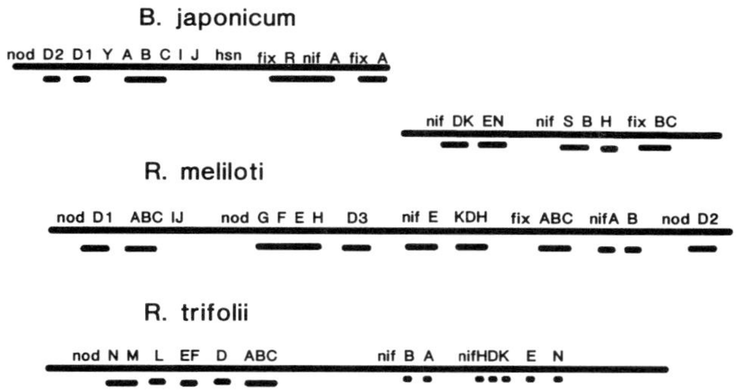

Fig. 2. Organization of nodulation and nitrogen fixation genes in Rhizobium and Bradyrhizobium. Transcriptional units are indicated by solid bars below nodulation (nod), nitrogen fixation (nif), or fixation-related (fix) genes. In *B. japonicum* the distance between nod and nif gene clusters has not been ascertained. Hsn refers to host-specific nodulation of siratro locus. Adapted from Stacey, 1988 and Long, 1989.

gene cluster has been reported to be located near the common *nod* region (Horvath *et al.*, 1986). This gene cluster is invovled in host specific nodulation (*hsn*) and has been designated *hsnABCD*. Analogous *hsn* genes, termed *nodEFG*, have also been reported in *R. leguminosarum* biovars *trifolii* (Schofield and Watson, 1986) and *viceae* (Shearman *et al.*, 1986; Spaink *et al.*, 1987). The *hsn* loci have been so named because mutations in these regions cannot be functionally complemented by analogous regions from other *Rhizobium* species. In addition to the *hsn* loci, the *nodD* genes of several *Rhizobium* species have also been reported to be determinants of host-specificity (Spaink *et al.*, 1987). To date, rhizobial host-range determinants have been defined by two means (Djordjevic *et al.*, 1987). Firstly, by the examination of nodulation mutants that have altered host-range properties; these mutants are those that cannot be complemented by DNA regions from other rhizobia. Secondly, *hsn* loci have been defined as those genes, which when transferred to a suitable strain, render the recipient capable of nodulating the donor's host plant. While many of the same nodulation gene loci have been found in *Bradyrhizobium japonicum*, two major organizational differences separate this bacterium from the fast-growers (Lambe and Hennecke, 1986; Nieuwkoop *et al.*, 1987; Sadowsky and Bohlool, 1983; Stacey, 1988). Firstly, in *B. japonicum*, the nodulation genes are not located on plasmids, but rather are distributed on the chromosome and secondly, the nitrogen fixation structural genes *nifDK* and *H* are physically separated a relatively large distance from the *nod* genes.

In addition to the two classes of nodulation loci discussed above, those involved with the rhizobial cell surface and cell surface polysaccharides are also reported to influence the specificity of symbioses. A detailed discussion on their role in symbiotic interactions can be found in several recent reviews (Djordjevic *et al.*, 1987; Long, 1989; Stacey, 1988).

Regulation of *nod* gene expression

Most information on nodulation gene regulation comes from *lacZ*-transcription and/or translation studies done with *R. meliloti* and *R. leguminosarum* biovars *trifolii* and *viceae* (Djordjevic *et al.*, 1987; Long, 1989; Peters *et al.*, 1986; Zaat *et al.*, 1987). Similar studies done with other fast-growers (Sadowsky *et al.*, 1988) and with *B. japonicum* (Kosslak *et al.*, 1988) indicate that the induction of many nod genes requires the presence of plant-derived signal molecules. The induction, which can be seen at the level of transcription, depends on the *nodD* gene product (Fisher *et al.*, 1987; Long, 1989). Current models suggest that the *nodD* gene product directly interacts with plant-derived signal molecules (Long, 1989). Moreover, the *nodD* protein has been shown to bind to putative promoter areas, termed the *nod* box sequence (Rostas *et al.*, 1986), of inducible nod genes (Fisher *et al.*, 1988; Hong *et al.*, 1987). Taken together, biochemical and genetical studies suggest that the *nodD* protein is a DNA binding, transcriptional activator (Long, 1989; Rostas *et al.*, 1986).

The identities of many of these plant-derived, *nod* gene-inducing, substances have been determined (Djordjevic *et al.*, 1987; Kosslak *et al.*, 1987; Peter *et al.*, 1986; Sadowsky *et al.*, 1988; Zaat *et al.*, 1987) and in the case of the fast-growing rhizobia, belong to classes of compounds called flavanols, flavones, and flavanones. While less is known about *nod* gene expression and regulation in *B. japonicum*, it has been shown that the inducing substances generally belong to flavanoid molecules of the isoflavone class (Kosslak *et al.*, 1987; Banfalvi *et al.*, 1988). Interestingly, isoflavones such as daidzein and genistein are capable of repressing *nod* gene induction in the clover (Djordjevic *et al.*, 1987) and pea (Firmim *et al.*, 1986) systems.

While earlier studies suggested that *nodD* genes from different rhizobial (or bradyrhizobial) species were functionally similar, recent studies (Banfalvi *et al.*, 1988; Horvath *et al.*, 1987; Spaink *et al.*, 1987) have shown that the *nodD* proteins from different rhizobia do not have identical behavior. Thus, in the strict sense, the *nodD* gene itself can also be viewed as a host-range determinant.

Hierarchy and types of host-range functions

As mentioned briefly above, rhizobial host-range determinants appear to function in concert with those genes controlling nodulation itself. For example, its quite easy to isolate mutants of several *Rhizobium* species that are impaired in their ability to nodulate only specific plant species; *e.g. B. japonicum* strains that still nodulate soybeans, but not cowpea, mungbean, or siratro (Hahn and Hennecke, 1988; Nieuwkoop *et al.*, 1987). Taken together, results from several studies now suggest that there is a hierarchy to nodulation determinants. While some host-range determinants affect nodulation at the legume genus level, such as the *hsn* loci of *R. meliloti* (Horvath *et al.*, 1986), others affect nodulation of a single genotype within a given legume species (Davis *et al.*, 1988; Heron *et al.*, 1989).

In the classic case of gene-for-gene interactions with plant pathogens, the avirulence loci are dominant alleles that function negatively to narrow host range. Several rhizobial host range determinants also appear to negatively function to limit host range. For example, Djordjevic and coworkers (1985) reported that a Tn5 mutation in a *R. leguminosarum* bv. *trifolii* gene region, caused an extended host-range phenotype. In addition, Heron *et al.* (1989) reported the isolation of *R. fredii* Tn5 mutants with extended host ranges. Similarly, Horvath and coworkers (1986) indicated that *hsnD* (*nod H*) mutants of *R. meliloti* have an extended host range which now include two *Vicia* species.

In addition to these types of negative factors, there are also several clear examples where host range is controlled by positively acting host-range determinants. In these cases, Tn5 insertions in specific gene regions cause narrowing of host-range. For example, Tn5 insertions in *hsn* or nod-1 loci of *B. japonicum* affect nodulation of siratro or cowpea, but not soybean (Hahn and Hennecke, 1988; Nieuwkoop *et al.*, 1987). Positively acting host-range determinants are also clearly exemplified in experiments where host range is extended by the transfer of new genes into an organism. For example, the host range of *R. meliloti* can be extended to include siratro when strains contain the *nodD1* gene from a broad host-range strain of Rhizobium, NGR234 (Horvath *et al.*, 1987).

Genotype-specific nodulation loci

While *hsn* loci have been localized in many legume nodulating bacteria, little information is available about genotype-specific nodulation (GSN) genes. GSN loci refers to those bacterial sequences which allow nodulation of specific plant genotypes within a given legume species. Such GSN genes have been found in *R. leguminosarum* bv. *viceae* strain Tom (Davis *et al.*, 1988) and, together with its pea host, illustrates the gene-for-gene interaction between legumes and a Rhizobium strain. While strain Tom can nodulate the primitive pea (*Pisum sativum*) cv. Afghanistan, the conventional European *R. leguminosarum* bv. *viceae* strains fail to nodulate this host (Brewin *et al.*, 1980).

The bacterial genes allowing strain Tom to nodulate Afghanistan pea have been located on the large Sym plasmid pRL5JI, and Gotz and coworkers (1985) have identified and cloned a 2.0-kb region of this plasmid that is responsible for Afghanistan pea nodulation. Recently, Davis *et al.* (1988) reported that a single gene in this region, named *nodX*, is necessary for the nodulation of cv. Afghanistan pea. The *nodX* locus is located downstream of the nodIJ genes in strain TOM (Fig. 3). Data from lacZ fusion studies indicated that *nodX* is moderately induced by pea root-exudate and by the flavanone hesperetin. Interestingly, there are no homologs to *nodX* in conventional *R. leguminosarum* bv. *viceae* strains. While mutations in *nodX* result in a nod$^-$ phenotype on Afghanistan pea, they do not affect nodulation of European pea genotypes. Moreover, transfer of *nodX* to conventional strains allows the recipients to nodulate Afghanistan peas. Thus, a positively acting host-range determinant controls the nodulation of a single pea genotype, and what makes this a good example of gene-for-gene interactions in symbioses is that host-controlled nodulation restriction in Afghanistan pea has been shown to be conditioned by a single recessive plant allele (Holl, 1975).

Interestingly, a similar gene-for-gene system also appears to be functioning to control nodulation of *B. japonicum* serocluster 123 isolates on soybean. As was pointed out previously (see Cregan *et al.*, this volume and Cregan and Keyser, 1986; Cregan *et al.*, 1989a; Keyser and Cregan, 1987; Sadowsky *et al.*, 1987), we have been examining several soybean plant introduction genotypes which specifically exclude nodulation of the relatively competitive members of *B. japonicum* serocluster 123. This serocluster is composed of strains in serogroups 123, 127, and 129 and some of the plant introduction (PI) genotypes restrict the nodulation of strains in an apparently serogroup-specific manner. For example, PI 377578 restricts the nodulation of all serogroup 123 strains and PI 417566 restricts nodulation of all tested serogroup 129 isolates. Some isolates, like DE3-1a, however, have a broad host-range and their nodulation is not restricted by any of the PI genotypes. Since this differential nodulation restriction was phenotypically strikingly similar to that which was found with the Afghanistan pea system, it suggested to us that specific plant alleles may be interacting, in a gene-for-gene manner, with particular bradyrhizobial genes.

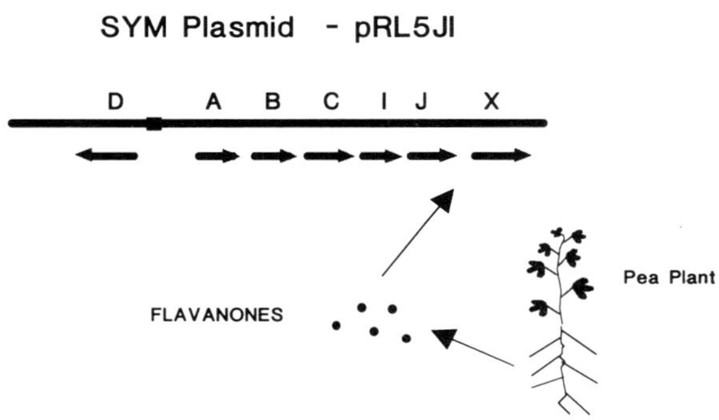

Fig. 3. Physical and genetic map of nodulation genes in *Rhizobium leguminosarum* bv. *viceae* strain TOM. Arrows on bars below the line refer to direction of transcription of individual nod genes. The nodX gene is involved in nodulation of Afghanistan pea and is located on the symbiotic plasmid pRL5JI. Adapted from Davis *et al.*, 1988.

To test this, we initially began by examining the interaction of serogroup 123 strains (like USDA 162 or a South Dakota isolate, SD6-1c) with the USDA 123 nodulation-restricting genotype, PI 377578. It should be noted that this genotype nodulates normally with inoculant quality strains, like USDA 110 or the serogroup 127 strain, MN1-1c. To identify those bradyrhizobial genes involved in the genotype-specific nodulation of soybeans, we used a nodulation complementation approach. We began by transferring cloned regions of the *B. japonicum* genome from strain USDA 110 to the serogroup 123 strains, SD6-1c and USDA 162. The two clones tested, pR29 and pR32 (Russell *et al.*, 1985) overlap the large common nodulation gene-containing cosmid pRJUT10 (see Nieuwkoop *et al.*, 1987). When the SD6-1c and USDA 162, pR29 and pR32 transconjugants were inoculated onto the PI genotypes (in soil), only the pR32 transconjugants nodulated the restrictive soybean host (Table 1). In most instances the SD6-1c and USDA(pR32) transconjugants nodulated the PI genotypes to an extent equal that of the nodulation unrestricted isolate MN1-1C.

To more precisely determine the location of the GSN locus, we subcloned 5 of the Hind III

Table 1. Nodulation of *Glycine max* PI 377578 by serocluster 123 strains SD6–1C, USDA 162, and their pR32 transconjugants in Monmouth soil

Isolate	Nodule number/Plant[a]
Uninoculated	3.6 ± 0.4
MN1-1C	19.9 ± 3.8
USDA 162	5.0 ± 1.0
USDA 162 (pR32)	16.4 ± 1.9
SD6-1C	2.0 ± 0.4
SD6-1C (pR32)	17.0 ± 1.2

[a] Values are means of at least five replicates ± standard error.

Table 2. Nodulation of *Glycine max* PI 377578 in Monmouth soil by SD6-1C and USDA 162 transconjugants containing plasmid pMJS12

Isolate	Soybean genotype	
	PI 377578	Williams
	Nodule number per plant[a]	
Uninoculated	1.5 ± 0.7	3.0 ± 1.8
SD6-1C	3.5 ± 2.2	14.3 ± 2.2
SD6-1C (pMJS12)	16.9 ± 3.7	19.5 ± 4.2
USDA 162	6.1 ± 1.1	11.3 ± 0.6
USDA 162 (pMJS12)	12.3 ± 0.8	12.1 ± 2.1
MN1-1C	18.5 ± 3.2	15.8 ± 1.3

[a] Values are means of at least three replicates (6–12 plants) ± standard error.

fragments from plasmid pR32 [called pMJS12, 9, 18, 24, and 22 (Fig. 4)] and individually mated them into USDA 162 and SD61c. Results of a subsequent nodulation study indicated that only subclone pMJS12 complemented the serogroup 123 isolates for nodulation of the restricted soybean host (Table 2). The 2.3-kb gene region cloned in pMJS12 is located about 2.9 kilobases transcriptionally downstream of nodD. Since the gene locus appears to affect soybean nodulation at the host genotype level, we have also termed this locus as a genotype-specific nodulation region. To determine if host-derived signal molecules also induce the soybean GSN region, we made *lacZ* transcriptional-fusions in and around the 2.3-kb HindIII fragment of pR32. Results indicated that the GSN gene region was moderately induced by soybean seed extract and the isoflavone genistein (data not shown). Fusion studies also indicated that this region is constitutively transcribed at low levels and transcribed in the same direction as *nodD*.

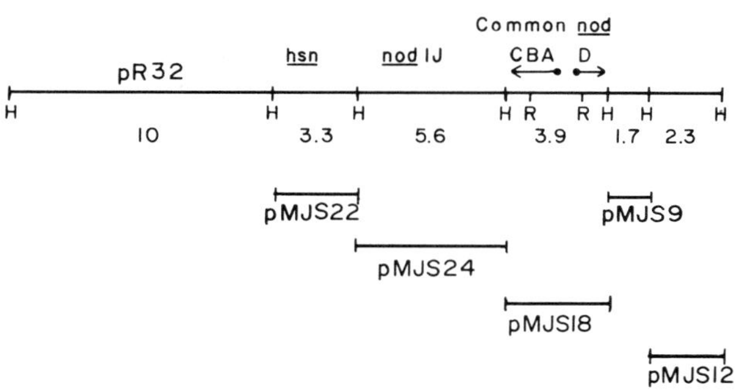

Fig. 4. Physical relationship of pMJS subclones to the nodulation complementing cosmid pR32. The HindIII fragments of pR32 were sub-cloned into cosmid pVK102 as described in the text. The location of the hsn, nodIJ, and nodDABC genes on pR32 are shown. Numbers below the line are fragment sizes in kilobases.

Interestingly, while the *R. leguminosarum* bv. *viceae* GSN gene *nodX* was present only in nodulation unrestricted strains, like strain TOM, the *B. japonicum* GSN gene region hybridizes to DNAs from both nodulation restricted and unrestricted strains. Hybridization, however, was to different sized *EcoR1* fragments in these two types of strains and suggested that this locus may be arranged differently in these phenotypically divergent organisms. While at present we are not sure of the significance of the differently sized hybridizing fragments, the observed hybridization profiles do have a strong correlation with nodulation restriction. We are currently sequencing the GSN gene region to learn more about its structure and possible mode of regulation.

In conclusion, our results indicate, that like *nodX* in strain TOM, the *B. japonicum* GSN gene region functions as a positively acting host-range determinant. Just as importantly, some of our recent studies have indicated that nodulation restriction in the plant introduction genotypes is conditioned in each, by a single dominant host allele. These results suggest that the genetic components of both symbiotic partners may interact to influence nodulation in the soybean-Bradyrhizobium symbiosis. Taken together, our results, and those of others (Davis *et al.*, 1988; Heron *et al.*, 1989) suggest that gene-for-gene interactions between a host plant and its microsymbiotic partner may be a relatively common mechanism for controlling nodulation specificity at the legume genotype level.

References

Banfalvi Z, Nieuwkoop A, Schell M, Besl L and Stacey G 1988 Regulation of nod gene expression in *Bradyrhizobium japonicum*. Mol. Gen. Genet. 214, 420–424.

Brewin N J, Beringer J E and Johnston A W B 1980 Plasmid mediated transfer of host range specificity between two strains of *Rhizobium leguminosarum*. J. Gen. Microbiol. 128, 1817–1827.

Cregan P B and Keyser H H 1986 Host restriction of nodulation by *Bradyrhizobium japonicum* strain USDA 123 in soybean. Crop Sci. 26, 911–916.

Cregan P B, Keyser H H and Sadowsky M J 1988 Soybean genotypes which reduce the competitiveness of strains of *Bradyrhizobium japonicum* serocluster 123. *In* Nitrogen Fixation: Hundred Years After. p. 782. Gustav Fischer Verlag, N.Y.

Cregan P B, Keyser H H and Sadowsky M J 1989a A soybean genotype that restricts nodulation of a previously unrestricted isolate of *Bradyrhizobium japonicum* serocluster 123. Crop Sci. 29, 307–312.

Cregan P B, Keyser H H and Sadowsky M J 1989b The use of host × strain incompatibilities in soybean to restrict nodulation of indigenous strains of *Bradyrhizobium japonicum*. *In* Proc. World Soybean Res. Conf. IV. Ed. A J Pascale. pp 1131–1137.

Crute I R 1986 Investigations of gene-for-gene relationships: the need for genetic analyses of both host and parasite. Plant Pathol. 35, 15–17.

Davis E O, Evans I J and Johnston A W B 1988 Identification of nodX, a gene that allows *Rhizobium leguminosarum* biovar *viceae* strain TOM to nodulate Afghanistan peas. Mol. Gen. Genet. 212, 531–535.

Debelle F, Rosenberg C, Vasse J, Maillet F, Martinez E, Denarie J and Truchet 1986 Assignment of symbiotic developmental phenotypes to common and specific nodulation (nod) genetic loci of *Rhizobium meliloti*. J. Bacteriol. 168, 1075–1086.

Denarie J, Boistard P, Casse-Delbart F, Atherley A G, Berry J O and Russell P 1981 Indigenous plasmids of *Rhizobium*. *In* International Review of Cytology, Supplement 13. Eds. G H Bourne and J F Danielle. pp 225–246. Academic Press. New York.

Djordjevic M A, Gabriel D W and Rolfe B G 1987 *Rhizobium* – the refined parasite of legumes. Annu. Rev. Phytopathol. 25, 145–168.

Djordjevic M A, Redmond J W, Batley M and Rolfe B G 1987 Clovers secrete specific phenolic compounds which either stimulate or repress nod gene expression in *Rhizobium trifolii*. EMBO J. 6, 1173–1179.

Djordjevic M A, Schofield P R and Rolfe B G 1985 Tn5 mutagenesis of *Rhizobium trifolii* host-specific nodulation genes results in mutants with altered host-range ability. Mol. Gen. Genet. 200, 463–471.

Egelhoff T T, Fisher R F, Jacobs T W, Mulligan J T and Long S R 1985 Nucleotide sequence of *Rhizobium meliloti* 1021 nodulation genes: nodD is read divergently from nodABC. DNA 4, 241–248.

Firmin J L, Wilson K E, Rossen L and Johnston A W B 1986 Flavanoid activation of nodulation genes in *Rhizobium* is reversed by other compounds present in plants. Nature 324, 90–92.

Fisher R F, Egelhoff T T, Mulligan J T and Long S R 1988 Specific binding of proteins from *Rhizobium meliloti* cell-free extracts containing nodD to DNA sequences upstream of inducible nodulation genes. Genes Dev. 2, 282–293.

Flor H H 1946 Genetics of pathogenicity in *Melamspora lini*. J. Agric. Res. 73, 335–357.

Flor H H 1956 The complementary genetic systems in flax and flax rust. Adv. Genet. 8, 29–54.

Flor H 1971 Current status of the gene-for-gene concept. Annu. Rev. Phytopath. 9, 275–296.

Gotz R, Evans I J, Downie J A and Johnston A W B 1985 Identification of the host-range DNA which allows *Rhizobium leguminosarum* strain TOM to nodulate cv. Afghanistan peas. Mol. Gen. Genet. 201, 296–300.

Hahn M and Hennecke H 1988 Cloning and mapping of a novel nodulation region from *Bradyrhizobium japonicum* by genetic complementation of a deletion mutant. Appl. Environ. Microbiol. 54, 55–61.

Heron D S, Ersek T, Krishan H B and Pueppke S G 1989 Nodulation mutants of *Rhizobium fredii* USDA 257. Molec. Plant-Microbe Interact. 2, 4–10.

Holl F B 1975 Host plant control of the inheritance of dinitrogen fixation in the *Pisum-Rhizobium* symbiosis. Euphytica 24, 767–770.

Hong G-F, Burn J E and Johnston A W B 1987 Evidence that DNA involved in the expression of nodulation (nod) genes in *Rhizobium* binds to the regulatory gene nodD. Nucl. Acids Res. 15, 9677–9690.

Horvath B, Bachem C W B, Schell J and Kondorosi A 1987 Host-specific regulation of nodulation genes in *Rhizobium* is mediated by a plant signal, interacting with the nodD gene product. EMBO J. 6, 841–848.

Horvath B, Kondorosi E, John M, Schmidt J, Torok I, Gyorgypal Z, Barabas I, Wieneke U, Schell J and Kondorosi A 1986 Organization, structure and symbiotic function of *Rhizobium meliloti* nodulation genes determining host specificity for alfalfa. Cell 46, 335–343.

Keen N T 1982 Specific recognition in gene-for-gene host parasite systems. Adv. Plant Pathol. 1, 35–82.

Keen N T and Staskawicz B 1988 Host-range determinants in plant phatogens and symbionts. Annu. Rev. Microbiol. 42, 421–440.

Keyser H H and Cregan P B 1987 Nodulation and competition for nodulation of selected soybean genotypes among *Bradyrhizobium japonicum* serogroup 123 isolates. Appl. Environ. Microbiol. 53, 2631–2635.

Kondorosi E, Banfalvi Z and Kondorosi A 1984 Physical and genetic analysis of a symbiotic region of *Rhizobium meliloti*: identification of nodulation genes. Mol. Gen. Genet. 193, 445–452.

Kondorosi A and Johnston A W B 1981 The genetics of *Rhizobium*. *In* International Review of Cytology, Supplement 13. Eds. G H Bourne and J F Danielle. pp 191–224. Academic Press, New York.

Kosslak R M, Bookland R, Barkei J, Paaren H E and Appelbaum E R 1987 Induction of *Bradyrhizobium japonicum* common nod genes by isoflavones isolated from *Glycine max*. Proc. Natl. Acad. Sci. USA 84, 7428–7432.

Lamb J W and Hennecke H 1986 In *Bradyrhizobium japonicum* the common nodulation genes nod-ABC are linked to Nif-A and Fix-A. Mol. Gen. Genet. 202, 512–517.

Leroux B, Yanofsky M F, Winans S C, Ward J E, Ziegler S F and Nester E W 1987 Characterization of the virA locus of *Agrobacterium tumefaciens*: a transcriptional regulator and host-range determinant. EMBO J. 6, 849–856.

Long S R 1984 The genetics of *Rhizobium* nodulation. In Plant Microbe Interactions. Eds. Kosuge and E W Nester. pp 265–306. Macmillan Pub. Co., New York.

Long S R 1989 *Rhizobium*-legume nodulation: life together in the underground. Cell 56, 203–214.

Mukherjee D, Lambert J, Cooper R and Kennedy B 1966 Inheritance of resistance to bacterial blight of soybeans. Crop Sci. 6, 324–326.

Nieuwkoop A J, Banfalvi Z, Deshmane N, Gerhold D, Schell M G, Sirotkin K M and Stacey G 1987 A locus encoding host range is linked to the common nodulation genes of *Bradyrhizobium japonicum*. J. Bacteriol. 169, 2631–2638.

Peters K, Frost J W and Long S R 1986 A plant flavone, luteolin, induces expression of *Rhizobium meliloti* nodulation genes. Science 233, 917–1008.

Rossen L, Johnston A W B and Downie A 1984 DNA sequence of the *Rhizobium leguminosarum* nodulation genes AB and C required for root hair curling. Nucl. Acids Res. 12, 9497–9508.

Rostas K, Kondorosi E, Horvath B, Simoncsits A and Kondorosi A 1986 Conservation of extended promoter regions of nodulation genes in *Rhizobium*. Proc. Natl. Acad. Sci. USA 83, 1757–1761.

Russell P, Schell M G, Nelson K K, Halverson L J, Sirotkin K M and Stacey G 1985 Isolation and characterization of a DNA region encoding nodulation functions in *Bradyrhizobium japonicum*. J. Bacteriol. 164, 1301–1308.

Sadowsky M J and Bohlool B B 1983 Possible involvement of a megaplasmid in nodulation of soybeans by fast-growing rhizobia from China. Appl. Environ. Microbiol. 46, 906–911.

Sadowsky M J and Bohlool B B 1985 Differential expression of the pea symbiotic plasmid, pJB5JI, in genetically dissimilar backgrounds. Symbiosis 1, 125–138.

Sadowsky M J, Olson E R, Foster V E, Kosslak R M and Verma D P S 1988 Two host-inducible genes of *Rhizobium fredii* and the characterization of the inducing compound. J. Bacteriol. 170, 171–178.

Sadowsky M J, Tully R E, Cregan P B and Keyser H H 1987 Genetic diversity in *Bradyrhizobium japonicum* serogroup 123 and its relation to genotype-specific nodulation of soybean. Appl. Environ. Microbiol. 53, 2624–2630.

Schofield P R and Watson J M 1986 DNA sequence of *Rhizobium trifolii* nodulation genes reveals a reiterated and potentially regulatory sequence preceding nodABC and FE. Nucl. Acids Res. 14, 2891–2903.

Shearman C A, Rossen L, Johnston A W B and Downie J A 1986 The *Rhizobium leguminosarum* nodulation gene nodF encodes a polypeptide similar to acyl carrier protein and is regulated by nodD plus a factor in pea root exudate. EMBO J. 5, 647–652.

Spaink H P, Wijffelman C A, Pees E, Okker R J H and Lugtenberg B J J 1987 *Rhizobium leguminosarum* nodulation gene nodD as a determinant of host specificity. Nature 328, 337–340.

Stacey G 1988 Genetics of symbiotic nitrogen fixation. *In* The Impact of Chemistry on Biotechnology-Multidisciplinary Discussions. Eds. M Phillips, S P Shoemaker, R D Middlekauff and R M Ottenbrite. pp 262–278. American Chemical Society, Washington, DC.

Staskawicz B J, Dahlbeck D and Keen N 1984 Cloned avirulence gene of *Pseudomonas syringae* pv. *glycinea* determines race-specific incompatibility on *Glycine max* (L.) Merr. Proc. Natl. Acad. Sci. USA 81, 6024–6028.

Staskawicz B, Dahlbeck D, Keen N and Napoli C 1987 Molecular characterization of cloned avirulence genes from race 0 and race 1 of *Pseudomonas syringae* pv. *glycinea*. J. Bacteriol. 169, 5789–5794.

Torok I, Kondorosi E, Stepkowski T, Posfai J and Kondorosi A 1984 Nucleotide sequence of *Rhizobium meliloti* nodulation genes. Nucl. Acids Res. 12, 9509–9524.

Vest G and Caldwell B E 1972 Rj4 – A gene conditioning ineffective nodulation in soybean. Crop Sci. 12, 692–693.

Yanofksy M F and Nester E W 1986 Molecular characterization of a host-range determining locus from *Agrobacterium tumefaciens*. J. Bacteriol. 168, 244–250.

Zaat S A, Wijffelman C A, Spaink H P, Van Brussel A A N, Okker R J H and Lugtenberg B J J 1987 Induction of the nodA promoter of *Rhizobium leguminosarum* Sym plasmid pRL1J1 by plant flavanones and flavones. J. Bacteriol. 169, 198–204.

Short papers

D. L. Keister and P. B. Cregan (Eds.), The rhizosphere and plant growth, 183.

PLSO BARC51

Extracellular polysaccharide-deficient mutants of Rhizobium strain CIAT899 induce chlorosis in beans

R.S. ARAUJO, G.A. BEATTIE and J. HANDELSMAN
Department of Plant Pathology, University of Wisconsin, Madison, WI 53706, USA

Key words: EPS, *Rhizobium leguminosarum* bv. *phaseoli*, Tn5 mutagenesis, toxin production

Extracellular polysaccharide (EPS) production is required for nodulation by some rhizobia but not for nodulation of beans by *Rhizobium leguminosarum* bv. *phaseoli*. In order to study the role of EPS in nodulation of beans by the wide host range Rhizobium strain CIAT899, we produced EPS-deficient mutants of this strain.

Ten EPS-deficient Tn*5* mutants of CIAT899 from a total of six separate mutageneses were identified by screening for altered colony morphology on solid medium. The 10 mutants produced colonies that were 1/4 the diameter of colonies produced by the wild-type parent, CIAT899, suggesting that these mutants were either deficient in EPS production or grew more slowly than the wild type. In order to differentiate between these possibilities, growth rates and EPS production by the mutants was measured in broth culture. All but one of the mutants had doubling times similar to that of the wild type, approximately 2.0 h, but they all produced less extracellular material than the wild type.

All of the EPS-deficient mutants induced pink nodules on bean plants, but the plants developed a severe interveinal chlorosis on the trifoliate leaves. The chlorosis was not reversed by the addition of nitrogen or micronutrients to the plants. When each EPS-deficient mutant was applied to plants in a 1:1 mixture with the wild type, CIAT899 predominated in the nodules.

Our results suggest a relationship between the loss of EPS production and acquisition of the chlorosis-inducing phenotype since all of the EPS-deficient mutants also induced chlorosis. That the Tn*5* is causal to both mutant phenotypes remains to be demonstrated.

The phenomenon described here is similar to that described by LaFavre *et al.* (1988) who showed that capsule-deficient variants of *Bradyrhizobium japonicum* strain USDA 76 induce chlorosis on soybeans, whereas capsule-producing variants do not. Chlorosis induction by USDA 76 is due to production of rhizobitoxine, an amino acid derivative, which inhibits methionine metabolism. The similarity between the soybean system and the data presented here regarding plant symptoms and the effects of EPS suggests that the EPS-deficient mutants of CIAT899 may produce rhizobitoxine. This is the first indication of toxin production among the fast-growing rhizobia.

Reference

LaFavre J S *et al.* 1988 Can. J. Microbiol. 34, 1017–1022.

D. L. Keister and P. B. Cregan (Eds.), The rhizosphere and plant growth, 184.

PLSO BARC52

Rhizosphere interactions between rhizobia and nonlegumes

G.L. BENDER, J. PLAZINSKI and B.G. ROLFE
Plant Microbe Interaction Group, Research School of Biological Sciences, Australian National University, Canberra, ACT, 2601 Australia

Key words: nonlegumes, rhizobia

The process of infection and nodule development in the Parasponia symbiosis with rhizobia differs significantly from that found in the legume symbiosis (Lancelle and Torrey, 1984, 1985; Bender *et al.*, 1987). Despite these differences, only minor genetic changes are needed in rhizobia to initiate Parasponia nodule development (Bender *et al.*, 1988). Transfer of the *nodDl* gene from Rhizobium strain NGR234, which nodulates Parasponia, to the clover-specific *R. leguminosarum* bv *trifolii* strain ANU843 and derivatives extends the host range of the recipients to include Parasponia. It is also possible to extend the host range of strain ANU843 to Parasponia by inducing single base pair changes in the resident *nodD* gene in this strain (McIver *et al.*, 1989). The *nodDl* gene enables the induction of *nod*:*lac* functions by root extracts from Parasponia and other nonlegumes including sunflower, cotton, wheat, rice and maize. The most active inducer in wheat root extracts has been isolated and identified and is unlike any signals isolated from legumes (K. Le Strange, pers. comm.). The induction of nodulation genes by nonlegume root extracts does not lead to nodulation, except for Parasponia. However, strain ANU845 (a Sym plasmid-cured derivative of strain ANU843) carying *nod* genes on a multicopy vector can induce a root hair curling response on maize and rice plants (Plazinski *et al.*, 1985). This is an important observation because root hair curling is an early plant response in the infection of many legumes by rhizobia.

References

Bender G L *et al.* 1987 Plant Sci. 51, 285–293.
Bender G L *et al.* 1988 Mol. Plant Microbe Interac. 1, 259–266.
Lancelle S A and Torrey J G 1984 Protoplasma 123, 26–37.
Lancelle S A and Torrey J G 1985 Can. J. Bot. 63, 25–35.
McIver J *et al.* 1989 Mol. Plant Microbe Interac. (*In press*).
Plazinski J *et al.* 1985 J. Bacteriol. 163, 812–815.

D. L. Keister and P. B. Cregan (Eds.), The rhizosphere and plant growth, 185.

PLSO BARC53

Photosynthetic N_2-fixing Rhizobia

J.M. ELLIS[1], M. HUNGRIA[2], R.W.F. HARDY[1] and A.R.J. EAGLESHAM[3]
[1]*Boyce Thompson Institute, Ithaca, NY 14853, USA* [2]*EMBRAPA-UAPNPBS, Km 47 antiga Ria-Sao Paulo, Seropedica, RJ, Brazil 23851 and* [3] *EniChem Americas Inc., Monmouth Junction, NJ 08852, USA*

Key words: bacteriochlorophyll, $^{14}CO_2$ uptake, O_2 uptake, stem nodule

An isolate from a stem nodule on the legume *Aeschynomene indica*, rhizobial strain BTAi 1, is pink when cultured in the light with a photoperiod of 16 h. The absorption spectra of whole and extracted cells of BTAi 1 indicate the presence of bacteriochlorophyll and carotenoid pigments. The objective of this work was to determine photosynthetic functionality of cultured cells. The non-pigmented rhizobial strain BTSr 3, from a stem nodule on *Sesbania rostrata*, was included as a check.

BTAi 1 failed to grow under anaerobic or C-free conditions, whether illuminated or not. Illumination had no significant effect on growth during the logarithmic phase, although $^{14}CO_2$ uptake was increased. However, after depletion of exogenous C, illuminated cells of BTAi 1 retained viability whereas dark-grown cells did not. Illumination had no effect on $^{14}CO_2$ uptake by, or on stationary-phase viability of, stain BTSr 3.

Rates of uptake of O_2 were significantly decreased upon illumination of light-grown cells of BTAi 1 (but not upon illumination of dark-grown cells). The above CO_2 and O_2 responses to light are indicative of photosynthesis in BTAi 1. Illumination had no effect on O_2-uptake by light-grown BTSr 3.

Although the combination of photosynthesis and N_2 fixation is common in the purple bacteria and cyanobacteria, BTAi 1 is the first photosynthetic rhizobium to have been described. We believe it should be assigned a new genus, and suggest that it be named *Photorhizobium thompsonianum*.

D. L. Keister and P. B. Cregan (Eds.), The rhizosphere and plant growth, 186.

PLSO BARC54

Physiological comparisons between root and stem nodules of *Aeschynomene scabra* and *Sesbania rostrata*

M. HUNGRIA[1], A.R.J. EAGLESHAM[2] and R.W.F. HARDY[3]
[1]*EMBRAPA-UAPNPBS, Km 47 antiga Rio-Sao Paulo, Seropedica, RJ, Brazil 23851,* [2]*EniChem Americas Inc., Monmouth Junction, NJ 08852, USA and* [3]*Boyce Thompson Institute, Ithaca, NY 14853, USA*

Key words: acetylene reduction, $^{14}CO_2$ uptake, GOGAT, GS

Species in the legume genera Aeschynomene and Sesbania can form nodules on both stems and roots. The cortical cells of stem nodules contain numerous chloroplasts, therefore it seems likely that energy for N_2 fixation originates *in situ*. The objective of this study was to verify that photosynthesis occurs in stem nodules, and to determine whether parameters of N_2 fixation differ between stem and root nodules.

In the greenhouse, stems and roots of *A. scabra* were inoculated with rhizobial strain BTAi 1, and those of *S. rostrata* with azorhizobial strain BTSr 3. Five weeks later, specific acetylene reduction activities, as well as GS and GOGAT activities were significantly higher in the stem nodules than in the root nodules of the respective species. All activities were highest in the Aeschynomene stem nodules which accumulated 24% more N than did root nodules.

Photosynthesis by stem nodules was verified by higher rates of uptake of $^{14}CO_2$ in the light than in the dark; $^{14}CO_2$ uptake activity of root nodules was unaffected by illumination, and in the dark was the same for root and stem nodules. The simulation of CO_2 uptake upon illumination was 11-fold for *S. rostrata* and 38-fold for *A. scabra*.

The data show that processes associated with N_2 fixation are more efficient in stem nodules than in root nodules, and more markedly so in *A. scabra*. This may be a function of energy production *in situ*. The high rates of $^{14}CO_2$ by *A. scabra* stem nodules are consistent with a photosynthetic capability in rhizobial strain BTAi 1 (Ellis *et al.*, 1990).

Reference

Ellis J M *et al.* 1990 *In* The Rhizosphere and Plant Growth. Eds. D L Keister and P B Cregan. p 185. Kluwer Academic Publishers, Dordrecht, The Netherlands.

D. L. Keister and P. B. Cregan (Eds.), The rhizosphere and plant growth, 187.

PLSO BARC55

Isolation and characterization of cowpea (*Vigna unguiculata*) lectin

R. LIU and E.L. SCHMIDT
Department of Soil Science, University of Minnesota, St. Paul, MN 55108, USA

Key words: cowpea, lectin isolation, lectin properties, methods

Successful initiation of the symbiotic relationship between legume plants and rhizobia depends on specific interactions between the partners. One current theory, the lectin binding hypothesis, holds that plant lectin binds complementary rhizobial exopolysaccharide (EPS) to accomplish the recognition. Cowpea is an important legume grain plant largely distributed in tropical regions. It is known that cowpea can be nodulated by various *B. japonicum* and rhizobia from wild legumes, as well as the cowpea rhizobia. Because the cowpea is relatively non-selective in symbiosis, it is of interest to examine its lectin(s). The lectin status of the cowpea is poorly understood with but one previous report available (Roberson and Strength, 1983). We report procedures leading to the extraction and isolation of a lectin from cowpea cultivar Bigboy, and summarize its properties. Separation of the cowpea lectin involved ammonium sulfate fractionation, acidification, gel filtration, ion exchange and affinity chromatography (fetuin agarose). Affinity chromatography is recognized as a very useful tool to purify lectins. We first tried acid-treated Sepharose 6B reported by Allen and Johnson (1976) since the crude extracts were strongly inhibited by D+ galactose. No adsorption peak was detected after eluting with 0.2 *M* D+ galactose. The fetuin-bound agarose gel reported by Sela *et al.* (1976) is a multi-lectin binding medium, which has been used for purification of several lectins (Donatucci *et al.*, 1987), which fall into glucose/mannose, N-acetyl glucosamine, and N-acetyl galactosamine/galactose families according to the Goldstein and Poretz (1986) classification. The fetuin agarose column yielded cowpea lectin in a sharp peak following elution with 0.25 *M* sodium chloride. Purified lectin presented a single band in native polyacrylamide gel electrophoresis. Periodic acid-Schiff's (PAS) strain indicated that the lectin is a glycoprotein. Sodium dodecyl sulfate (SDS) gel electrophoresis revealed components with molecular weights in the range of 20k to 65k. Sugar binding specificity tests showed minor inhibition by glucose, mannose, and N-acetyl glucoseamine and much stronger inhibition by galactose. The lectin agglutinated rabbit erythrocytes but not human type A, B, O, AB or dog erythrocytes.

References

Allen H J and Johnson E A Z 1976 Carbohydrate Research 50, 121–131.
Donatucci D A *et al.* 1987 J. Nutr. 117, 2154–2160.
Goldstein I J and Poretz R D 1986 *In* The Lectins: Properties, Functions and Application in Biology and Medicine. Eds. I E Liener, N Sharon and I J Goldstein. pp 34–248. Academic Press Inc., Orlando.
Roberson B J and Strength D R 1983 Preparative Biochemistry 13, 45–56.
Sela B *et al.* 1976 J. Biol. Chem. 250, 7535–7538.

D. L. Keister and P. B. Cregan (Eds.), The rhizosphere and plant growth, 188.

PLSO BARC56

Rapid plate assay for hydrolytic enzymes of Rhizobium

N. SALEH-RASTIN[1], M.A. PETERSEN[2], S.E. COLEMAN[2] and D.H. HUBBELL[2]
[1]*Department of Soil Science, University of Tehran, Karadj, Iran, and* [2]*Department of Soil Science, IFAS 0151, University of Florida, Gainesville, FL 32611-0151, USA*

Key words: hydrolytic enzymes, plate assays, Rhizobium

It has been suggested that extracellular hydrolytic enzymes function in the penetration phase of legume root infection by Rhizobium. It is now generally accepted that rhizobia produce cellulase, hemicellulase and pectinase enzymes. Studies of the genetics and physiology of these enzymes in relation to the infection process are currently hampered by the absence of rapid and reliable *in vitro* assays for the enzymes. Such studies are critical to assessment of the importance of these enzymes in the infection process. This work was initiated as a comparative study that would result in rapid and reproducible assays for the qualitative demonstration of these enzymes in rhizobia.

Strains of *Rhizobium leguminosarum* biovar *trifolii* assayed were noninfective strains 0402 and 0435-2 and infective strains 0403, 0435 and CIAT 79. Controls used were *Xanthomonas campestris* bv *campestris* (pectinase), *X. campestris* bv *vesicatoria* (cellulase) and *Pseudomonas aeruginosa* (protease). The rhizobia were maintained on yeast extract-mannitol (YEM) agar and control bacteria on trypticase soy agar. Bacteria to be assayed for enzyme activity were grown in induction media which included YEM salts, 0.1% yeast extract, 0.1% mannitol and the following inducers: (1) 0.1% carboxymethylcellulose (CMC) for cellulase; (2) 0.1% pectin or polygalacturonic acid for pectinase and (3) 0.1% gelatin for protease. Pure enzymes and heat-killed bacteria were used as controls. For the assay procedure, cell were washed from the induction media with 0.1 *M* phosphate citric acid buffer, pH 5 and spot inoculated onto plates containing a 1 mm layer (5 mL) of assay medium prepared with 1% GelRite in the same buffer. The following substrates were used: (1) 1% CMC for cellulase; (2) 0.001% pectin or polygalacturonic acid for pectinase; and (3) 0.1% gelatin for protease. The plates were incubated for 1 hr in a moist chamber, the cells washed from the plates and flooded with developing media consisting of 1% hexadecyltrimethyl ammonium bromide for 20 min or 0.1% Congo red followed by bleaching with 1 *M* NaCl for 15 min and rinsed in 0.05 *M* HCl to stabilize the clear zones of hydrolysis.

All strains tested were positive for cellulase. Optimum pectinolytic activity occurred when the bacteria were grown in the presence of 0.1% mannitol + 0.5% glycerol using 0.1% polygalacturonic acid or pectin as substrates. Optimum proteolytic activity was observed with strains 0403 and 0402 at pH 7. Although this method is not quantitative, it provides a rapid method for estimating the potential for infectivity of strains of Rhizobium.

D. L. Keister and P. B. Cregan (Eds.), The rhizosphere and plant growth, 189.

PLSO BARC57

Evaluation of *Rhizobium meliloti* strains indigenous to Sonoran Desert soils by plasmid profile and intrinsic antibiotic resistance

M. SHISHIDO and I.L. PEPPER[1]
[1]*Department of Soil and Water Science, University of Arizona, Tucson, AZ 85721, USA*

Key words: alfalfa, nitrogen fixation, rhizobia

Introduction

In Arizona, alfalfa (*Medicago sativa* L.) is not normally inoculated and yet yields are high due to biological nitrogen fixation by indigenous strains of *Rhizobium meliloti*. The objectives of this study were 1.) to determine the dominant *R. meliloti* strains indigenous to the Southwest, 2.) to evaluate environmental factors influencing the distribution of these dominant strains, and 3.) to assess the effectiveness of those indigenous strains on a locally adapted alfalfa cultivar.

Materials and methods

Five sampling locations were selected from the established alfalfa production areas of Arizona. Root nodules were randomly sampled from individual alfalfa plants from fields which had never received commercial inoculants. Sites were sampled during the winter of 1987 and the summer of 1988. Nodules were surface sterilized, and N fixing organisms isolated. Identification of *R. meliloti* strains was by plasmid profile analysis and intrinsic antibiotic resistance. Dominant strains were considered to be those showing more than 20% nodule occupancy at a particular location. The effectiveness of four dominant strains, indigenous strains and one commercial strain, 102F77b (Nitragin Co. Milwaukee, Wisconsin) was determined in a Leonard jar study.

Results and discussion

Four dominant strains were identified. Strain AZTCYJ occupied 31.5% of all nodules analyzed and had 3 plasmids. Strain AZSC, occupied 14% of all nodules analyzed and had 4 plasmids. Strain AZY, occupied 11% of all nodules analyzed, and also had 4 plasmids. Strain AZJT, occupied 4% of all nodules analyzed, and had 1 plasmid. Strain AZSC was the only dominant strain found at Safford site where the ESP was 18.5%. Thus, AZSC may well be tolerant of high sodium content. Strain AZY was found only in Yuma, and was more common in summer than the winter (80% *vs* 30%). In Yuma maximum air temperatures frequently exceed 45°C and soil temperatures at 10 cm depth approach 40°C. Thus strain AZY may be tolerant of high temperature. Strain AZTCYJ was the one most commonly found in this research and occupied 31.5% of all nodules. This strain seems to be well adapted to the semi-arid desert southwest, except for environmental extremes such as soils with high sodium in Safford and the high temperature area around Yuma. All the dominant indigenous *R. meliloti* strains were equally as effective as the commercial strain, at least in association with the alfalfa cultivar, Lew.

The full version of this paper is published in *Soil Biology and Biochemistry* 22 (1990), 11–16.

D. L. Keister and P. B. Cregan (Eds.), The rhizosphere and plant growth, 190.

PLSO BARC58

Enhancing rhizobial colonization in the rhizosphere of legumes by a systemic fungicide

M.A. SIDDIQI and M. ALEXANDER
Laboratory of Soil Microbiology, Department of Agronomy, Cornell University, Ithaca, NY 14853, USA

Key words: *Bradyrhizobium japonicum*, competition, *Rhizobium meliloti*

Inoculated rhizobia and bradyrhizobia do not extensively colonize the rhizosphere of legumes. To become established, rhizobia must compete with the indigenous rhizosphere microflora. A study was designed to find means to enhance colonization of the legume rhizosphere by using a basipetally translocated fungicide, aliette.

Strains of *Rhizobium meliloti* and *Bradyrhizobium japonicum* resistant to 500 μg of streptomycin and 200 μg of aliette per ml were used. Alfalfa and soybeans were grown in environmental chambers at 27°C. The populations of the two rhizobia on the root system were determined on agar amended with streptomycin and aliette.

There was no difference in the density of *R. meliloti* in the rhizosphere of alfalfa grown from seeds treated with aliette at 7 or 3 μg per seed, but the populations in both the cases were larger than on plants not receiving aliette. The greatest nodule abundance was associated with use of high aliette concentrations combined with a large inoculum size. A foliar spray of aliette in addition to seed coating enhanced early colonization and nodulation. Application of aliette to seed or foliage often increased colonization of the rhizosphere by resistant rhizobia during different stages of plant growth. Soybeans derived from seeds coated with aliette bore more nodules and had greater nitrogenase activity than plants receiving a foliar spray although differences in plant weights were not found. The percentage of nodules formed by aliette-resistant *B. japonicum* was enhanced by coating the seeds with aliette. Aliette caused an increase in nodule weight in soybeans inoculated with *B. japonicum*. The results indicate that aliette-resistance may give rhizobia a selective advantage in the rhizosphere during early stages of plant growth, eventually leading to a greater number of nodules and enhanced nitrogenase activity.

D. L. Keister and P. B. Cregan (Eds.), The rhizosphere and plant growth, 191.

PLSO BARC59

Mapping and subcloning of genes involved in nodulation competitiveness from *Rhizobium leguminosarum* bv. *trifolii* T24

E.W. TRIPLETT, M.J. SCHINK and K.L. NOELDNER
Department of Agronomy and the Center for the Study of Nitrogen Fixation, 1575 Linden Dr., University of Wisconsin-Madison, Madison, WI 53706, USA

Key words: competition, nodulation, symbiotic nitrogen fixation

Introduction

Trifolitoxin consists of 10 amino acids with a thiazoline ring and a UV-absorbing chromophore of unknown structure (Triplett *et al.*, 1988). It is produced by the ineffective strain, *R. leguminosarum* bv. *trifolii* T24 and is bacteriostatic toward *R. leguminosarum* and *R. fredii* (Triplett and Barta, 1987). Production of trifolitoxin by T24 limits nodulation of clover roots by trifolitoxin-sensitive rhizobia (Schwinghamer and Belkengren, 1968; Triplett and Barta, 1987). The trifolitoxin production and resistance genes (*tfx*) have been cloned from a genomic library of T24 DNA (Triplett, 1988).

Materials and methods

Stability of trifolitoxin production in *tfx* transconjugants and exconjugants was determined by purifying each strain to single colonies up to 10 consecutive times on non-selective media. Trifolitoxin production was determined during each purification.

Results and dicussion

Restriction mapping of pTFX1::Tn5 mutants showed that the *tfx* region in pTFX1 was 4.4 kb in size. This region was subcloned with the resulting plasmid referred to as pTFX2. The trifolitoxin-resistance region in *tfx* was subcloned with the resulting plasmid referred to as pTFX4. The *tfx* region was inserted into the genome of *R.1.* bv. *trifolii* TA1 by marker exchange. Two isogenic, recombinant exconjugants of TA1 were produced in which one strain contains *tfx* with a Tn5 insertion 0.5 kb away from the *tfx* region while the other contains a Tn5 insertion inside *tfx*. As a result, one of these strains, TA1::10–15, produces trifolitoxin while the other, TA1::12–10, does not. These two isogenic strains were each co-inoculated on clover roots with the trifolitoxin-sensitive strain, *R.1..* bv. *trifolii* 2046. When TA1::10–15 was co-inoculated with 2046 at an inoculum ratio of 1.6:1, the trifolitoxin-producing exconjugant occupied 91% of the nodules while 2046 occupied 41%. When the trifolitoxin-minus exconjugant, TA1::12–10, was co-inoculated with 2046, only 19% of the nodules were occupied by the TA1 derivative while 100% of the nodules were occupied by 2046. Clover plants inoculated with either exconjugant were Fix^{+}. These experiments represent the first genetic construction of an effective, competitive strain of Rhizobium.

References

Schwinghamer E A and Belkengren R P 1968 Arch. Mikrobiol. 64, 130–145.
Triplett E W 1988 Proc. Natl. Acad. Sci. (USA) 85, 3810–3814.
Triplett E W and Barta T M 1987 Plant Physiol. 85, 335–342.
Triplett E W *et al.* 1988 *In* Nitrogen Fixation: Hundred Years After. Eds. H Bothe, F J de Bruijn and W E Newton. Gustav Fischer, Stuttgart. 788 p.

D. L. Keister and P. B. Cregan (Eds.), The rhizosphere and plant growth, 192.

PLSO BARC60

Competition for nodulation of soybean by *Bradyrhizobium japonicum* 123 and 138 in Missouri soil

R.E. ZDOR and S.G. PUEPPKE
Department of Plant Pathology, University of Missouri, Columbia, MO 65211, USA

Key words: Bradyrhizobium, competition, nodulation, soybean

Seed inoculation often fails to increase soybean yields due to highly competitive native populations of rhizobia. We have sought to characterize indigenous serogroups capable of nodulating soybeans in Missouri soil and to measure the competitiveness of peat-formulations of *B. japonicum* strains 123 (serogroup 123) and 138 (serogroup c1) in the soil.

Soybean seeds were planted in a field of Mexico silt loam soil and nodule isolates were serotyped. Serogroups c1, 38-115, 76, 122, c3, and 94 were found to occupy such nodules. Serogroup c1 was the most prevalent, occupying 20% of all nodules. Serogroup 123 occupied only 13% of all nodules, and about 17% of the nodules were occupied by unknown serogroups.

Containers of field soils were used for competition experiments in a growth chamber. Immunofluorescence was used to assess nodule occupancy by strains 123 and 138 after single or dual strain inoculants. Inoculation with either strain 123 or 138 in unsterile soil increased nodule occupancy by the inoculant serogroup 13- and 2-fold, respectively. In dually inoculated plants grown in unsterile soil, strain 138 was more competitive than strain 123, occupying twice as many nodules as strain 123 (Table 1). In contrast, strain 123 was more competitive than strain 138 in sterile soil, occupying 74% of all nodules. This dominance was even more pronounced on lateral roots. It appears that this soil lacks populations of highly competitive serogroups 123 and soil sterilization influences the competitiveness of strain 123.

Table 1. Competition of strains 123 and 138 on soybeans grown in Missouri field soil

Root inoculation		Mean % of nodules occupied by:			
	Soil	123	138	123/138	Neither
Primary	Unsterile	21	33	14	32
	Sterile	67	14	19	0
Lateral	Unsterile	20	60	10	10
	Sterile	86	9	5	0
All	Unsterile	20	44	12	23
	Sterile	74	12	14	0

D. L. Keister and P. B. Cregan (Eds.), The rhizosphere and plant growth, 193.

PLSO BARC61

Effect of acidified soil on symbiosis in soybeans

G.R. CLINE and KARAN KAUL
Community Research Service, Kentucky State University, Frankfort, KY 40601, USA

Key words: acidity, aluminum, nitrogen fixation, soybean

To improve the tolerance of soybean plants to acidic soils, it is important to determine which acidic soil infertility factors are most detrimental and whether the plant or the symbiotic relationships with *Bradyrhizobium japonicum* is more adversely affected. The objectives of this study were (1) to compare the effects of acidity and Al toxicity on soybeans grown in soil and (2) to determine if the plant or the symbiosis is more susceptible to acidic soil conditions.

A completely randomized 3 soil acidity × 5 N treatment factorial experimental design was used with each treatment combination replicated in three pots containing three plants each. The three soil acidity treatments consisted of nonacidified Maury silt loam soil and soil acidified with 7.5 g kg^{-1} $Al_2(SO_4)_3$ or 1.7 g kg^{-1} S_2. The N treatments included a nonamended control, addition of 100 mg N kg^{-1} soil as NH_4NO_3, and inoculation with commercial *B. japonicum* or the locally isolated KY1 strain plus the addition of 0.05 mg kg^{-1} Mo to the soil as $NaMoO_3$.

Treatment of soil with $Al_2(SO_4)_3$ resulted in a soil pH of 4.8, 30 μM soluble Al, and 0.77 $cmol_c$ $kg-^{-1}$ KCl-Al. Soil amended with S_2 was more acidic, pH 4.6, but only contained about half as much Al, 14 μM soluble Al and 0.37 $cmol_c$ KCl-Al.

Soil acidification with either acidulent did not significantly reduce shoot or root dry weights of plants receiving inorganic N, but acidification significantly ($\alpha = 0.05$) reduced shoot and root dry weights, nodule dry weights and number, and foliar N content of inoculated plants. Thus, under acidic soil conditions the adversely-affected symbiosis appeared to be the primary cause of reduced plant growth. Inoculated plants from S_2-amended soil were significantly ($\alpha = 0.05$) smaller and less well nodulated than those grown in $Al_2(SO_4)_3$-amended soil which was less acidic but contained more soluble Al than the former soil. Thus, acidity appeared to be more detrimental to plant growth than Al toxicity. Addition of Mo to inoculated plants was not beneficial.

Addition of the two soil acidulents also increased electrical conductivity (2.6 mS cm^{-1}), Ca (1.6 mM), and SO_4-ions (1.9 mM). However, these conditions did not appear to adversely affect the plant or bacterium since dry weights of N-fertilized plants were not significantly reduced by soil acidification and growth of KY1 *B. japonicum* in pure culture was not affected when subjected to similar conditions.

It was concluded that (1) acidity appeared to affect soybeans more so than did Al toxicity and (2) the induced acidified soil conditions affected the symbiosis more than the host plant.

D. L. Keister and P. B. Cregan (Eds.), The rhizosphere and plant growth, 194.

PLSO BARC62

Distribution of P metabolites and compartmentation in soybean nodules as studied by electron microscopy and ^{31}P NMR spectroscopy

D. ROLIN, P. PFEFFER, J. SCHIMDT, R. BOSWELL, P. COOKE and S. JONES
Eastern Regional Research Center, USDA, ARS Philadelphia, PA 19118, USA

Key words: electron microscopy, nitrogen fixation, nuclear magnetic resonance, phosphate

Introduction

In a previous study (Rolin *et al.*, 1989) we established the optimum parameters for obtaining ^{31}P spectra of functioning soybean nodules. The purpose of the present study was to investigate nodule compartmentation using *in vivo* ^{31}P NMR in conjunction with electron microscopy to establish the distribution of the metabolites in nodules.

Results and discussion

The root nodules consist of a heterogeneous cell population. The cortex surrounding the central matrix represents 50% of the total volume of the nodule. The active, N_2 fixing tissue is a central core of enlarged host parenchyma cells, which are packed with the endosymbiotic bacteria. The cells of the central tissue form a compact mass with the uninfected and infected cells. The infected host cells of the nodule central tissue are densely packed within the peribacteroid membrane.

During *in vivo* ^{31}P NMR experiments, all of the examined tissues remained alive for at least 20h when the perfusion medium contained 50m*M* glucose and was saturated with O_2. Eighty-seven percent of the total mobile Pi in the 9 week-old nodules was in the acidic compartment whose Pi shift corresponds to pH 5.5. A ^{31}P spectrum of the carefully excised cortex layer showed that 59% of the total mobile Pi of the nodules resided in the vacuole of the cortical cells. The spectrum of 7 week-old intact nodules shows that 77% of the total mobile Pi is present in the acidic compartment. The central matrix of the nodule was separated from the cortical cells by careful excision with a scalpel. Only 34% of the total mobile Pi in the matrix was found to be present in the vacuole of the uninfected cells, the only acidic compartment in the central matrix. In addition, it may be possible to estimate the relative quantity of mobile Pi in the bacteroid. In the *in vivo* spectrum of isolated bacteroids, the resonance at 2.30 ppm represents the cytoplasmic Pi of the bacteroids corresponding to a pH of 6.85. The bacteroids have a relatively more acidic cytoplasm than the observed pH 7.4 of the host plant cells. However, these pH dependent shift differences are not resolved in the nodule spectra due to broader peak widths. The relative distribution of cytoplasmic and vacuolar Pi in nodules was compared in 4 to 12 week-old nodules. In the earlier growth stages (4 week old), most of the Pi was found in the cytoplasm (61%). After 7 weeks, the majority of the Pi (77%) was stored in the vacuole of the host cells (cortical plus uninfected cells). Since most of the Pi of nodules at 4 weeks was present in the metabolically active cytoplasm, absorption of Pi probably did not greatly exceed the cell's needs. Throughout the growth period from 4 to 12 weeks the vacuolar Pi of the uninfected cells remained relatively constant at 32% of the total Pi. However, Pi was preferentially accumulated in the vacuole of the cortical cells (61% of the total Pi for the 12 week-old nodules) at the expense of the cytoplasmic Pi.

Reference

Rolin D *et al.* 1989 Plant Physiol. 89, 1238–1246.

Part 3

Rhizosphere interactions and plant rest control

D. L. Keister and P. B. Cregan (Eds.), The rhizosphere and plant growth, 197–209.

PLSO BARC19

Exudate molecules initiating fungal responses to seeds and roots

ERIC B. NELSON
Department of Plant Pathology, Cornell University, Ithaca, NY 14853, USA

Key words: Pythium, rhizosphere, spermosphere, spore germination

Abstract

Plant pathogenic fungi survive in soils in a quiescent state. In order for many root-pathogen interactions to be initiated, dormant propagules must be activated by molecules present in seed and root exudates. Without the release of such stimulatory molecules, the majority of root infections do not occur. Currently, little is known about the specific molecules involved in stimulating propagule germination and initiating root-pathogen interactions. Although certain molecules can be shown to elicit germination responses *in vitro*, responses of propagules reared on conventional culture media do not always reflect the responses of those formed on plant tissues in soil. Consequently, it is not possible to extend conclusions from laboratory determinations of the role of specific exudate molecules in stimulating fungal propagule germination to soil systems. The interaction of *Pythium* species with germinating seeds has served as a model system to answer questions about propagule behavior and the role of exudate stimulant molecules in establishing root-fungus interactions. The potential role of both volatile and water-soluble molecules in stimulating propagule germination are discussed.

Introduction

In agricultural soils, nutrient availability is a limiting factor for microbial growth and activity. Consequently, most microorganisms, including pathogenic fungi, exist in a state of exogenous dormancy or fungistasis (Lockwood, 1977). Propagules of soilborne fungi that arise from fungal thalli developing on and in infected plant tissues during pathogenesis, as well as those propagules released into soil upon decomposition of plant residues, are susceptible to fungistasis. However, fungistatic suppression can be readily overcome once propagules are supplied with appropriate organic and inorganic stimuli (Lockwood, 1977; 1988). Stimulants most commonly introduced into the soil environment that alleviate fungistasis are components of seed and root exudates released during seed germination and root development. Since the majority of host-parasite interactions in soil do not occur without the release of such stimuli (Curl and Truelove, 1986; Mitchell, 1976), an understanding of the molecules involved in the activation of quiescent fungal propagules by seed and root exudates is of prime importance in understanding the initiation and control of seed and root infections by soil-borne pathogens.

Although there have been numerous studies of seed and root exudation, much of the published work is of a descriptive nature. Little attention has been given to the biological activity of specific exudate molecules, particularly as they relate to the establishment of seed and root infections. As a result, there is a fairly good understanding of the types of molecules present in seed and root exudates but very little knowledge about how and when specific exudate components interact with root-infecting pathogens.

This review will focus on our current knowledge of exudate molecules potentially involved in activating fungal propagules in soil. I will not discuss research with mycorrhizal fungi, although it should be noted that considerable work has been done in this area. I wish to focus on initial responses of fungal propagules (*i.e.* activation and germination) to seed and root exudates since responses subsequent to propagule germination such as chemotaxis will be covered in a separate review as part of this symposium.

Seed and root exudation

Our knowledge of the nature of seed and root exudates has largely developed over the past 30–35 years as a result of much descriptive work concerning exudate composition and sites of production (*e.g.* Curl and Truelove, 1986; Hale *et al.*, 1971; 1978). As might be expected, those compounds typically found in plants as normal cell constituents and metabolic by-products are generally present in soluble seed or root exudates. Commonly described compounds present in seed and root exudates include sugars, amino acids, organic acids, flavonoids, sterols and proteins. Additionally, a great amount of the plant photosynthate that is transported to roots is deposited in the rhizosphere as insoluble or volatile components (Barber and Martin, 1976). Both the quantity and quality of materials released from germinating seeds and developing plant roots may be affected by factors altering the physiology of the plant (Curl and Truelove, 1986). For example, factors such as stage of plant development, soil pH, soil moisture, temperature, presence of microorganisms, *etc.* have been shown to influence the deposition of certain exudate components into the rhizosphere. Exudates released from germinating seeds are also normal products of metabolism and their production can be influenced by many of the same factors affecting root exudation.

Compared with soluble seed or root exudates, there have been relatively few investigations into the production and composition of volatile seed and root exudates. The principle volatile compounds released from germinating seeds are aldehydes, alcohols, ethylene, CO_2 and volatile fatty acids (Gorecki *et al.*, 1985; Stotzky and Schenck, 1976; Vancura and Stotzky, 1976). Due to technical difficulties in studying volatile exudates from roots, these have received much less attention than seed volatiles. Much of the earlier work with root volatiles has been reviewed previously (Linderman and Gilbert, 1975). Most of the volatile compounds that have been described from plant roots are low molecular weight alcohols, aldehydes, and organic acids. In some specific systems, more unique compounds and processes of evolution have been characterized and have been implicated in establishing specificity in root pathogen interactions (Coley-Smith and King, 1970).

Exudate stimulation of fungal propagule germination

In order for host-pathogen interactions to be initiated, dormant propagules of soilborne pathogens must germinate. Soluble and volatile exudates from germinating seeds and developing roots are the primary stimuli for such responses although other nutrient inputs will stimulate propagule germination of some soil fungi under certain conditions (Cook and Baker, 1983). Propagules of nearly all of the common genera of soilborne pathogens respond to seed or root exudates (Table 1). However, surprisingly little is known about the specific exudate molecules that elicit these responses.

Soluble components

The specific soluble molecules involved in stimulating propagule germination of soilborne fungal pathogens are largely unknown but are thought to involve primarily sugars and amino acids. Propagules of *Phytophthora*, *Macrophomina phaseolina* and *Verticillium dahliae* will respond to amino acids but not sugars (El-Hamalawi and Erwin, 1986; Mircetich *et al.*, 1968; Schreiber and Green, 1963; Smith, 1969) while propagules of *Pythium* and *Fusarium* spp. reared on synthetic culture media will respond to both sugars and amino acids. Endoconidia and chlamydospores of *Thielaviopsis basicola*, on the other hand, do not germinate in response to either sugars or amino acids but in response to C_{16}–C_{18} fatty acids (Papavizas and Adams, 1969; Papavizas and Kovacs, 1972). Palmitic, stearic, oleic, linoleic and palmitoleic acids were the major active fatty acid components detected after 8 days from the rhizosphere of a number of plant species grown in sand culture. None, however, were released from roots grown in solution culture.

Frequently studied host-pathogen interactions involve responses of *Pythium* and *Fusarium* spp. to germinating seeds and seedling roots. For example, sporangia of *Pythium ultimum* and sporangia and oospores of *P. aphanidermatum* germinate within 1.5 to 3 hr after exposure to bean or pea seed exudates (Lifshitz *et al.*, 1986; Stanghellini and Burr, 1973; Stanghellini and Hancock, 1971a, 1971b). Likewise, exudates collected from three-week-old pea roots increase germination of encysted zoospores of *P. aphanidermatum* (Chang-Ho, 1970) where germination is evident within 1 hr of exposure. Exudates collected from 6-day-old red pine seedlings stimulate the germination of sporangia and promote tube growth of *P. ultimum* (Agnihotri and Vaartaja, 1967a) and *P. irregulare* (Agnihotri and Vaartaja, 1970) and stimulate the germination of encysted zoospores of *P. afertile* (Agnihotri and Vaartaja, 1968). Agnihotri and Vaartaja (1967a, 1967b, 1970) found mixtures of sugars and amino acids to be most stimulatory to *P. ultimum* and *P. irregulare* while Chang-Ho (1970) found mixtures of glucose and organic acids to be most stimulatory to zoospore cysts of *P. aphanidermatum*. Stanghellini and Hancock (1971a) reported that at least 30 μg glucose/g soil were needed to induce maximum germination of sporangia of *P. ultimum* in soil. Schlub and Schmitthenner (1978) found as little as 12 μg glucose equivalents capable of inducing high levels of germination but found no correlation between increased carbohydrate exudation from germinating soybean seeds and increased acitvity of *P. ultimum*.

Chlamydospores of *Fusarium solani* f.sp. *pisi* germinate maximally in response to pea seeds 20–42 hr after planting (Cook and Flentje, 1967; Short and Lacy, 1974). Similarly, germination of chlamydospores of *F. oxysporum* f.sp. *vasinfectum* and *F. oxysporum* f.sp. *tracheiphilum* begins within 8 hr of exposure to cotton or cowpea roots with maximum germination occurring within 20–24 hr (Smith, 1977). Chlamydospores of *F. solani* f.sp. *phaseoli* within the first two millimeters of the surface of bean seeds begin to germinate within 4–5 hr after planting (Stanghellini and Hancock, 1971a); maximum germination occurs within 16–24 hr after planting (Schroth and Snyder, 1961). Glucose, sucrose, fructose, asparagine, aspartic acid, glutamine, glutamic acid, glycine and phenylalanine have been reported to be responsible for stimulating chlamydospore germination in *F. solani* f.sp. *phaseoli* and f.sp. *pisi* both *in-vitro* and in soil (Loria and Lacy, 1979; Schroth and Snyder, 1961; Schroth *et al.*, 1963).

Volatile components

Over the past several years, the importance of volatile seed and root exudates in soil ecology has become more clear (Gorecki *et al.*, 1985; Griebel and Owens, 1972; Linderman and Gilbert, 1975; Norton and Harman, 1985; Schenck and Stotzky, 1975; Stenhagen *et al.*, 1987; Vancura and Stotzky, 1976). As a result, the role of seed- and root-derived volatiles in stimulating fungal spore germination and initiating host-pathogen interactions is increasingly being investigated (Catska, 1979, 1980; Catska *et al.*, 1975; Coley-Smith and King, 1970; Harman *et al.*, 1978; Nelson, 1987).

One of the most thoroughly studied host-pathogen systems in which fungal propagules require a volatile exudate stimulus released from roots in order to germinate is the interaction of *Sclerotium cepivorum* with *Allium* species (Coley-Smith and King, 1970). During growth of species of Allium, water soluble allyl and propyl cysteines and their respective sulphoxides are released from root tips. These compounds are then degraded by soil bacteria to give rise to volatile alkyl sulphides which stimulate sclerotial germination (King and Coley-Smith, 1969). These compounds and their precursors are limited to speices of Allium and helps explain why sclerotia of *S. cepivorum* will not germinate in response to exudates from other plant genera (Coley-Smith and Holt, 1966).

In other studies, germination of sporangia of *P. ultimum*, *P. irregulare* and *P. sylvaticum* is evident within 2 hr after exposure to volatiles from germinating cotton and other plant seeds (Nelson, 1987). Maximum germination occurs following 8 hr of exposure to volatiles. Ethanol is

Table 1. Germination of propagules of plant pathogenic fungi in response to seed and root exudates

Organism	Propagule	Exudate source	Stimulant molecule(s)	Soil assay[a]	References
Alternaria alternata	conidium	seeds of *Pisum sativum*	unknown volatiles	N	Harman *et al.*, 1978
A. solani	conidium	seeds of *Cuminum cyminum*	unknown	N	Chaturverdi *et al.*, 1974
Aphanomyces euteiches	oospore	roots of various plant species	unknown	Y	Scharen, 1960
Aspergillus flavus	conidium	roots of *Arachis hypogaea*	glucose + unknown N source	Y	Griffin, 1969
A. ruber	conidium	seeds of *Pisum sativum*	unknown voltiles	N	Harman *et al.*, 1978
Botrytis cinerea	conidium	seeds of various plant species	unknown volatiles	N	Catska *et al.*, 1975; Catska, 1980
Ceratocystis fimbriata	conidium	roots of *Ipomoea batatas*	Ca^{++}	N	Kojima and Uritani, 1978
Colletotrichum capsici	conidium	seeds of *Cuminum cyminum*	unknown	N	Chaturverdi *et al.*, 1974
Curvularia lunata	conidium	seeds of *Cuminum cyminum*	unknown	N	Chaturverdi *et al.*, 1974
Fusarium oxysporum	chlamydospore	roots of *Arachis hypogaea*	glucose + unknown N source	Y	Griffin, 1969
	macroconidia	seeds of *Lens esculenta*	ethanol	N	Catska, 1979
F. oxysporum f. sp. *cucumerinum*	chlamydospore	seeds and seedlings of *Pisum sativum*	unknown	Y	Elad and Baker, 1985; Lifshitz *et al.*, 1986; Sneh *et al.*, 1984
F. oxysporum f. sp. *cubense*	microconidium	roots of *Musa* sp.	unknown	N	Buxton, 1962
F. oxysporum f. sp. *melonis*	chlamydospore	roots of *Cucumis melo*	unknown	Y	Sivan and Chet, 1989
F. oxysporum f. sp. *pisi*	microconidium chlamydospore	roots of *Pisum sativum*	unknown	Y, N	Buxton, 1957; Kommedahl, 1966; Whalley and Taylor, 1973
F. oxysporum f. sp. *tracheiphilum*	chlamydospore	roots of *Gossypium hirsutum*	unknown	Y	Smith, 1977
F. oxysporum f. sp. *vasinfectum*	chlamydospore	seeds and roots of *Cassia obtusifolia*; roots of *Gossypium hirsutum* and *G. barbadense*	unknown	Y, N	Brown and Curl, 1987; Sivan and Chet, 1989; Smith, 1977
F. solani	chlamydospore	roots of *Pisum sativum*	unknown	Y	Jackson, 1957
F. solani f. sp. *phaseoli*	chlamydospore; microconidium	seeds and roots of various plant species	sugars, amino acids, lipids	Y, N	Cook and Snyder, 1965; Schroth and Snyder, 1961; Schroth and Hendrix, 1962; Schroth *et al.*, 1963; Schroth and Cook, 1964; Stanghellini and Hancock, 1971a; Youssef *et al.*, 1975
F. solani f. sp. *pisi*	chlamydospore macroconidium	seeds and roots of *Pisum sativum*	sugars and unknown volatile compounds; ethanol	Y	Cook and Flentje, 1967; Harman *et al.*, 1987; Kraft, 1974; Loria and Lacy, 1979; Short and Lacy, 1974; Smucker and Erickson, 1987; Whalley and Taylor, 1973

Macrophomina phaseolina	sclerotium	roots of *Pinus lambertiana*	amino acids	Y	Cloud and Rupe, 1988; Short and Wyllie, 1978; Smith, 1969
Phytophthora cinnamomi	chlamydospore	roots of *Persea americana*	amino acids, organic acids	N	Mircetich *et al.*, 1968
P. infestans	oospore	roots of *Solanum tuberosum*	unknown	Y	Smoot *et al.*, 1958
P. megasperma f. sp. *glycinea*	oospore	roots of *Glycine max*	unknown	N	Jimenez and Lockwood, 1982
P. megasperma f. sp. *medicaginis*	oospore	roots of various plant species	amino acids	N	El-Hamalawi and Erwin, 1986; Förster *et al.*, 1983; Kuan and Erwin, 1980
P. palmivora	sporangium zoospore cyst	roots of various plant species	unknown	N	Turner, 1983
Puccinia punctiformis	teliospore	roots of *Cirsium arvense*	aplotaxene	N	French *et al.*, 1988
Pythium aphanidermatum	oospore	seeds of *Phaseolus vulgaris*; roots of various plant species	unknown	Y	Elad and Chet, 1987; Stanghellini and Burr, 1973
	zoospore cyst	roots of *Pisum sativum* and *Phaseolus aureus*	sugars, amino acids	N	Chang-Ho, 1970; Kraft and Erwin, 1967
P. irreguare	sporangium	seeds of *Pinus resinosa*	sugars, amino acids	Y	Agnihotri and Vaartaja, 1970
P. mamillatum	oospore	roots of *Brassica campestris*	unknown	Y	Barton, 1957
P. splendens	sporangium	roots of *Cucumis sativus*	unknown	Y	Ko and Ho, 1983
P. ultimum	oospore	seeds of *Pisum sativum*; roots of *Gossypium hirsutum*	unknown	Y	Johnson and Arroyo, 1983; Norton and Harman, 1985
	sporangium	seeds of various plant species	glucose, ethanol	Y	Gindrat, 1976; Lifshitz *et al.*, 1986;
				Y	Nelson, 1987; Schlub and Schmitthenner, 1978; Stanghellini and Hancock, 1971a, 1971b
Rhizoctonia fragariae	pseudoconidium	roots of *Fragaria* sp.	unknown	N	Hussain and McKeen, 1963
R. solani	sclerotium	seeds of *Pisum sativum*	unknown volatiles	Y	Norton and Harman, 1985
Sclerospora sorghi	oospore	roots of various plant species	unknown	Y	Pratt, 1978
Sclerotium cepivorum	sclerotium	roots of various *Allium* species	alkyl sulfides	Y	Coley-Smith *et al.*, 1968; Coley-Smith and King, 1969; Coley-Smith and King, 1970; King and Coley-Smith, 1968; King and Coley-Smith, 1969
S. rolfsii	sclerotium	seeds of *Pisum sativum*	unknown volatiles	Y	Norton and Harman, 1985
Thielaviopsis basicola	endoconidium chlamydospore	roots of various plant species	fatty acids	Y	Papavizas and Adams, 1969; Papavizas and Kovacs, 1972
Ustilago maydis	chlamydospore	seeds of *Zea mays*	ethanol	N	Catska, 1979
Verticillium albo-atrum/ *V. dahliae*	microsclerotium	roots of various plant species	amino acids	Y, N	Farley *et al.*, 1971; Olsson and Nordbring-Hertz, 1985 Schreiber and Green, 1963
	conidium	seeds and roots of various plant species	unknown	N	Catska *et al.*, 1975; Catska, 1979; Catska, 1980; Schreiber and Green, 1963

[a] Y = germination was assessed in a soil system; N = germination was assessed in an in-vitro system; Y, N = both types of assays have been performed.

the primary volatile stimulant evolved from germinating cotton seeds and is active at nanomolar to subnanomolar concentrations (Nelson, 1987). Higher concentrations on the other hand, appear to be inhibitory to germination. Acetaldehyde, another major component of both seed volatiles (Gorecki *et al.*, 1985; Nelson, 1987; Stotzky and Schenck, 1976), is not stimulatory to sporangium germination of *P. ultimum* at concentrations greater than ~1 n*M*.

Increased sporangium germination in response to seed volatiles can lead to increases in soil populations of *Pythium* spp. Norton and Harman (1985) observed up to 60 fold increases in soil populations of *P. ultimum* after exposing amended sterile soil to volatiles from aged pea seeds. In the presence of other soil microorganisms, however, soil populations of *P. ultimum* declined after exposure. In contrast, acetaldehyde and ethanol can increase general soil respiration without increases or decreases in soil microbial populations (Griebel and Owens, 1972). Griebel and Owens demonstrated that ethanol and acetaldehyde could be effectively metabolized by quiescent macroconidia of *Fusarium solani* without giving rise to propagule germination.

Linderman and Gilbert (1975) point out that volatiles are ideal candidates for pathogen stimulatory molecules since they can diffuse away from their source in either a vapor or liquid phase, are less susceptible to inactivation than water-soluble materials, can induce responses in extremely low concentrations and concentrations eliciting responses need not persist for extended periods of time. Volatile molecules may be particularly important in stimulating 'rapid response' pathogens such as *Pythium* spp. Despite the potential importance of volatiles in soil, much still needs to be learned about their production, biological activity and role in regulating pathogen behavior.

Temporal exudation and pathogen behavior

Much of the previous work on pathogen responses to exudates has focussed on responses of pathogens to water-soluble exudates collected from a few days to several weeks after seed germination with no attempt to link this response to the temporal behavior of the pathogen in the spermosphere or rhizosphere. For example, *Pythium* spp. respond to a germinating seed within 2 hr after exposure, but, the activity of exudates released 2 hr after the initiation of imbibition has rarely been investigated. Simple chemical analysis of seed or root exudates collected *in vitro* and the demonstration that bulk exudates stimulate germination of culture-produced propagules is not reliable information for determining the role of particular exudate components in initiating host-pathogen interactions in soil. The significance of exudates in regulating soilborne pathogen activity can only be determined when specific temporal behavioral patterns of the pathogen can be related to the time and site of production of specific exudate components. In many cases, we have no knowledge of the temporal behavior of soilborne pathogens in response to plants. Likewise, no research has definitively linked the exudation or lack of exudation of specific molecules to specific pathogen behavior. Consequently, no sound conclusions can be made from information currently available regarding the relationship of seed or root exudate components in regulating propagule behavior and initiating root-pathogen interactions. As a result, predicting the activities of soilborne fungal pathogens and associated microorganisms has not been entirely possible.

Pythium-seed interactions: A model system

Soilborne *Pythium* species are among the most responsive fungi to germinating seeds and developing seedlings (Lifshitz *et al.*, 1986; Loper, 1988; Nelson, 1988; Nelson *et al.*, Stanghellini and Hancock, 1971a). Sporangia of *P. ultimum* are activated and germinate within 1-1.5 hr of exposure to germinating seeds (Stanghellini and Hancock, 1971a,b). Within 6–12 hrs after planting in Pythium-infested soil, nearly all seeds are heavily colonized and rotted. Because of these rapid responses, Pythium-seed interactions serve as ideal model systems for examining various responses of soilborne pathogens to plants and plant exudates since temporal behavior as it relates to plant infection can be readily investigated within a narrow time frame. Despite the evidence that significant elements of Pythium-seed interactions are initiated within hours after sowing seeds, few studies on seed or root exudation have addressed the exudation of specific component molecues during the initial stages of seed germination when exogenously dormant propagules are being activated and colonization of host tissues is occurring. In most studies, the activity of exudates collected after several days and sometimes weeks of plant growth have been examined for their effects on propagule germination. Although Short and Lacy (1976a) examined carbohydrate exudation from germinating pea seeds within the first few hours of seed germination, no attempt was made to link exudation dynamics to pathogen behavior in the spermosphere.

Exudate molecules involved in stimulating sporangium germination of *Pythium ultimum*

Conclusions drawn from many of the previous studies of propagule germination among *Pythium* spp. have suggested that carbohydrates and amino acids are the primary exudate components responsible for initiating Pythium-seed or Pythium-root interactions in nature. For example, Agnihotri and Vaartaja (1967a, 1967b) found mixtures of sugars and amino acids to be most stimulatory to *P. ultimum*. Stanghellini and Hancock (1971a) reported that glucose was a primary stimulant of sporangium germination in soil. However, a major problem in relating observed *in vitro* propagule behavior to that in soils under natural conditions, is that it has not been known whether culture-produced propagules behave as they would if they were produced on diseased plant tissue in soil. Without this knowledge, conclusions about the stimulatory activity of specific exudate molecules cannot readily be evaluated in the laboratory with the goal of extending these concepts to natural conditions.

Recently, we have shown that sporangia of *P. ultimum* behave much differently in their response to sugar and amino acid exudate components when produced on synthetic media commonly used for the cultivation of *Pythium* spp. than when produced on diseased plant tissue (Nelson and Craft, 1989). Sporangia produced on common synthetic culture media germinate readily in response to individual sugars and amino acids present in cotton seed exudates (Table 2) but fail to respond to these same molecules when produced on metabolically active plant tissue (diseased seeds and radicles) (Tables 3 and 4). Regardless of how sporangia are reared, they remain fully germinable in response to unfractionated cotton seed exudate either collected *in vitro* or extracted from natural soils. This suggests that in nature, sugars and amino acids may not be important soluble exudate stimulants for *Pythium* spp. in soil. Interestingly, a mineral salts medium amended with soy lecithin gives rise to sporangia that behave similarly to those produced on living plant tissue. Although the role of complex lipid sources in regulating propagule responses are not understood, it appears that sporangia produced on such media physiologically resemble sporangia produced on plant tissues and thus may more closely reflect expected responses of propagules produced in soil.

Soluble stimulants

Utilizing bioassays with sporangia produced on plant-amended or lecithin-amended media, we have discovered that moderately non-polar water soluble molecules present in cotton seed exudate can stimulate sporangium germination. These stimulants are released as early as 2 hr after imbibition is initiated and reach a maximum by 4–6 hr in *in vitro* systems. Molecules conferring activity are low molecular weight compounds that are stable to autoclaving but unstable to extreme acid conditions (pH 2.5). Active molecules can be partitioned into ethyl acetate, n-butanol and chloroform but not into carbon tetrachloride. Individual components of exudate fractions conferring activity can readily be resolved by reverse phase liquid chromatography. However, the identity of individual components are currently unknown.

Soluble exudate stimulants can be extracted from non-sterilized soil in which cotton seeds had been germinating by leaching soil columns with water or less polar solvents such as methanol, ethanol, or acetone. For example, acetone extractions can provide up to 125 times more recovered activity than water extractions.

Table 2. Germination of culture-produced sporangia of *Pythium ultimum* (P4) in response to cotton seed exudate and exudate components

Culture medium[b]	% Sporangia germinated after addition of:[a]				
	H_2O	D-Glucose	Sucrose	L-Asparagine	Exudate[c]
GEA	0.0 a	78.7 a	16.2 b	7.9 b	96.0 a
CMA	0.0 a	76.0 a	77.0 a	31.8 a	87.8 a
SM	0.0 a	71.1 a	73.2 a	4.9 b	95.2 a
LBA	0.0 a	68.6 a	21.5 b	1.5 b	98.0 a
V8	0.0 a	35.4 b	10.5 bc	12.5 ab	93.5 a
SM + L	0.0 a	0.8 c	2.3 c	1.5 b	98.0 a

[a] Determined 3 hr after addition of 10 μLaliquots of germination stimulant; sugars and amino acids added at concentrations of 10 m*M*.
[b] GEA = grass extract agar; CMA = Difco cornmeal agar; SM = glucose-asparagine mineral salts agar; LBA = lima bean agar; V8 = V8 juice agar; SM + L = SM = 0.1% α-phosphatidyl choline (soy lecithin).
[c] Equivalent of exudate from one 24 hr-germinated seed added to each sporangial disk.
Numbers in each column followed by the same letter are not significantly ($P = 0.05$) different. (From Nelson and Craft, 1989).

Table 3. Influence of sugars and amino acids on germination of sporangia of *Pythium ultimum* produced on three different media[a]

Germination stimulant[c]	% Germinated sporangia[b]		
	CMA	SM + L	A + CS
Sugars			
D-Fructose	0.5[d]	0.0	0.4
D-Galactose	16.4	0.0	0.0
D-Glucose	75.6	2.7	0.0
D-Lactose	0.0	0.0	0.0
D-Maltose	64.9	0.0	0.8
Sucrose	66.6	0.0	0.6
Amino Acids			
L-Alanine	21.1	0.0	0.0
L-Arginine	2.0	0.0	0.2
L-Asparagine	37.6	0.6	0.5
Dl-Aspartic Acid	5.0	0.8	0.3
L-Glutamic Acid	37.2	0.0	0.7
L-Serine	43.9	0.0	0.3
H_2O	0.0	0.0	0.0
Seed Exudate[e]	93.0	96.8	90.4

[a] CMA = 3% Difco cornmeal agar; SM + L = a glucose-asparagine mineral salts agar amended with 0.1% α-phosphatidyl choline (soy lecithin); A + CS = 3% agarose amended with five surface-sterilized germinable cotton seeds.
[b] Germination determined 3 hr after addition of stimulant.
[c] Ten μL of 10 m*M* solutions of each stimulant added to agar disks containing sporangia.
[d] Numbers represent percent germinated sporangia assessed 3 hr after addition of stimulant.
[e] Equivalent of exudate from one 24 hr-germinated seed added to each sporangial disk (10 μL) (From Nelson and Craft, 1989).

Acetone and ethanol are the most effective stimulant extractants while non-polar solvents such as chloroform and carbon tetrachloride are not effective extractants of stimulant activity from soils. Aqueous soil leachates yield the highest level of stimulatory activity when seed exudates are extracted 48 hr after planting. Significant levels can be detected after 6 hr from non-sterile soil, although it is currently not clear whether stimulants recovered within the first 6 hr

Table 4. Influence of plant tissue amendments to an agarose medium on responses of sporangia of *P. ultimum* to cotton seed exudate and exudate components

Production medium[b]	% Germinated sporangia (3 hr) after addition of:[a]				
	H_2O	D-Glucose	Sucrose	L-Asparagine	Exudate[c]
Controls					
CMA	0.0 a	73.9 a	56.0 a	16.3 a	96.1 a
SM + L	0.0 a	0.0 c	0.0 d	1.1 bc	93.1 a
Tissue amendments[d]					
Wheat	0.0 a	1.0 bc	1.6 bcd	0.0 c	99.0 a
Barley	0.0 a	5.9 bc	0.0 d	1.1 bc	94.0 a
Oat	0.0 a	2.7 bc	2.3 bc	3.0 b	98.0 a
Snapbean	0.0 a	0.0 c	0.6 cd	0.0 c	97.5 a
Cotton	0.0 a	4.0 bc	0.5 cd	0.7 bc	96.0 a
Soybean	0.0 a	3.7 bc	3.2 b	1.1 bc	93.4 a

[a] Germination assessed 3 hr after the addition of 10 μL of 10 mM solution of the stimulant.
[b] CMA = 3% Difco cornmeal agar (broad-response control); SM + L = a glucose-asparagine mineral salts agar amended with 0.1% α-phosphatidyl choline (soy lecithin) (narrow-response control).
[c] Equivalent of exudate from one 24 hr-germinated seed added to each sporangial disk.
[d] Seeds of the various plant species surface sterilized and embedded in molten 3% agarose; 5 germinable seeds added to each plate.
Numbers in each column followed by the same letter are not significantly ($P = 0.05$) different. (From Nelson and Craft, 1989).

of seed germination are the same as those recovered at 48 hr. Studies are in progress to characterize these non-polar exudate stimulants and to determine their production and persistence in natural soils.

Volatile stimulants

Although propagules of *P. ultimum* have been shown to respond to volatiles in *in vitro* systems (Nelson, 1987), the stimulatory activity of these molecules in soil systems has not yet been clearly demonstrated. The activity of *P. ultimum* in sterile soils exposed to pea seed volatiles can be increased (Norton and Harman, 1985). However, in natural soils, populations of *Pythium* spp. decreased, presumably due to the action of antagonists. Nonetheless, in Pythium-conducive soils, exudate stimulants must be present for sufficent periods of time to allow for propagule activation and germination. Currently, the major limitation in demonstrating the activity of these molecules in soil systems is with our ability to detect them. A purge and trap technique followed by thermal desorption similar to that described by Stenhagen *et al.* (1987) may provide the most sensitive means of detection of pathogen stimulatory volatiles in soil.

Exudate stimulant metabolism and biological control

Since many seed- and root-infecting pathogens are highly dependent on exudate molecules to initiate plant infections, interference with the production and activity of exudate stimulants is a promising approach by which to interfere with pathogen activity and achieve biological control. Few biological control studies have focussed on this imporrtant step in pathogenesis. Although Ahmad and Baker (1988) observed reductions in sporangium germination of *P. ultimum* in the presence of Trichoderma-treated seeds as compared with non-treated seeds, the mechanism by which this phenomenon occurred was not elucidated. It is believed to occur through the metabolism of pathogen-stimulatory exudates. In a study by Gorecki *et al.*, (1985), pea seeds evolved significantly lower levels of ethanol and acetaldehyde during germination when treated with *Enterobacter cloacae* or *Trichoderma harzianum* as compared with untreated seeds. Ethanol has since been shown to be an effective activator of propagule germination of *P. ultimum* (Nelson, 1987).

Elad and Chet (1987) found a significant correlation between the ability of various bacterial

Table 5. Stimulatory activity of soil-extracted exudates from germinating untreated and *Enterobacter*-treated cotton seeds[a]

Seed treatment	% Germinated sporangia[b]		
	0 hr	6 hr	48 hr
Control (No Seeds)	0.0	2.1	4.4
Untreated	0.0	10.4	91.5
E. cloacae-treated[c]	0.0	6.6	8.8

[a] Aqueous leachates from 100 seeds germinating in 80 cc soil.
[b] Determined 3 hr after exposure to exudate.
[c] Seed populations of strain EcCT-501: 3.0×10^9 cells/seed.

strains to inhibit Pythium seed rot of cucumber and their ability to inhibit oospore germination of *P. aphanidermatum*. Effective bacterial biocontrol agents inhibited oospore germination by as much as 57% while ineffective biocontrol strains inhibited oospore germination by only 13–20%. There was no direct interaction between the bacteria and oospores and the ability of effective strains to inhibit oospore germination was not related to the production of inhibitory metabolites. They speculated that bacterial strains metabolized exudate components responsible for stimulating oospore germination. Recently, we confirmed this observation with *P. ultimum* (Table 5). Bacterial strains effective in suppressing Pythium seed rot of cotton eliminated the stimulatory activity of cotton seed exudates while ineffective strains only slightly reduced the stimulatory activity. In addition, soluble exudates, extracted from soil in which cotton seeds treated with an effective bacterial biocontrol agent were germinating, were no longer stimulatory to *P. ultimum* sporangium germination. Similarly, in *in vitro* systems, *E. cloacae* will eliminate the stimulatory activity of cotton seed volatiles to sporangia of *P. ultimum* (Table 6). This reduction in activity is apparently related to reduced levels of ethanol present. However, the production of undetected toxic volatiles cannot be ruled out. Unfortunately, repeated attempts to demonstrate this phenomenon in soils have failed. The ability of biocontrol agents to metabolize exudate signal molecules may indeed be an important step in biocontrol processes. However, further research is needed to more thoroughly explore this hypothesis.

Table 6. Stimulatory activity of volatiles released untreated and *Enterobacter*-treated cotton seeds germinating in water and in sterile sand[a]

Seed treatment	% Germinated sporangia[b]	
	H_2O[c]	Sand[c]
Control (No Seeds)	0.5	0.8
Untreated	43.7	2.8
E. cloacae-treated[d]	3.2	0.8

[a] 20 seeds germinating in a sealed flask (51 mL total volume).
[b] Germination determined after 8 hr of continuous exposure to headspace above germinating seeds.
[c] 1.2 mL water added to 20 seeds germinating in absence of sand; 7 mL water added to 20 seeds germinating in 30 g sand.
[d] Seed populations of *E. cloacae* strain E6: 1.9×10^8 cells/seed.

Conclusions

In order to successfully develop new and more effective control strategies for soilborne fungal diseases, a more thorough understanding of the host-pathogen interaction is necessary. We need to develop a better understanding of the succession of pathogen events leading up to plant infection as well as the factors regulating these activities. Since seed and root exudates are the major driving force behind the spermosphere and rhizosphere activities of soil-borne plant pathogens as well as associated microorganisms, knowledge of when and how exudates initiate and regulate pathogen response to plants is essential for an understanding of the interactions of soil-borne pathogens with seeds, roots and other spermosphere and rhizosphere microorganisms. This knowledge is also essential for an understanding of how to manipulate the spermosphere and rhizosphere environment in order to reduce or eliminate seed or root infection by pathogenic fungi.

References

Agnihotri V P and Vaartaja O 1967a Root exudates from red pine seedlings and their effects on *Pythium ultimum*. Can. J. Bot. 45, 1031–1040.
Agnihotri V P and Vaartaja O 1967b Effects of amendments,

soil moisture contents, and temperatures on germination of *Pythium* sporangia under the influence of soil mycostasis. Phytopathology 57, 1116–1120.

Agnihotri V P and Vaartaja O 1968 Seed exudates from *Pinus resinosa* and their effects on growth and germination of *Pythium afertile*. Can. J. Bot. 46, 1135–1141.

Agnihotri V P and Vaartaja O 1970 Effect of seed exudates of *Pinus resinosa* on the germination of sporangia and on the population of *Pythium irregulare* in the soil. Plant and Soil 32, 246–249.

Ahmad J S and Baker R 1988 Implications of rhizosphere competence of *Trichoderma harzianum*. Can. J. Microbiol. 34, 229–234, 1988.

Barber D A and Martin J K 1976 The release of organic substances by cereal roots into soil. New Phytol. 76, 69–80.

Barton R 1957 Germination of oospores of *Pythium mamillatum* in response to exudates from living seedlings. Nature 180, 613–614.

Brown S L and Curl E A 1987 Rhizosphere effect of herbicide-stressed sicklepod (*Cassia obtusifolia*) on chlamydospores of *Fusarium oxysporum* f.sp. *vasinfectum*. Plant Dis. 71, 919–922.

Buxton E W 1957 Some effects of pea root exudates on physiologic races of *Fusarium oxysporum* Fr. f. *pisi* (Linf.) Snyder and Hanson. Trans. Br. Mycol. Soc. 40: 145–154.

Buxton E W 1962 Root exudates from banana and their relationship to strains of the *Fusarium* causing Panama wilt. Ann. Appl. Biol. 50, 269–282.

Catska V 1979 Volatile and gaseous metabolites of germinating seeds and their role in fungus-plant relationships. *In* Soilborne Plant Pathogens. Eds. B Schippers and W Gams. pp 241–245. Academic Press, New York.

Catska V 1980 Effect of volatile and gaseous metabolites of germinating pea seeds on micromycetes. Folia Microbiol. 25, 174–176.

Catska V, Afifi A F and Vancura V 1975 The effect of volatile and gaseous metabolites of swelling seeds on germination of fungal spores. Folia Microbiol. 20, 152–156.

Chang-Ho Y 1970 The effect of pea root exudate on the germination of *Pythium aphanidermatum* zoospore cysts. Can. J. Bot. 48, 1501–1514.

Chaturverdi S N, Siradhana B S and Muralla R N 1974 The influence of cumin seed exudates on fungal spore germination. Plant and Soil 39, 49–56.

Cloud G L and Rupe J C 1988 Differences in the germination of *Macrophomina phaseolina* sclerotia in the presence of soybean and sorghum seed. Phytopathology 78, 1693.

Coley-Smith J R and Holt R W 1966 The effect of species of *Allium* on germination in soil of sclerotia of *Sclerotium cepivorum* Berk. Ann. Appl. Biol. 58, 273–278.

Coley-Smith J R and King J E 1969 The production by species of *Allium* of alkyl sulphides and their effect on germination of sclerotia of *Sclerotium cepivorum* Berk. Ann. Appl. Biol. 64, 289–301.

Coley-Smith J R and King J E 1970 Response of resting structures of root-infecting fungi to host exudates: An example of specificity. *In* Root Diseases and Soil-Borne Pathogens. Eds. T A Toussoun, R V Bega and P E Nelson. pp 130–133. Univ. Calif. Press, Berkeley.

Coley-Smith J R, Dickinson D J, King J E and Holt R W 1968 The effect of species of *Allium* on soil bacteria in relation to germination of sclerotia of *Sclerotium cepivorun* Berk. Ann. Appl. Biol. 62, 103–111.

Cook R J and Baker K F 1983 The Nature and Practice of Biological Control. The American Phytopathological Society, St. Paul, MN, 539 p.

Cook R J and Flentje N T 1967 Chlamydospore germination and germling survival of *Fusarium solani* f. *pisi* in soil as affected by soil water and pea seed exudation. Phytopathology 57, 178–182.

Cook R J and Snyder W C 1965 Influence of host exudates on growth and survival of germlings of *Fusarium solani* f. *phaseoli* in soil. Phytopathology 55, 1021–1025.

Curl E A and Truelove B 1986 'The Rhizosphere', Springer-Verlag, New York, 288 p.

Elad Y and Chet I 1987 Possible role of competition for nutrients in biocontrol of Pythium damping-off by bacteria. Phytopathology 77, 190–195.

Elad Y and Baker R 1985 Influence of trace amounts of cations and siderophore-producing pseudomonads on chlamydospore germination of *Fusarium oxysporum*. Phytopathology 75, 1047–1052.

El-Hamalawi Z A and Erwin D C 1986 Components in alfalfa root extract and root exudate that increase oospore germination of *Phytophthora megasperma* f.sp. *medicaginis*. Phytopathology 76, 508–513.

Farley J D, Wilhelm S and Snyder W C 1971 Repeated germination and sporulation of microsclerotia of *Verticillium albo-atrum* in soil. Phytopathology 61, 260–264.

Förster H, Ribeiro O K and Erwin D C 1983 Factors affecting oospore germination of *Phytophthora megasperma* f.sp. *medicaginis*. Phytopathology 73, 442–448.

French R C, Turner S K, Sonnett P E, Pfeffer P and Piotrowski E 1988 Properties of an extract from Canada thistle roots that stimulates germination of dormant teliospores of Canada thistle rust (*Puccinia punctiformis*). J. Agric. Food Chem. 36, 1043–1047.

Gindrat D 1976 Components of unbleached commercial chitin stimulate *Pythium ultimum* in sugar beet spermosphere. Phytopathology 66, 312–316.

Gorecki R J, Harman G E and Mattick L R 1985 The volatile exudates from germinating pea seeds of different viability and vigor. Can. J. Bot. 63, 1035–1039.

Griebel G E and Owens L D 1972 Nature of the transient activation of soil microorganisms by ethanol or acetaldehyde. Soil Biol. Biochem. 4, 1–8.

Griffin G J 1969 *Fusarium oxysporum* and *Aspergillus flavus* spore germination in the rhizosphere of peanut. Phytopathology 59, 1214–1218.

Hale M G, Foy C L and Shay F J 1971 Factors affecting root exudation. Adv. Agron. 24, 89–109.

Hale M G, Moore L D and Griffin G J 1978 Root exudates and exudation. *In* Interactions Between Non-Pathogenic Soil Microorganisms and Plants. Eds. Y. R. Dommergues and S. V. Krupa. pp 163–203. Elsevier Publ. Co., Amsterdam.

Harman G E, Nedrow B and Nash G 1978 Stimulation of fungal spore germination by volatiles from aged seeds. Can. J. Bot. 56, 2124–2127.

Husain S S and McKeen W E 1963 Interactions between

strawberry roots and *Rhizoctonia fragariae*. Phytopathology 53, 541–545.
Jackson R M 1957 Fungistasis as a factor in the rhizosphere phenomenon. Nature 180, 96–97.
Jimenez B and Lockwood J L 1982 Germination of oospores of *Phytophthora megasperma* f.sp. *glycinea* in the presence of soil. Phytopathology 72, 662–666.
Johnson L F and Arroyo T 1983 Germination of oospores of *Phythium ultimum* in the cotton rhizosphere. Phytopathology 73, 1620–1624.
King J E and Coley-Smith J R 1968 Effects of volatile products of *Allium* species and their extracts on germination of sclerotia of *Sclerotium cepivorum* Berk. Ann. Appl. Biol. 61, 407–414.
King J E and Coley-Smith J R 1969 Production of volatile alkyl sulphides by microbial degradation of synthetic alliin and alliin-like compounds, in relation to germination of sclerotia of *Sclerotim cepivorum* Berk. Ann. Appl. Appl. Biol. 64, 303–314.
Ko W and Ho W 1983 Screening soils for suppressiveness to *Rhizoctonia solani* and *Pythium splendens*. Ann. Phytopath. Soc. Japan 49, 1–9.
Kojima M and Uritani I 1978 Studies on factors in sweet potato root which induce spore germination of *Ceratocystis fimbriata*. Plant Cell Physiol. 19, 91–97.
Kommedahl T 1966 Relation of exudates of pea roots to germination of spores in races of *Fusarium oxysporum* f. *pisi*. Phytopathology 56, 721–722.
Kraft J M 1974 The influence of seedling exduates on the resistance of peas to Fusarium and Pythium root rot. Phytopathology 64, 190–193.
Kuan T L and Erwin D C 1980 Predisposition effect of water saturation of soil on Phytophthora root rot of alfalfa. Phytopathology 70, 981–986.
Lifshitz R, Windham M T and Baker R 1986 Mechanism of biological control of preemergence damping-off of pea by seed treatment with *Trichoderma* spp. Phytopathology 76, 720–725.
Linderman R G and Gilbert R G 1975 Influence of volatiles of plant origin on soil-borne plant pathogens. *In* Biology and Control of Soil-Borne Plant Pathogens. Ed. G W Bruehl. pp 90–99. The American Phytopathological Society, St. Paul, MN.
Lockwood J L 1977 Fungistasis in soils. Biol. Rev. 52, 1–43.
Lockwood J L 1988 Evolution of concepts associated with soilborne plant pathogens. Ann. Rev. Phytopathol. 26, 93–121.
Loper J E 1988 Role of fluorescent siderophore production in biological control of *Pythium ultimum* by a *Pseudomonas fluorescens* strain. Phytopathology 78, 166–172.
Loria R and Lacy M L 1979 Mechanism of increased susceptibility of bleached pea seeds to seed and seedling rot. Phytopathology 69, 573–575.
Mircetich S M, Zentmyer G A an Kendrick J B Jr. 1968 Physiology of germination of chlamydospores of *Phytophthora cinnamomi*. Phytopathology 58, 666–671.
Mitchell J E 1976 The effect of roots on the activity of soilborne plant pathogens. *In* Encyclopedia of Plant Physiology, New Series, Vol. 4, Physiological Plant Pathology. Eds. R Heitefuss and P H Williams. pp 104–128. Springer-Verlag, New York.
Nelson E B 1987 Rapid germination of sporangia of *Pythium* species in response to volatiles from germinating seeds. Phytopathology 77, 1108–1112.
Nelson E B 1988 Biological control of Pythium seed rot and preemergence damping-off cotton with *Enterobacter cloacae* and *Erwinia herbicola* applied as seed treatments. Plant Dis. 72, 140–142.
Nelson E B, Chao W L, Norton J M, Nash G T and Harman G E 1986 Attachment of *Enterobacter cloacae* to hyphae of *Pythium ultimum*: Possible role in the biological control of Pythium pre-emergence damping-off. Phytopathology 76, 327–335.
Nelson E B and Craft C M 1989 Comparative germination of culture-produced and plant-produced sporangia of *Pythium ultimum* in response to soluble seed exudates and exudate components. Phytopathology 79, 1009–1013.
Norton J M and Harman G E 1985 Responses of soil microorganisms to volatile exudates from germinating pea seeds. Can. J. Bot. 63, 1040–1045.
Olsson S and Nordbring-Hertz B 1985 Microsclerotial germination of *Verticillium dahliae* as affected by rape rhizosphere. FEMS Microbiol. Ecol. 31, 293–299.
Papavizas G C and Adams P B 1969 Survival of root-infecting fungi in soil. XII. Germination and survival of endoconidia and chlamydospores of *Thielaviopsis basicola* in fallow soil and in soil adjacent to germinating bean seed. Phytopathology 59, 371–378.
Papavizas G C and Kovacs M F Jr 1972 Stimulation of spore germination of *Thielaviopsis basicola* by fatty acids from rhizosphere soil. Phytopathology 62, 688–694.
Pratt R G 1978 Germination of oospores of *Sclerospora sorghi* in the presence of growing roots of host and non-host plants. Phytopathology 68, 1606–1613.
Scharen A L 1960 Germination of oospores of *Aphanomyces euteiches* embedded in plant debris. Phytopathology 50, 274–277.
Schenck S and Stotzky G 1975 Effect on microorganisms of volatile compounds released from germinating seeds. Can. J. Microbiol. 21, 1622–1634.
Schlub R L and Schmitthenner A F 1978 Effects of soybean seed coat cracks on seed exudation and seedling quality in soil infested with *Pythium ultimum*. Phytopathology 68, 1186–1191.
Schreiber L R and Green R J Jr 1963 Effect of root exudates on germination of conidia and microsclerotia of *Verticillium albo-atrum* inhibited by the soil fungistatic principle. Phytopathology 53, 260–264.
Schroth M N and Cook R J 1964 Seed exudation and its influence on pre-emergence damping-off of bean. Phytopathology 54, 670–673.
Schroth M N and Hendrix F F Jr 1962 Influence of nonsusceptible plants on the survival of *Fusarium solani* f.sp. *phaseoli* in soil. Phytopathology 52, 906–909.
Schroth M N and Snyder W C 1961 Effect of host exudates on chlamydospore germination of the bean root rot fungus, *Fusarium solani* f. *phaseoli*. Phytopathology 51, 389–393.
Schroth M N, Toussoun T A and Snyder W C 1963 Effect of

certain constituents of bean exudate on germination of chlamydospores of *Fusarium solani* f. *phaseoli* in soil. Phytopathology 53, 809–812.

Short G E and Lacy M L 1974 Germination of *Fusarium solani* f. sp. *pisi* chlamydospores in the spermosphere of pea. Phytopathology 64, 558–562.

Short G E and Lacy M L 1976a Carbohydrate exudation from pea seeds: Effects of cultivar, seed age, seed color, and temperatures. Phytopathology 66, 182–187.

Short G E and Lacy M L 1976b Factors affecting pea seed and seedling rot in soil. Phytopathology 66, 188–192.

Short G E and Wyllie T D 1978 Inocuum potential of *Macrophomina phaseolina*. Phytopathology 68, 742–746.

Sivan A and Chet I 1989 The possible role of competition between *Trichoderma harzianum* and *Fusarium oxysporum* on rhizosphere colonization. Phytopathology 79, 198–203.

Smith S N 1977 Comparison of germination of pathogenic *Fusarium oxysporum* chlamydospores in host rhizosphere soils conducive and suppressive to wilts. Phytopathology 67, 502–510.

Smith W H 1969 Germination of *Macrophomina phaseolina* sclerotia as affected by *Pinus lambertiana* root exudate. Can. J. Microbiol. 15, 1387–1391.

Smoot J J, Gough F J, Lamey H A, Eichenmuller J J and Gallegly M E 1958 Production and germination of oospores of *Phytophthora infestans*. Phytopathology 48, 165–171.

Smucker A J M and Erickson A E 1987 Anaerobic stimulation of root exudates and disease of peas. Plant and Soil 99, 423–433.

Sneh B, Dupler M, Elad Y and Baker R 1984 Chlamydospore germination of *Fusarium oxysporum* f.sp. *cucumerinum* as affected by fluorescent and lytic bacteria from a Fusarium-suppressive soil. Phytopathology 74, 1115–1124.

Stanghellini M E and Burr T J 1973 Germination *in vivo* of *Pythium aphanidermatum* oospores and sporangia. Phytopathology 63, 1493–1496.

Stanghellini M E and Hancock J G 1971a The sporangium of *Pythium ultimum* as a survival structure in soil. Phytopathology 61, 157–164.

Stanghellini M E and Hancock J G 1971b Radial extent of the bean spermosphere and its relation to the behavior of *Pythium ultimum*. Phytopathology 61, 165–168.

Stenhagen G, Alborn H and Lundborg T 1987 Variation of root and microflora exudates in genotypes of barley. *In* Allelochemicals: Role in Agriculture and Forestry. Ed. G R Waller. pp 76–88. American Chemical Society Symposium Series 330, American Chemical Society, Washington.

Stotzky G and Schenck S 1976 Observations on organic volatiles from germinating seeds and seedlings. Am. J. Bot. 63, 798–805.

Turner P D 1963 Influence of root exudates of cacao and other plants on spore development of *Phytophthora palmivora*. Phytopathology 53, 1337–1339.

Vancura V and Stotzky G 1976 Gaseous and volatile exudates from germinating seeds and seedlings. Can. J. Bot. 54, 518–532.

Whalley W M and Taylor G S 1973 Influence of pea-root exudates on germination of conidia and chlamydospores of physiologic races of *Fusarium oxysporum* f. *pisi*. Ann. Appl. Biol. 73, 269–276.

Youssef Y A, Mankarios A T and Abo Sbadi M R 1975 Root exudates from *Phaseolus vulgaris* and their effects on spore germination and growth of *Fusarium solani* (Mart.) Sacc. Egypt. J. Phytopathol. 7, 21–32.

D. L. Keister and P. B. Cregan (Eds.), The rhizosphere and plant growth, 211–219.

PLSO BARC20

Beneficial and deleterious effects of HCN-producing pseudomonads on rhizosphere interactions

B. SCHIPPERS, A.W. BAKKER, P.A.H.M. BAKKER and R. VAN PEER
Willie Commelin Scholten Phytopathological Laboratory (Department of Plant Pathology, University of Utrecht) Javalaan 20, 3742 CP Baarn, The Netherlands

Key words: beneficial rhizosphere pseudomonads, cyanide negative mutants, cyanide production by pseudomonads, deleterious rhizosphere pseudomonads, disease suppression by pseudomonads, induced resistance, plant growth promotion, *Pseudomonas fluorescens*, *Pseudomonas putida*, Transposon mutagenesis

Abstract

Rhizobacteria live around roots but also inside the cortical root tissues by utilizing organic substances released from root cells into the intercellular spaces and the root environment. The effects of metabolites of these rhizosphere-inhabiting bacteria on root physiology and plant development have hardly been studied. However, recent studies indicate that, depending on environmental factors and plant species, certain strains of rhizosphere *Pseudomonas* spp. and some of their metabolites such as HCN may inhibit or enhance plant establishment or inhibit development of plant disease. Cultural practices such as cropping frequency, no tillage, and soilless cultivation, as well as edaphic factors seem to determine these rhizosphere interactions.

Beneficial rhizosphere pseudomonads

Root colonizing bacteria in the genus Pseudomonas have captured special attention over the last 10 years for their beneficial and deleterious effects on plant development. Their plant growth promoting effects are due to the suppression of soilborne plant parasitic microorganisms and of deleterious rhizosphere microorganisms (Kloepper *et al.*, 1988; Schippers, 1988; Weller, 1988). Some Pseudomonas strains promote plant growth in the absence of pathogens, for example by the enhancement of phosphate uptake (Lifshitz *et al.*, 1987).

Among the pathogenic microorganisms suppressed are *Gaeumannomyces graminis* var. *tritici* (Ggt) causing 'take all' of wheat, *Fusarium oxysporum* causing wilt in several crops, *Thielaviopsis basicola* causing root rot of tobacco, *Pythium* spp. and *Rhizoctonia solani* causing seedling decay, and deleterious rhizosphere microorganisms causing reduced plant growth.

Different mechanisms and different Pseudomonas metabolites were shown to be involved in the suppression of pathogenic microorganisms. Siderophore-mediated competition for iron is claimed to be a major factor in the suppression of *F. oxysporum* (Baker *et al.*, 1986) and Pythium (Becker and Cook, 1988; Loper, 1988) and of deleterious rhizosphere microorganisms (Leong, 1986; Schippers *et al.*, 1987). Antibiosis by phenazin(s) is partly responsible for the suppression of *Ggt* (Thomashow and Weller 1988; Weller, 1988), whereas the antibiotics pyoluteorin and pyrrolnitrin are involved in the suppression of *Pythium* spp. and *Rhizoctonia solani*, respectively (Howell and Stipanovic, 1979, 1980). Cyanide seems to play a role in the suppression of root rot in tobacco caused by *T. basicola* (Ahl *et al.*, 1986). Defense mechanisms in carnation against *F. oxysporum* f. sp. *dianthi* were shown to be stimulated by an as yet unknown characteristic(s) of *Pseudomonas* sp. strain WCS417 (Van Peer, unpublished).

Deleterious rhizosphere pseudomonads

Rhizosphere pseudomonads have also been suspected to retard plant growth (Schippers *et al.*, 1987; Suslow, 1982; Suslow and Schroth, 1982). Much of the experimental evidence for the deleterious effects caused by rhizosphere microorganisms, including *Pseudomonas* spp., has been obtained indirectly from increased plant growth and yield in a variety of crops following seed or tuber bacterization with plant growth promoting rhizobacteria (PGPR) (Burr and Ceasor, 1984; Hemming, 1986; Schippers *et al.*, 1987). It was shown also in soilless plant cultivation that a deleterious rhizosphere microflora of *Pseudomonas* spp., inhibiting plants in the early stages of growth, may easily develop in a variety of horticultural crops including lettuce, tomato, cucumber and carnation. Inoculation of the substrate with a plant growth promoting *Pseudomonas* sp. strain WCS417 prevented the development of such a deleterious microflora (Van Peer and Schippers, 1989).

Information on the metabolites involved in the deleterious activity of pseudomonads has only been recently presented. (Schippers *et al.*, 1988) (Table 1). *Pseudomonas* spp. have been suspected to inhibit potato root development by their production of hydrogen cyanide (Bakker and Schippers, 1987; Bakker *et al.*, 1989). Cyanide seems also to be involved in the deleterious activity of *Pseudomonas* spp. causing growth inhibition in lettuce and bean (Alström and Burns, 1989). Cyanide produced by rhizosphere pseudomonads thus is postulated to have a detrimental effect on plant establishment in several crops (Schippers *et al.*, 1987) but a beneficial one in others by the suppression of root pathogens (Défago *et al.*, 1990).

Cyanide production by pseudomonads

The ability to produce HCN is most common among the rhizosphere pseudomonads. Approximately 50% of the pseudomonads isolated from potato and wheat rhizospheres were able to produce HCN *in vitro*. The rate of cyanide production in batch cultures differed widely among the different isolates tested (Bakker and Schippers, 1987).

Several environmental factors influence the rate of HCN production by *Pseudomonas* spp. (Castric, 1975, 1983; Knowles, 1976). Production depends highly on the amino acid composition of the substrate. Glycine was shown to be the direct precursor of microbial cyanide production and exerts the strongest effect of the amino acids tested (Askeland and Morison, 1983; Bakker *et al.*, 1987; Knowles, 1976; Voisard *et al.*, 1989). Proline also stimulates microbial cyanide production but to a lesser extent. In combination with glycine, proline reduces the stimulating effect of the latter (Bakker *et al.*, 1987). Glycine and proline have been found in root exudates (Bakker *et al.*, 1989; Curl and Truelove, 1986; Rovira and Davey, 1974). The rate of microbial HCN production in the rhizosphere of different crops possibly varies widely due to differences in the amino acid composition of their root exudates. Environmental factors affecting root exudation rate and exudate composition such as light intensity and soil water potential may probably also affect HCN-production in the rhizosphere.

The availability of ferric iron in the environment is another important factor affecting HCN

Table 1. Metabolites produced by Pseudomonas involved in plant growth retardation

Plant	Metabolite	Author
Canola seedlings	polygalacturonic acid hydrolase	Campbell *et al.*, 1987
Wheat seedlings	polypeptide phytotoxin	Frederickson and Elliot, 1985
Potato	HCN	Bakker and Schippers, 1987
Lettuce	HCN	Alström and Burns, 1989

production by pseudomonads (Askeland and Morrison, 1983; Castric, 1975; Knowles and Bunch, 1986). The *in vitro* production of HCN per certain amount of *Pseudomonas* biomass was stimulated by $FeCl_3$ as well as by the addition of iron bound to the siderophore of *Pseudomonas* sp. strain WCS361. Sodium chloride did not have this effect (Bakker *et al.*, 1987). Production of HCN by Pseudomonas strain A14 was stimulated by addition of iron bound to its own siderophore but not by iron bound to pseudobactin 358, the siderophore of *P. putida* WCS358 which cannot be utilized by A14 (Bakker *et al.*, 1989). Addition of vermiculite, a clay mineral rich in Fe^{3+}, to a batch culture of the HCN-producing *P. fluorescens* strain CHAo induced cyanogenesis whereas illite, a clay mineral poor in Fe^{3+} iron, did not (Keel *et al.*, 1989).

Inorganic phosphate concentrations in soil may also influence HCN production by *Pseudomonas* spp. Castric (1975) reported that whereas normal cell growth of *P. aeruginosa* in a synthetic medium occured at phosphate concentrations from 3 to 300 μM, cyanide production was inhibited at the higher concentrations (150–300 μM). He also observed that although HCN production by *P. aeruginosa* growing in a synthetic medium required aerobiosis, this process operated efficiently at low dissolved oxygen concentrations. The ability of *P. aeruginosa* to produce HCN was rapidly lost in the absence of a carbon source and glycine at 180 μM oxygen or more, whereas at 1.0 μM this loss did not occur (Castric, 1983). Cyanide production in the rhizosphere by pseudomonads may thus depend on many soil and plant characteristics.

Effects of cyanide on root metabolism and root development

In plant roots, HCN inhibits cytochrome oxidase thereby negatively affecting root energy metabolism (Lambers, 1980). When roots are exposed to cyanide, electrons released by oxidation of NADH follow the 'alternative cyanide resistant respiratory pathway' to oxygen instead of the cyanide sensitive cytochrome oxidase pathway (Lambers, 1980). We observed that 5 μM KCN caused an inhibition of the cytochrome oxidase respiration of approximately 40% in potato roots (Bakker and Schippers, 1987). In absence of an inhibitor of the alternative respiration path (SHAM), 5 μM KCN did not decrease root respiration. This indicates an increased contribution of the alternative path in total root respiration. Kuiper (1983) demonstrated that the growth rate of *Plantago major* (plantain) was negatively correlated with the relative contribution of the alternative path to the total root respiration.

Défago *et al.* (1990) observed in a gnotobiotic system that the cyanide producing *P. fluorescens* strain CHAo, but not its HCN^- mutant CHA5, stunted the roots and stimulated root hair formation in tobacco. A significant decrease in root (plant) fresh (dry) weight was not observed in this system. The negative effect of HCN on root energy metabolism thus may affect plant growth differentially, depending on plant species and environmental factors.

Cytochrome oxidase is not the only enzyme sensitive to HCN. A variety of other enzymes, especially Fe-containing (*e.g.* peroxidases) and Cu-containing enzymes, were shown to be sensitive to HCN as well (Lambers, 1985; Solomonson, 1981). Peroxidase activity has been demonstrated to be linked with processes such as suberization and lignification of cell walls (Symons *et al.*, 1985) and with hypersensitive reactions of plant tissue after infection by pathogens and after tissue wounding (Arora and Bajaj, 1985). Thus, HCN might influence these processes.

Suppression of disease by Pseudomonas HCN

Défago and co-workers in Switzerland recently obtained evidence that *Pseudomonas fluorescens* strain CHAo, isolated from soil suppressive to black root rot of tobacco caused by *Thielaviopsis basicola*, efficiently suppresses disease caused by *T. basicola* on tobacco, cotton and cherry trees and by *Gaeumannomyces graminis* var. *tritici* on wheat in natural non-sterile soil (Défago and Haas 1990; Stutz *et al.*, 1986). Strain CHAo produces several extracellular metabolites: HCN, siderophore(s) of the pyoverdine type, salicylic acid, pyoluteorin, mono acetyl phloroglucinol, 2,4-diacetyl phloroglucinol and additional unidentified compounds (Défago *et al.*, 1990).

In a gnotobiotic system containing an artificial Fe^{3+}-rich vermiculite soil, the wild type strain CHAo offered ±70% protection of tobacco plants, whereas CHA5, a mutant deficient in HCN-production, gave ±35% protection. Complementation of CHA5 with the hcn^+ genes cloned in pVK 100 restored HCN production and suppression of black root rot (Voisard *et al.*, 1989). *P. fluorescens* strain P3, which does not produce HCN, synthesized HCN and gave improved protection of tobacco when a derivative of transposon Tn5 carrying the hcn^+ genes of strain CHAo had been inserted into the genome of *P. fluorescens* P3 (P3:: Tn hcn). P3:: Tn hcn did not reach the suppression capacity of CHAo. Wild type and genetically modified strains colonized the roots in similar densities (Voisard *et al.*, 1989). In the gnotobiotic system, protecting strains did not reduce the number of *T. basicola* propagules although the fungus is very sensitive to HCN on agar (Ahl *et al.*, 1986). It is assumed, however, that the antagonistic effect on the fungus is restricted to the vicinity of the root surface (Voisard *et al.*, 1989). Based on these results, it is concluded that HCN produced by CHAo contributes significantly to suppression of black root rot, but other metabolites, presumably antibiotic compounds, may also be involved in the disease suppression (Voisard *et al.*, 1989).

When the iron-rich vermiculite in the gnotobiotic system is replaced by the clay mineral illite, which is poor in iron, CHAo protected plants from disease poorly. However, when the illite was supplemented with $FeCl_3$ or Fe EDDHA, CHAo gave good protection (Keel *et al.*, 1989). The availability of iron in soil apparently is essential for the disease suppressive activity of the *P. fluorescens* strain CHAo.

A siderophore negative transposon mutant of CHAo (CHA400), unable to use Fe^{3+} from Fe EDDHA, gave similar protection as strain CHAo in vermiculite and in illite amended with $FeCl_3$ but protected plants poorly in illite amended with Fe EDDHA, in contrast to the wild type strain CHAo. The siderophore-negative mutant CHA400 and strain CHAo do not differ significantly in their protection capacity in raw soil or in gnotobiotic systems containing vermiculite. Competition for iron therefore is excluded as a mechanism of suppression in these rhizosphere interactions (Défago *et al.*, 1990). Disease suppression in this system is rather linked to a high availability than to a low availability of Fe^{3+}.

Cyanide also seems to play a role in the suppression of take all disease in wheat and of *Rhizoctonia solani* in cotton. Strain CHAo increased the weight of wheat plants infected with Ggt dramatically and reduced the number of running hyphae of the pathogen in natural, partially sterilized soil, but the HCN-negative mutant CHA5 had no such effect (Défago *et al.*, 1990). The ability of the bacteria to produce HCN, to suppress root rot, and to stimulate root hair formation are positively correlated (Défago *et al*, 1990). Cyanide stress induces cyanide resistant respiration and production of phytoalexins in potato slices (Alves *et al.*, 1979; Laties, 1982). Défago and co-workers therefore postulate that bacterial HCN modifies the metabolism of tobacco in a way that induces some plant defense mechanism against the pathogen (Défago *et al.*, 1990; Voisard *et al.*, 1989).

Recently it was demonstrated by van Peer (unpublished), that colonization of the roots of carnation by *Pseudomonas* sp. strain WCS417 in soilless culture significantly suppresses disease caused by *F. oxysporum* f. sp. *dianthi*. Strain WCS417 and the pathogenic fusarium were inoculated onto the roots and into the vascular tissue of the stem, respectively, and shown to be separated spatially during the experiment. These experimental results strongly suggest that induction of plant defense mechanisms are involved in the suppression of the disease (Van Peer, unpublished). WCS417, however, does not have the ability to produce HCN (Van Peer, unpublished). In this system, traits of Pseudomonas other than HCN production may be involved in the induction of plant defense mechanisms.

Suppression of potato growth by microbial cyanide

Production of hydrogen cyanide by rhizosphere pseudomonads is suspected to be partly responsible for the inhibition of potato plant growth and for the reduction of potato tuber yields in high frequency potato cropping (Bakker and Schippers, 1987; Schippers *et al.*, 1987). This suspicion originates from the observed HCN-induced inhibition of potato root cytochrome oxidase respiration (Bakker and Schippers, 1987), from the high percentage of HCN-producing pseudomonads in the rhizosphere, from the strong dependence of the Pseudomonas-HCN production on the availability of iron, and from the known presence of particular amino acids, such as glycine in potato root exudates. The strong inhibition of potato root development within one week after planting in soil frequently cropped to potato and its counteraction by root bacterization with *P. putida* WCS358 (which does not produce HCN), but not by its siderophore negative Tn*5* transposon mutant (P.A.H.M. Bakker *et al.*, 1987), fits the observed Fe-dependence of the HCN production. Reduced tuber yields in high frequency potato cropping could also be diminished by seed tuber bacterization with strain WCS358, but not by its siderophore negative Tn*5* mutant (Bakker *et al.*, 1986). Finally, most HCN-producing pseudomonads isolated from potato rhizosphere were unable to utilize the Fe^{3+}-pseudobactin 358 of strain WCS358 (Bakker *et al.*, 1990).

The possible role of HCN in inhibiting potato growth has been visualized in a working hypothesis (Fig. 1) (Bakker and Schippers, 1987; Bak-

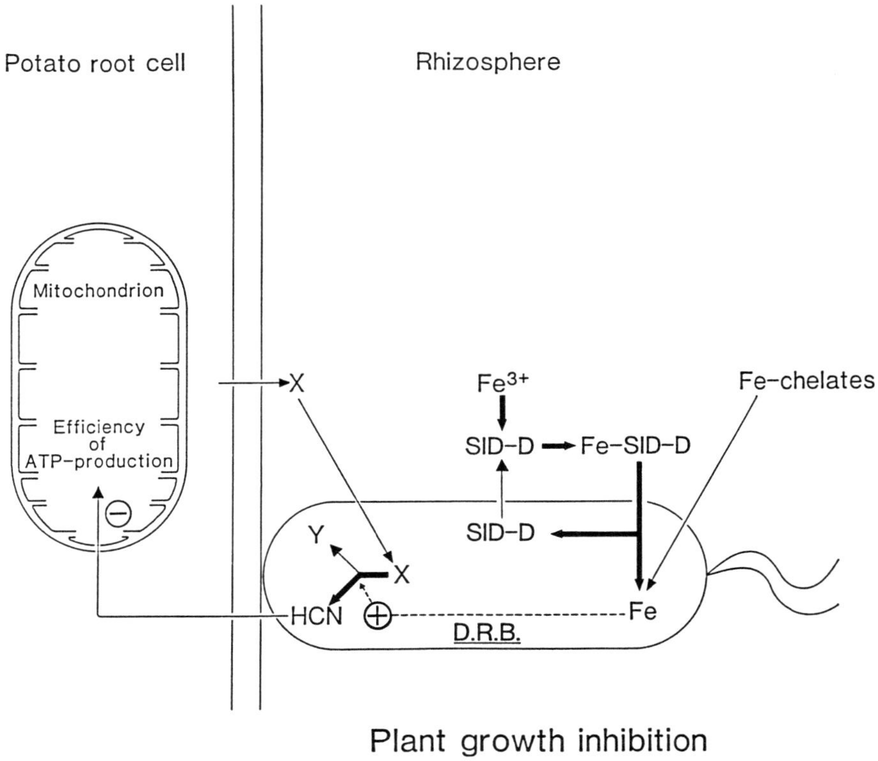

Fig. 1. Hypothetical mechanism of plant growth inhibition due to cyanide production by deleterious rhizobacteria (DRB) in the rhizosphere of potato in short rotation soil.

The presence of substrate (X) for cyanide production and the high iron availability stimulate cyanide production by the deleterious rhizobacteria (DRB).
SID-D : siderophore of DRB
Fe-SID-D : siderophore of DRB bound to Fe^{3+}
Y : non-deleterious microbial product
(After: Bakker, A.W. *et al.*, 1989).

ker *et al.*, 1989). This hypothetical model shows how HCN, by increasing the relative contribution of the alternative respiration path to the total root respiration, may suppress plant growth. Effects of HCN on the activity of other enzymes *e.g.* peroxidases in the roots may be involved as well and need further studies. The model also illustrates how the production of HCN may be counteracted by siderophore mediated competition for Fe^{3+} (Fig. 2). In this respect, HCN can be replaced by other phytotoxic microbial metabolites of which the production is iron-dependent.

This hypothesis was investigated experimentally in a gnotobiotic system using artificial soil low in iron availability, sterile potato plants grown from meristem culture, and HCN-producing *Pseudomonas* sp. strain A47. Addition of the Fe^{3+}-siderophore complex of strain A47 to this system significantly reduced potato plant growth and reduction may be due to increased HCN production in the rhizosphere. These results suggest that HCN production by A47, induced by increased Fe^{3+} availability, inhibits potato plant growth. An experimental approach in which the sensitivity of the potato plant for HCN can be manipulated and in which HCN-negative mutants are included, however, is needed to further investigate whether microbially produced HCN indeed inhibits potato plant growth.

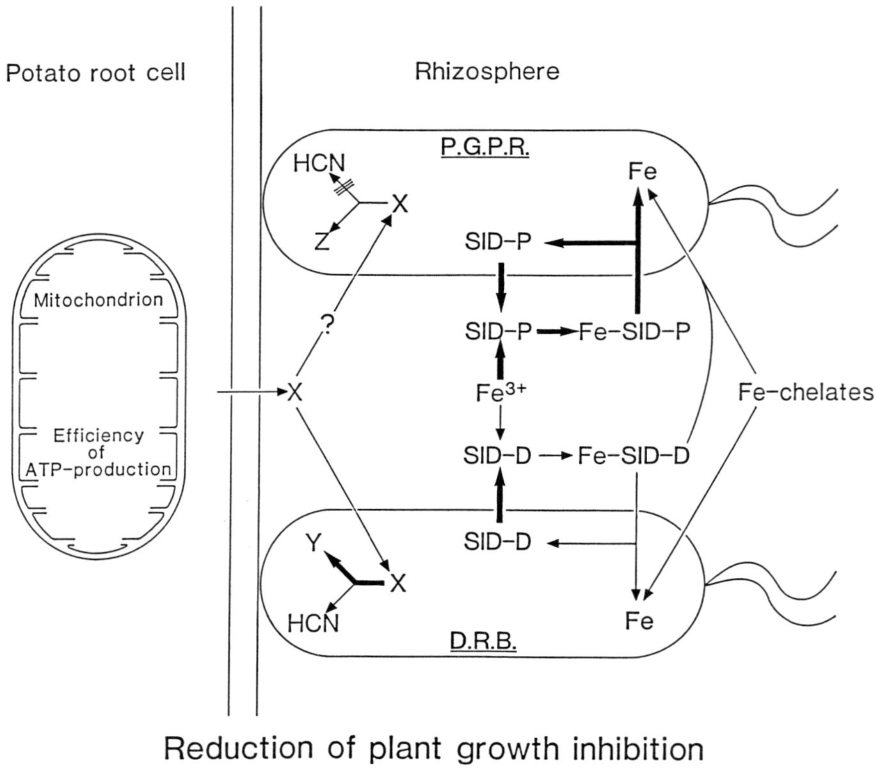

Fig. 2. Hypothetical mechanism of plant growth stimulation by bacterization with plant growth-promoting rhizobacteria (PGPR) in short potato-rotations.

Due to competition for iron (free ferric iron and certain iron-chelates) between PGPR and DRB, iron availability for the DRB decreases. This will reduce the rate of cyanide production.
SID-P : siderophore of PGPR
Fe-SID-P : siderophore of PGPR bound to Fe^{3+}
Z : non-deleterious microbial product
(After: Bakker, A.W. *et al.*, 1989).

Sites of activity of rhizosphere pseudomonads

Intercellular penetration of the root cortex by rhizosphere bacteria, often in high densities, has been demonstrated in many different crops (Darbyshire and Greaves, 1971; Jacobs *et al.*, 1985; Old and Nicolson, 1978; Patriquin and Dobereiner, 1978). In some of these cases *Pseudomonas* spp. are predominant (Kleeberger *et al.*, 1983; Van Peer, unpublished). Particularly with plants grown in nutrient films or on rockwool (soilless cultures), they sometimes represent more than 50% of the total aerobic bacterial population inside the roots compared to 2–5% in soil grown tomatoes (Van Peer *et al.*, unpublished).

A strong negative correlation ($r = 0.8$) was observed between plant weight and the density of pseudomonads inside the healthy looking roots of tomato plants grown in soilless cultures. Inoculation of the nutrient solution with a plant growth promoting *Pseudomonas* sp. strain WCS417 displaced these root inhabiting pseudomonads. This was accompanied by a significant increase in tomato seedling growth (Van Peer and Schippers, 1989). When applied to young seedlings of tomato or when introduced into the nutrient solution before planting, a significantly higher percentage of pseudomonads isolated from inside the roots (30%), compared to isolates obtained from the root surface (4%), inhibited the growth of tomato seedlings. Pseudomonas isolates obtained from inside the roots, recolonized the root cortex in higher numbers compared to Pseudomonas isolates obtained from the root surface (Van Peer, unpublished).

Populations of bacteria inside the roots are most dense in the outermost cortex layers and in the intercellular spaces of the cortex near the endodermis, as could be observed after staining of the tomato roots with 2,3,5-triphenyl-tetrazolium chloride (Van Peer, unpublished). Nutrients coming from the central cylinder and partly released as root exudates in soil probably have the highest concentration in the cortex near the endodermis. Bacterial cells were also found intracellularly in the cortex. Both the root surface and the cortex of tomato contain pseudomonads that have the potential of producing HCN. A positive correlation between the potential to produce HCN and to inhibit tomato development in soilless cultures, however, could not be demonstrated. Other microbial metabolites may thus be involved (Alström and Burns, 1989; Van Peer, unpublished).

Our observations, particularly in soilless cultures, suggest that deleterious and plant growth promoting activities of rhizosphere pseudomonads may take place inside the cortical tissue of roots as well as at the root surface. Metabolites of bacteria invading the root, however, may have a greater effect on root physiology than those of bacteria on the root surface. Their metabolities are trapped in the root tissue and are less vulnerable to microbial degradation and to adsorption to soil components. The Pseudomonas strains inhibiting the growth of canola seedlings and of winter wheat seedlings were also observed to colonize the intercellular spaces of root tissue (Campbell *et al.*, 1987; Frederikson and Elliott, 1985). Défago *et al.* (1990), using immunofluorescence staining of the root tissue, showed that also *P. fluorescens* strain CHAo protecting tobacco roots against *T. basicola*, invaded the outer cortex layers.

Conclusions

It seems likely that HCN production by rhizosphere pseudomonads is a rather common phenomenon. The ecological significance for the producers is subject to speculation (Schippers *et al.*, 1989).

The contribution of cyanide production by *Pseudomonas fluorescens* strain CHAo, to the suppression of several soilborne plant pathogenic fungi, has been demonstrated elegantly by using defective mutants and subsequent complementation (Voisard *et al.*, 1989). It is also evident, that cyanide production by strain CHAo affects root morphology in tobacco (Défago *et al.*, 1990). These observations suggest that HCN may be involved in the induction of plant resistance. This supposition, however, needs further study.

Although there is circumstantial evidence that microbial cyanide, by affecting root energy metabolism and the activity of many enzymes, may inhibit plant development, convincing evidence has not yet been presented. Also the actual production of cyanide in the rhizosphere needs further experimental evidence. Whether it significantly influences plant development apparently depends on soil characteristics such as the availability of iron from soil particles or microbial iron siderophore complexes, root exudate composition, plant susceptibility to HCN, and on a specific relationship between particular HCN-producing Pseudomonas strains and the plant root *e.g.* penetration of the root tissue. This specific relationship and the effects of microbial HCN on the root physiology deserve further study. Beneficial effects of pseudomonads on plant development by the suppression of disease may mask negative effects of microbial metabolites on plant development that take place simultaneously. These deleterious effects therefore may only become apparent in the absence of plant parasitic pathogens that are suppressed by HCN.

References

Ahl P, Voisard C and Défago G 1986 Iron-bound siderophores, cyanide and antibiotics involved in suppression of *Thielaviopsis basicola* by a *Pseudomonas fluorescens* strain. J. Phytopath. 116, 121–134.

Alström S and Burns R G 1989 Cyanide production by rhizobacteria as a possible mechanism of plant growth inhibition. Biol. Fertil. Soils 7, 232–238.

Alves L M, Heisler E G, Kissinger J C, Patterson J M and Kalan E G 1979 Effects of controlled atmospheres on production of sesquiterpenoid stress metabolites by white potato tuber. Pl. Physiol. 63, 359–362.

Arora Y K and Bajaj K L 1985 Peroxidase and polyphenol oxidase associated with induced resistance of mung bean to *Rhizoctonia solani* Kühn. Phytopathol. Z. 114, 325–331.

Askeland R A and Morrison S M 1983 Cyanide production by *Pseudomonas fluorescens* and *Pseudomonas aeruginosa*. Appl. Environ. Microbiol. 45, 1802–1807.

Baker R, Elad Y and Sneh B 1986 Physical, biological and host factors in iron competition in soils. *In* Iron, Siderophores, and Plant Diseases. Ed. T R Swinburne. pp. 77–84. Plenum, New York.

Bakker A W, Bakker P A H M and Schippers B 1989 Deleterious cyanide-producing rhizosphere pseudomonads as a factor limiting potato root growth and tuber yield in high frequency potato-cropping soil. *In* Effects of Crop Rotation on Potato Production in the Temperate Zones. Eds. J Vos, C D van Loon and G J Bollen. pp. 153–162. Kluwer Academic Publishers, Dordrecht.

Bakker A W, Cosse A A and Schippers B 1987 The role of HCN-producing *Pseudomonas* spp. in yield reductions in short potato rotations. Meded. Fac. Landbouww. Rijksuniv. Gent 52, 1111–1117.

Bakker A W and Schippers B 1987 Microbial cyanide production in the rhizosphere in relation to potato yield reduction and *Pseudomonas* spp.-mediated plant growth stimulation. Soil Biol. Biochem. 19, 249–256.

Bakker P A H M, Bakker A W, Marugg J D, Weisbeek P J and Schippers B 1987 A bioassay for studying the role of siderophores in potato growth stimulation by *Pseudomonas* spp. in short potato rotations. Soil Biol. Biochem. 19, 443–449.

Bakker P A H M, Lamers J G, Bakker A W, Marugg J D, Weisbeek P J and Schippers B 1986 The role of siderophores in potato tuber yield increase by *Pseudomonas putida* in a short rotation of potato. Neth. J. Pl. Pathol. 92, 249–256.

Bakker P A H M, Van Peer R and Schippers B 1990 Specificity of siderophores and siderophore receptors and biocontrol by *Pseudomonas* spp. *In* Biological Control of Soil-borne Plant Pathogens. Ed. D Hornby. pp. 131–142. CAB International.

Becker J O and Cook R J 1988 Role of siderophores in suppression of *Phythium* species and production of increased growth response of wheat by fluorescent pseudomonads. Phytopathology 78, 778–782.

Burr T J and Caesor A 1984 Beneficial plant bacteria. CRC Crit. Rev. Pl. Sci. 2, 1–20.

Campbell J N, Cass D D and Peteya D J 1987 Colonization and penetration of intact canola seedling roots by an opportunistic fluorescent *Pseudomonas* sp. and the response of host tissue. Phytopathology 77, 1166–1173.

Castric P A 1975 Hydrogen cyanide, a secondary metabolite of *Pseudomonas aeruginosa*. Can. J. Microbiol. 21, 613–618.

Castric P A 1983 Hydrogen cyanide production by *Pseudomonas aeruginosa* at reduced oxygen levels. Can. J. Microbiol. 29, 1344–1349.

Curl E A and Truelove B 1986 The Rhizosphere. Springer-Verlag, Berlin/Heidelberg/New York/Tokyo.

Darbyshire J F and Greaves M P 1971 The invasion of pea roots, *Pisum sativum* L., by soil microorganisms, *Acanthamoeba palestinensis* (Reich) and *Pseudomonas* spp. Soil Biol. Biochem. 3, 151–155.

Défago G, Berling C H, Burger U, Haas D, Kahr G, Keel C, Voisard C, Wirthner Ph and Wütrich B 1990 Suppression of black root rot of tobacco by a *Pseudomonas* strain: potental applications and mechanisms. *In* Biological Control of Soil-borne Plant Pathogens. Ed. D Hornby. pp. 93–108. CAB International.

Défago G and Haas D 1990 Pseudomonads as antagonists of soilborne plant pathogens: mode of action and genetic analysis. Soil Biochem. 6, 249–291.

Frederickson J K and Elliott L F 1985 Effects on winter wheat seedling growth by toxin-producing rhizobacteria. Plant and Soil 83, 399–409.

Hemming B C 1986 Microbial-iron interactions in the plant rhizosphere. An overview. J. Pl. Nutr. 9, 505–521.

Howell C R and Stipanovic R D 1979 Control of *Rhizoctonia solani* on cotton seedlings with *Pseudomonas fluorescens* and with an antibiotic produced by the bacterium. Phytopathology 69, 480–482.

Howell C R and Stipanovic R D 1980 Suppression of *Phytium ultimum*-induced damping-off cotton seedlings by *Pseudomonas fluorescens* and its antibiotic, pyoluteorin. Phytopathology 70, 712–715.

Jacobs M J, Bugbee W M and Gabrielson D A 1985 Enumeration, localisation, and characterization of endophytic bacteria within sugar beet roots. Can. J. Bot. 63, 1262–1265.

Keel C, Voisard C, Berling C H, Kahr G and Défago G 1989 Iron sufficiency, a prerequisite for suppression of tobacco black root rot by *Pseudomonas fluorescens* strain CHAO under gnotobiotic conditions. Phytopathology 79, 584–589.

Kleeberger A, Castorph H and Klingmuller W 1983 The rhizosphere microflora of wheat and barley with special reference to gram-negative bacteria. Arch. Microbiol. 136, 306–311.

Kloepper J W, Lifshitz R and Schroth M N 1988 *Pseudomonas* inoculants to benefit plant production. ISI Atlas of Science: Anim. Pl. Sci. 60–63.

Knowles C J 1976 Microorganisms and cyanide. Bacteriol. Rev. 40, 652–680.

Knowles C J and Bunch A W 1986 Microbial cyanide metabolism. Adv. Microbiol. Physiol. 27, 73–111.

Kuiper D 1983 Genetic differentiation in *Plantago major*: Growth and root respiration and their role in phenotypic adaptation. Physiol. Pl. 57, 222–230.

Lambers H 1980 The physiological significance of cyanide-resistant respiration in higher plants. Pl. Cell Environ. 3, 293–302.

Lambers H 1985 Respiration in Intact Plants and Tissues: Its Regulation and Dependence on Environmental Factors, Metabolism and Invaded Organism. Encyclopedia Pl. Physiol. Vol. 18, 418–465.

Laties G G 1982 The cyanide-resistant, alternative path in higher plant respiration. Ann. Rev. Pl. Physiol. 33, 519–555.

Leong J 1986 Siderophores: their biochemistry and possible role in the biocontrol of plant pathogens. Annu. Rev. Phytopath. 24, 187–209.

Lifshitz R, Kloepper J W and Kozlowski M, *et al.* 1987 Growth promotion of canola (rapeseed) seedlings by a strain of *Pseudomonas putida* under gnotobiotic conditions. Can. J. Microbiol. 33, 390–395.

Loper J E 1988 Role of fluorescent siderophore production in biological control of *Pythium ultimum* by a *Pseudomonas fluorescens* strain. Phytopathology 78, 166–172.

Old K M and Nicolson T H 1978 The root cortex as part of a microbial continuum. *In* Microbial Ecology. Eds. M W Loutit and J A R Miles. pp 291–294. Springer-Verlag, Berlin/Heidelberg/New York.

Patriquin D G and Dobereiner J 19878 Light microscopy observations of tetrazolium-reducing bacteria in the endorhizosphere of maize and other grasses in Brasil. Can. J. Microbiol. 24, 734–742.

Rovira A D and Davey C B 1974 Biology of the rhizosphere. *In* The Plant Root and Its Environment. Ed. E W Carson. pp 153–204. Univ. Press Virginia, Charlottesville.

Schippers B 1988 Biological control of pathogens with rhizobacteria. Philosophical Transactions of the Royal Society, B 318, 283–293.

Schippers B, Bakker A W and Bakker P A H M 1987 Interactions of deleterious and beneficial rhizosphere microorganisms and the effect of cropping practices. Annu. Rev. Phytopath. 25, 339–358.

Schippers B, Bakker P A H M, Bakker A W and Van Peer R 1988 Crop losses due to deleterious rhizobacteria and their prevention by bacterization. Proceedings Brighton Crop Protection Conference 'Pest and Diseases': 611–614.

Schippers B, Bakker P A H M, Salentijn E and Hoekstra W P M 1989 Antimicrobial metabolite production: a competitive advantage to a soil microorganism? *In* Risk Assessment in Agricultural Biotechnology. Proc. Int. Conf. UC, Davis, 1988. (*In press*).

Solomonson L P 1981 Cyanide as a metabolite inhibitor. *In* Cyanide in Biology. Eds. B Vennesland *et al.* pp. 11–28. Academic Press, London.

Stutz E W, Défago G and Kern H 1986 Naturally occuring fluorescent pseudomonads involved in suppression of black root rot of tobacco. Phytopathology 76, 181–185.

Suslow T V 1982 Role of root-colonizing bacteria in plant growth. *In* Phytopathogenic Prokaryotes Eds. M S Mount and G H Lacy. Vol. 1, pp 187–223. Academic Press, New York and London.

Suslow T V and Schroth M N 1982 Role of deleterious rhizobacteria as minor pathogens in reducing crop growth. Phytopathology 72, 111–115.

Symons P C, Kolattukudy P E and Bienfait H F 1985 Iron deficiency decreases suberization in bean roots through a decrease in suberinspecific peroxidase activity. Pl. Physiol. 78, 115–120.

Thomashow L S and Weller D 1988 Role of a phenazine antibiotic from *Pseudomonas fluorescens* in biological control of *Gäumannomyces graminis* var. *tritici*. J. Bacteriol. 170, 3499–3508.

Van Peer R and Schippers B 1989 Plant growth responses to bacterization with selected *Pseudomonas* spp. strains and rhizosphere microbial development in hydroponic cultures. Can. J. Microbiol. 35, 456–463.

Voisard C, Keel C, Haas D and Défago G 1989 Cyanide production by *Pseudomonas fluorescens* helps suppress black root rot of tobacco under gnotobiotic conditions. EMBO J. 8, 351–358.

Weller D M 1988 Biological control of soilborne plant pathogens in the rhizosphere with bacteria. Annu. Rev. Phytophath. 26, 379–407.

D. L. Keister and P. B. Cregan (Eds.), The rhizosphere and plant growth, 221–228.

PLSO BARC21

Induction of rhizosphere competence in the biocontrol fungus Trichoderma

RALPH BAKER
Plant Pathology and Weed Science, Colorado State University, Fort Collins, CO 80523, USA

Key words: cucumber, increased growth response, mechanisms of biocontrol, minor pathogens, mutation, pea, radish, tomato

Abstract

Apparently, wild types of *Trichoderma* spp. do not possess the attribute of rhizosphere competence (RC), *i.e.* they cannot colonize the rhizosphere adjacent to developing roots from a seed treatment. To enhance RC, strains of *T. harzianum* were exposed to N-methyl-N-nitro-N-nitrosoguanidine and mutants selected for benomyl resistance. When benomyl was added to soil, these mutants applied to seeds colonized the rhizosphere 8 cm below the seed at densities approaching 10^6 cfu/g rhizosphere soil. Unexpectedly, RC was induced when benomyl was not added to the soil. There is no evidence that benomyl resistance is related to RC; however, a perfect correlation was observed between increased cellulase production by mutants and RC. This suggested that mutants may utilize the mucigel on roots as a substrate. Mutation to RC and faster growth rate, comparable to root elongation, was induced in a slow growing species such as *T. polysporum*. Mutants with RC also were more efficient biocontrol agents and induced greater increased growth responses from a seed treatment than wild-type parents.

Introduction

Rhizosphere competence is defined here as the ability of a microorganism applied by seed treatment to colonize the rhizosphere of developing roots. This attribute is valuable in the enhancement of a biocontrol agent since not only the seed may be protected from soilborne plant pathogens but the roots as well. Because the mechanisms underlying such an attribute are not well understood and even those agents that are reported to be rhizosphere competent are not very efficient in doing so (Weller, 1988), the development of strains that colonize such sites could be a valuable asset in increased biocontrol of soilborne pathogens.

Rhizosphere competence in *Trichoderma* spp.

Trichoderma spp. are not rhizosphere-competent (Papavizas, 1985); however, the report by Mendez-Castro and Alexander (1983) suggested an approach for establishing a previously rhizosphere-incompetent microorganism on roots. By this method a fungus resistant to a relatively broad spectrum fungicide that inhibited competitors for sites on roots could become rhizosphere-competent by addition of the fungicide to the soil. Strains of *Trichoderma harzianum* Rifai were mutated by use of *N*-methyl-N-nitro-*N*-nitrosoguanidine (NTG) and selected for benomyl resistance (Chang *et al.*, 1986). The mutants were not only competent when the benomyl was added to the soil (Ahmad and Baker, 1987b) but also when the fungicide was not added (Fig. 1).

Population densities approached 10^6 cfu/cm root. This was noticed at various soil reactions and temperatures (Fig. 2).

The total length of all hyphae was measured 1–8 mm along root surfaces (Fig. 3).

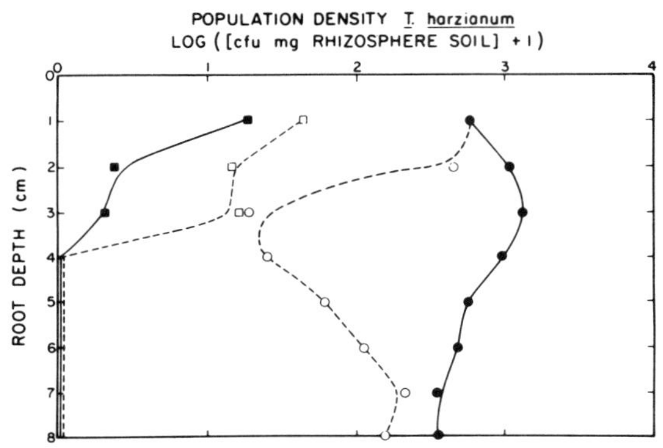

Fig. 1. Population densities of the wild type (WT) parent and mutant strain T-95 of *Trichoderma harzianum* in the rhizosphere soil of cucumber plants: ■---■ isolate WT with benomyl in the soil, □---□ isolate WT without benomyl, ●---● isolate T-95 with benomyl, ○---○ T-95 without benomyl (from Ahmad and Baker, 1987b by permission).

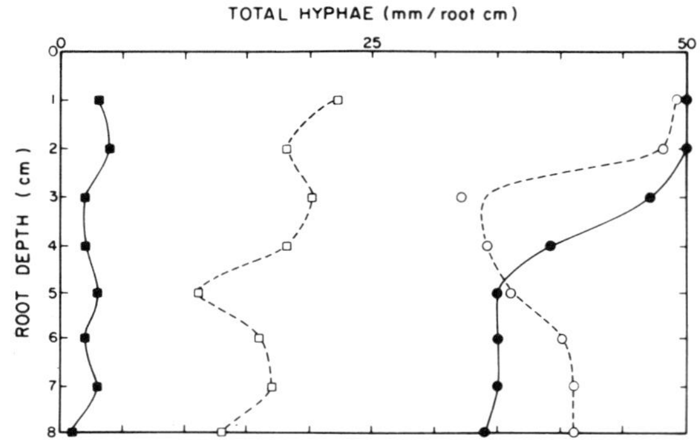

Fig. 3. Total hyphae observed per root centimeter depth of cucumber plants from treated (*Trichoderma harzianum*, T-95) and untreated seeds: ■---■ untreated seeds with benomyl in the soil, □---□ untreated seeds without benomyl, ●---● strain T-95 treated seeds with benomyl, ○---○ T-95 treated seeds without benomyl (from Ahmad and Baker, 1987b, by permission).

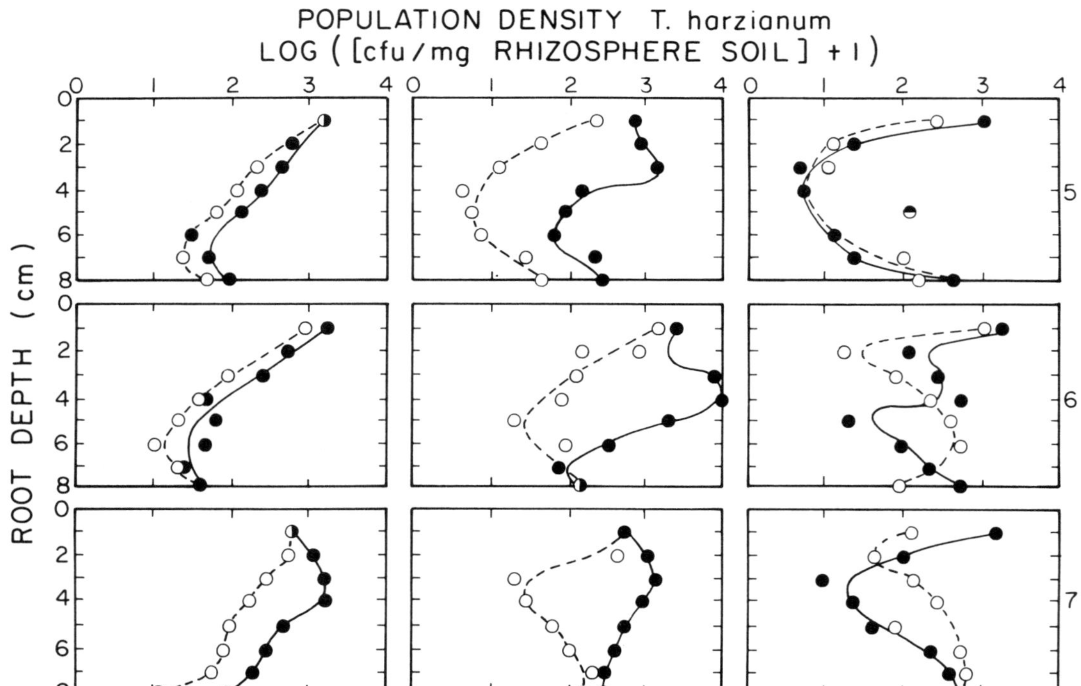

Fig. 2. Population densities of *Trichoderma harzianum* T-95 in rhizosphere soil of cucumber plants grown at three pH levels and three temperatures: ●---with benomyl in soil, ○---○ without benomyl (from Ahmad and Baker, 1987b by permission).

When wild-type parents were applied to seed, hyphal length was approximately 10–25 but was reduced to <5 mm/cm root when benomyl was incorporated into the soil. Seed treatment with the mutants resulted in hyphal lengths from 35–50 mm/cm root.

In most of the experiments when cfu were plotted at centimeter units along the root, a C-shaped curve was observed (Figs. 1, 2). The high population densities on the root immediately below the seed could be explained by exudates from the seed providing more substrates for growth of the fungus than those from roots at lower depths. Also, exudates may be more abundant at root tips (Schroth and Hildebrand, 1964); however, this could not explain the lower counts on the mid-portions of the root. Population density increase below 6–8 cm in comparison with mid-portion rhizosphere depth could be due to the 'flush' of exudates resulting from photosynthates as green cotyledon and shoots emerged above the soil level. However, evidence for this hypothesis was not demonstrated (Ahmad and Baker, 1987b). Elements of the C-shaped curve appeared to be present when the experiment was performed in the dark. Also, instead of sampling sections of roots after development for 8 days, 1-cm sections of root tips were sampled each day for 8 days with illumination. There were no significant differences among cfu below the 2-cm rhizosphere depth, suggesting that quantity of substrates for growth was constant on the root tips and that no increase occurred when shoots emerged and photosynthesis was initiated. It is possible that lower populations in the mid-portion of the root is due to lysis of and/or production of chlamydospores by the maturing hyphae.

The results described above involved mutation of strains of *T. harzianum* (Ahmad and Baker, 1987b). Mutants resistant to benomyl also were developed for *T. viride* Rifai, *T. polysporum* Rifai, and *T. koningii* Rifai (Ahmad and Baker, 1988c). Induction of rhizosphere competence in the last two species was complicated by the factor that they had approximately half the linear growth rate in culture in comparison with cucumber rates of root extension. By mutation, however, these were induced to increased hyphal growth rates comparable with root elongation.

Why are benomyl resistant mutants rhizosphere-competent?

Fungal genes and mechanisms responsible for resistance to benzimidazoles have been identified for some fungi, namely, exclusion of the chemical by reduced membrane permeability in Sporobolomyces, detoxification by breakdown or binding in Verticillium, or decreased affinity for tubulin to the fungicide in Aspergillus (Delp, 1980). Considerable speculation and extrapolation would be necessary to formulate hypotheses to explain rhizosphere competence based on these mechanisms of benomyl resistance. Although it has been difficult to quantify the impact of antibiotics produced by the rhizosphere microflora on fungal activity (Baker, 1984), it is possible that the benomyl-resistant mutants could tolerate this type of antagonism to a greater extent than the wild types. Indeed, in preliminary tests when conidia were observed in the rhizosphere of root tips of cucumber over 8 days, germination of the wild type was below 10% at 1–2 cm below the seed and 0% below that depth whereas 23–37% germination occurred in the mutant T-95. This phenomenon, however, may not be wholly explained by tolerance to antibiosis in the mutants. In another preliminary experiment, conidia of either the wild type of T-95 were added to raw soil at various concentrations of glucose-asparagine (4:1) and resultant germination observed. Higher germination of the mutant compared with the wild type was observed at glucose concentrations of 0.4–0.8 $\mu g\ g^{-1}$ soil. Therefore, conidia of the mutants may have a lower exogenous nutrient requirement to overcome soil fungistasis than wild types but how this is associated with benomyl resistance is not readily apparent.

The prolific hyphae observed on root tips when mutant T-95 was applied as a seed treatment (Ahmad and Baker, 1987b) suggested that substrates supporting growth on the rhizoplane might be involved. Foster, Rovira and Cook (1983) identified the mucigel on the surface of roots as the remains of epidermal cell walls and these are rich in cellulose. *Trichoderma* spp. are well-known cellulose decomposers. When mutants or their parents were added to soil containing either wheat straw segments or cellophane disks, the mutants colonized the substrates at a higher frequency (Ahmad and Baker, 1987a).

The competitive saprophytic ability *sensu* Garrett (1970) was determined for the mutants and their wild type parents and quantified by a competitive saprophytic ability index (CSA) as follows (Ahmad and Baker, 1987a):

$$\text{CSA index} = \sum_{n}^{i=1} [\ln(1/1 - C_i)/t_i(\log p_i)]/n \quad (1)$$

where C = frequency of isolation of a specific strain of Trichoderma from units of cellulose substrates, t = time of incubation, p = population density of conidia added to soil, and n = number of treatments. Also, a rhizosphere competence (RC) index was developed:

$$\text{RC index} = \sum_{n}^{i=1} [\log(p_i + 1) \cdot \ln(d + 1)]/n \quad (2)$$

where p = population density per milligram of rhizosphere soil, d = root depth, and n = total root depth. The interactions of data accumulated from equations (1) and (2) and cellulase activity on cellobiose, carboxymethyl cellulose and cotton linters as soil carbon sources by the strains were analyzed by correlation statistics (Fig. 4). High correlations were apparent among these factors. Therefore, the hypothesis was advanced that the rhizosphere-competent mutants utilized the mucigel of the root more efficiently than wild types.

Growth of the mutants and wild types on glucose and various cellulose substrates as sole carbon sources was compared (Ahmad and Baker, 1988a). Both mutants and wild types grew poorly on glucose, galactose, cellobiose, or xylose as a sole source of carbon. The rhizosphere-competent mutants, however, produced significantly higher biomass than wild types when grown on cotton linters, microcrystalline cellulose, wood cellulose, or xylan. Typical results are in Figure 5.

Since simple sugars are present in root exudates (Schroth and Hildebrand, 1964), it is possible that cellulose decomposers would be influenced by catabolic repression (Montenecourt, 1983). Strain T-95 had significantly higher production of mycelial dry weight on wood cellulose, however, when glucose was added at concentrations of 0.5 to 3% (Ahmad and Baker, 1988a) (Fig. 6).

In contrast, T-12B produced significantly less growth when glucose was added to the wood cellulose substrate. If such observations have ecological significance, strain T-95 should have a RC index greater than T-12B and, indeed, this was the case (Ahmad and Baker, 1987b).

Is benomyl resistance necessary for induction of rhizosphere competence? There was a perfect

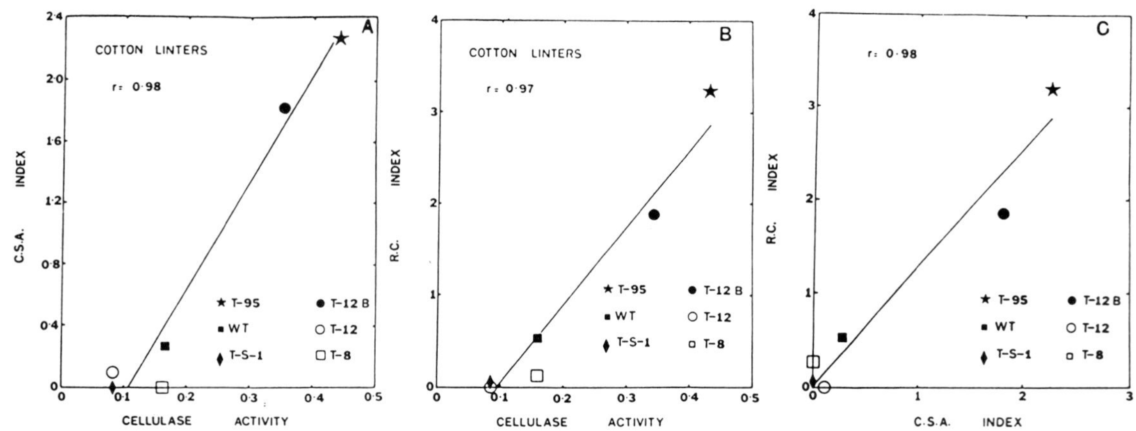

Fig. 4. Relation of **A**, competitive saprophytic ability (CSA) index of *Trichoderma* spp. and strains to their cellulase activity, **B**, rhizosphere competence (RC) index of *Trichoderma* spp. and strains to the cellulase activity when cotton linters were the sole source of carbon, and C, RC index to CSA index of *Trichoderma* spp. and strains (Ahmad and Baker, 1987a by permission).

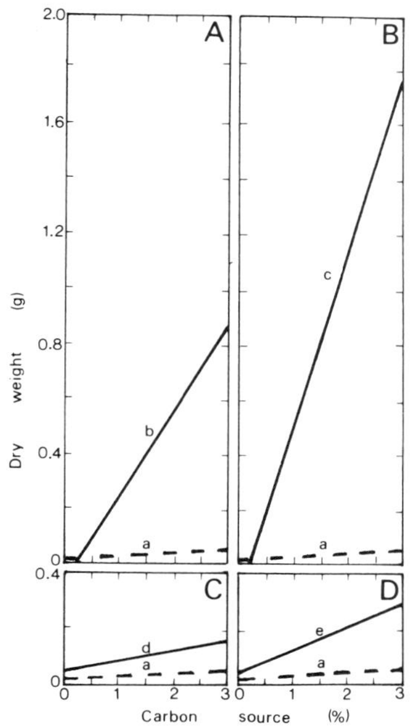

Fig. 5. Dry weight of mycelium of *Trichoderma harzianum* mutant strains, **A**, T-95, **B**, T-12B, and wild type parents **C**, WT, and **D**, T-12 grown for 6 days in Czapek-Dox broth with glucose (– – –) or wood cellulose (———) as sole carbon source in increasing concentrations. Lines labelled by the same letters are not significantly different ($P = 0.05$) (from Ahmad and Baker, 1988a by permission).

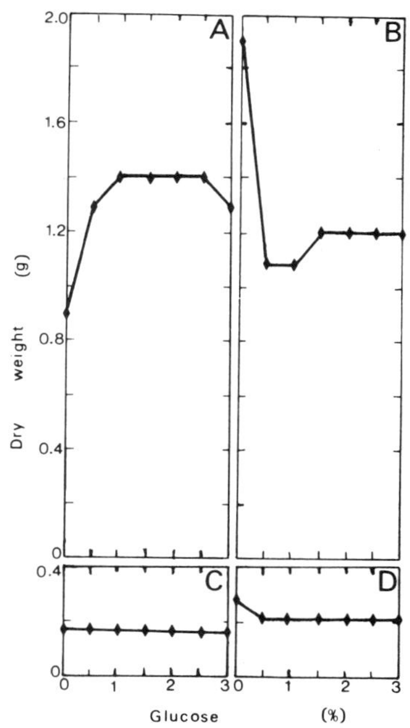

Fig. 6. Dry weight of mycelium of *Trichoderma harzianum* mutant strains, **A**, T-95, **B**, T-12B and wild type parents, **C**, WT, and **D**, T-12 grown for 6 days in Czapek-Doz broth with glucose in increasing concentrations as carbon source and 3% wood cellulose at each concentration (from Ahmad and Baker, 1988a by permission).

correlation between superior growth on cellulose and rhizosphere-competence. Strains were found, however, that were benomyl-resistant but were not competent and these did not grow well on cellulose substrates (Ahmad and Baker, 1988c). This suggests that benomyl-resistance is often associated with increased cellulase activity in mutations induced by NTG but might not be necessary for the attribute of rhizosphere competence. It does not eliminate the possibility, however, that both benomyl resistance and enhanced utilization of cellulose substrates are required for competence; *i.e.* a second site mutation. Further research involving mutations for increased cellulase production without benomyl resistance (Cuskey *et al.*, 1983) should provide better evidence of the mechanisms involved. Benomyl resistance, however, is a desirable attribute for biocontrol agents used in agriculture, especially in the ornamental industry where benzimidazoles are applied routinely (Baker and Scher, 1987).

Implications

What benefits may accrue from induction of rhizosphere competence? Hypothetically, the activities associated with biocontrol (Papavizas, 1985) and increased growth responses induced by *Trichoderma* spp. (Baker, 1988) could be enhanced by such competence.

Rhizosphere competence should increase the efficiency of biocontrol. Cucumber seeds were treated with the parent strain or a rhizosphere-competent mutant (T-95) and the inoculum density of *Pythium ultimum* Trow measured in centimeter sections along 8 cm of the developing seedling roots (Ahmad and Baker, 1988b). The roots of non-treated controls, or those treated with the wild type with or without the addition of benomyl to the soil, grew only 7 cm during 8 days and were discolored, whereas seeds treated with T-95 produced 8-cm lengths of roots and displayed no obvious symptoms (Fig. 7).

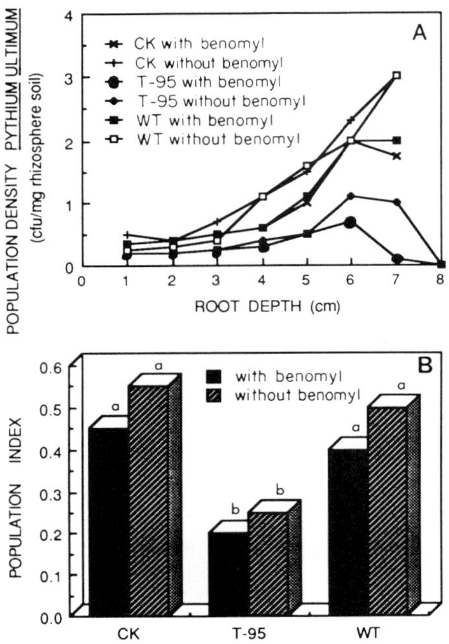

Fig. 7. Spatial distribution (**A**) and population index (Ahmad and Baker, 1987a) of *Pythium ultimum* in rhizosphere soil of 8-day old cucumber seedlings grown from seeds treated with *Trichoderma harzianum* mutant strain T-95 and its wild type parent (WT). Bars labelled by the same letters are not significantly different ($P = 0.05$). CK = control (from Ahmad and Baker, 1988b by permission).

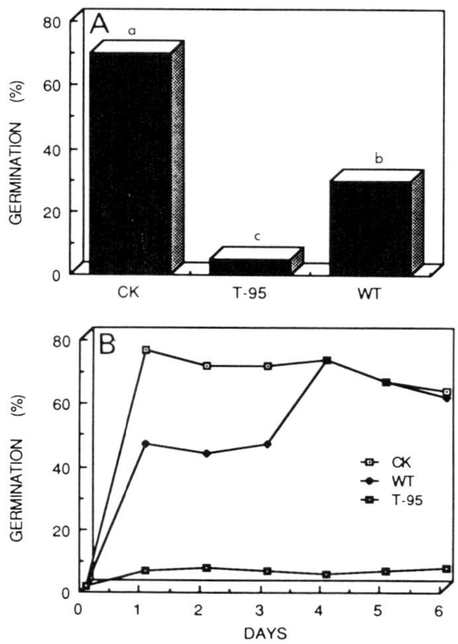

Fig. 9. Percent germination of sporangia of *Pythium ultimum* placed on **A**, cucumber seed, and **B**, rhizosphere soil of cucumber seedlings treated with the *Trichoderma* rhizosphere-competent mutant T-95 and its wild type parent (WT). Bars labelled by the same letters are not significantly different ($P = 0.05$). CK = control (from Ahmad and Baker, 1988b by permission).

Inoculum densities in rhizospheres of *P. ultimum* were significantly reduced by seed treatment with T-95 in comparision with those treated with the parent or non-treated controls. Indeed, *P. ultimum* was not detected in the 7 to 8 cm portion of the rhizosphere when seeds were treted with T-95.

The report by Lifshitz *et al.* (1986) may provide an explanation for the apparent protection of roots from infection by *P. ultimum* by seed treatment with a rhizosphere-competent mutant. No experimental evidence was obtained for competition as a biocontrol mechanism for suppression of Pythium damping-off by seed treatment with *Trichoderma* spp. Mycoparasitism (Baker, 1987) was not likely because it does not occur

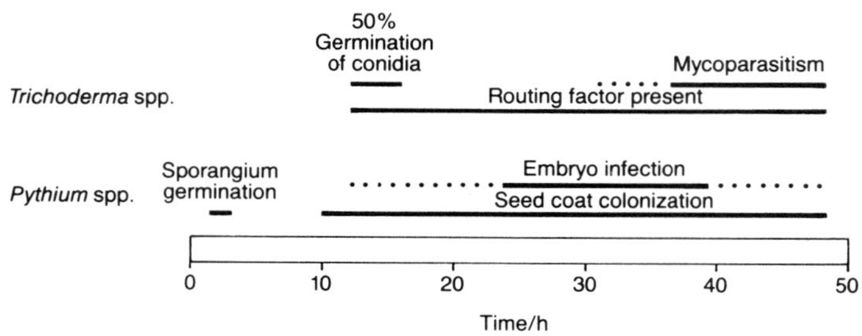

Fig. 8. Time course of interactions of pea germination, *Pythium* spp. and *Trichoderma* spp. applied to seed (Lifshitz *et al.*, 1986). The routing factor produced by *Trichoderma* spp. is present early enough to provide protection for the embryo during the period of susceptibility. Mycoparasitism occurs too late and operates infrequently in systems involving *Pythium* spp. even *in vitro* (from Baker, 1989 by permission).

early enough in the seed germination process to be effective (Fig. 8). The only mechanism that provides an explanation for the phenomena appears to be an antibiotic-like reaction detectable at microscopic levels (Lifshitz *et al.*, 1986). Since the initial reaction of *P. ultimum* to the approach of *Trichoderma* spp. is a split-second retreat of the protoplasm in the hyphal tip, the presumed antibiotic has been called 'the routing factor.' Ahmad and Baker (1988b) observed significant differences in sporangium germination of *P. ultimum* in rhizospheres of cucumber roots originating from seed treatment with a rhizosphere-competent mutant compared with that observed with seed not treated or treated with the parent wild type (Fig. 9). This phenomenon could be explained by operation of the routing factor.

Most attention on implications was focused on effects of mutants on phenomena associated with the increased growth response of plants induced by *Trichoderma* spp. (Baker, 1988). When rhizosphere-competent mutants were applied to seed or when a peat-bran culture preparation (Hadar *et al.*, 1979) was added to soil, cucumber, pea, radish, and tomato produced significantly higher fruit weight and/or dry weights (Fig. 10) than those treated with wild-types or in controls (Ahmad and Baker, 1988b). Of most interest, however, was the observation that seed treatment with mutants increased the incidence of emergence and resultant plant growth significantly in comparison with these strains applied to soil in a peat-bran culture. In contrast, there was no significant difference between the two types of application of the wild types.

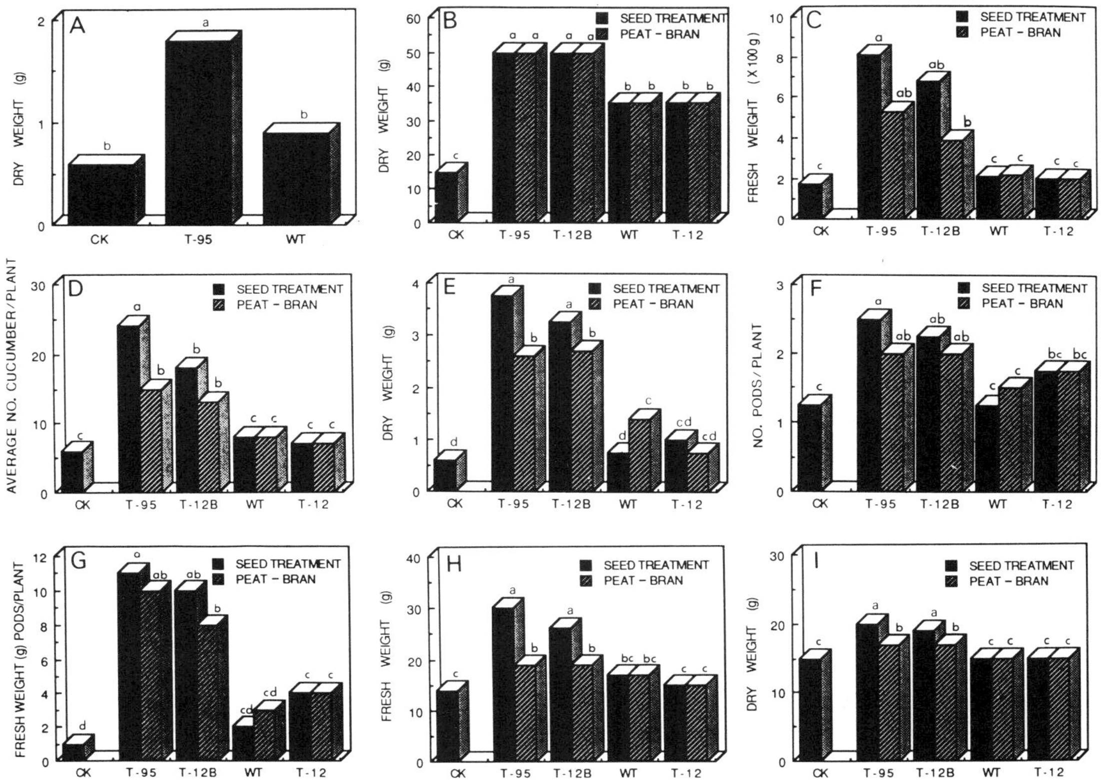

Fig. 10. Effect of caryopses or seed treatment and peat-bran application of *Trichoderma harzianum* strains on (**A**) dry weight of 2-week-old barley shoots, (**B**) dry weight of 8-week-old cucumber shoots, (**C**) fresh weight of cucumbers per plant, (**D**) average number of cucumbers per plant (**E**) dry weight of 8-week-old pea shoots, (**F**) average number of pods per pea plant, (**G**) fresh weight of pods, (**H**) fresh weight of 8-week-old radish roots, and (**I**) dry weights of 9-week-old tomato shoots. Bars in each figure labelled by the same letters are not significantly different ($P = 0.05$) (from Ahmad and Baker, 1988b by permission).

Hypotheses and evidence for the phenomena were reviewed (Baker, 1988, 1989). Increased growth responses induced by bacteria are thought to be the result of control of minor pathogens (Kloepper and Schroth, 1981). Important minor pathogens may be species of *Pythium* (Salt, 1978). As reviewed above, rhizosphere-competent mutants reduced inoculum density and sporangium germination of *P. ultimum* in the rhizosphere of roots in comparison with wild types or controls (Ahmad and Baker, 1988b) (Fig. 7, 8). *Trichoderma* spp. also produce a growth-promoting factor (Windham *et al.*, 1986). Therefore, rhizosphere-competent *Trichoderma* spp. active at root tips, where infection courts characteristically invaded by *Pythium* spp. are protected and where absorption of growth factors occurs, could contribute to the observed increased growth response.

These studies are the first evidence that the rhizosphere competence of microorganisms can be improved by genetic manipulation. The mechanisms responsible for the phenomenon might be applied to systems other than *Trichoderma* spp. For example, utilization of the mucigel substrate could increase the biocontrol efficiency of bacteria as well as fungi. Certainly, the attribute of rhizosphere-competence is a significant component in current attempts to enhance the activity of biocontrol agents (Baker and Scher, 1987).

References

Ahmad J S and Baker R 1987a Competitive saprophytic activity of rhizosphere-competent mutants of *Trichoderma harzianum*. Phytopathology 77, 358–362.

Ahmad J S and Baker R 1987b Rhizosphere competence of *Trichoderma harzianum*. Phytopathology 77, 181–189.

Ahmad J S and Baker R 1988a Growth of rhizosphere-competent mutants of *Trichoderma harzianum* on carbon substrates. Can. J. Microbiol. 34, 807–814.

Ahmad J S and Baker R 1988b Implication of rhizosphere competence of *Trichoderma harzianum*. Can. J. Microbiol. 34, 229–234.

Ahmad J S and Baker R 1988c Rhizosphere competence of benomyl-tolerant mutants of *Trichoderma* spp. Can. J. Microbiol. 34, 694–696.

Baker R 1984 Biological control of plant pathogens: Definitions. *In* Biological Control in Agricultural IPM Systems. Eds. M A Hoy and D C Herzog. pp 24–39. Academic Press. New York.

Baker R 1988 Mycoparasitism: ecology and physiology. Can. J. Phytopathology 9, 370–379.

Baker R 1988 *Trichoderma* spp. as growth stimulants. CRC Critical Rev. Biotech. 7, 97–106.

Baker R 1989 Improved *Trichoderma* spp. for promoting crop productivity. Trends Biotech. 7, 34–38.

Baker R and Scher F M 1987 Enhancing the activity of biological control agents. *In* Innovative Approaches to Plant Disease Control. Ed. I Chet. pp 1–17. John Wiley & Sons. New York.

Chang Y C, Chang Y C, Baker R, Kleifeld O and Chet I 1986 Increased growth of plants in the presence of the biological control agent, *Trichoderma harzianum*. Plant Dis. 70, 145–148.

Cuskey, S M, Montenecourt B S and Eveleigh D E 1983 Screening for cellulolytic mutants. *In* Liquid Fuel Developments. Ed. D L Wise. pp 31–47. CRC Press. Boca Raton, FL.

Delp C J 1980 Coping with resistance to plant disease control agents. Plant Dis. 64, 652–657.

Foster R C, Rovira A D and Cock T W 1983 Ultrastructure of the Root-Soil Interface. The Amer. Phytopathol. Soc., St. Paul, MN, 157 p.

Garrett S D 1970 Pathogenic Root-infecting Fungi. Cambridge Univ. Press, London, 294 p.

Hadar Y, Chet I and Henis Y 1979 Biological control of *Rhizoctonia solani* damping-off with wheat bran culture of *Trichoderma harzianum*. Phytopathology 69, 64–68.

Kloepper J W and Schroth M N 1981 Relationship of *in vitro* antibiosis of plant growth-promoting rhizobacteria to plant growth and the displacement of root microflora. Phytopathology 71, 1020–1024.

Lifshitz R, Windham M T and Baker R 1986 Mechanism of biological control of preemergence damping-off of pea by seed treatment with *Trichoderma* spp. Phytopathology 76, 720–725.

Mendes-Castro F A and Alexander M 1983 Method of establishing a bacterial inoculum on corn roots. Appl. Environ. Microbiol. 45, 258–264.

Montenecourt B S 1983 *Trichoderma reesei* cellulases. Trends Biotechnol. 1, 156–161.

Papavizas G C 1985 *Trichoderma* and *Gliocladium*: biology, ecology, and potential for biocontrol. Annu. Rev. Phytopathol. 23, 23–54.

Salt G H 1978 The increasing interest in minor pathogens. *In* Soilborne Plant Pathogens. Eds. B Schippers and W Gams. pp 289–312. Academic Press. New York.

Schroth M N and Hildebrand D C 1964 Influence of plant exudates on root-infecting fungi. Annu. Rev. Phytopathol. 2, 389–393.

Weller D M 1988 Biological control of soilborne plant pathogens in the rhizosphere with bacteria. Annu. Rev. Phytopathol. 26, 379–407.

Windham M T, Elad Y and Baker R 1986 A mechanism for increased plant growth induced by *Trichoderma* spp. Phytopathology 76, 518–521.

D. L. Keister and P. B. Cregan (Eds.), The rhizosphere and plant growth, 229–236.

PLSO BARC22

Mechanisms of biocontrol of soil-borne plant pathogens by Rhizobacteria

I. CHET[1], A. ORDENTLICH[1], R. SHAPIRA[2] and A. OPPENHEIM[2]
[1]*The Hebrew University, Otto Warburg Center of Biotechnology, Faculty of Agriculture, Rehovot, Israel and* [2]*The Hebrew University, Hadassah Medical School, Jerusalem, Israel*

Key words: biological control, chitinase, cloning, parasitism, antibiosis, competition, rhizosphere, *Serratia marcescens*

Abstract

Bacterial antagonism, responsible for biological control, may operate by antiobiosis, competition or parasitism. Parasitism relies on lytic enzymes for the degradation of cell walls of pathogenic fungi. *Serratia marcescens* was found to be an efficient biocontrol agent of *Sclerotium rolfsii* and *Rhizoctonia solani* under greenhouse conditions. Populations of 10^5 or 10^6 colony forming units g^{-1} soil were the most effective. Drench and drip application of *S. marcescens* suspension were more effective in controlling *S. rolfsii* than spraying, mixing in soil or seed coating. The highest population density of the bacteria in the rhizosphere was found on the proximal portion of the root, decreasing significantly until the tips, where it increased again. The isolated *Serratia*, found to possess chitinolytic activity, was able to release N-acetyl D-glucosamine from cell walls of *S. rolfsii*. The gene coding for chitinase was cloned into *Escherichia coli* and the enzyme was uniquely excreted from the bacterium into its growth medium. When *S. rolfsii* was sprayed by partially purified chitinase produced by the cloned gene, rapid and extensive bursting of the hyphal tips was observed. This chitinase preparation was effective in reducing disease incidence caused by *S. rolfsii* in beans and *R. solani* in cotton, under greenhouse conditions. A similar effect was obtained when a viable *E. coli* cell, containing the plasmid with the chitinase gene (pLCHIA), was applied. It appears that genetic engineering of the lytic enzymes, such as chitinase, which play an important role in plant disease control, may improve the efficacy of biocontrol agents.

Introduction

The rhizosphere region is heavily populated by different microorganisms including both beneficial and deleterious ones. The rhizosphere is the first-line defence for roots against attack by pathogenic fungi (Weller, 1988). Therefore, there is an excellent opportunity to find rhizosphere-competent bacteria in the rhizosphere which are potential biocontrol agents.

A successful biocontrol agent efficiently suppresses the pathogen and reduces disease incidence. Biocontrol agents act against pathogens by such forms of antagonism as competition, antibiosis and parasitism. Activity is not restricted to only one of these, and, indeed, an efficient biocontrol agent may affect pathogens by a combination of mechanisms (Chet, 1987).

Competition

Competition between the biocontrol bacteria and the pathogen can result in displacement of the pathogen. Microorganisms compete with each other for food and essential elements in the soil and around the rhizosphere (Baker, 1981).

Kloepper *et al.* (1980) were the first to demonstrate the importance of siderophores (a low molecular weight, high affinity Fe^{3+} chelator that transports iron into cells of microorganisms [Leong, 1986; Neilands, 1981]) in the mechanism of biological control. Siderophores mediate the limited amount of iron in the rhizosphere, deprive pathogens of iron and suppress their growth. Many reports have been published showing siderophore involvement in the suppression of plant pathogenic fungi (Baker, 1985; Baker *et al.*, 1986; Bakker *et al.*, 1986; 1987; Becker and Cook, 1988; Cook and Weller, 1987; Elad and Baker, 1985a,b; Kloepper *et al.*, 1980; Kleopper and Schroth, 1981; Loper, 1988; Scher, 1986; Scher and Baker, 1982; Schippers *et al.*, 1987; Sneh *et al.*, 1984; Weller and Cook, 1986; Weller *et al.*, 1988; Wilkinson *et al.*, 1981).

Elad and Chet (1987) found that competition for nutrients between germinating oospores of *P. aphanidermatum* and bacteria was significantly correlated with suppression of disease in the greenhouse. Oospore germination in rhizospheres of wheat, tomato, cucumber, melon, bean and cotton plants was decreased in the presence of these bacteria. Broadcast application or seed coating with these bacteria reduced disease incidence in cucumbers, planted in peat-vermiculite mixture or in soil, by 60–75% compared with the untreated control. Damping-off caused by *P. ultimum* in cucumbers was reduced by up to 94%. Combining low levels of *Pseudomonas putida* or *P. cepacia* with the fungicide prothiocarb resulted in a positive additive effect. Bacteria applied to cucumber seeds became established along the roots. Populations were 100-fold higher at root tips than at their upper parts. On the other hand, except for the root tips, the total population of bacteria was decreased in the presence of the beneficial ones. Disease control was achieved in bean, pepper, melon, tomato and cotton plants as well.

Antibiosis

Antibiosis plays an important role in disease suppression by certain bacteria. Suppression of pathogens by antibiotic-producing bacteria have been shown (Colyer and Mount, 1984; Gutterson *et al.*, 1986; Howell and Stipanovic, 1979, 1980; James and Gutterson, 1986; Kerr, 1980). Rothrock and Gottlieb (1984) suppressed *Rhizoctonia solani* in soil by applying purified Geldanomycin, an antibiotics produced by *Streptomyces hygroscopicus* var. *geldonus*.

Recently, Thomashow and Weller (1988) showed that *Pseudomonas fluorescens* 2-79 (NRRL B-15132) and its rifampin-resistant derivative 2-79RN_{10} are suppressive to take-all, a major root disease of wheat caused by *Gaeumannomyces graminis* var. *tritici*. Strain 2-79 produces the antibiotic phenazine-1-carboxylate, which is active *in vitro* against *G. graminis* var. *tritici* and other fungal root pathogens. Mutants defective in phenazine synthesis (Phz^-), were generated by Tn5 insertion and then compared with the parental strain to determine the importance of the antibiotic in take-all suppression on wheat roots. Six independent, prototrophic Phz^- mutants were noninhibitory to *G. graminis* var. *tritici in vitro* and provided significantly less control of take-all than strain 2-79 on wheat seedlings. Antibiotic synthesis, fungal inhibition *in vitro* and suppression of take-all on wheat were coordinately restored in two mutants complemented with cloned DNA from a 2-79 genomic library. These results support the importance of the phenazine antibiotic in disease suppression in the rhizosphere.

Excellent comprehensive reviews were recently dedicated to antiobiosis and competition (Fravel, 1988; Schippers, 1988; Weller, 1988) but not much was done on parasitism. In this chapter we would like to emphasize the bacterial lytic activity as a potential mechanism for biological control.

Parasitism and lysis

Microorganisms capable of lysing other organisms are widespread in natural ecosystems. These microorganisms seem to play an important role in the microbiological equilibrium in the soil and can serve as a powerful tool for biological control (Henis and Chet, 1975).

Cell walls of many fungi contain chitin. Chet *et al.* (1967), Chet and Huttermann (1980) and later Sivan and Chet (1989) analyzed the chemical composition of several antagonistic fungi (Fig. 1).

Various bacteria and fungi have been shown to produce hydrolytic enzymes after induction by the appropriate substrates (Bull, 1967; Bull and Chester, 1966; Ridout *et al.*, 1986). Mitchell and Alexander (1961) demonstrated biological con-

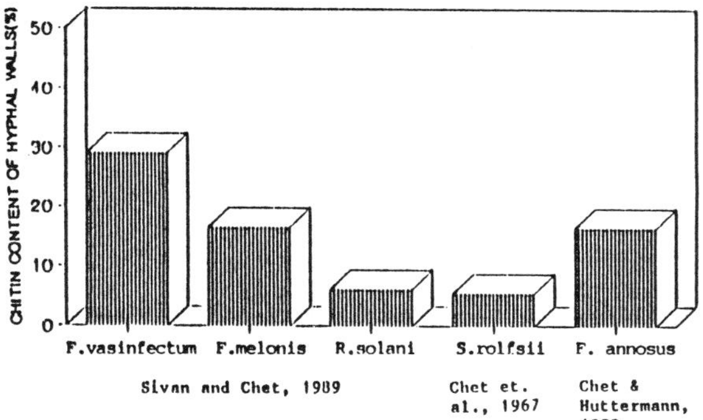

Fig. 1. Chitinase content of hyphal cell walls of *Fusarium oxysporum* f. sp. *vasinfectum*, *F. oxysporum* f. sp. *melonis*, *Rhizoctonia solani*, *Sclerotium rolfsii* and *Heterobasidion annosum*.

trol of *Fusarium* spp. and *Pythium* sp. by bacteria that degrade the cell wall of these plant pathogens. Mitchell and Hurwitz (1965) protected tomato plants against damping off caused by *Pythium aphanidermatum* (Edson) Fitzpatrick with lytic Arthrobacter. Koths and Gunner (1967) and Sneh (1981) also demonstrated biological control of disease caused by Fusarium on carnation by using a chitinolytic *Arthrobacter* sp. *Enterobacter cloacae* (Jordan) Hormaeche and Edwards controlled Pythium in peas, sugar beet and cucumber, probably by degrading the mycelium (Hadar *et al.*, 1983). Campbell (1983) and Campbell and Ephgrave (1983) showed that hyphal lysis, induced by *Bacillus* sp. in soil, was involved in the biological control of *Gaeumannomyces graminis* (Sacc.) v. Arx and Olivier.

Ordentlich *et al.* (1987) isolated a bacterium, identified as *Serratia marcescens*, from soil. This bacterium was found to suppress *Sclerotium rolfsii* in the greenhouse (up to 75% disease reduction). Populations of 10^5 or 10^6 colony forming units g^{-1} soil were the most effective in disease control. Drench and drip application of *S. marcescens* suspension were more effective in controlling *S. rolfsii* than spraying, mixing in soil or coating seeds with the bacterium. This bacterium significantly reduced (by 50%) the incidence of damping-off caused by *Rhizoctonia solani* in beans, but was not effective against *Pythium aphanidermatum* in cucumbers. A natural mutant of *S. marcescens*, resistant to the antibiotic rifampicin, was isolated. The mutant, as effective as the wild type, was used to study rhizosphere colonization. The highest population density of the bacteria was found on the proximal portion of the root, decreasing significantly until the tips, where it increased again. Recently Scher *et al.*, (1988) found very good root colonization by Serratia.

Serratia marcescens was grown on cell wall of *Sclerotium rolfsii* and its components, chitin and laminarin, as a sole carbon source (Fig. 2). A culture filtrate, possessing chitinolytic activity, was obtained and incubated with different substrates, including *S. rolfsii* cell wall and dry and washed mycelium. A release of N-acetyl-D-glucosamine at rates of 1–5.2 chitinase units indicated substrate degradation. Degradation of hyphae of the plant pathogen was observed by light and scanning electron microscopy. Crude chitinase caused very swift swelling at the hyphal tips of the fungus. Sixty three percent of the cells of the hyphal tips were lysed (Ordentlich *et al.*, 1988).

We have tried to prove the involvement of chitinase in the control of *Sclerotium rolfsii* by *Serratia marcescens* by genetic engineering techniques. This microorganism is capable of secreting several chitinolytic enzymes (Fuchs *et al.*, 1986). The gene *chi*A, coding for the major chitinase produced by *S. marcescens*, was recently cloned and sequenced. *Escherichia coli* cells harboring a plasmid carrying the *chi*A gene, were found to synthesize chitinase following prolonged incubation (Jones *et al.*, 1986; Sundheim, 1987; Sundheim, *et al.*, 1988).

Based on these findings, Shapira *et al.* (1989) constructed a system for the efficient expression of chitinase in *E. coli*. A DNA fragment, carrying the *chi*A gene from *S. marcescens*, was subcloned into the plasmid pBR322. Cleavage by various restriction enzymes established the similarity of this DNA fragment to the published chitinase sequence. This plasmid, pCHIA, includes a 2 kbp segment upstream of the *chi*A gene and, presumably, carries the *S. marcescens* regulatory elements. The very low levels of chitinase observed in an *E. coli* culture carrying this plasmid are probably due to the inefficient transcription of *chi*A in the heterologous bacteria. In order to obtain high levels of chitinase expression, we introduced the leftward operators promoter of bacteriophage λ oLpL, upstream of the *chi*A gene. The resulting plasmid, pLCHIA,

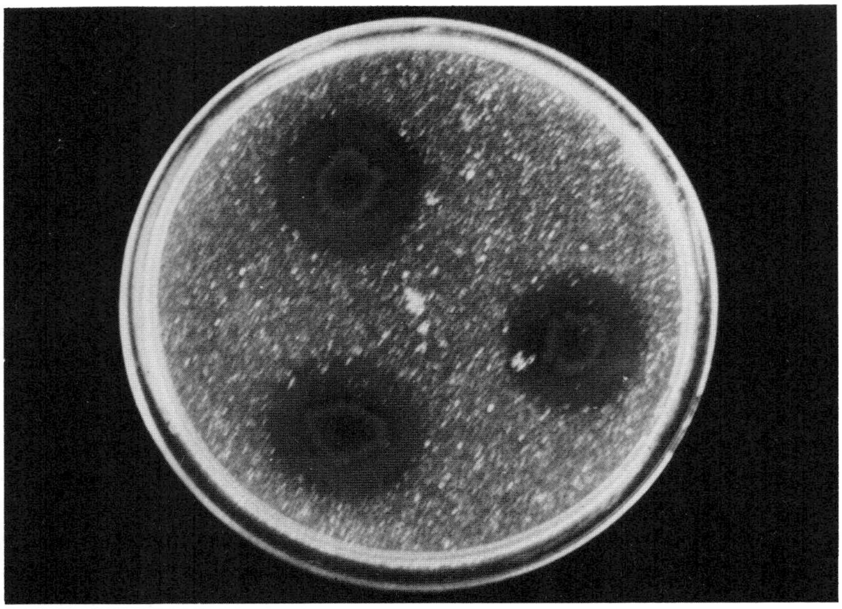

Fig. 2. Clear halos produced by *Serratia marcescens* colonies when plated on media plates containing colloidal chitin.

was introduced into *E. coli* cells expressing a thermolabile λ repressor. At 42°C, the repressor is inactivated and oLpL is no longer repressed. When the cells were grown at this temperature on chitin-containing plates (0.3%), a clearing halo was produced, indicating that chitinase was secreted from the cells to the surrounding medium (Shapira *et al.*, 1988).

Fermentation conditions for the production of high levels of chitinase were established. Cells carrying the plasmid pLCHIA were grown at 30°C and then shifted to 42°C for 3 hours to induce chitinase synthesis. Intact cells were removed by centrifugation and the supernatant was found to contain chitinase activity (Ordentlich *et al.*, 1988). Since *E. coli* secretes very few proteins, the medium was expected to contain mainly the secreted chitinase protein. Indeed, analysis on sodium dodecyl sulphate-polyacrylamide gel revealed that chitinase constitutes the major protein species of the supernatant. A similar culture of the same bacterial strain carrying the plasmid pCHIA, which lacks the pL promoter, produced 26 times less chitinase (Shapira *et al.*, 1988).

It was previously shown that a non-purified supernatant from *S. marsescens* culture, which is a mixture of various hydrolytic enzymes including several chitinases, is capable of degrading fungal tips (Ordentlich *et al.*, 1988). The hyphal tips of *S. rolfsii*, the sites of hyphal growth, were shown to contain exposed chitin (Gutterson *et al.*, 1986; Jones *et al.*, 1986). We therefore tested whether the chitinase produced by *E. coli* can similarly degrade hyphal tips of this fungus. When *S. rolfsii* was sprayed by partially purified chitinase produced by the cloned gene described above, rapid and extensive bursting of the hyphal tips was observed (Fig. 3). These results suggest that the purified *chi*A gene product, produced by *E. coli*, could serve in the control of pathogenic fungi.

A large number of reports dealing with biological control of soil-borne pathogens propose that chitinolytic activities play a major role in plant protection (Barrows-Broaddus and Kerr, 1981; Chet, 1987; Elad *et al.*, 1982; Morrissey *et al.*, 1976; Ordentlich *et al.*, 1988). However, to our knowledge, there is no direct proof demonstrating that purified chitinase can protect growing plants from fungal infection. The availability of a large supply of chitinase enabled us to test this hypothesis directly. Bean seeds were sown in soil artificially infested with 40 mg sclerotia of *S. rolfsii* per Kg soil and grown under standard

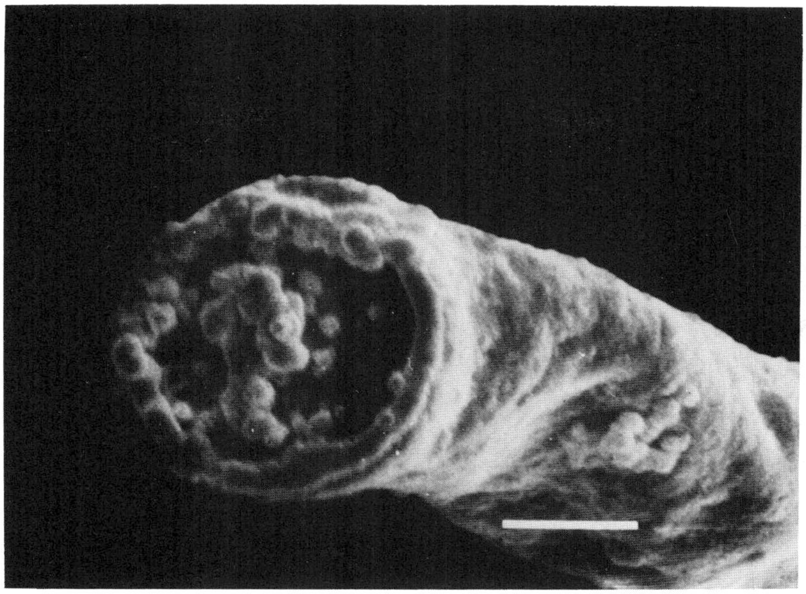

Fig. 3. Scanning electron micrograph showing the effect of the partially purified chitinase on hyphal tips of *Scelrotium rolfsii.*

greenhouse conditions (Ordentlich *et al.*, 1987). The pots were irrigated daily with tap water containing a diluted chitinase preparation (0.04 units per pot – one chitinase unit was defined as the amount of enzyme that liberates one micromole of N-acetyl-D-glucosamine from chitin per minute). Control pots were irrigated with tap water or tap water containing a protein preparation obtained from equivalent amounts of culture medium of the low producing *E. coli* strain A5186. We found that chitinase can effectively reduce the number of diseased plants. During the course of the experiment, over 50% of the plants irrigated with tap water showed severe disease symptoms. The daily addition of chitinase to the irrigation water significantly reduced the incidence of infection. The protein did not decompose immediately in soil. We repeatedly found about a 3-fold reduction in the number of diseased plants. Plants which were irrigated daily with the chitinase preparation were significantly protected from *S. rolfsii* infection throughout. Boiling the chitinase preparation destroyed its enzymatic and fungicidal activity. The protein preparation isolated from *E. coli* carrying the plasmid pCHIA, lacking the pL promoter, contained low levels of chitinase activity and was rather ineffective in protecting bean seedlings (Fig. 4). These results demonstrate that chitinase may be instrumental in the biocontrol of *S. rolfsii* by the bacterium.

Similar experiments, carried out at 40°C with whole viable *E. coli* cells containing the plasmid pLCHIA, showed that these cells were also effective in inhibiting *S. rolfsii*, albeit to a lower degree. When the same experiment was carried out at 30°C, a temperature at which the pL promoter is repressed, no protection was observed (Shapira *et al.*, 1988).

In addition, a λ phage DNA library of *Serratia marcescens* was constructed and a clone, carrying the gene coding for chitobiase (E.C. 3.2.1.29), was isolated and characterized. Deletion analysis limited the cloned region to 4.5 kbp, *i.e.* capable of efficient expression of chitobiase. *E. coli* cells, harboring a plasmid carrying the cloned gene, express chitobiase constitutively (Kless *et al.*, 1989). It seems, however, that the key enzyme in this biocontrol system is the chitin, not chitobiase. The ability to protect plants by applying lytic enzyme opens the way for developing of an alternative to the common chemical control of soil-borne plant pathogenic fungi. These findings presented here may allow the development of a

Fig. 4. Protection of bean seedlings by chitinase. From left to right: irrigation with tap water, tap water containing chitinase preparation excreted from strain A5186, and tap water containing protein preparation excreted from strain A5187.

model system for testing other genes and gene products that may participate in biological control.

References

Baker R 1981 Eradication of plant pathogens by adding organic amendments to soil. *In* Handbook of Pest Management in Agriculture. Vol. 2. Ed. D Pimental. pp 137–157. Chemical Rubber Company Press, Boca Raton, Florida.

Baker R 1985 Biological control of plant pathogens: definitions. *In* Biological Control in Agricultural IPM Systems. Eds. M A Hoy and D C Herzog, pp 25–29. Academic Press, Orlando.

Baker R, Elad Y and Sneh B 1986 Physical, biological and host factors in iron competition in soils. *In* Iron, Siderophores and Plant Diseases. Ed. T R Swinburne. pp 77–84. Plenum Press, New York.

Bakker P A H M, Bakker A W, Marugg J D, Weisbeek P J and Schippers B 1987 Bioassay for studying the role of siderophores in potato growth stimulation by *Pseudomonas* spp. in short potato rotations. Soil. Biol. Biochem. 19, 443–449.

Bakker P A H M, Lamers J G, Bakker A W, Marugg J D, Weisbeek P J and Schippers B 1986 The role of siderophores in potato tuber yield increase by *Pseudomonas putida* in a short rotation of potato. Neth. J. Plant Pathol. 92, 249–256.

Barrows-Broaddus J and Kerr T K 1981 Inhibition of *Fusarium moniliforme* var. *subglutinans*, the causual agent of pitch canker, by the soil bacterium *Arthrobacter* sp. Can. J. Microbiol. 27, 20–27.

Becker O and Cook R J 1988 Role of siderophores in suppression of *Pythium* species and production of increased growth response of wheat by fluorescence pseudomonads. Phytopathology 78, 778–782.

Bull A T 1967 The enzyme degradation of β-glucan. Int. Biodeterior. Bull. 3, 3–12.

Bull A T and Chester C G C 1966 The biochemistry of laminarin and the nature of laminarinase. Adv. Enzymol. 28, 325–364.

Campbell R 1983 Ultrastructural studies of *Gaeumannomyces graminis* in the water films on wheat roots and the effect of clay on the interactions between this fungus and antagonistic bacteria. Can. J. Microbiol. 29, 39–45.

Campbell R and Ephgrave J M 1983 Effect of bentonite clay on the growth of *Gaeumannomyces graminis* var. *tritici* and on its interaction with antagonistic bacteria. J. Gen. Microbiol. 129, 771–777.

Chet I 1987 *Trichoderma* – application, mode of action, and potential as a biocontrol agent of soil-borne plant pathogenic fungi. *In* Innovative Approaches to Plant Disease Control. Ed. I Chet. pp 137–160. John Wiley and Sons, New York.

Chet I, Henis Y and Mitchell R 1967 Chemical composition of hyphal and sclerotial walls of *Sclerotium rolfsii* Sacc. Can. J. Microbiol. 13, 137–141.

Chet I and Huttermann A 1980 Chemical composition of hyphal walls of *Fomes annosus*. Eur. J. Forest Pathol. 10, 65–70.

Colyer P D and Mount M S 1984 Bacterization of potatoes

with *Pseudomonas putida* and its influence on postharvest soil rot diseases. Plant Dis. 68, 703–706.

Cook R J and Weller D M 1987 Management of take-all in consecutive crops of wheat or barley. *In* Innovative Approaches to Plant Disease Control. Ed. I Chet. pp 41–76, John Wiley and Sons, New York.

Elad Y and Baker R 1985a Influence of trace amounts of cations and siderophore-producing pseudomonads on chlamydospore germination of *Fusarium oxysporum*. Phytopathology 75, 1047–1052.

Elad Y and Baker R 1985b The role of competition for iron and carbon in suppression of chlamydospore germination of Fusarium spp. by *Pseudomonas* spp. Phytopathology 75, 1053–1059.

Elad Y, Chet I and Henis Y 1982 Degradation of plant pathogenic fungi by *Trichoderma harzianum*. Can. J. Microbiol. 28, 719–725.

Elad Y and Chet I 1987 Possible role of competition for nutrients in biocontrol of *Pythium* damping-off by bacteria. Phytopathology 77, 190–195.

Fravel D R 1988 Role of antibiosis in the biocontrol of plant diseases. Annu. Rev. Phytopathol. 26, 75–91.

Fuchs R L, McPherson S A and Drehos D J 1986 Cloning of a *Serratia marcescens* gene encoding chitinase. Appl. Environ. Microbiol. 51, 504–509.

Gutterson N I, Layton T J, Zeigle J S and Warren G J 1986 Molecular cloning of genetic determinants for inhibition of fungal growth by a fluorescent pseudomonad. J. Bacteriol. 165, 695–701.

Hadar Y, Harman G E, Taylor A G and Norton J M 1983 Effects of pregerminating of pea and cucumber seeds and of seed treatment with *Enterobacter cloacae* on rots caused by *Pythium* spp. Phytopathology 73, 1322–1325.

Henis Y and Chet I 1975 Microbial control of plant pathogens. Adv. Appl. Microbiol. 19, 85–111.

Howell C R and Stipanovic R D 1979 Control of *Rhizoctonia solani* on cotton seedlings with *Pseudomonas fluorescens* and with an antibiotic produced by the bacterium. Phytopathology 69, 480–482.

Howell C R and Stipanovic R D 1980 Suppression of *Pythium ultimum*-induced damping off of cotton seedlings by *Pseudomonas fluorescens* and its antibiotic, pyoluteorin. Phytopathology 70, 712–715.

James D W Jr and Gutterson N I 1986 Multiple antibiotics produced by *Pseudomonas fluorescens* HV37a and their differential regulation by glucose. Appl. Environ. Microbiol. 52, 1183–1189.

Jones J D G, Grady K L, Suslow T V and Bedbrook J R 1986 Isolation and characterization of genes encoding two chitinase enzymes from *Serratia marcescens*. Embo J 5, 467–473.

Kerr A 1980 Biological control of crown gall through production of agrocin 84. Plant Dis. 64, 25–30.

Kless H, Sitrit Y, Chet I and Oppenheim A B 1989 Cloning of the gene coding for chitobiase of *Serratia marcescens*. Mol. Gen Genet. 217, 471–473.

Kloepper J W, Leong J, Teintze M and Schroth M N 1980 Enhanced plant growth by siderophores produced by plant growth-promoting rhizobacteria. Nature 286, 835–836.

Kloepper J W, Leong J, Teintze M and Schroth M N 1980 *Pseudomonas* siderophores: A mechanism explaining disease suppressive soils. Curr. Microbiol. 4, 317–320.

Kloepper J W and Schroth M N 1981 Relationship of *in vitro* antibiosis of plant growth-promoting rhizobacteria to plant growth and the displacement of root microflora. Phytopathology 71, 1020–1024.

Koths J S and Gunner H R 1967 Establishment of a rhizosphere microflora on carnation as a means of plant protection in steamed greenhouse soils. Am. Soc. Hortic. Sci. 91, 617–626.

Leong J 1986 Siderophores: their biochemistry and possible role in the biocontrol of plant pathogens. Annu. Rev. Phytopathol. 24, 187–209.

Loper J E 1988 Role of fluorescent siderophore production in biological control of *Pythium ultimum* by a *Pseudomonas fluorescens* strain. Phytopathology 78, 166–172.

Mitchell R and Alexander M 1961 The mycolytic phenomenon and biological control of *Fusarium* in soil. Nature 190, 109–110.

Mitchell R and Hurwitz R 1965 Suppression of *Pythium deharyanum* by lytic rhizosphere bacteria. Phytopathology 55, 156–158.

Morrissey R F, Dugan E P and Koths J S 1976 Chitinase production by an *Arthrobacter* sp. lysing cells of *Fusarium roseum*. Soil Biol. Biochem. 8, 23–28.

Neilands J B 1981 Microbial iron compounds. Annu. Rev. Biochem. 50, 715–731.

Ordentlich A, Elad Y and Chet I 1987 Rhizosphere colonization by *Serratia marcescens* for the control of *Sclerotium rolfsii*. Soil Biol. Biochem. 19, 747–751.

Ordentlich A, Elad Y and Chet I 1988 The role of chitinase of *Serratia marcescens* in biocontrol of *Sclerotium rolfsii*. Phytopathology 78, 84–88.

Ridout C J, Coley-Smith J R and Lynch J M 1986 Enzyme activity and electrophoretic profile of extracellular proteins induced in *Trichoderma* spp. by cell walls of *Rhizoctonia solani*. J. Gen. Microbiol. 132, 2345–2352.

Rothrock C S and Gottlieb D 1984 Role of antibiosis in antagonism of *Streptomyces hygroscopicus* var. *geldanus* to *Rhizoctonia solani* in soil. Can. J. Microbiol. 30, 1440–1447.

Scher F M 1986 Biological control of *Fusarium* wilts by *Pseudomonas putida* and its enhancement by EDDHA. *In* Iron, Siderophores and Plant Diseases. Ed T R Swinburne. pp 109–117, Plenum Press, New York.

Scher F M and Baker R 1982 Effect of *Pseudomonas putida* and a synthetic non chelator on induction of soil suppressiveness to Fusarium wilt pathogens. Phytopathology 72, 1567–1573.

Scher F M, Kloepper J W, Singleton C, Zaleska I and Laliberte M 1988 Colonization of soybean roots by *Pseudomonas* and *Serratia* species. Relationship to bacterial motility, chemotaxis and generation time. Phytopathology 78, 1055–1059.

Schippers B 1988 Biological control of pathogens with rhizobacteria. Phil. Trans. R. Soc. Lond. B 318, 283–293.

Schippers B, Bakker A W and Bakker P A H M 1987 Interactions of deleterious and beneficial rhizosphere microorganisms and the effect of cropping practices. Annu. Rev. Phytopathol. 25, 339–358.

Shapira R, Ordentlich A, Chet I and Oppenheim A B 1989 Control of plant diseases by chitinase expressed from cloned DNA in *Escherichia coli*. Phytopathology 79, 1246–1249.

Sivan A and Chet I 1989 Cell wall composition of *Fusarium oxysporum*. Soil Biol. Biochem. 21, 869–876.

Sneh B 1981 use of rhizosphere chitinolytic bacteria for biological control of *Fusarium oxysporum* f. sp. *dianthi* in carnation. Phytopathology Z. 100, 251–256.

Sneh B, Dupler M, Elad Y and Baker R 1984 Chlamydospore germination of *Fusarium oxysporum* f. sp. *cucumerinum* as affected by fluorescent and lytic bacteria from Fusarium-suppressive soil. Phytopathology 74, 1115–1124.

Sundheim L 1987 Cloning and conjugational transfer of chitinase encoding genes. J. Agric. Sci. Fin. 59, 209–215.

Sundheim L, Poplawsky A and Ellingboe A H 1988 Molecular cloning of two chitinase genes from *Serratia marcescens* and their expression in *Pseudomonas* species. Physiol. Mol. Plant Pathol. 33, 483–491.

Thomashow L S and Weller D M 1988 Role of a phenazine antibiotic from *Pseudomonas fluorescens* in biocontrol of *Gaeumannomyces graminis* var. *tritici*. J. Bacteriol. 170, 3499–3500.

Weller D M 1988 Biological control of soilborne plant pathogens in the rhizosphere with bacteria. Annu. Rev. Phytopathol. 26, 379–407.

Weller D M and Cook R J 1986 Suppression of root diseases of wheat by fluorescent pseudomands and mechanisms of action. *In* Iron, Siderophores, and Plant Diseases. Ed. T R Swinburne. pp 99–107, Plenum Press, New York.

Weller D M, Howie W J and Cook R 1988 Relationships between *in vitro* inhibition of *Gaeumannomyces graminis* var. *tritici* and suppression of take-all of wheat by fluorescent pseudomonads. Phytopathology 78, 1094–1100.

Wilkinson H T, Miller R D and Millar R L 1981 Infiltration of fungal and bacterial propagules into soil. Soil Sci. Soc. Am. J. 45, 1034–1039.

D. L. Keister and P. B. Cregan (Eds.), The rhizosphere and plant growth, 237–243.

PLSO BARC23

Biocontrol of soilborne plant pathogens with fungi

D.R. FRAVEL and A.P. KEINATH
Biocontrol of Plant Diseases Laboratory, USDA, ARS, Beltsville, MD 20705, USA

Key words: antibiosis, biocontrol mechanisms, competition, mycoparasitism

Abstract

Determining mechanisms by which antagonistic fungi control plant pathogens is an important step in developing biorational approaches to disease management. These approaches should increase the probability of successful, consistent biocontrol in the field. Mechanisms of biocontrol include competition, parasitism, and antibiosis, all of which function in the rhizosphere. Competition may be the mechanism most prevalent in natural systems, but it is often difficult to observe experimentally. Parasitism can be an efficient control mechansm and has the potential to be very target specific, as in control of *Sclerotinia* species by Sporidesmium. Although antibiosis has not been definitively demonstrated to be the sole control mechanism of any biocontrol fungus, there is evidence for involvement of antibiotics, antibiotic-like compounds, or enzymes in control mediated by numerous fungi including isolates of Trichoderma, Gliocladium, and Talaromyces. In general, these mechanisms are not mutually exclusive; different mechanisms may act in combination to produce biocontrol. Knowledge of mechanisms can be used to design screening procedures for potential antagonists, evaluate management practices which favor antagonists, select genetically improved antagonists, and assist in registration of antagonists.

Introduction

Understanding the mechanisms through which the biocontrol of plant diseases is achieved is critical to the improvement and wider use of biocontrol methods. These mechanisms generally are classified as competition, parasitism/predation, and antibiosis (Baker, 1968; Cook and Baker, 1983). Other mechanisms, such as cross protection, do not readily fit into these categories. This classification should provide a framework to aid our understanding of biocontrol, but should not circumscribe our thinking. For example, several antagonists are known to have multiple modes of action which act against the same or different pathogens. Furthermore, different mechanisms may operate under different environmental conditions. Identification of one mode of action does not always 'tell the whole story'.

Competition

Competition may be the mechanism most prevalent in natural systems. For example, an organism introduced into sterile soil will proliferate and establish itself to a greater extent than if it is introduced into a natural soil. It is difficult to introduce successfully an organism into soil of an undisturbed ecosystem (Garrett, 1956). This difference in performance of the organism is due to the more intense competition in the more complex ecosystem. Competition among microorganisms in soil and the rhizosphere is primarily for nutrients. Arguments concerning the importance of competition for space and oxygen continue. Organisms presumably do not compete for water in the rhizosphere or soil since the availability of water is determined by the energy status of water (water potential), which microorganisms do not affect significantly (Clark, 1965). Because nutrients are nearly always associated with substrates (*e.g.* roots, seeds, organic debris, *etc.*) competition for food may manifest itself as a competition for space, such as in the physical exclusion of a pathogen from a host rhizosphere due to previous colonization by an antagonist. Competition is often difficult to document since particular competitors or the objects of competition usually have not been identified and there may be several competitors involved.

One of the most familiar examples of competition as a mechanism of biocontrol is the use of the fungus *Phanerochaete gigantea* (Fr.:Fr.) Rattan *et al.* (=*Peniophora gigantea* Fr.:Karst.) to control root rot of pine caused by *Heterobasidion annosum* (Fr.:Fr.) Bref. [=*Fomes annosus* (Fr.:Fr.) Cooke] (Rishbeth, 1975). Heterobasidion infects healthy pine roots through root grafts from colonized tree stumps. Conidia of the antagonist *P. gigantea* are applied, usually in chain saw oil, to freshly cut stumps. Although the primary mechanism is competition for space by physical exclusion of the pathogen, *P. gigantea* also produces an antibiotic-like substance against *H. annosum* (Ikediugwu *et al.*, 1970). This substance may be involved in the observed hyphal interference.

Competition between an antagonist and pathogen may be regulated by edaphic factors. For example, the chloride ion concentration of California soils is important in competition between the pathogen *Pythium ultimum* Trow and *P. oligandrum* Drechsler for organic substrates (Martin and Hancock, 1986). Soils suppressive to *P. ultimum* had higher levels of Cl^- than conducive soils. *Pythium oligandrum* had a higher degree of saprophytic activity (measured as colonization of dried cotton leaves) than *P. ultimum* in these soils because *P. oligandrum* was significantly more tolerant of Cl^- than *P. ultimum*. *Pythium oligandrum* also has been reported to be a mycoparasite of *P. ultimum* in dual cultures. However, both species were rarely recovered from the same leaf debris in natural systems. The ability of *P. oligandrum* to act as a primary colonizer of leaf tissue prevented *P. ultimum* from occupying that tissue and reproducing, leading to an increase in inoculum.

In another system involving two *Pythium* species, both competitive ability and mycoparasitism may be involved. *Pythium nunn* Lifshitz, Stanghellini and Baker is a more aggressive antagonist in soil than *P. oligandrum* since *P. nunn* can displace *P. ultimum* from organic debris (Paulitz and Baker, 1988). After 12 hours *P. ultimum* and *P. nunn* colonized 54% and 78%, respectively, of bean leaf fragments added to naturally infested soil. By 72 hours, more than 90% of the bean tissue was colonized by *P. nunn* alone. This reduction in colonization of tissue by *P. ultimum* could result in a reduction of the inoculum density of the pathogen.

Fusarium oxysporum Schlecht is sensitive to competition from antagonists. Germination of chlamydospores of *F. oxysporum* f. sp. *melonis* and f. sp. *vasinfectum* was significantly reduced in the rhizospheres of melon and cotton, respectively, by *Trichoderma harzianum* Rifai (Sivan and Chet, 1989). However, this inhibition was negated by the addition of seed exudates, which suggested that competition for nutrients occurred. Several different *formae speciales* of *F. oxysporum*, causing wilts of vegetables and ornamentals, were inhibited in suppressive soils by indigenous populations of nonpathogenic *F. oxysporum* and *F. solani* (Alabouvette *et al.*, 1979). Besides interactions between two species, interactions between a pathogen and other members of the native soil microflora can result in biocontrol. Fungi which naturally recolonized fumigated soil inhibited reproduction of *F. oxysporum* f. sp. *radicis-lycopersici* and decreased disease incidence on tomato (Marois and Mitchell, 1981).

In addition to the ability to compete with pathogens, antagonists must be able to compete with the indigeous microflora in the rhizosphere and soil. Because of the inherent 'biological buffering' of the natural soil system, any antagonist introduced into the system must possess certain competitive attributes. In one of the few studies designed to measure specific competitive attributes, three components of growth through soil were measured for *Gliocladium roseum* Bain., *Thelavia terricola* (Gilman and Abbot) Emmons, and *Trichoderma* spp. (Stack *et al.*, 1987). The three components were percentage of carrier particles supporting hyphal growth, the number of hyphae radiating from a particle, and the length of these hyphae. All three components were influenced by the carbon and nitrogen sources and the C:N ratio of the nutrient solution.

Because of the complex system of biological checks and balances in soil, it is very unusual for an antagonist introduced into an ecosystem to become established, yet the antagonists *Sporidesmium sclerotivorum* Uecker, Ayres and Adams, *Talaromyces flavus* (Klöcker) Stolk and Samson, and other antagonists can become established. These competitive antagonists are discussed further in this article.

Parasitism

Parasitism of fungi by fungi is generally termed mycoparasitism or hyperparasitism. There are at least 50 different species of fungi which are known to parasitize other fungi (Lumsden, 1981). Fungi in all taxa, ranging from the Chytridimycetes to the Basidiomycetes, are known to parasitize and be parasitzed by other fungi. Some of these mycoparasites, such as *Trichoderma* spp., have very broad host ranges while others, such as many of the Chytrids, are extremely host specific. Only a few of these have been studied in the context of biological control. Trichoderma and Gliocladium species appear to have potential for biocontrol and the literature has been reviewed recently (Papavizas, 1985). Other mycoparasites reported to have potential for biocontrol are *Verticillium biguttatum* Gams (Jager and Velvis, 1988), *Coniothyrium minitans* Campbell (Huang, 1977; Huang, 1980; Tribe, 1957), *Laetisaria arvalis* Burdsall (Hoch and Abawi, 1979; Martin *et al.*, 1986), *P. nunn*, *T. flavus*, and *S. sclerotivorum*.

Sporidesmium sclerotivorum is a nearly ideal biocontrol agent. In nature, *S. sclerotivorum* behaves as an obligate parasite of *Sclerotinia minor* Jagger, *S. sclerotiorum* (Lib.) de Bary, *S. trifolium* Eriks., *Sclerotium cepivorum* Berk., and *Botrytis cinerea* Pers. (Ayers and Adams, 1981). Macroconidia of the antagonist germinate in response to chemicals released by sclerotia of the pathogen and germ tubes infect the sclerotia. Inside the sclerotium, Sporidesmium penetrates the intercellular matrix which is composed primarily of glucans. In response to the mycoparasite, the pathogen produces increased levels of glucanase which results in the degradation of sclerotial glucans to glucose, which is utilized by the mycoparasite (Adams and Ayres, 1983). After infection of the host sclerotium, Sporidesmium mycelia grow out into the surrounding soil where they can infect healthy sclerotia up to 3 cm away. Since the pathogen occurs in aggregate distribution in the field, new infections of the pathogen by Sporidesmium are likely. The interaction between the two fungi is density dependent. Sporidesmium will produce approximately 1.5×10^4 new macroconidia per infected sclerotium (Adams *et al.*, 1984).

As indicated previously, *P. nunn* and *T. flavus* are good soil competitors. Both fungi also are known to parasitize several plant pathogens. Hyphae of *P. nunn* coiled around and subsequently lysed the hyphae of the hosts *P. ultimum* and *P. vexans* de Bary (Lifshitz *et al.*, 1984). *P. nunn* formed appressoria-like structures and parasitized the hyphae of *P. aphanidermatum* (Edson) Fitzp., *Rhizoctonia solani* Kühn, *Phytophthora cinnamomi* Rands, and *P. parasitica* Dast. *Talaromyces flavus* parasitizes hyphae of *R. solani* (Boosalis, 1956), sclerotia of *S. sclerotiorum* (McLaren *et al.*, 1986), and microsclerotia of *V. dahliae* (Fahima *et al.*, 1988). Since both *P. nunn* and *T. flavus* are also known to produce compounds which mediate antibiosis, both antagonists may actually be necrotrophic mycoparasites.

Species of Trichoderma and Gliocladium may also control pathogens through a combination of parasitism and antibiosis (Papavizas, 1985). Both genera are known to parasitize directly a number of pathogens *in vitro*. Both genera also produce toxic metabolites and mycolytic enzymes such as glucanases. In mycoparasitism, the host may first be weakened by toxic metabolites and/or enzymes prior to disorganization and death of the tissue. Further, it may not be necessary to kill the pathogen to prevent disease. By incubating various concentrations of *G. virens* with sclerotia of *Sclerotium rolfsii* Sacc. in field soil, Papavizas found that living sclerotia colonized by *G. virens* were unable to infect bean stems (personal communication). These sclerotia did not germinate on artificial media, although sclerotia from the same treatment germinated following surface sterilization.

Antibiosis

While convincing evidence for the roles of competition and parasitism in biocontrol by fungi has been obtained, definitive evidence establishing the importance of antibiosis has been more elusive (Fravel, 1988). Antibiosis is antagonism mediated by specific or nonspecific metabolites of microbial origin, by lytic agents, enzymes, volatile compounds, or other toxic substances (Jackson, 1965). Most antibiotics are produced by soil microorganisms. Hence, it has been assumed that antibiotics confer a selective advantage to the producers in the rhizosphere and soil. However, the lack of methods to evaluate in a meaningful way the production and function of compounds mediating antibiosis has been a major stumbling block to the acquisition of basic information on this mechanism with fungal biocontrol agents (Papavizas and Lumsden, 1980).

The earliest evidence for the involvement of antibiotics in biocontrol was a 1956 report that a gliotoxin-producing strain of *T. viride* Pers. Gray (subsequently thought to be *G. virens* Miller, Giddens and Foster) gave better control of Pythium-induced damping-off than did a viridin-producing isolate, while a strain that produced neither antibiotic gave the least control (Wright, 1956b). This work directed attention toward the production of antibiotics in microsites. Wright demonstrated that gliotoxin was produced on the coats of pea seeds inoculated with *T. viride* (=*G. virens*) and of noninoculated pea seeds planted in natural soil (Wright, 1956a, 1956c). Gliotoxin and viridin are also produced by *G. virens* on sclerotia of Rhizoctonia borne on potato tubers (Aluko and Herring, 1970).

There are conflicting reports about the role of another antibiotic, gliovirin, in the biocontrol of *R. solani* by *G. virens*. A gliotoxin-producing isolate of *G. virens* did not limit either damping-off or root rot of peanut caused by *R. solani* as effectively as a non-gliotoxin-producing isolate (Jones *et al.*, 1984). In contrast, mutants of *G. virens* unable to parasitize *R. solani* were still able to suppress Rhizoctonia-induced damping-off of cotton in field soil, indicating that mycoparasitism was not the major mechanism of biocontrol of *R. solani* by *G. virens* (Howell, 1987). The results of chromatography indicated that one nonparastiic mutant produced gliovirin, while another produced viridin.

Volatile metabolites from *T. harzianum* identified as alkyl pyrones suppressed *R. solani*-induced damping-off of lettuce when added to a peat-soil mixture (Claydon *et al.*, 1987). These metabolites are powerful inhibitors of several fungi *in vitro*. While these compounds were fungistatic to most fungi, they acted as paramorphogens (*i.e.* they altered the spatial organization of the fungal biomass) for some fungi, particularly *R. solani* and *R. cerealis* van der Hoeven. Filtrates of various mutants of *T. harzianum* are also known to suppress the white-rot pathogen *S. cepivorum* (Papavizas *et al.*, 1982).

The involvement of enzymes in biocontrol obscures the distinction between parasitism and antibiosis. For example, production of a cell wall-degrading enzyme by an antagonist could be involved simultaneously in parasitism and antibiosis. Other enzymes may act only as antibiotics.

For example, *T. flavus* isolate Tf1 controlled Verticillium wilt of eggplant (Marios *et al.*, 1982) and has shown the potential to suppress Verticillium wilt of potato (Davis *et al.*, 1986; Fravel *et al.*, 1986) under commercial production conditions. *Talaromyces flavus* Tf1 produces glucose oxidase which reacts with glucose to produce hydrogen peroxide (Kim *et al.*, 1988). Neither glucose nor glucose oxidase alone kills the microsclerotia in soil, while the simultaneous addition of glucose and glucose oxidase results in reduction of numbers of viable microsclerotia buried in soils. In addition, relatively high concentrations of hydrogen peroxide are required to kill microsclerotia in soil. Antibiosis is therefore suspected to be invovled in biocontrol, but the role of glucose oxidase has not been definitively demonstrated.

In the future, other currently unidentified compounds which mediate antibiosis may be identified as enzymes. For example, water extracts from 'germling' preparations of *Trichoderma* spp. and *Gliocladium virens* affected growth of *R. solani* (Lewis and Papavizas, 1987). These extracts induced leakage of soluble protein, carbohydrate, amino acids, and salts from hyphae of the pathogen. Autoclaved extracts were 75–80% less effective in causing leakage than nonheated extracts.

Conclusions

One often overlooked advantage of understanding mechanisms which mediate biocontrol is that this knowledge is useful in designing screening procedures to discover new antagonists. For example, rapid growth on agar plates has been used as a primary screening criterion. While this method may identify potential antagonists which operate primarily through competition, it would not detect those that function by mycoparasitism or antibiosis. Specifically, Sporidesmium, a slow-growing obligate parasite, could not have been discovered by this method. A more reasonable way to screen for mycoparasites is to bury propagules of the intended host and recover organisms colonizing these propagules. One approach that we have used to screen for new antagonists is to create suppressive soils by growing successive crops of susceptible plants in the presence of the pathogen. When disease incidence begins to decline, then we bury propagules of the pathogen to determine what colonizes them (potential mycoparasites) and recover root microflora (potential competitors). When isolating root-colonizing microorganisms as potential antagonists, it is important to consider pathogen mechanisms also. For example, *R. solani* can infect at any location on the root while *V. dahliae* infects primarily at root tips. Hence, for *V. dahliae* it may be more appropriate to select organisms which move with the growing root tip, or, more likely, to seek antagonists with modes of action, such as mycoparasitism, which do not require protection of the infection court.

Knowledge of biocontrol mechanisms is also necessary for developing management strategies which favor the antagonist. For example, since competition for nutrients may be of primary importance for antagonists which act primarily via competition, crop choice and organic amendments may affect biocontrol. Successive croppings may be necessary to build up populations of the antagonist, in the case of the mycoparasite Sporidesmium.

Characteristics which govern biocontrol must also be identified in order to improve genetically any antagonist. Knowledge of biocontrol mechanisms is necessary for the registration of these organisms with EPA. Registration should be less cumbersome for antagonists which are target specific and which do not produce toxic metabolites.

Any control agent, biological or chemical, must be in the 'right place at the right time' to effect control. Thus, parasitism is least likely to function successfully in the rhizosphere because plant infection will have occurred by the time mycoparastism is established. Parasitism is more effective when a strategy of reducing the amount of initial inoculum in nonrhizosphere soil is appropriate, such as destruction of sclerotia of Sclerotinia and Verticillium. Competition and antibiosis may be the most important biocontrol mechanisms in the rhizosphere, since these mechanisms can prevent infection of the root by actively growing hyphae or germ tubes. While competition is also important in nonrhizosphere soil, as with Fusarium, antibiosis can only function in nonrhizosphere soil if nutrients are available. In the future, more emphasis should be placed on elucidating these mechanisms of biocontrol.

References

Adams P B and Ayres W A 1983 Histological and physiological aspects of infection of sclerotia of two *Sclerotinia* species by two mycoparasites. Phytopathology 73, 1072–1076.

Adams P B, Marois J J and Ayres W A 1984 Population dynamics of the mycoparasite, *Sporidesmium sclerotivorum*, and its host, *Sclerotinia minor*, in soil. Soil. Biol. Biochem. 16, 627–633.

Alabouvette C, Rouxel F and Louvet J 1979 Characteristics of *Fusarium* wilt-suppressive soils and prospects for their utilization in biological control. *In* Soil-borne Plant Pathogens. Eds. B Schippers and W Gams. pp 165–182. Academic Press, London.

Aluko M O and Hering T F 1970 The mechanism associated

with the antagonistic relationship between *Corticium solani* and *Gliocladium virens*. Trans. Br. Mycol. Soc. 55, 173–179.

Ayers W A and Adams P B 1981 Mycoparasitism and its application to biological control of plant diseases. *In* Biological Crop Production, Beltsville Symposium in Agricultural Research Vol. 5. Ed. G C Papavizas. pp 91–103. Allenheld Osmun, Totowa.

Baker R 1968 Mechanisms of biological control of soil-borne pathogens. Ann. Rev. Phytopathol. 6, 263–294.

Boosalis M G 1956 Effect of soil temperature and green-manure amendments of unsterilized soil on parasitism of *Rhizoctonia solani* by *Penicillium vermiculatum* and *Trichoderma* sp. Phytopathology 46, 473–478.

Clark F E 1965 The concept of competition in microbiology. *In* Ecology of Soil-Borne Plant Pathogens. Eds. K F Baker and W C Snyder. pp 339–347. Univ. of Calif. Press, Berkeley, CA.

Claydon N, Allan M, Hanson J R and Avent A G 1987 Antifungal alkyl pyrones of *Trichoderma harzianum*. Trans. Br. Mycol. Soc. 88, 503–513.

Cook R J and Baker K F 1983 The Nature and Practice of Biological Control of Plant Pathogens. The Amer. Phytopathol. Soc., St. Paul, MN, 539 p.

Davis J R, Fravel D R, Marois J J and Sorenson L H 1986 Effect of soil fumigation and seedpiece treatment with *Talaromyces flavus* on wilt incidence and yield, 1983. Biol. Cult. Tests 1, 18.

Fahima T, Katan J and Henis Y 1988 Biological control of Verticillium wilt of eggplant by *Talaromyces flavus*. Phytoparasitica 16, 67.

Fravel D R 1988 Role of antibiosis in the biocontrol of plant diseases.

Fravel D R, Davis J R and Sorenson L H 1986 Effect of *Talaromyces flavus* and metham on Verticillium wilt incidence and potato yield 1984–1985. Biol. Cult. Tests 1, 17.

Garrett S D 1956 Biology of Root-Infecting Fungi. Cambridge Univ. Press, Cambridge, England, 293 p.

Hoch H C and Abawi G S 1979 Biological control of Pythium root rot of table beet seed with *Corticium* sp. Phytopathology 69, 417–419.

Howell C R 1987 Relevance of mycoparasitism in the biological control of *Rhizoctonia solani* by *Gliocladium virens*. Phytopathology 77, 992–994.

Huang H C 1977 Importance of *Coniothyrium minitans* in survival of sclerotia of *Sclerotinia sclerotiorum* in wilted sunflower. Can. J. Bot. 55, 289–295.

Huang H C 1980 Control of Sclerotinia wilt of sunflower by hyperparasites. Can. J. Plant Pathol. 2, 26–32.

Ikediugwu F E O, Dennis C and Webster J 1970 Hyphal interference by *Peniophora gigantea* against *Heterobasidium annosum*. Trans. Br. Mycol. Soc. 54, 307–309.

Jackson R M 1965 Antibiosis and fungistasis of soil microorganisms. *In* Ecology of Soil-Borne Plant Pathogens. Eds. K F Baker and W C Snyder. pp 363–369. Univ. of Calif. Press, Berkeley, CA.

Jager G and Valvis H 1988 Inactivation of sclerotia of *Rhizoctonia solani* on potato tubers by *Verticillium biguttatum*, a soil-borne mycoparasite. Neth. J. Plant Pathol. 94, 225–231.

Jones R W, Pettit R E and Taber R A 1984 Lignite and stillage: Carrrier and substrate for application of fungal biocontrol agents to soil. Phytopathology 74, 1167–1170.

Kim K K, Fravel D R and Papavizas G C 1988 Identification of a metabolite produced by *Talaromyces flavus* as glucose oxidase and its role in the biocontrol of *Verticillium dahliae*. Phytopathology 78, 488–492.

Lewis J A and Papavizas G C 1987 Permiability changes in hyphae of *Rhizoctonia solani* incited by germling preparations of *Trichoderma* and *Gliocladium*. Phytopathology 77, 699–703.

Lifshitz R, Dapler M, Elad Y and Baker R 1984 Hyphal interactions between a mycoparasite, *Pythium nunn*, and several soil fungi. Can. J. Microbiol. 30, 1482–1487.

Lumsden R D 1981 Ecology of mycoparasitism. *In* The Fungal Community. Ed. D T Wicklow and G C Carroll. pp 295–318. Marcel Dekker, Inc., New York.

Marois J J, Johnston S A, Dunn M T and Papavizas G C 1982 Biological control of Verticillium wilt of eggplant in the field. Plant Dis. 66, 1166–1168.

Marois J J and Mitchell D J 1981 Effects of fungal communities on the pathogenic and saprophytic activities of *Fusarium oxysporum* f. sp. *radicis-lycopersici*. Phytopathology 71, 1251–1256.

Martin F N and Hancock J G 1986 Association of chemical and biological factors in soils suppressive to *Pythium ultimum*. Phytopathology 76, 1221–1231.

Martin S B, Abawi G S and Hoch H C 1986 The relationship of population densities of the antagonist, *Laetisaria arvalis*, to seedling diseases of table beet incited by *Pythium ultimum*. Can. J. Microbiol. 32, 156–159.

McLaren D L, Huang H C, and Rimmer S R 1986 Hyperparasitism of *Sclerotinia sclerotiorum* by *Talaromyces flavus*. Can. J. Plant Pathol. 8, 43–48.

Papavizas G C 1985 *Trichoderma* and *Gliocladium*: biology, ecology, and potential for biocontrol. Annu. Rev. Phytopathol. 23, 23–54.

Papavizas G C, Lewis J A and Abd-El Moity T H 1982 Evaluation of new biotypes of *Trichoderma harzianum* for tolerance to benomyl and enhanced biocontrol capabilities. Phytopathology 72, 126–132.

Papvizas G C and Lumsden R D 1980 Biological control of soilborne fungal propagules. Annu. Rev. Phytopathol. 18, 389–413.

Paulitz T C and Baker R 1988 Interactions between *Pythium nunn* and *Pythium ultimum* on bean leaves. Can. J. Microbiol. 8, 947–951.

Rishbeth J 1975 Stump inoculation: A biological control of *Fomes annosus*. *In* Biology and Control of Soil-borne Pathogens. Ed. G W Bruehl. pp 158–162. Amer. Phytopathol. Soc., St. Paul, MN.

Sivan A and Chet I 1989 The possible role of competition between *Trichoderma harzianum* and *Fusarium oxysporum* on rhizosphere colonization. Phytopathology 79, 198–203.

Stack J P, Kenerley C M and Pettit R E 1987 Influence of carbon and nitrogen sources, relative carbon and nitrogen concentrations, and soil moisture on the growth in non-sterile soil of soilborne fungal antagonists. Can. J. Microbiol. 33, 626–631.

Tribe H T 1957 On the parasitism of *Sclerotinia trifoliorum* by *Coniothyrium minitans*. Trans. Br. Mycol. Soc. 40, 489–499.

Wright J M 1956a Biological control of a soil-borne Pythium infection by seed inoculation. Plant and Soil 8, 132–140.

Wright J M 1956b The production of antibiotics in soil. III. Production of gliotoxin in wheat straw buried in soil. Ann. Appl. Biol. 44, 461–466.

Wright J M 1956c The production of antibiotics in soil. IV. Production of antibiotics in coats of seeds sown in soil. Ann. Appl. Biol. 44, 561–566.

D. L. Keister and P. B. Cregan (Eds.), The rhizosphere and plant growth, 245–251.

PLSO BARC24

Role of antibiotics and siderophores in biocontrol of take-all disease of wheat

LINDA S. THOMASHOW and DAVID M. WELLER
USDA, Agricultural Research Service, Root Disease and Biological Control Research Unit, Pullman, WA 99164-6430, USA

Key words: antibiotics, biocontrol, *Gaeumannomyces graminis*, phenazine-1-carboxylic acid, Pseudomonas, siderophores, take-all, wheat

Abstract

Both antibiotics and siderophores have been implicated in the control of soilborne plant pathogens by fluorescent pseudomonads. In *Pseudomonas fluorescens* 2-79, which suppresses take-all of wheat, the importance of the antibiotic phenazine-1-carboxylic acid was established with mutants deficient or complemented for antiobiotic production and by isolation of the antibiotic from the roots of wheat colonized by the bacteria. Genetic and biochemical studies of phenazine synthesis have focused on two loci; the first is involved in production of both anthranilic acid and phenazine-1-carboxylic acid, and the second encodes genes involved directly in phenazine synthesis. Because the antibiotic does not account fully for the suppressiveness of strain 2-79, additional mutants were analyzed to evaluate the role of the fluorescent siderophore and of an antifungal factor (Aff, identified as anthranilic acid) that accumulates when iron is limiting. Whereas strains producing only the siderophore conferred little protection against take-all, Aff^+ strains were suppressive, but much less so than phenazine-producing strains. Iron-regulated nonsiderophore antibiotics may be produced by fluorescent pseudomonads more frequently than previously recognized, and could be partly responsible for beneficial effects that were attributed in the past to fluorescent siderophores.

Introduction

Certain strains of fluorescent pseudomonads, when introduced on seed or planting material, control plant diseases caused by soilborne pathogens or promote plant growth by suppressing deleterious rhizosphere microorganisms (Schippers *et al.*, 1987; Weller, 1988). These beneficial strains compete aggressively for sites on roots or in the rhizosphere where nutrients are available. Occupation of these sites allows introduced bacteria both to preempt establishment of pathogens on the root (niche exclusion; Suslow, 1982) and to reduce the amount of substrate available to pathogens (Foster, 1986; Weller, 1988). Competition for sites and nutrients is an underlying mechanism that confers a low to moderate level of suppressiveness to most introduced bacteria, but it does not account fully for their effectiveness as biocontrol agents.

Beneficial pseudomonads also antagonize pathogens by producing one or more of a variety of metabolites that include antibiotics (Fravel, 1988; Weller, 1988) siderophores (Leong, 1986; Schippers *et al.*, 1987) and other substances such as cyanide (Voisard *et al.*, 1989). The improvement and eventual commercialization of fluorescent pseudomonads as biocontrol agents depends in part on understanding and exploiting the mechanisms involved in these antagonistic interactions among bacteria, pathogens and their plant hosts.

Mechanisms in the control of take-all

Take-all, caused by *Gaeumannomyces graminis* var. *tritici*, is an important root and crown disease of wheat worldwide. Take-all decline, the spontaneous diminution in disease and increase in yield that occurs with wheat monoculture (Shipton, 1975), provides a natural model for biological control of take-all. Take-all decline results from the accumulation of a microflora antagonistic to *G. g.* var. *tritici* (Cook and Rovira, 1976; Gerlagh, 1968; Shipton, 1975), and fluorescent pseudomonads are an important component of the suppressive microflora. Roots of wheat grown in a take-all suppressive soil have larger populations of fluorescent pseudomonads antagonistic to *G. g.* var. *tritici* than roots grown in soils conducive to the disease. Further, strains from suppressive soils provide greater protection against take-all than do strains from conducive soils (Weller *et al.*, 1988). This strong association between inhibition of *G. g.* var. *tritici in vitro* and suppression of take-all, which differs markedly from the more usual lack of correlation (reviewed in Fravel, 1988) between antagonism *in vitro* and in soil, supports the idea (Weller *et al.*, 1988) that strains from roots grown in suppressive soils already have been selectively enriched by the infected root-pathogen complex for rhizosphere competence, antagonism to *G. g.* var. *tritici* and other uncharacterized traits essential for biocontrol activity. Such strains are a logical starting point for analyses of the mechanisms important in take-all suppression.

Role of phenazine antibiotics

P. fluorescens 2-79, isolated originally from a suppressive soil, inhibited *G. g.* var. *tritici* on both King's medium B and potato dextrose agar, used to detect siderophore and antibiotic production, respectively. Chemically derived mutants defective in either activity also were less suppressive of take-all, suggesting that both antibiotics and siderophores have a role in biocontrol (Weller *et al.*, 1988). A greenish-yellow antibiotic purified from cultures of strain 2-79 initially was identified as a dimer of phenazine-1-carboxylic acid (PCA) (Gurusiddaiah *et al.*, 1986) but later was shown to exist as the monomer (Brisbane *et al.*, 1987). The antibiotic was inhibitory to a variety of fungi, including the wheat pathogens *G. g.* var. *tritici*, *Rhizoctonia solani* and several *Pythium* spp. (Gurusiddaiah *et al.*, 1986).

The relationship between PCA production and take-all suppression in 2-79 was further investigated using transposon mutants defective in synthesis of the antibiotic (Phz^-). These mutants were noninhibitory to the pathogen *in vitro* and much less suppressive of take-all than 2-79 (Thomashow and Weller, 1988). Genetic analysis of six mutants revealed that each contained a single Tn*5* insertion and that the insertions in two of the mutants were closely linked. These mutants, when complemented *in trans* with homologous DNA from a 2-79 genomic library, regained the ability to produce PCA (although only at about half the wild-type level) and to inhibit the pathogen *in vitro*. Introduction of the wild-type sequences into the mutant genomes fully restored PCA production *in vitro* and suppressiveness of take-all.

Phz^- mutants were detected on roots in numbers similar to the parental strain in both steamed and natural soils, indicating that differences in suppressiveness were not due to failure of the mutants to colonize the root and the rhizosphere. Although antibiotic production apparently does not contribute to competitiveness in the short term (11–13 days), its role in long-term colonization remains uncertain. Results from this and other (Schippers *et al.*, 1987; Voisard *et al.*, 1989) studies suggest that uncharacterized and unselected rhizospere competence traits need not be lost during genetic manipulations *in vitro*. The loss of such traits, if frequent, would limit the value of the molecular approach in genetic analyses of biocontrol determinants as well as future strain improvements.

Isolation of antibiotic from the rhizosphere of wheat

Although many microorganisms from the soil produce inhibitory substances *in vitro*, there is little direct evidence that they do so in nature (Williams, 1982). Further, almost all of the evidence for the involvement of antibiotics in biocontrol is indirect or circumstantial (reviewed in Fravel, 1988; Weller and Thomashow, 1989). Because of the fundmental uncertainty about the role of antibiotics in natural habitats, it was important to demonstrate directly that PCA production on roots of wheat was responsible for suppression of take-all.

Washings from the roots and rhizosphere soil of seedlings colonized by 2-79 were analyzed for the presence of PCA after fractionation by high pressure liquid chromatography. The antibiotic was identified by its chromatographic retention time and UV-visible spectrum, both of which were indistinguishable from those of purified PCA. In growth chamber studies, seedlings colonized by strain 2-79 in steamed soil yielded about tenfold more PCA (up to 350 ng/g root with adhering soil) than seedlings in natural soils (28–43 ng) in the presence or absence of *G. g.* var. *tritici*. The population size of 2-79 on roots from steamed soil (1.4×10^8 colony-forming units (cfu)/g root) also was larger than on roots from natural soil (approximately 1.6×10^7 and 2×10^6 cfu/g root in the presence and absence, respectively, of the pathogen). Seedlings grown in a virgin field plot without introduced *G. g.* var. *tritici* were directly comparable both in terms of PCA yield and 2-79 population with seedlings grown without introduced pathogen in the growth chamber. No antibiotic was detected from roots of seedlings colonized by a Phz^- mutant or grown from nontreated seed. Finally, in both steamed and natural soils, roots from which PCA was recovered had significantly less disease than roots from which no PCA was isolated, indicting that suppression of take-all is related directly to the presence of the antibiotic in the rhizosphere.

The finding that quantities of PCA sufficient to reduce disease were present on roots in the absence of the pathogen was somewhat unexpected. Previous studies (Weller, 1983) had documented the proliferation of strain 2-79 on roots infected with *G. g.* var. *tritici*, leading to the hypothesis (Weller *et al.*, 1988) that PCA production occurs mainly within take-all lesions enriched by the leakage of nutrients, and functions to limit secondary infections by the fungus. This model is consistent with earlier observations (Cook *et al.*, 1986) that microorganisms in take-all suppressive soil limit disease progress rather than initiation, an idea originally put forth by Shipton (1975). Recently, however, an inverse relationship between the number of take-all lesions and the population size of 2-79 present on roots was demonstrated (Bull, 1987), suggesting that if large enough populations are established early enough, even lesion initiation may be limited. No such effect was observed when Phz^- mutants were introduced. The finding that populations of 2-79 on uninfected roots can produce biologically relevant amounts of PCA strongly implicates the antibiotic in reducing infection incidence.

That significant take-all control occurred in the presence of as little as 25–30 ng of PCA per gram of root is consistent with efficient delivery of the antibiotic to the pathogen whether in lesions or prior to their development, but a more complex targeting mechanism probably is needed in the latter case. Bacteria grow on roots as discrete microcolonies in sites enriched in root exudates, lysates or mucilage (Foster, 1986), and the antibiotic also is likely to be more concentrated in these areas. If the passages by which fungi gain access to the root are favored sites for bacterial growth, as proposed by Foster (1986), then occupation of such sites by PCA-producing antagonists would position them strategically for antibiotic delivery.

Relative importance of antibiotics and siderophores

Disease ratings of seedlings treated with Phz^- mutants of 2-79 are intermediate between those of the parental strain and the nontreated control (Thomashow and Weller, 1988), indicating that PCA is not the sole factor responsible for suppressiveness. Siderophores produced by fluorescent pseudomonads have been implicated as a control mechanism for many pathogens (reviewed in Leong, 1986; Schippers *et al.*, 1987) including *G. g.* var. *tritici* (Kloepper *et al.*, 1980; Weller *et al.*, 1988; Wong and Baker, 1984), although not all studies have supported their importance in take-all control (*e.g.* Brisbane and Rovira, 1988). To evaluate the importance of the fluorescent siderophore produced by strain 2-79, Hamdan (1988) characterized mutants deficient in production of PCA (Phz^-), the fluorescent siderophore (Flu^-) or both factors (Phz^-Flu^-). These studies revealed that two Phz^- mutants, 2-79-B46 and 2-79-782 (Thomashow and Weller, 1988) also failed to produce an antifungal factor (Aff) that accumulated in iron-limited cultures of 2-79 but was distinct from the fluorescent siderophore. Mutants at this locus therefore were $Phz^-Aff^-Flu^+$ as compared to phenazine mutants at other loci which were $Phz^-Aff^+Flu^+$. Moreover, $Phz^-Aff^+Flu^-$ mutants were inhibitory to *G. g.* var. *tritici* on King's medium B whereas $Phz^-Aff^-Flu^+$ strains were not, indicating that the Aff factor, rather than the fluorescent siderophore, is responsible for inhibition by strain 2-79 *in vitro* under iron-limiting conditions.

Iron availability in soil increases as a function of decreasing pH, and siderophore production occurs only in environments low in available iron. Suppressiveness associated with siderophore activity therefore is expected to be greatest in high pH soils low in available iron (reviewed in Baker *et al.*, 1986). In contrast, strain 2-79 requires iron for phenazine production (Gurusiddaiah *et al.*, 1986; Weller *et al.*, 1988) and activity of the antibiotic *in vitro* increases with decreasing pH (Brisbane and Rovira, 1988). Because of the potential impact of pH and iron availability on factors contributing to suppressiveness, Hamdan's strains were compared for ability to control take-all in two soils, one of high pH (7.6) and low extractable iron (6.75 $\mu g/g$) and the other of low pH (5.5) and higher extractable iron (93.7 $\mu g/g$).

In both soils, seedlings treated with Phz^- mutants were significantly more diseased than those treated with strain 2-79 regardless of the Flu or Aff phenotype (Hamdan, 1988). The relationship between soil pH and suppressiveness due to PCA was examined further with strain 2-79 or a $Phz^+Aff^+Flu^-$ mutant in steamed soil adjusted to pH values from 4.9 to 8.0. Both strains suppressed take-all over the full pH range tested (B.H. Ownley and D.M. Weller, personal communication). These results provide additional evidence for the importance of PCA as a major determinant of suppressiveness, but conflict with *in vitro* data and theoretical considerations (Brisbane *et al.*, 1987; Brisbane and Rovira, 1988) suggesting that PCA might be ineffective in alkaline environments. Perhaps the simplest explanation for this apparent discrepancy is that the pH in the microsites where the bacteria and the pathogen interact may be significantly lower than that of the surrounding bulk soil, or even of the rhizosphere in general.

Comparisons of a $Flu^-Phz^+Aff^+$ mutant with 2-79, and of $Flu^-Phz^-Aff^+$ and $Flu^+Phz^-Aff^+$ mutants, revealed no significant difference in suppressiveness between the paired strains. Further, a $Flu^+Phz^-Aff^-$ mutant was no more protective than one deficient in all three factors at either pH 7.6 or 5.5 (Hamdan, 1988). Collectively, these results based on the use of genetically defined mutants suggest that the fluorescent siderophore has a minor role, if any, in control of take-all. This conclusion differs from that of a previous report (Weller *et al.*, 1988) implicating the siderophore in disease suppression. However, the earlier study used less-chacterized, chemically derived nonfluorescent mutants that also were somewhat reduced in inhibition of *G. g.* var. *tritici in vitro* in potato dextrose agar, suggesting that they may have been pleiotropically altered in antibiotic production.

Seedlings treated with $Aff^+Flu^+Phz^-$ or $Aff^+Flu^-Phz^-$ strains were less diseased than when treated with Aff^- counterparts of these strains. The differences were much smaller than those between Phz^+ and Phz^- strains and were significant only in the pH 7.6 soil, however (Hamdan, 1988), indicating that Aff contributes much less than PCA to suppressiveness, and then only under certain conditions. The ability of Aff^+ strains to provide protection in soil at pH 7.6 but not at pH 5.5 may be related to the greater availability in the latter soil of iron, which negatively controls accumulation of the Aff product *in vitro*.

Although factors negatively regulated by iron apparently are of much less importance than the phenazine antibiotic in control of take-all by 2-79, iron-sensitive mechanisms contribute substantially to pathogen suppression or plant growth promotion by fluorescent pseudomonads in other systems (Leong, 1986; Schippers *et al.*, 1987). Siderophores generally have been considered responsible for these phenomena because beneficial effects were correlated with their presence or presumed presence, and these effects were reversed upon addition of iron. A few recent studies (Loper, 1988; Schippers *et al.*, 1987) using well-defined mutants have clearly documented the importance of siderophores in biocontrol, but the methods in many of the earlier studies would not have distinguished between effects due to siderophores and those due to antibiotics such as Aff, both of which would have been present under the same conditions. The ability to produce nonsiderophore inhibitors in response to iron limitation may be more broadly distributed among fluorescent pseudomonads than previously realized. Nonfluorescent Tn*5* mutants of *P. fluorescens* M4-80R and *P. putida* L30b-80 remained inhibitory to *G. g.* var. *tritici* in iron-limited media (Hamdan, 1988), and an antibiotic fungistatic to *Pythium ultimum* in iron-limited cultures of *Pseudomonas* sp. NZ130 recently was described (Gill and Warren, 1988). If such substances are of common occurrence, then the relative contribution of siderophores to disease control may need to be reevaluated.

Genetics and biochemistry of phenazine production

Genetic and biochemical studies of phenazine production have focused on two loci, the first of which is required for production of both PCA and the antifungal factor Aff. The ability of single Tn*5* insertions at this locus to cause multiple phenotypic changes serves as an important reminder that even 'clean' transposon mutants can be pleiotropically altered, and the effects can be both relevant and not immediately apparent.

The Aff factor from 2-79 cultures was chromatographically and spectrally indistinguishable from anthranilic acid. Further, anthranilate was inhibitory to *G. g.* var. *tritici* at concentrations comparable to those in 2-79 culture supernates. Anthranilate is the first intermediate in tryptophan biosynthesis and is derived from chorismate, which also is a phenazine precursor. Although the enzyme responsible for the initial step in phenazine synthesis shares features in common with anthranilate synthase (AS), anthranilate previously was thought not to be a phenazine precursor (reviewed in Turner and Messenger, 1986). However, recent studies of AS in *P. aeruginosa* do not support this conclusion. *P. aeruginosa* has two pairs of related but distinguishable genes for the two heterodimeric subunits of AS. One pair, now designated *phn*AB, complements AS mutants of *Escherichia coli* (Crawford *et al.*, 1986) and is required for production of the phenazine pyocyanin, whereas the other pair, *trp*E and *trp*G, mainly functions in tryptophan biosynthesis (D. Essar, pers. comm.). These results indicate that anthranilate is an early intermediate in phenazine synthesis, and suggest that the Tn*5* insertions in Phz^-Aff^- mutants of 2-79 (which do not have a tryptophan requirement) may be in structural or regulatory genes for *phn*AB. Presumably, anthranilate synthesized by the products of these genes then is converted to PCA by a pathway that requires iron for its full expression or activity. Whether iron also has a direct positive or negative influence on the *phn*AB locus *per se* is presently uncertain.

The second locus of interest was identified originally in *P. aureofaciens* 30-84, a take-all suppressive strain that produes two 2-hydroxyphenazine derivatives as well as PCA. The phenazines also are important determinants of suppressiveness in strain 30-84 (L.S. Pierson, pers. comm.). The 30-84 cosmid pLSP259 complemented several different Phz^- mutants of both 30-84 and 2-79, including a 30-84 mutant that produced only PCA and not the 2-hydroxy derivatives. This suggested that pLSP259 might contain genes for at least part of a phenazine biosynthetic pathway. Structural and functional analyses subsequently defined contiguous regions of 2.8 kb and 1.9 kb required for production of PCA and 2-hydroxy-phenazine-1-carboxylic acid, respectively, and an unlinked region necessary for production of 2-hydroxyphenazine. When a subcloned 9.2 kb *Eco*R1 fragment containing the 2.8 and 1.9 kb regions was introduced into *E. coli*, PCA and 2-hydroxyphenazine-1-carboxylic acid were produced (L.S. Pierson, pers. comm.). Further, a 6.3 kb *Eco*R1 fragment subcloned from pPHZ173, a 2-79 cosmid homologous to pLSP259, also was expressed in *E. coli*, again resulting in PCA production. These results provide strong evidence that the cloned sequences from 30-84 and 2-79 contain genes involved directly in phenazine biosynthesis, and demonstrate the feasibility of mobilizing and expressing such genes in nonproducer organisms.

Future directions

Understanding the mechanisms involved in disease suppression is an essential first step toward improved biological control with existing strains as well as the selection or development of new agents. In strains 2-79 and 30-84, efforts now are focused on identifying the biochemical and genetic mechanisms that regulate phenazine synthesis and the environmental factors that may limit antibiotic production or activity. The goal of these studies is to find ways to manipulate the bacteria, the environment, or both to increase the level or consistency of performance. New and potentially superior biocontrol strains already are being identified on the basis of phenazine production, and hybridization probes based on phenazine biosynthetic genes could increase the speed and reliability of the screening process. Finally, traditional and molecular genetic techniques are available both to modify the kinds and amounts of phenazines by existing strains and to introduce and express phenazine biosynthetic genes in nonproducer organisms that have other desirable attributes. Such engineered strains will have immediate value as research tools and may represent a key to future generations of improved biocontrol agents.

References

Baker R, Elad Y and Sneh B 1986 Physical, biological and host factors in iron competition in soils. *In* Iron, Siderophores and Plant Diseases. Ed. T R Swinburne. pp 77–84. Plenum, New York.

Brisbane P G and Rovira A D 1988 Mechanisms of inhibition of *Gaeumannomyces graminis* var. *tritici* by fluorescent pseudomonads. Plant Pathology 37, 104–111.

Brisbane P G, Janik L J, Tate M E and Warren R F O 1987 Revised structure for the phenazine antibiotic from *Pseudomonas fluorescens* 2-79 (NRRL B-15132). Antimicrob. Agents Chemother. 31, 1967–1971.

Bull C T 1987 Wheat root colonization by disease-suppressive or nonsuppressive bacteria and the effect of population size on severity of take-all caused by *Gaeumannomyces graminis* var. *tritici*. M.S. Thesis, Washington State Univ., Pullman.

Cook R J and Rovira A D 1976 The role of bacteria in the biological control of *Gaeumannomyces graminis* by suppressive soils. Soil Biol. Biochem. 8, 269–273.

Cook R J, Wilkinson H T and Alldredge J R 1986 Evidence that microorganisms in suppressive soil associated with wheat take-all decline do not limit the number of lesions produced by *Gaeumannomyces graminis* var. *tritici*. Phytopathology 76, 342–345.

Crawford I P, Wilde A, Yelverton E M, Figurski D and Hedges R W 1986 Structure and regulation of the anthranilate synthase genes in *Pseudomonas aeruginosa*: II. Cloning and expression in *Escherichia coli*. Mol. Evol. Biol. 3, 449–458.

Foster R C 1986 The ultrastructure of the rhizoplane and rhizosphere. Annu. Rev. Phytopathol. 24, 211–234.

Fravel D R 1988 Role of antibiosis in the biocontrol of plant diseases. Annu. Rev. Phytopathol. 26, 75–91.

Gerlagh M 1968 Introduction of *Ophiobolus graminis* in new polders and its decline. Neth. J. Plant Pathol. 74, Suppl. 2. 97 pp.

Gill P R and Warren G J 1988 An iron-antagonized fungistatic agent that is not required for iron assimilation from a fluorescent rhizosphere pseudomonad. J. Bacteriol. 170, 163–170.

Gurusiddaiah S, Weller D M, Sarkar A and Cook R J 1986 Characterization of an antibiotic produced by a strain of *Pseudomonas fluorescens* inhibitory to *Gaeumannomyces graminis* var. *tritici* and *Pythium* spp. Antimicrob. Agents Chemother. 29, 488–495.

Hamdan H 1988 The fluorescent siderophore of *Pseudomonas fluorescens*: role in suppression of *Gaeumannomyces graminis* var. *tritici* and genetic analysis of siderophore production. PhD Thesis, Washington State Univ., Pullman.

Kloepper J W, Leong J, Teintze M and Schroth M N 1980 *Pseudomonas* siderophores: A mechanism explaining disease suppressive soils. Curr. Microbiol. 4, 317–320.

Leong J 1986 Siderophores: their biochemistry and possible role in the biocontrol of plant pathogens. Annu. Rev. Phytopathol. 24, 187–209.

Loper J E 1988 Role of fluorescent siderophore production in biological control of *Pythium ultimum* by a *Pseudomonas fluorescens* strain. Phytopathology 78, 166–172.

Schippers B, Bakker A W and Bakker P A H M 1987 Interactions of deleterious and beneficial rhizosphere microorganisms and the effect of cropping practice. Annu. Rev. Phytopathol. 25, 339–358.

Shipton P J 1975 Take-all decline during cereal monoculture. *In* Biology and Control of Soilborne Plant Pathogens. Ed. G W Bruehl. pp 137–155. American Phytopathological Society, St. Paul, MN.

Suslow T V 1982 Role of root-colonizing bacteria in plant growth. *In* Phytopathogenic Prokaryotes Vol. 1. Eds. M S

Mount and G H Lacy. pp 187–223. Academic Press, New York.

Turner J M and Messenger A J 1986 Occurrence, biochemistry and physiology of phenazine pigment production. Adv. Microbial Physiol. 27, 211–275.

Thomashow L S and Weller D M 1988 Role of a phenazine antibiotic from *Pseudomonas fluorescens* in biological control of *Gaeumannomyces graminis* var. *tritici*. J. Bacteriol. 170, 3499–3508.

Voisard C, Keel C, Haas D and Defago G 1989 Cyanide production by *Pseudomonas fluorescens* helps suppress black root rot of tobacco under gnotobiotic conditions. EMBO J. 8, 351–358.

Weller D M 1983 Colonization of wheat roots by a fluorescent pseudomonad suppressive to take-all. Phytopathology 73, 1548–1553.

Weller D M 1988 Biological control of soilborne plant pathogens in the rhizosphere with bacteria. Annu. Rev. Phytopathol. 26, 379–407.

Weller D M, Howie W J and Cook R J 1988 Relationship between *in vitro* inhibition of *Gaeumannomyces graminis* var. *tritici* and suppression of take-all by fluorescent pseudomonads. Phytopathology 78, 1094–1100.

Weller D M and Thomashow L S 1990 Antibiotics: evidence for their production and sites where they are produced. *In* New Directions in Biological Control. Eds. R Baker and P Dunn. A. R. Liss, Inc., New York.

Williams S T 1982 Are antibiotics produced in soil? Pedobiologia 23, 427–435.

Wong P T W and Baker R 1984 Suppression of wheat take-all and Ophiobolus patch by fluorescent pseudomonads from a Fusarium-suppressive soil. Soil Biol. Biochem. 16, 397–403.

D. L. Keister and P. B. Cregan (Eds.), The rhizosphere and plant growth, 253–261.

PLSO BARC25

Factors influencing siderophore-mediated biocontrol activity of rhizosphere *Pseudomonas* spp.

JOYCE E. LOPER and CAROL A. ISHIMARU
U.S. Department of Agriculture, Agricultural Research Service, Horticultural Crops Research Laboratory, Corvallis, OR 97330, USA

Key words: aerobactin, *Erwinia carotovora*, iron, pseudobactin, pyoverdine(s)

Abstract

Variable efficacy of biological control agents between different sites of application poses a significant obstacle to agronomic application of biocontrol strategies. Theoretically, any factor influencing the activity or establishment of a biocontrol agent will influence its field performance. The sensitivity of the indigenous pathogen population to a biocontrol agent may be one factor varying among agricultural fields. Certain *Pseudomonas* spp. strains control potato seed piece decay and soft rot diseases caused by *Erwinia carotovora*, due to the production of fluorescent siderophore(s). Sensitivity to biocontrol by *Pseudomonas* spp. varies among the target *E. carotovora* strains evaluated (Xu and Gross, 1986b). Studies were initiated to identify the siderophores produced by *E. carotovora* and their potential roles in interactions with *Pseudomonas* spp. All of 23 *E. carotovora* strains tested produced a catechol siderophore but only strain W3C105 produced the hydroxamate siderophore, aerobactin. We postulate that iron-acquisition systems of target *E. carotovora* strains may influence their sensitivities to siderophore-mediated biocontrol by *Pseudomonas* spp.

Introduction

Historically, the control of plant disease has been achieved through the use of resistant plant varieties and agricultural practices, including the application of synthetic fungicides and other agrichemicals. However, growing ecological problems, including health risks associated with human exposure, ground water contamination, and the selection of pathogens with chemical resistance are associated with the use of agrichemicals. The extensive monoculture of specific plant cultivars selects for the predominance of new virulent pathogen races among pathogen populations. There is an urgent need for improvement, complementation, and at least partial replacement of these approaches by new methods of disease control. Biocontrol, utilizing introduced or manipulating indigenous microorganisms to suppress disease, is receiving increased attention due to concern over the use of pesticides in agriculture, and because it may provide control for diseases that cannot be managed by traditional strategies. Notable among these biocontrol agents are rhizosphere inhabitants that are applied to soil, seed, or roots of agricultural crops for the suppression of soilborne plant diseases. Fluorescent pseudomonads, especially certain *Pseudomonas fluorescens* and *Pseudomonas putida* strains, have received prominant attention due to their abilities to establish large rhizosphere population sizes on the emerging root systems of treated plants. The fluorescent pseudomonads are characterized by their production of yellow-green pigments that fluoresce under ultraviolet irradiation and function as siderophores.

Siderophores are biosynthetic compounds that

are produced by microbial cells under iron-limiting conditions, chelate the ferric ion (Fe^{3+}) with high specific activity, and serve as vehicles for the transport of iron(III) into the cell. Iron is an essential element for living organisms by virtue of its two stable valences that act as cofactors in various oxidative-reductive enzymatic reactions. It is assumed generally that organisms have evolved these high affinity iron uptake systems because the naturally abundant ferric ion is virtually insoluble in oxygenated environments, and thus unavailable for microbial growth. The fluorescent pigments/siderophores, termed pyoverdines or pyoverdins, represent one class of siderophores produced by *Pseudomonas* spp. (Demange *et al.*, 1987). Structural characterization of the pyoverdines has been impeded since only pseudobactin, the pyoverdine produced by *P. fluorescens* strain B10 (Teintze *et al.*, 1981), has been crystallized successfully for x-ray structure determination. Pseudobactin and other pyoverdines, which have been characterized by FAB mass spectrometry, nuclear magnetic resonance and circular dichroism, have several structural similarities (Demange *et al.*, 1987; Teintze *et al.*, 1981). All are chromopeptides, possessing a peptide chain of 6 to 10 amino acids bound to a chromophore derived from 2,3-diamino-6,7-dihydroxyquinoline. However, structural and functional differences also are observed among the pyoverdines. Thus, a pyoverdine produced by one Pseudomonas strain may or may not be utilized as an iron-carrier by another Pseudomonas strain (Buyer and Leong, 1986). The functional strain specificities of the pyoverdines are due, at least in part, to the selective recognition of outer membrane siderophore receptors, since the introduction of a gene for the membrane receptor is accompanied by the uptake of a previously unrecognized ferricpyoverdine (Magazin *et al.*, 1986). The specificity of a membrane receptor for a given ferricsiderophore is the basis of the siderophore mechanism theory in biocontrol. An organism that cannot utilize as a source of iron the prevalent ferricsiderophores in a given environment may be subject to iron-starvation. Although siderophores with direct toxicity have been described (Knusel *et al.*, 1969), the major biocontrol activity associated with siderophores is iron-depletion of the target pathogen.

Evidence for a role of siderophores in biocontrol of soilborne diseases

The pyoverdines provide the most thorough case study for a discussion of the role of siderophores in biocontrol. Moreover, the successes and failures of biocontrol strategies mediated by pyoverdines accent a need for the identification of factors impacting the underlying mechanisms in biocontrol. The following is a general presentation of the evidence supporting a role for pyoverdines in biocontrol of soilborne diseases.

Effect of pyoverdines on soilborne phytopathogens in culture

A ferricpyoverdine and the corresponding free-ligand often have differential effects on the growth and development of phytopathogens. For example, pseudobactin inhibits growth of the bacterial phytopathogen, *Erwinia carotovora*, on agar plates while ferric pseudobactin does not (Kloepper *et al.*, 1980). Presumably, the ferricpyoverdine, already saturated with ferric ions, does not deplete the medium of available iron. Similarly, the pyoverdine of *Pseudomonas tolaasii* decreases mycelial growth of the fungal phytopathogen, *Pythium ultimum* while the ferricpyoverdine does not (Meyer *et al.*, 1987). Chlamydospore germination and subsequent germ tube elongation of *Fusarium oxysporum* are inhibited with EDDHA [ethylene diaminedi(o-hydroxyphenylacetic acid)], a chelator with a high specific activity for the ferric ion, but not with Fe-EDDHA (Scher and Baker, 1982). These results suggest that certain siderophores are antagonistic through chelation of iron from the environment of the target pathogen. However, in other cases, a free-ligand may stimulate the germination and growth of a phytopathogen while the corresponding ferricpyoverdine has a neutral or toxic effect. For example, the ferricpyoverdine of *P. fluorescens* strain CHA0 inhibits endoconidial germination and radial growth of the soilborne fungus, *Thielaviopsis basicola*, while the corresponding pyoverdine does not (Ahl *et al.*, 1986). The ferricpyoverdine of CHA0 may inhibit *T. basicola* directly or indirectly, by making iron more available and thus more toxic. Iron-availability clearly plays a role in the growth and development of certain soilborne phytopathogens in culture, although no universal pattern describes the role of iron in these processes.

Evidence for a role of pyoverdines in biocontrol of soilborne diseases

Despite its relative abundance in soil, most iron is bound in insoluble complexes that are unavailable to microbial growth (Lindsay, 1979). Due to the limited availability of iron in rhizosphere and bulk soils, certain phytopathogens are subject to iron-competition by siderophore-producing antagonists. Iron competition is implicated in the biological control by *Pseudomonas fluorescens* and *Pseudomonas putida* of wilt diseases caused by *Fusarium oxysporum* (Scher and Baker, 1982) and of damping-off of cotton caused by *Pythium ultimum* (Loper, 1988). Iron competition also is implicated in plant growth promotion (Leong, 1986; Schippers *et al.*, 1987), which will be considered the biocontrol of minor soilborne pathogens in this discussion. Data suggesting the involvement of siderophores in biological control of soilborne plant pathogens have been obtained from experiments 1) evaluating the control achieved with applications of purified siderophores or synthetic iron-chelators, or 2) comparing the biocontrol activities of parental and derivative strains that are deficient in siderophore biosynthesis (Sid^-).

The addition of purified pseudobactin to soils planted with potatoes results in quantitatively similar levels of plant growth promotion as does inoculation with the producing strain, *P. fluorescens* B10, although no effect is observed with addition of ferric pseudobactin (Leong, 1986). Similarly, addition of EDDHA to soil suppresses Fusarium wilt symptoms, presumably due to inhibition of chlamydospore germination and germ tube elongation in rhizosphere and bulk soil (Simeoni, 1987). These experiments suggest that pseudobactin or specific synthetic iron chelators such as EDDHA can suppress certain soilborne plant diseases. However, they provide no evidence for the *in situ* production of pyoverdines in concentrations adequate for disease suppression.

Indirect approaches, comparing the biocontrol activity of Sid^+ parental and Sid^- derivative strains, are useful for evaluating the potential roles of siderophores in biocontrol of soilborne diseases. For example, derivatives of *P. fluorescens* and *P. putida* strains, deficient in pyoverdine production (Sid^-) and obtained following chemical, ultraviolet (Kloepper and Schroth, 1981) or Tn*5* (Bakker *et al.*, 1987; Schippers *et al.*, 1987) mutagenesis, do not promote plant growth as do parental strains. Several Sid^- Tn*5* mutants of *P. fluorescens* 3551 are less effective than the parental strain in biocontrol of a damping-off disease of cotton, caused by *Pythium ultimum* (Loper, 1988). These results suggest a role for pyoverdine production in the biocontrol of *P. ultimum* by strain 3551, and provide indirect evidence for pyoverdine production in the spermosphere. Assumptions and limitations of the mutagenesis approach for evaluating the role of pyoverdines in biocontrol have been reviewed recently (Loper, 1989).

Pyoverdines are not universally implicated in the biocontrol effected by rhizosphere pseudomonads, although by definition, all fluorescent pseudomonads produce these pigments. The relative contributions of pyoverdines and other metabolites to biocontrol activity have been compared in several systems. Pyoverdines contribute to, but do not account for, the biocontrol of take-all of wheat or black root rot of tobacco by *Pseudomonas* spp. (Ahl *et al.*, 1986; Hamdan *et al.*, 1988; Weller *et al.*, 1988). Plant pathogens also may exhibit differential sensitivities to pyoverdines of different Pseudomonas strains. For example, seed treatments with strains 3551 (Loper, 1988) or HV37a (Suslow *et al.*, 1986) control a damping-off disease of cotton caused by *Pythium utimum*. However, the pyoverdine of strain 3551 appears important in biocontrol activity while that of strain HV37a does not. The role of pyoverdines in the suppression of plant diseases by fluorescent pseudomonads appears to vary with the plant host, pathogen, and Pseudomonas strain evaluated.

Detection of siderophores in soil

Knowledge of the nature and concentrations of siderophores present in the rhizosphere and bulk soil is critical to an understanding of their roles in biological control. The types of hydroxamate siderophores produced by many soil microbes, especially soil fungi, are known (Waid, 1975). The soil concentrations of hydroxamate siderophores, estimated typically by bioassay of soil extracts with one or more indicator organisms, range from 0–240 μg/Kg soil in various experiments (Bossier *et al.*, 1988). Concentrations of siderophores are consistently greater in rhizosphere than in non-rhizosphere soils (Akers, 1983b; Nelson *et al.*, 1988). Only one hydroxamate siderophore, schizokinen, has been detected in soil by analytical chemical techniques (Akers, 1983a).

The bioassay indicator strains used for the detection of hydroxamate and catechol siderophores in soil are not useful for the detection of pyoverdines. However, Bakker *et al.* (1988) demonstrated indirectly the *in situ* production of pyoverdines by examining interactions of two Pseudomonas strains and their Sid^- derivatives in the potato rhizosphere. An *in situ* crossfeeding experiment between parental and Sid^- derivative strains demonstrated that a Sid^- mutant, which could utilize the pyoverdine produced by a coinoculated wildtype strain, established a greater population size in the potato rhizosphere than a Sid^- mutant that could not. Although these data do not provide a quantitative estimate of the concentrations of pyoverdines in the soil, they do suggest that fluorescent siderophores are produced in the potato rhizosphere in adequate concentrations to influence bacterial growth.

Inconsistent efficacy of biocontrol agents

Biocontrol strategies based on the use of rhizosphere antagonists have been successful in greenhouse and field experiments. However, despite notable successes, the field of biocontrol has offered little to modern agriculture in the form of effective products or practices. Variable efficacy of biocontrol agents among agricultural fields poses the greatest single obstacle to agronomic application of biocontrol strategies. This variability appears ubiquitous among biocontrol agents, including the rhizosphere *Pseudomonas* spp. that are the subject of this review. For example, statistically significant growth or yield responses were achieved with plant growth promoting *Pseudomonas* spp. strains in only 32 of 69 field trails in California in the late 1970s and early 1980s (Suslow, 1982). Inconsistent field efficacy of *Pseudomonas* spp. for the biocontrol of take-all of wheat in Washington state also has been observed (Weller, 1988). The synopsis of these field trials, which were exploratory in nature, may present an unduly pessimistic view of the performance of rhizosphere pseudomonads. Nevertheless, it is clear that field performance of these biocontrol agents is variable, sometimes in an unexplained or unpredictable manner. Theoretically, any physical, chemical or biological factor influencing the establishment of or expression of a critical phenotype by a biocontrol agent may affect the biocontrol activity of that agent. Similarly, factors influencing the sensitivity of the indigenous pathogen population to competition or antagonism by a bicontrol agent will affect the success of biocontrol. Such factors, which potentially influence pyoverdine-mediated biocontrol, are discussed below.

Variable sensitivities of indigenous pathogen populations to biocontrol

Theoretically, pathogens differing in siderophore production and utilization patterns may also differ in their sensitivities to iron-deprivation imposed by the presence of a non-utilized iron chelator. This phenomenon is exemplified by the differential sensitivities of two *Escherichia coli* stains to iron-regulated antibiosis by *Pseudomonas fluorescens* strain B10 (Fig. 1). *E. coli* strain AN194, which produces and utilizes the catechol siderophore, enterochelin (Ent^+), as an iron-scavenging compound, is not sensitive to iron-regulated antibiosis by *P. fluorescens* strain B10 (Kloepper *et al.*, 1990). However, strain AN193, which is deficient in enterochelin production (Ent^-), is sensitive to antibiosis. Further, the production and utilization of multiple siderophores may augment the number and types of habitats exploited by an organism. For example, the production of the hydroxamate siderophore, aerobactin, by clinical isolates of *E. coli* is highly correlated with virulence while

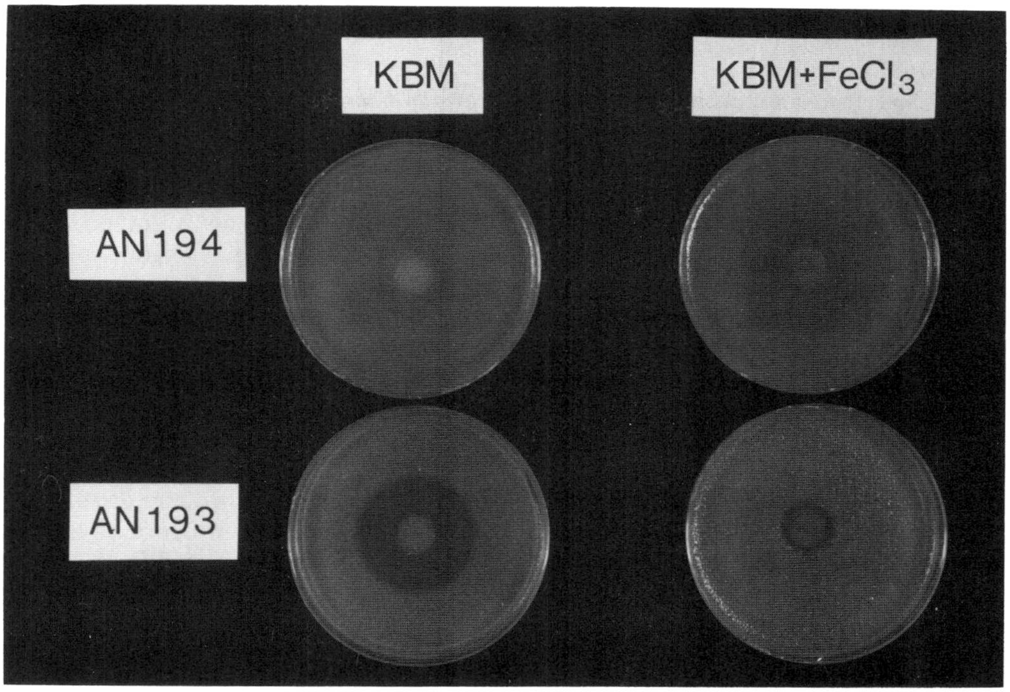

Fig. 1. Antagonism of *E. coli* by *Pseudomonas fluorescens* strain B10. *P. fluorescens* was grown for 48 h on King's medium B (colony in center of plates), killed with chloroform vapors and overlaid with 5 mL of molten King's medium B (King *et al.*, 1954) seeded with 10^6 colony forming units of the indicator, *E. coli* strain AN193 (Ent^-) or strain AN194 (Ent^+). The antagonism was reversed by the addition of $FeCl_3$ (0.1 m*M*) to the overlay medium.

enterochelin production is not (Williams, 1979). Although enterochelin has a greater affinity for iron than does aerobactin, its lability and slow diffusion rate presumably account for the relatively subtle role of enterochelin in virulence. Aerobactin, on the other hand, is expressed and excreted rapidly at an early stage in cell growth and effectively competes with other iron chelators, such as human transferrin, that are present in mammalian tissues (Bindereif and Neilands, 1985; Neilands, 1981). Theoretically, differential sensitivities of phytopathogens to siderophore-mediated antagonism by biocontrol agents also may be related to differential high-affinity iron uptake systems.

Siderophore production determines biocontrol activity of fluorescent pseudomonads against certain soilborne plant pathogens, yet little is known about the role of iron in the ecology or virulence of target pathogens. Although the siderophores produced by many phytopathogens are known (Neilands and Leong, 1986), their importance as virulence factors or as ecological determinants is unclear. Agrobactin, a hydroxamate siderophore produced by *Agrobacterium tumefaciens*, is not required for pathogenicity (Ong *et al.*, 1979). Comparison of Sid^- derivative and Sid^+ parental *Pseudomonas syringae* strains provided no evidence of a determinative role of pyoverdine production to the growth, survival, or pathogenicity of *P. syringae* on bean (Loper and Lindow, 1987) or cherry (Cody and Gross, 1987). However, the production of a catechol siderophore, chrysobactin, by *Erwinia chrysanthemi* is a virulence factor on african violet (*Saintpaulia ionantha*) (Enard *et al.*, 1988; Expert and Toussant, 1985; Persmark *et al.*, 1989). The ability of a phytopathogen to utilize, as a source of iron, siderophores produced by other soil microbes may improve its iron status in rhizosphere and bulk soil. Multiple siderophore receptors, for which the corresponding siderophore production phenotype could not be demonstrated, occur in *E. coli* (Braun *et al.*,

1987) and *E. herbicola* (Berner *et al.*, 1988). The possession of such receptors by soilborne phytopathogens could enhance their competitiveness for iron in a soil environment. However, the siderophore production and utilization systems of soilborne pathogens that are targets of iron-mediated biocontrol strategies and an ecological role for these systems are unknown.

Recently, a study was initiated to evaluate the role of siderophores in the sensitivity of *Erwinia carotovora* strains to biocontrol by *P. fluorescens* strains. *Erwinia carotovora* subsp. *carotovora* and subsp. *atroseptica* cause soft rot diseases of storage tissues of several crop plants, and seed piece decay and black leg disease of potato. Biocontrol of potato soft rot (Coyler and Mount, 1984; Kloepper, 1983) and seed piece decay (Xu and Gross, 1986a) is achieved by inoculation of seed pieces with specific *Pseudomonas* spp. prior to planting. Pyoverdines produced by the *Pseudomonas* spp. strains are implicated in the biocontrol of seed piece decay (Xu and Gross, 1986b). Of particular interest to the present discussion, *E. carotovora* strains differ in their sensitivities to biocontrol by various Pseudomonas strains (Xu and Gross, 1986b). *Erwinia carotovora* subsp. *carotovora* strain W3C105 is generally less sensitive to biocontrol than *E. c.* subsp. *atroseptica* strain W3C37 (Fig. 2). Strains W3C105, W3C37, and 21 other *E. carotovora* strains isolated from Oregon and Washington state produced catechol(s) when grown under iron limiting conditions in our laboratory. Preliminary results suggest that the catechol is similar functionally to enterochelin, a catechol siderophore common to many enteric bacteria. However, of the 23 *E. carotovora* strains evaluated, only W3C105, which was relatively insensitive to biocontrol by *Pseudomonas* spp. strains, produced a hydroxamate siderophore. The hydroxamate was identified as aerobactin by chemical analyses, including NMR spectroscopy (Ishimaru and Loper, unpublished results). It is clear that aerobactin is not strictly required for pathogenicity of *E. carotovora*, since only one of 23 tested strains produced this hydroxamate. However, aerobactin may play a role in the ecology of *E. carotovora* subsp. *carotovora* W3C105, including its sensitivity to siderophore-mediated biocontrol strategies.

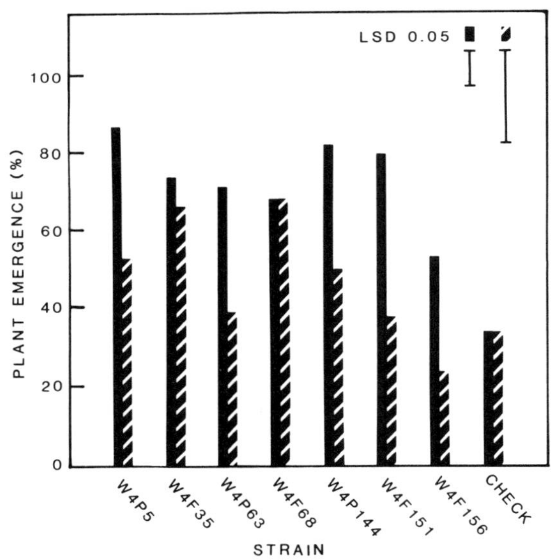

Fig. 2. Effects of antagonistic fluorescent pseudomonads on pre-emergence seed piece decay caused by *Erwinia carotovora* subsp. *atroseptica* strain W3C37 (solid bars) or *E. c.* subsp. *carotovora* strain W3C105 (diagonally shaded bars). Potato seed pieces were treated with *Pseudomonas* spp. (10^9 cfu per seed piece) and *E. carotovora* (10^6 cfu per seed piece) before planting in pots containing field soil. Check seed pieces were treated with *E. carotovora* only. Emergence data are means of five replicate pots (three seed pieces per pot) observed 2 wk after planting. Reprinted with permission from Xu and Gross, 1986.

Definition of an ecological role for aerobactin, achieved by comparing parental and aerobactin deficient derivative strains in competition experiments, is a subject of future experiments. To this end, we have cloned the genes encoding aerobactin biosythesis from *E. carotovora* subsp. *carotovora* W3C105 and have determined their structural and functional relationship to *E. coli* aerobactin biosynthesis genes. Introduction of specific mutations into cloned genes and subsequently into the genome of W3C105 will result in aerobactin-deficient derivatives that can be evaluated in ecological studies.

Variable production and utilization of siderophores

The production of pyoverdines by rhizosphere *Pseudomonas* spp. in culture is highly dependent on temperature, carbon and nitrogen source, pH, and concentrations of micronutrients, including but not limited to Fe^{3+} (Elad and Baker, 1985; reviewed by Loper and Schroth, 1986).

Table 1. Effect of strain and media composition on catechol production by *E. carotovora* subsp. *carotovora*

Strain	Media	DHBA equivalents[a] (μM)
E.c.c. W3C105	TMS[b]	50
	TMS + casamino acids[c]	35
	TMS + glucose[d]	37
	TMS + glucose + casamino acids	0
E.c.c. SR319	TMS	20
	TMS + casamino acids	22
	TMS + glucose	0
	TMS + glucose + casamino acids	0
E.c.c. CC106	TMS	20
	TMS + casamino acids	0
	TMS + glucose	20
	TMS + glucose + casamino acids	0

[a] dihydroxybenzoic acid (DBHA) equivalents determined by the Arnow reaction and adjusted for cell density (A_{510}/A_{620}).
[b] Tris-minimal-succinate medium (Simon and Tessman, 1963).
[c] deferrated casamino acids (0.3% w/v).
[d] glucose (0.5% w/v).

Physical and chemical factors also influence the siderophore production of a target pathogen. For example, catechol production of three *E. carotovora* strains were differentially influenced by nutritional components of a culture medium (Table 1). In soil, the availability of carbon substrates is a major factor limiting the production of siderophores (Bossier *et al.*, 1988). The efficacy of biocontrol with fluorescent pseudomonads is dependent on iron availability in the soil as mediated by soil pH (Elad and Baker, 1985), and the presence (vermiculite) or absence (illite) of specific clay minerals (Keel *et al.*, 1988). Iron-competition between biocontrol strains and target pathogens must be evaluated in the context of the multiple interacting factors influencing the *in situ* siderophore production and utilization patterns of both organisms.

Conclusions

Considering the variable and complex physical, chemical, and biological environment characterizing the plant rhizosphere, it is not surprising that a given biocontrol agent is not universally successful. Nevertheless, an understanding of how and why biocontrol succeeds, and more importantly, why it fails, is critical to the future agronomic application of biocontrol strategies. This understanding will have predictive value, facilitating an identification of field conditions under which a biocontrol strategy will succeed or fail. The composition of indigenous pathogen populations, with respect to sensitivity or resistance to biocontrol, may be one biological factor determining the success or failure of biocontrol strategies.

The presence of resistant individuals within a sensitive pathogen population limits the success of control strategies based on chemical application or plant host resistance, but has received little attention in biocontrol strategies. One notable exception is the variable sensitivity observed among *Agrobacterium tumefaciens* strains to antagonism by *Agrobacterium radiobacter* strain K84. The prevalence in an agricultural field of *A. tumefaciens* strains with sensitivity to strain K84 determines the success of strain K84 as a biocontrol agent (Moore and Warren, 1979). The present discussion expands our thinking of variable sensitivities of pathogen populations beyond *A. radiobacter* to other biocontrol agents. The intrinsic high-affinity iron acquisition system(s) of a target pathogen may determine its sensitivity to biocontrol mediated by microbial siderophores.

References

Ahl P, Voisard C and Defago G 1986 Iron bound-siderophores, cyanic acid and antibiotics involved in suppression of *Thielaviopsis basicola* by a *Pseudomonas fluorescens* strain. J. Phytopathology 116, 121–134.

Akers H A 1983a Isolation of the siderophore schizokinen from soil of rice fields. Appl. Environ. Microbiol. 45, 1704–1706.

Akers H A 1983b Multiple hydroxamic acid microbial iron chelators (siderophores) in soils. Soil. Sci. 135, 156–159.

Bakker P A H M, Weisbeek P J and Schippers B 1988 Siderophore production by plant growth-promoting *Pseudomonas* spp. J. Plant Nutr. 11, 925–933.

Bakker P A H M, Bakker A W, Marugg J D, Weisbeek P J and Schippers B 1987 Bioassay for studying the role of siderophores in potato growth stimulation by *Pseudomonas* spp. in short potato rotations. Soil. Biol. Biochem. 19, 443–449.

Berner I, Konetschny-Rapp, Jung G and Winkelmann G 1988 Characterization of ferrioxamine E as the principal siderophore of *Erwinia herbicola* (*Enterobacter agglomerans*). Biol. Metals 1, 51–56.

Bindereif A and Neilands J B 1985 Aerobactin genes in clinical isolates of *Escherichia coli*. J. Bacteriol. 161, 727–735.

Bossier P, Hofte M and Verstraete W 1988 Ecological significance of siderophores in soil. Adv. Microb. Ecol. 10, 385–414.

Braun V, Hantke K, Eick-Helmerich K, Koster W, Prebler U, Sauer M, Schaffer S, Schoffler H, Staudenmaier H and Zimmerman L 1987 Iron transport systems in *Escherichia coli*. *In* Iron Transport systems in Microbes, Plants and Animals. Eds. G. Winkelmann, D. Van der Helm, J B Neilands. pp 35–51. VCH Chemie, Weinheim, FRG.

Buyer J S and Leong J 1986 Iron transport-mediated antagonism between plant growth-promoting and plant-deleterious *Pseudomonas* strains. J. Biol. Chem. 261, 791–794.

Cody Y S and Gross D C 1987 Outer membrane protein mediating iron uptake via pyoverdin$_{pss}$, the fluorescent siderophore produced by *Pseudomonas syringae* pv. *syringae*. J. Bacteriol. 169, 2207–2214.

Coyler P D and Mount M S 1984 Bacterization of potatoes with *Pseudomonas putida* and its influence on postharvest soft rot diseases. Plant Disease 68, 703–706.

Demange P, Wendenbaum S, Bateman A, Dell A and Abdallah M A 1987 Bacterial siderophores: Structure and physicochemical properties of pyoverdins and related compounds. *In* Iron Transport in Microbes, Plants and Animals. Eds. G Winkleman, D Van der Helm, and J B Neilands. pp 167–187. VCH Chemie, Weinheim, FRG.

Elad Y and Baker R 1985 Influence of trace amounts of cations and siderophore-producing pseudomonads on chlamydospore germination of *Fusarium oxysporum*. Phytopathology 75, 1047–1052.

Enard C, Diolez A and Expert D 1988 Systemic virulence of *Erwinia chrysanthemi* 3937 requires a functional iron assimilation system. J. Bacteriol. 170, 2419–2426.

Expert D and Toussaint A 1985 Bacteriocin-resistant mutants of *Erwinia chrysanthemi*: Possible involvement of iron acquisition in phytopathogenicity. J. Bacteriol. 163, 221–227.

Hamdan H, Thomashow L S and Weller D M 1988 Relative importance of fluorescent siderophore and phenazine antibiotic by *Pseudomonas fluorescens* 2-79 in suppression of take-all. Phytopathology 78, 1522 (abstr).

Keel C, Voisard C, Berling C H, Kahr G and Defago G 1988 Influence of vermiculite, illite and iron on suppression of tobacco black root rot by *Pseudomonas fluorescens* strain CHA0. *In* Abstracts of Papers, Fifth International Congress of Plant Pathology, August 20–27, Kyoto, Japan (abstr.).

King E O, Ward M K and Raney D E 1954 Two simple media for the demonstration of pyocyanin and fluorescin. J. Lab. Clin. Med. 44, 301–307.

Kloepper J W 1983 Effect of seed piece inoculation with plant growth-promoting rhizobacteria on populations of *Erwinia carotovora* on potato roots and in daughter tubers. Phytopathology 73, 217–219.

Kloepper J W, Leong J, Teintze M and Schroth M N 1980 Enhanced plant growth by siderophores produced by plant growth-promoting rhizobacteria. Nature 286, 885–886.

Kloepper J W and Schroth M N 1981 Relationship of *in vitro* antibiosis of plant growth-promoting rhizobacteria to plant growth and displacement of root microflora. Phytopathology 71, 1020–1024.

Knusel F, Schiess B and Zimmermann W 1969 The influence exerted by sideromycins on poly-U-directed incorporation of phenylalanine in the S-30 fraction of *Staphylococcus aureus*. Arch. Mikrobiol. 68, 99–106.

Leong J 1986 Siderophores: Their biochemistry and possible role in the biocontrol of plant pathogens. Annu. Rev. Phytopath. 24, 187–209.

Lindsay W L 1979 Chemical Equilibria in Soils. Wiley, New York.

Loper J E 1988 Role of fluorescent siderophore production in biocontrol of *Pythium ultimum* by a *Pseudomonas fluorescens* strain. Phytopathology 78, 166–172.

Loper J E 1989 Molecular and biochemical bases for activities of biological control agents: The role of siderophores. *In* New Directions in Biological Control. pp 735–747. UCLA Symposia Vol. 112. Eds. R. Baker and P. Dunn. Alan R. Liss, Inc., New York.

Loper J E and Schroth M N 1986 Importance of siderophores in microbial interactions in the rhizosphere. *In* Iron, Siderophores and Plant Diseases. Ed. T R Swinburne. pp 85–98. Plenum Publishing Corporation, New York.

Loper J E and Lindow S E 1987 Lack of evidence for *in situ* fluorescent pigment production by *Pseudomonas syringae* pv. *syringae* on bean leaf surfaces. Phytopathology 77, 1449–1454.

Magazin M D, Moores J C and Leong J 1986 Cloning of the gene coding for the outer membrane receptor protein for ferric pseudobactin, a siderophore from a plant growth-promoting *Pseudomonas* strain. J. Biol. Chem. 261, 795–799.

Meyer J-M, Halle F, Hohnadel D, Lemanceau P and Ratefiarivelo H 1987 Siderophores of *Pseudomonas*-biological properties. *In* Iron Transport in Microbes, Plants and Animals. Eds. G. Winkleman, D. Van der Helm and J B Neilands. pp 189–205. VCH Chemie, Weinheim, FRG.

Moore L W and Warren G 1979 *Agrobacterium radiobacter* strain 84 and biological control of crown gall. Annu. Rev. Phytopathol. 17, 163–179.

Neilands J B 1981 Microbial iron compounds. Annu. Rev. Biochem. 50: 715–731.

Neilands J B and Leong S A 1986 Siderophores in relation to plant growth and disease. Annu. Rev. Plant Physiol. 37, 187–208.

Nelson M, Cooper C R, Crowley D E, Reid C P P and Szaniszlo P J 1988 An *Escherichia coli* bioassay of individual siderophores in soil. J. Plant. Nutr. 11, 915–924.

Ong S A, Peterson T and Neilands J B 1979 Agrobactin, a siderophore from *Agrobacterium tumefaciens*. J. Biol. Chem. 254, 1860–1865.

Persmark M, Expert D and Neilands J B 1989 Isolation, characterization and synthesis of chrysobactin, a compound with siderophore activity from *Erwinia chrysanthemi*. J. Biol. Chem. 264, 3187–3193.

Scher F M and Baker R 1982 Effect of *Pseudomonas putida* and a synthetic iron chelator on induction of soil suppressiveness to Fusarium wilt pathogens. Phytopathology 72, 1567–1573.

Schippers B, Bakker A W and Bakker P A H M 1987 Interactions of deleterious and beneficial rhizosphere microorganisms and the effect of cropping practices. Annu. Rev. Phytopath. 25, 339–358.

Simeoni L A, Lindsay W L and Baker R 1987 Critical iron level associated with biological control of Fusarium wilt. Phytopathology 77, 1057–1061.

Simon E H and Tessman I 1963 Thymidine-requiring mutants of phage T4. Proceedings of the National Academy of Sciences of the United States of America 50, 526–532.

Suslow T V 1982 Role of root colonizing bacteria in plant growth. *In* Phytopathogenic Prokaryotes, Vol. 1. Eds. M S Mount and G H Lacy. pp 187–223. Academic Press, Inc., London.

Suslow T V, Matsubara D and Davies M 1986 Role of ferric antagonized inhibition (FAI) in root colonization and fungal inhibition by a strain of *Pseudomonas fluorescens*. Phytopathology 76, 1069 (abstr.).

Teintze M, Hossain M B, Barnes C L, Leong J and van der Helm D 1981 Structure of ferric pseudobactin, a siderophore from a plant growth promoting *Pseudomonas*. Biochemistry 20, 6446–6457.

Waid J S 1975 Hydroxamic acids in soils systems. *In* Soil Biochemistry. Eds. E A Paul and A D McLaren. pp 65–101. Marcel Dekker, New York.

Weller D M 1988 Biological control of soilborne plant pathogens in the rhizosphere with bacteria. Annu. Rev. Phytopathol. 26, 379–407.

Weller D M, Howie W J and Cook R J 1988 Relationship between *in vitro* inhibition of *Gaeumannomyces graminis* var. *tritici* and suppression of take-all of wheat by fluorescent pseudomonads. Pythopathology 78, 1094–1100.

Williams P H 1979 Novel iron uptake system specified by ColV plasmids: An important component in the virulence of invasive strains of *Escherichia coli*. Infect. Immun. 26, 925–932.

Xu G-W and Gross D C 1986a Field evaluations of the interactions among fluorescent pseudomonads, *Erwinia carotovora*, and potato yields. Phytopathology 76, 423–430.

Xu G-W and Gross D C 1986b Selection of fluorescent pseudomonads antagonistic to *Erwinia carotovora* and suppressive of potato seed piece decay. Phytopathology 76, 414–422.

D. L. Keister and P. B. Cregan (Eds.), The rhizosphere and plant growth, 263–269.

PLSO BARC26

Rhizosphere interactions and siderophores

JEFFREY S. BUYER and LAWRENCE J. SIKORA
Soil Microbial Systems Laboratory, Natural Resources Institute, USDA-ARS, Beltsville, MD 20705, USA

Key words: biocontrol, Gaeumannomyces, iron, monoclonal antibodies, Pseudomonas, siderophores

Abstract

Certain plant growth-promoting pseudomonads inhibit deleterious and pathogenic rhizosphere bacteria and fungi by producing siderophores. Properties of a siderophore transport system which might provide a competitive advantage under iron stress conditions include ability to utilize other organisms' siderophores, higher Fe(III) stability constant, faster kinetics of dissolution of Fe(III) minerals, more efficient transport system, and resistance to degradation. In order to determine the concentration and localization of siderophores in the rhizosphere monoclonal antibodies (Mabs) to ferric pseudobactin, the siderophore of *Pseudomonas putida* B10, have been developed. Several Mabs cross reacted differently with various pseudobactins. A growth medium has been developed for the study for siderophore-mediated rhizosphere interactions in the laboratory.

Introduction

Siderophores are low molecular weight iron-chelating agents produced by virtually all bacteria and fungi under iron-limiting conditions (Neilands, 1981). The effect of siderophores on rhizosphere interactions and biocontrol of plant pathogens was first demonstrated in 1980 (Kloepper *et al.*, 1980a; Kloepper *et al.*, 1980b). The evidence for siderophore-mediated biocontrol of deleterious and pathogenic microorganisms has been reviewed (Leong, 1986; Neilands and Leong, 1986; Schippers *et al.*, 1987). In this article we present first a conceptual model for siderophore interactions which includes factors often overlooked in siderophore-mediated rhizosphere interactions and suggest fruitful areas for further research. We then present some results from our current research on the use of monoclonal antibodies to study rhizosphere siderophores and on a growth medium which may be useful for studying siderophore-mediated rhizosphere interactions in the laboratory.

Conceptual model

Introduction

The model for siderophore-mediated rhizosphere interactions is limited to two microorganisms, each producing their own siderophore, but could be expanded to include more species (Fig. 1). Each siderophore (Sid) chelates Fe(III), forming the ferric siderophore (FeSid), which is taken up by the cells. The dotted lines represent ligand exchange by which one Sid displaces another Sid chelating an Fe^{3+} ion. The dashed lines indicate transport by one organism of the FeSid produced by the other organism. Degradation pathways, including biological and chemical mechanisms, are represented by the arrows at the bottom of the figure. Degradation undoubtedly would also occur with the free ligand but for convenience was drawn in only for the iron complex.

Under iron-limiting conditions the only known mechanism for microbial iron uptake is via a FeSid transport system. Therefore, the limiting

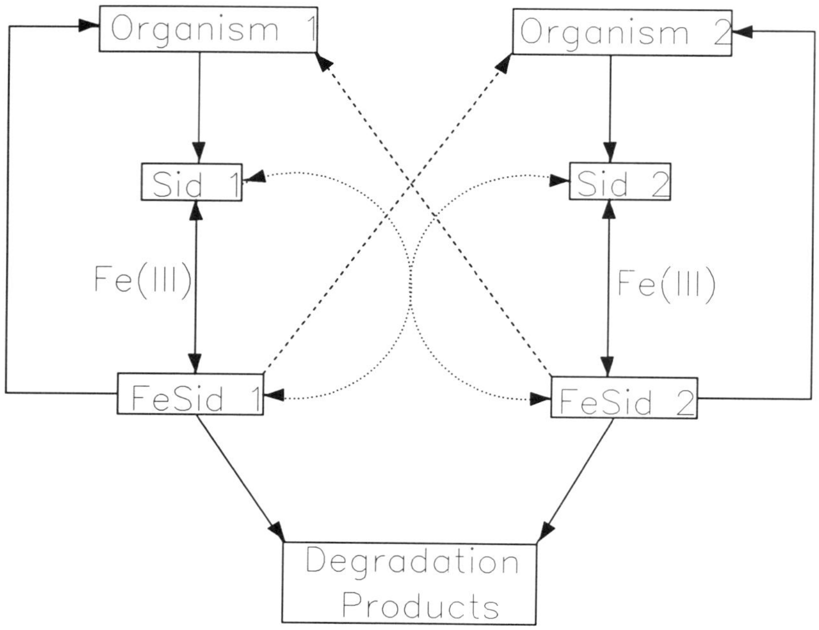

Fig. 1. Model of siderophore-mediated interactions.

substrate for growth should be FeSid, and any condition which lowers the concentration of FeSid should reduce the growth rate of the organism. We will attempt to outline the various aspects of siderophore-mediated iron transport that could result in interactions, such as antibiosis, as well as those factors giving one organism a competitive advantage under iron-limiting conditions over another organism without a direct interaction.

Formation and equilibrium constants

The literature on siderophore-mediated biocontrol is replete with unproven assumptions concerning the role of Fe(III) formation constants in competition for iron. The formation constant for FeSid may be defined as follows:

$$Fe^{3+} + Sid = FeSid \quad K_f = [FeSid]/[Fe^{3+}][Sid] \tag{1}$$

However, all Sids lose protons (H^+) upon chelation of Fe^{3+}, so the equilibrium constant K is defined as in equation 2.

$$Fe^{3+} + H_nSid = FeSid + nH^+$$
$$K = [FeSid][H^+]^n/[Fe^{3+}][H_nSid] \tag{2}$$

K is therefore pH dependent, with the degree of dependence controlled by n and by the acidity (pKa) of the protons involved. With 6 protons lost upon chelation enterobactin is highly pH dependent, such that the ratio $[FeEnt^{3-}]/[Fe^{3+}][H_6Ent]$ equals 10^{26} at pH 6.0 (Hider, 1984). While this represents extremely strong chelation, it is much less than the formation constant (Eq. 1) of 10^{52}. Trihydroxamate siderophores will be much less pH dependent because they lose only three protons upon chelation. Therefore, stability and formation constants cannot be used as a measure of the ability of Sids to compete with each other except at very high pH, where $[H^+]$ is very low. Instead, equilibrium constants must be measured at various pH levels, or K_f and pKa's must be determined in order to calculate the actual stability of the complex at any particular pH. For rhizosphere interactions the rhizosphere pH must be used.

Equilibrium constants and kinetics

Consider first the case of two Sids competing for a pool of Fe(III). Because of the high affinity of Sids for Fe(III) these reactions go virtually to completion. In the absence of ligand exchange each Sid will chelate Fe(III) independently of the other, at a rate determined by the kinetics of each reaction, until all the Fe(III) is chelated or until all Sid is converted to FeSid. Therefore competition for iron will take place according to the kinetics of dissolution and chelation of Fe(III), and by the quantity of each Sid initially present, rather than by equilibria.

Now consider the effect of ligand exchange:

$$FeSid_1 + Sid_2 = Sid_1 + FeSid_2 \qquad (3)$$

If this reaction is allowed to proceed to equilibrium, and if the Sids are present in equal concentration, the ratio of $FeSid_1$ to $FeSid_2$ will be directly proportional to the ratio of K_1 to K_2. Therefore, ligand exchange will tend to promote competition on the basis of FeSid equilibria rather than kinetics. This will only be effective if ligand exchange is fast compared to the rates of dissolution and chelation. The half-life for exchange between two hydroxamate siderophores was 220 hours at pH 7.4 (Tufano and Raymond, 1981), which might well be too slow to achieve equilibrium.

The relative concentrations of the two Sids must be known in order to apply equilibria calculations. If K_1 is 10 times greater than K_2, but Sid_2 is present at 10 times the concentration of Sid_1, the FeSids will be present in equal concentrations at equilibrium.

We conclude that the concentrations of Sids produced by each organism and the kinetics of FeSid formation and ligand exchange are as important as the relative equilibrium constants. We should also remember that the rhizosphere is heterogeneous both in time and space, with steep gradients in pH, oxygen, and substrate availability. It would be naive then to assume that a slow chemical reaction such as ligand exchange between hydroxamate siderophores will reach equilibrium under these conditions.

Siderophore transport

Certain properties of the FeSid transport system may contribute to competition for iron. Iron uptake may be characterized by the Michaelis-Menten parameters K_M and V_{max}. An organism with a more efficient system, *i.e.* lower K_M and higher V_{max}, may be expected to compete more effectively at low FeSid concentrations. Reported values of K_M and V_{max} are in the range 0.1–0.23 μM and 2-140 pmol min^{-1} mg^{-1} dry weight, respectively (Braun and Winkelmann, 1987; Ecker *et al.*, 1986; Frost and Rosenberg, 1973; de Weger *et al.*, 1988).

Furthermore, many bacteria have been shown to transport FeSids produced by other organisms. If organism 1 can utilize the iron from $FeSid_2$ then clearly Sid_2 will never induce iron starvation of organism 1, regardless of differences in equilibrium constants or kinetics of dissolution and chelation. This mechanism has been shown to be operative in antibiosis between rhizosphere pseudomonads. Using bioassays iron-inhibited antagonism was demonstrated to be due to the inability of one pseudomonad to use another pseudomonad's siderophore for iron transport, while pseudomonads which were able to use another pseudomonad's siderophore were not inhibited (Buyer and Leong, 1986). These results have been confirmed by another laboratory working with different pseudomonads (de Weger *et al.*, 1988), suggesting this mechanism of antibiosis to be a widespread phenomenon.

This argument assumes that a resistant organism uses its own siderophore and the exogenous siderophore at similar rates. If, due to differences in K_M or V_{max}, organism 1 used FeSid2 at a lower rate than $FeSid_1$ (Hohnadel and Meyer, 1988), growth of organism 1 might still be inhibited even though plate bioassays indicated that organism 1 utilized $FeSid_2$ for iron transport. A combination of antibiosis studies and characterization of transport rates will be required to resolve this question.

Antagonism is not the only possible result of this mechanism. If a biocontrol pseudomonad is introduced into a rhizosphere containing other pseudomonads which are possibly deleterious or pathogenic and which are able to utilize the biocontrol agent's siderophore for iron transport, these other pseudomonads might well increase in population because the concentration of FeSid available to them would increase. This could result in decreased plant growth and crop yield.

Finally, organisms which require less iron for growth and metabolism will be better suited to a low-iron environment. We know of no studies concerning the iron content or requirements of rhizosphere microorganisms.

Conclusions

It has been frequently assumed that pseudomonads inhibit pathogenic fungi under iron-limiting conditions because pseudobactin-type siderophores have higher affinity for Fe(III) than fungal siderophores. This assumption has not been substantiated by data on the equilibrium or even stability constants of the particular siderophores involved. Indeed, certain trihydroxamate siderophores such as Ferrichrome A and Ferrioxamine E (Carrano *et al.*, 1979) have K_f values as great or greater than those reported for siderophores isolated from fluorescent pseudomonads (Meyer and Abdallah, 1978; Torres *et al.*, 1986). We have attempted to demonstrate that other factors, including kinetics, concentration of siderophores, and FeSid transport systems may be more significant than equilibria. There is an obvious need for further research in this area.

Monoclonal antibodies to ferric pseudobactin

Monoclonal antibodies (Mabs) have been used extensively for the detection of organisms and components of organisms since the introduction of the method by Kohler and Milstein (1975). The specificity of the immunological reaction and the virtually unlimited quantities of the antibodies combine to provide a detection mechanism which can be used both for research and commercial purposes. Mabs have been used to detect bacterial infection of plants (De Boer *et al.*, 1988), viruses (Underwood, 1985), fungi (Ouellette and Benhamou, 1987; Wright *et al.*, 1986), A antigens of gram negative bacteria (Quinn *et al.*, 1984), and structural proteins (Mutharia and Hancock, 1984). Antigens that are too small to elicit an immune response alone are attached to a larger molecule and screens are used to isolate the Mab which reacts specifically with the small antigen. The detection limit of a Mab system depends on several factors including type of assay and affinity of the Mab for the antigen.

The detection of siderophores has most often involved the use of iron deficient media and growth rates of auxotrophic mutants. Most of the assays using microorganisms test for the presence of a class of siderophores, *i.e.* hydroxamates or catechols as opposed to a specific siderophore. Few assays detect a particular siderophore partly because the microbial transport systems are not specific enough and the structure of many siderophores are unknown. Powell *et al.* (1980) and Bossier and Verstraete (1986) used a siderophore-auxotrophic bacterium, Arthrobacter JG-9, in a bioassay to determine the relative quantities of hydroxamate siderophores in soil. The detection limit was 5 μg in ferrioxamine B equivalents Kg^{-1} soil. Powell *et al.* (1983) used an enterobactin defective *Escherichia coli* K-12 strain to detect ferrichrome-type siderophores in soil. The sensitivity in a 1:1 soil-water extract was approximately 78 n*M*. Nelson *et al.* (1988) used a panel of *E. coli* auxotrophic mutants to determine 4 types of siderophores in bulk soil and in rhizosphere soil. The types were a, enterobactin, b, ferrichromes, c, coprogen, ferrioxamines, and rhodotorulic acid, and d, unidentified siderophores. The detection limit was in the range of picomoles g^{-1} soil, using one hundred g of soil to perform the assay.

Because Mabs have greater potential to detect specific siderophores at lower concentrations than bioassays we used Mab technology similar to Oi and Herzenberg (1980) to raise Mabs reactive to ferric pseudobactin produced by *Pseudomonas putida* B10 (Teintze *et al.*, 1981). Chromic pseudobactin was attached to bovine serum albumin via a linker and injected into Balb/c mice over a 4 month period. Mouse spleen cells were fused with mouse myeloma cell line P3×63-AG8.653 at a ratio of 4:1 in the presence of 44% polyethylene glycol. The resulting hybridomas were grown in RPMI-1640 media with HAT selection for the first 14 days in 96 well microtiter plates at 37°C in a 5% CO_2 atmosphere. After 14 days the hybridomas were grown in the same medium without aminopterin. Screening was performed by coating 96 well microtiter plates with chromic pseudobactin linked to KLH. ELISA with alkaline phosphatase-conjugated anti-mouse immunoglobulin was used for screening and isotyping.

Other ferric siderophores were obtained by growing other pseudomonads in low-iron media and amending the supernatant with iron. Cross-reactivity was tested by preincubating the iron-amended culture supernatant for 1 hour with Mabs followed by overnight incubation of the mixture in chromic pseudobactin coated plates. ELISA was performed as described above. Pseudomonads were chosen that produced siderophores of known structure.

Three Mabs were raised that reacted with ferric pseudobactin (Table 1). All three Mabs were IgG1. Detection limits ranged from 10^{-10} to 10^{-12} moles of ferric pseudobactin. Several other ferric pseudobactins were tested to determine cross-reactivity. The results (Table 1) suggest that none of the Mabs were completely specific for ferric pseudobactin and that cross-reactivity to the ferric siderophore from *P. aeruginosa* 15692 was particularly high. The common chromophore was not the epitope or all siderophores would have reacted. Cross-reactivity may have been due to similarities in amino acid sequence and stereochemistry or to the stereochemistry of the ligand around the iron center.

These Mabs may prove useful for the detection of ferric pseudobactin in rhizosphere samples. We are currently characterizing the cross-reactivity of other siderophores with these Mabs and purifying other Mabs to ferric pseudobactin in order to obtain a more specific assay.

Table 1. Cross-reactivity of Mabs to ferric pseudobactins

Strain	5D4-C3[a]	4B5-E6	4B5-E8
P. putida B10[b]	+	+	+
P. fluorescens-putida 7SR1[c]	–	–	–
P. fluorescens-putida A214[d]	–	–	–
P. aeruginosa ATCC 15692[e]	+	P	+
P. chlororaphis ATCC 9446[e]	P	–	P
P. putida ATCC 12633[e]	–	–	–
P. fluorescens ATCC 13525[e]	–	–	–
P. fluorescens ATCC 17400[e]	P	–	P

[a] +, reactivity similar to B10; P, partially reactive; –, not reactive.
[b] Teintze *et al.*, 1981; [c] Yang and Leong, 1984; [d] Buyer *et al.*, 1986; [e] Demange *et al.*, 1987.

A growth medium for studying siderophore-mediated interactions

We have recently introduced Rhizosphere Medium (RSM) (Buyer *et al.*, 1989) as a first step towards modeling rhizosphere interactions in the laboratory. The inorganic composition of RSM is based on soil solution so that the metabolism and physiology of the organisms grown might be more similar to that in the rhizosphere. Sucrose and casamino acids were chosen as carbon and nitrogen sources which support rapid growth and siderophore production by a wide variety of bacteria and fungi.

RSM is sufficiently defined to permit the use of metal chelate buffers (Chaney and Bell, 1987). In these systems, metals and chelating agents are combined in suitable proportions such that the free metal activities ($pM = -\log(M^{n+})$) may be controlled and varied over a large range. The computer program Geochem-PC is used to calculate the pM values. For studies on siderophores it is desirable to control the free ferric ion activity. By using metal chelate buffers we can fix pFe at the desired level without starving the organisms for other metals, which is the inevitable result of simply adding large amounts of strong chelating agents. Complex media such as King's B and potato dextrose agar would not allow for the pM calculations. Schwyn and Neilands (1987) introduced a colorimetric test utilizing Chrome Azurol S (CAS) for the production of siderophores in plates or liquid media. Certain complex media such as King's B and potato dextrose agar interfere with this assay. RSM-CAS plates were blue, indicating little or no interference by the casamino acids. *G. graminis* was remarkable for the lack of siderophore production on these plates when compared to a number of other fungi including three Fusarium and two Ustilago species (Buyer *et al.*, 1989). This may explain the greater sensitivity of *G. graminis* to pFe^{3+} compared to the three Fusarium species when using metal chelate buffers as described above.

Variations in media components are known to

result in either production of different siderophores (Teintze and Leong, 1981) or in production of different amounts of siderophore. These may reflect contamination of media components by iron, variations in the iron status of the organism, regulation of the siderophore biosynthesis genes, or alterations in physiology and metabolism due to the carbon or nitrogen source. The latter suggests that the carbon and nitrogen sources available to biocontrol agents and rhizosphere pathogens might have an important influence on siderophore-mediated antibiosis and the resulting biocontrol of plant disease.

We have prepared a chemically defined minimal medium based on RSM by substituting NH_4Cl for casamino acids. Various carbon sources were substituted for sucrose and antibiosis between *P. putida* B10 and *G. graminis* was observed (Buyer, Sikora, and Kratzke, submitted for publication). Antibiosis was observed only when certain carbon sources were used and not with others. These results suggest that the carbon sources present in the rhizosphere may be crucial to biocontrol of *G. graminis* by *P. putida* B10.

Because the inorganic composition of RSM is based on that of the rhizosphere, it may be a first step towards the study of the chemistry and biology of the rhizosphere under well defined conditions. RSM may prove superior to other media for such purposes as screening potential biocontrol agents, producing and detecting siderophores, and examining the effects of metals on microbial growth. The most important factor in biocontrol by siderophores is undoubtedly the availability of iron, since this regulates siderophore production. Our laboratory model suggests that the carbon and nitrogen sources available to the biocontrol agent may also be significant.

References

Bossier P and Verstraete W 1986 Detection of siderophores in soil by a direct bioassay. Soil Biol. Biochem. 18, 481–486.

Braun V and Winkelmann G 1987 Microbial iron transport structure and function of siderophores. Progress in Clinical Biochemistry and Medicine 5, 67–99.

Buyer J S and Leong J 1986 Iron transport-mediated antagonism between plant growth-promoting and plant-deleterious *Pseudomonas* strains. J. Biol. Chem. 261, 791–794.

Buyer J S, Wright J M and Leong J 1986 Structure of Pseudobactin A214, a siderophore from a bean-deleterious *Pseudomonas*. Biochemistry 25, 5492–5499.

Buyer J S, Sikora L J and Chaney R L 1989 A New Growth Medium for the Study of Siderophore-Mediated Interactions. Biol. Fertil. Soils 8: 97–101.

Carrano C J, Cooper S R and Raymond K N 1979 Coordination Chemistry of Microbial Iron Transport Compounds. 11. Solution Equilibria and Electrochemistry of Ferric Rhodotorulate Complexes. J. Am. Chem. Soc. 101, 599–604.

Chaney R L and Bell P F 1987 Complexity of iron nutrition: Lessons for plant-soil interaction research. J. Plant Nutr. 10, 963–994.

De Boer S H, Wieczorek A and Karnmer H 1988 An ELISA test for bacterial ring rot of potato with a new monoclonal antibody. Plant Disease 72, 874-878.

Demange P, Wendenbaum S, Bateman A, Dell A and Abdallah M 1987 Bacterial siderophores: structure of pyoverdins and related compounds. *In* Iron Transport in Microbes, Plants and Animals. Eds. G Winkelmann, D van der Helm and J B Neilands. pp 167–187. V. C. H., Weinheim, Federal Republic of Germany.

Ecker D J, Matzanke B F and Raymond K N 1986 Recognition and transport of ferric enterobactin in *Escherichia coli*. J. Bacteriol. 167, 666–673.

Frost G E and Rosenberg H 1973 The inducible citrate-dependent iron transport system in *Escherichia coli* K-12. Biochim. Biophys. Acta 330, 90–101.

Hider R C 1984 Siderophore mediated absorption of iron. Structure and Bonding 58, 25–87.

Hohnadel D and Meyer J 1988 Specificity of pyoverdine-mediated iron uptake among fluorescent *Pseudomonas* strains. J. Bact. 170, 4865–4873.

Kloepper J W, Leong J, Teintze M and Schroth M N 1980a *Pseudomonas* siderophores: a mechanism explaining disease-suppressive soils. Current Microbiology 4, 317–320.

Kloepper J W, Leong J, Teintze M and Schroth M N 1980b Enhanced plant growth by siderophores produced by plant growth-promoting rhizobacteria. Nature 286, 885–886.

Kohler G and Milstein C 1975 Continuous cultures of fused cells secreting antibody of predefined specificity. Nature 256, 495–497.

Leong J 1986 Siderophores: their biochemistry and possible role in the biocontrol of plant pathogens. Ann. Rev. Phytopath. 24, 187–209.

Mutharia L M and Hancock R E W 1984 Surface localization of *Pseudomonas aeruginosa* outer membrane porin protein F by using monoclonal antibodies. Infect. Immun. 42, 1027–1033.

Meyer J M and Abdallah M A 1978 The fluorescent pigment of *Pseudomonas fluorescens*: biosynthesis, purification and physicochemical properties. J. Gen. Microbiol. 107, 319–328.

Neilands J B 1981 Microbial iron compounds. Annu. Rev. Biochem. 50, 715–731.

Neilands J B and Leong S A 1986 Siderophores in relation to plant growth and disease. Annu. Rev. Plant Physiol. 37, 187–208.

Nelson M, Cooper C R, Crowley D E, Reid C P P and Szaniszlo P J 1988 An *Escherichia coli* bioassay of individual siderophores in soil. J. Plant Nutr. 11, 915–924.

Oi V T and Herzenberg L A 1980 Immunoglobulin-producing hybridoma cell lines. *In* Selected Methods in Cellular Immunology. Eds. B B Mishell and S M Shiigii. pp 351–372. Freeman, San Francisco.

Ouellette G B and Benhamou W 1987 Use of monoclonal antibodies to detect molecules of fungal plant pathogens Can. J. Plant Path. 9, 167–176.

Powell P E, Cline G R, Reid C P P and Szaniszlo P J 1980 Occurrence of hydroxamate siderophores in soil. Nature 287, 833–834.

Powell P E, Szaniszlo P J and Reid C P P 1983 Confirmation of occurrence of hydroxamate siderophores in soil by a novel *Escherichia coli* bioassay. Appl. Environ. Microbiol. 46, 1080–1083.

Quinn R, Campbell A M and Phillips A P 1984 A monoclonal antibody specific for the A antigen of *Brucella* spp. J. Gen. Microbiol. 130, 2285–2289.

Schippers B, Bakker A W and Bakker P A H M 1987 Interactions of deleterious and beneficial rhizosphere microorganisms and the effect of cropping practices. Annu. Rev. Phytopathol. 25, 339–58.

Schwyn B and Neilands J B 1987 Universal chemical assay for the detection and determination of siderophores. Anal. Biochem. 160, 47–56.

Teintze M, Hossain M B, Barnes C L, Leong J and van der Helm D 1981 Structure of ferric pseudobactin, a siderophore from a plant growth promoting *Pseudomonas*. Biochemistry 20, 6446–6457.

Teintze M and Leong J 1981 Structure of pseudobactin A, a second siderophore from plant growth promoting *Pseudomonas* B10. Biochemistry 20, 6457–6462.

Torres L, Perez-Ortin J E, Tordera V and Beltran J P 1986 Isolation and characterization of an Fe(III)-chelating compound produced by *Pseudomonas syringae*. Appl. Env. Microbiol. 52, 157–160.

Tufano T P and Raymond K N 1981 Coordination Chemistry of Microbial Iron Transport Compounds. 21. Kinetics and mechanisms of iron exchange in hydroxamate siderophore complexes. J. Am. Chem. Soc. 103, 6617–6624.

Underwood P A 1985 Practical considerations of the ability of monoclonal antibodies to detect antigenic differences between closely related variants. J. Immunolog. Meth. 85, 309–323.

de Weger L A, van Arendonk J J C N, Recourt K, van der Hofstad G A J M, Weisbeek P J and Lugtenberg B 1988 Siderophore-mediated uptake of Fe^{3+} by the plant growth-stimulationg *Pseudomonas putida* strain WCS358 and by other rhizosphere microorganisms. J. Bacteriol. 170, 4693–4698.

Wright S F, Foster J G and Bennet, O L 1986 Production and use of monoclonal antibodies for identification of strains of *Rhizobium trifolii*. Appl. Environ. Microbiol. 52, 119–123.

Yang C-C and Leong J 1984 Structure of Pseudobactin 7SR1, a siderophore from a plant-deleterious *Pseudomonas*. Biochemistry 23, 3534–3540.

D. L. Keister and P. B. Cregan (Eds.), The rhizosphere and plant growth, 271–278.

PLSO BARC27

Genetics of iron transport in plant growth-promoting *Pseudomonas putida* WCS358

J. LEONG, W. BITTER, M. KOSTER, J.D. MARUGG and P.J. WEISBEEK
Department of Molecular Cell Biology, University of Utrecht, Padualaan 8, 3584 CH Utrecht, The Netherlands

Key words: genetics, iron transport, *Pseudomonas putida*, receptors, siderophores

Abstract

Root-colonizing *Pseudomonas putida* WCS358 enhances growth of potato in part by producing under iron-limiting conditions a yellow-green, fluorescent siderophore designated pseudobactin 358. This siderophore efficiently complexes iron(III) in the rhizosphere, making it less available to certain endemic microorganisms, including phytopathogens, thus inhibiting their growth. At least 15 genes distributed over five gene clusters are required for the biosynthesis of pseudobactin 358. The transcriptional organization and the iron-regulated expression of a major gene cluster (A) involved in the biosynthesis of pseudobactin 358 and transport of iron(III) via pseudobactin 358 were analyzed. The DNA sequence of the gene encoding the outer membrane receptor protein for ferric pseudobactin 358 has been determined; the mature protein consists of 772 amino acids (86.01 kilodaltons) with a signal sequence of 47 amino acids, which is extremely long for prokaryotes. The receptor protein, which appears to be specific for ferric pseudobactin 358, shares strong homology with four regions of TonB-dependent receptor proteins of *Escherichia coli*, which suggests the presence of a TonB-like protein in strain WCS358 required for iron(III) transport. Strain WCS358 also possesses a low-affinity iron uptake system for ferric pseudobactin 358 and ferric pseudobactins from many other fluorescent pseudomonads. Genes coding for the biosynthesis of pseudobactin 358 and the high-affinity receptor protein for ferric pseudobactin 358 are regulated transcriptionally by iron(III).

Introduction

The genus Pseudomonas comprises a vast group of bacteria which are found in a variety of natural environments (soils, fresh and seawater) and in many different associations with plants and animals. Their ecological diversity is a reflection of their simple nutritional requirements and their ability to metabolize a wide range of organic compounds. The fluorescent pseudomonads, including the species *P. aeruginosa*, *P. fluorescens*, *P. putida*, and *P. syringae*, constitute a major group among rhizosphere microorganisms. They produce and excrete a yellow-green, fluorescent, water-soluble pigment, whch has a very high affinity for ferric iron, and which functions as an iron transport agent (siderophore) (Neilands, 1984). They also produce a variety of other secondary metabolites, some of which are antibiotics, phytotoxins or plant growth hormones (Gross and Cody, 1985; Leisinger and Margraff, 1979).

From many investigations in recent years it has become clear that fluorescent pseudomonads play an important role in the protection of plants against deleterious microorganisms (Burr and Caesor 1984). Due to the production of siderophores and antibiotics, many rhizosphere pseudomonads are antagonistic to phytopathogenic fungi and bacteria. It has been proposed that the siderophores promote plant growth by depriving deleterious rhizosphere microorganisms of iron, thereby inhibiting their growth (Leong, 1986; Schippers *et al.*, 1987). Harmful, nonparasitic microorganisms increase with increasing cropping frequencies and thereby decrease crop yield. Certain beneficial fluorescent pseudomonads can reverse this effect by inhibiting the growth of these deleterious microorganisms in the rhizosphere.

Beneficial fluorescent *Pseudomonas* spp. strains have been isolated from potato roots on the basis of their ability to inhibit a variety of pathogenic and saprophytic rhizosphere fungi and bacteria (like *Fusarium tabacinum*, *Rhizoctonia solani*, *Erwinia carotovora*, and *Streptomyces scabies*). This paper describes the use of molecular genetic techniques to study siderophore biosynthesis and high-affinity iron(III) assimilation in the selected isolates *Pseudomonas fluorescens* WCS374 and *Pseudomonas putida* WCS358.

Pseudobactin 358

Structure of pseudobactin 358

The production of yellow-green, fluorescent siderophores is characteristic for the group of the fluorescent pseudomonads. These siderophores are only produced when cells are grown under iron-limited conditions, and they display strong affinity for ferric ion. The chemical structures of several siderophores produced by fluorescent pseudomonads have been reviewed (Leong, 1986). The structure of pseudobactin 358, the siderophore produced by *P. putida* WCS358 has recently been elucidated (G A J M van der Hofstad, personal communication) and is given in Figure 1. In the nonapeptide the α-hydroxyaspartic acid and the δ-N-hydroxyornithine constitute together with the *o*-dihydroxy aromatic group derived from the quinoline derivative the three bidentate iron(III)-chelating groups. The structure of pseudobactin 358 resembles the other fluorescent siderophores very much (Leong, 1986).

The structural differences in fluorescent siderophores are believed to be important in iron uptake since outer membrane receptor proteins for ferric siderophores appear to recognize specific features of the ferric siderophore. Pseudobactin 358 could be utilized by only a few of several hundred rhizosphere fluorescent pseudomonads that were tested.

Organization of pseudobactin 358 *biosynthetic genes*

Genes required for the biosynthesis of yellow-green, fluorescent pseudobactin 358 were identified by complementing non-fluorescent transposon Tn*5* mutants of strain WCS358 defective in the biosynthesis of pseudobactin 358 with a gene bank of DNA from strain WCS358 maintained in *Escherichia coli*. Approximately 15 genes distributed over five gene clusters were identified in this manner (Marugg *et al.*, 1986). This minimum number of genes seems reasonable considering the structural complexity of pseudobactin 358 and the fact that only a fraction of the non-fluorescent mutants could be complemented. In a similar way Moores *et al.* (1984), identified a minimum of 12 genes arranged in four clusters that were required for biosynthesis of pseudobactin, the native siderophore of plant growth-promoting *Pseudomonas* B10.

The transcriptional organization and the iron-regulated expression of a major gene cluster (A) in strain WCS358 were analyzed in more detail (Fig. 2). The mapping of the Tn*5* integration sites of 11 siderophore-defective mutants, together with the complementation analysis, demonstrates that at least a region of 33.5 kb is required for the biosynthesis of the siderophore. At least five transcriptional units are present in this region (IA, IB, II, III and IV; Fig. 2), some of which are relatively large (Marugg *et al.*, 1988). The polypeptides that were encoded by these transcriptional units were determined by analysis of *E. coli* minicells. The sizes of the encoded polypeptides shown in Figure 2 are minimal ones.

With the exception of the 85-kilodalton protein, which is probably the outer membrane receptor protein for ferric pseudobactin 358 (see following section), the biosynthetic functions of genes in this cluster are not known. However, three mutants in this region IB synthesized mutated non-fluorescent siderophores that con-

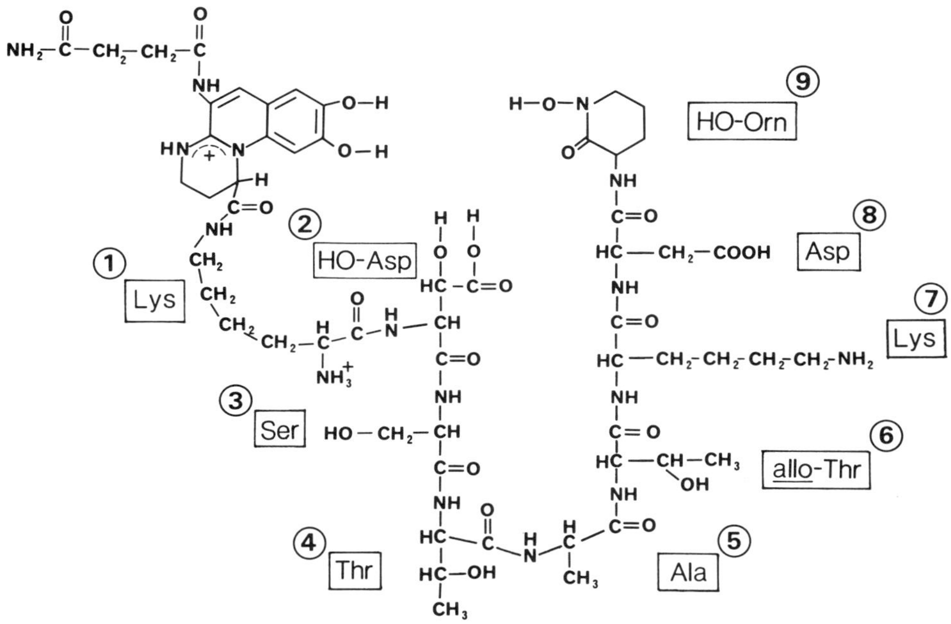

Fig. 1. Proposed chemical structure of pseudobactin 358, the siderophore of *Pseudomonas putida* WCS358.

tained the complete wild-type peptide (G A J M van der Hofstad, personal communication). These results, together with the overall structure of the siderophore, *i.e.* a short peptide chain attached to a fluorescing group, suggest a biosynthetic pathway in which the biosynthesis of the fluorescing group is preceded by the enzymatic synthesis of the peptide part and that specific mutants in cluster A may be defective in certain steps of the synthesis of the hydroxyquinoline derivative.

Regulation

Pseudobactin 358 is produced only under iron-limiting condtions. Under these conditions, certain outer membrane proteins are also expressed specifically (Weger *et al.*, 1986). By RNA-RNA hybridization the expression of specific genes within cluster A is regulated by iron(III) at the transcriptional level. This effect of iron(III) can be mediated via a regulatory protein (activator or repressor) which allows initiation of transcription strictly under iron-limited conditions. This implies that the promoter or operator regions of the respective genes contain specific features that are recognized by such a regulatory protein in the presence or absence of iron. Since the positions and direction of transcription of four transcriptional units in cluster A have been determined, the promoter regions on the DNA were localized as depicted in Figure 2.

These promoter regions were analyzed in broad host-range promoter probe vectors by constructing transcriptional fusions with the structural genes for β-galactosidase (*lac*Z) and catechol-2,3-dioxygenase (*xyl*E) (CDO). The promoter region of transcriptional unit III was localized to a 189-base pair *Hae*III-*Bam*HI fragment; this promoter caused the synthesis of β-galactosidase or CDO in *P. putida* WCS358 only when cells were grown under iron-limited conditions. This confirms that regulation by iron occurs at the transcriptional level. This promoter and other iron-regulated siderophore promoters are not expressed in *E. coli*. The nucleotide sequence of the promoter on the 189-bp *Hae*III-*Bam*HI fragment present within a 0.4 kb *Hin*dIII-*Bam*HI

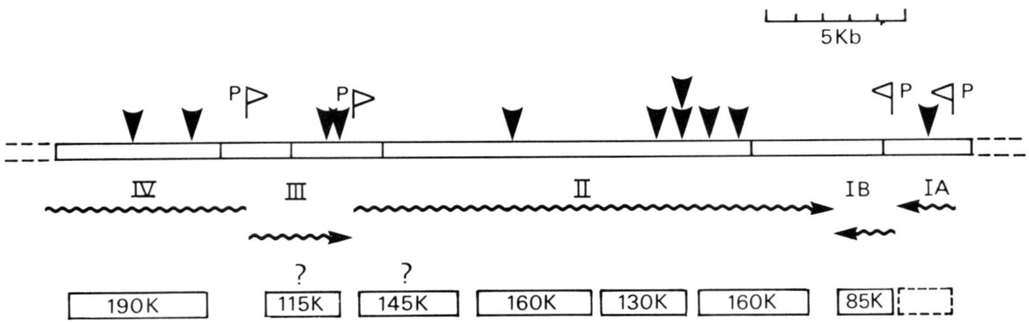

Fig. 2. Genetic organization of cluster A. The localization and orientation of the transcriptional units (IA, IB, II, III, IV) are indicated by the wavy lines. Arrowheads on the *Eco*RI restriction map represent Tn*5* insertions in the siderophore-defective mutants. Flags (with the letter P) represent possible promoter regions. The proposed location and the sizes of the polypeptide products are indicated by the closed boxes in the lower part of the figure. The size of the protein represented by the hatched box is unknown.

fragment was determined (Fig. 3). By nuclease S1 mapping, the transcriptional start site was mapped within this fragment, which showed that the necessary information for iron-regulated expression is present within 73 bp upstream of the transcription start site. Deletion analysis revealed that the −73 to −67 region is required for proper promoter functioning. This region may be involved in recognition or binding of regulatory protein(s) at the siderophore promoter.

We are using this and other iron-regulated promoter fusions to identify iron regulatory genes in strain WCS358.

Iron uptake and ferric siderophore receptors

The inability of *P. fluorescens* WCS374 to transport iron(III) via pseudobactin 358 was exploited to identify a gene coding for the 90-kilodalton outer membrane receptor protein for ferric pseudobactin 358. By mobilization of a gene bank of DNA from strain WCS358 to strain WCS374, a unique cosmid clone, pMR, was identified which made strain WCS374 competent to utilize pseudobactin 358. The essential genetic information of pMR was localized by subcloning to a 5.3-kb *Bgl*II fragment (pAK21/pAK22) (Fig. 4). Transposon Tn*5* mutagenesis limited the functional gene to a region of approximately 2.5 kb consistent with the observed molecular weight of the receptor protein (Fig. 4). The gene is flanked on both sides by pseudobactin 358 biosynthetic genes and is on a separate transcriptional unit (Fig. 5). Strain WCS374 harboring the putative receptor gene was able to transport Fe(III) via pseudobactin 358, whereas strain WCS374 carrying pMR-derivatives with Tn*5* in the receptor

```
HindIII             210             HaeIII                 250
AAGCTT--<203 bp>--TGCACGAGGGGGGCCTAAATCCTTGGCAGTGCTCAATCGTC

                                                  +1 300
TAACTACGGAGGAAATTGAAGAGCTGATTTCTTAGGAGATTCCCTGTGCAAGAGCTGCT

                                             350
CGATACAGTCGTATCGTTGTCCACCCAAAAACGTAAGGCGTTGGCAGCCTTGCTCAAGA

                                     400          BamHI
AACAAGGCGTCAACCTGTATGGGGTCACGCCGATCTTCCCGAGGGATCC------
    SD?
```

Fig. 3. Nucleotide sequence of the promoter region present on a 0.4-kb *Hin*dIII-*Bam*HI fragment. The sequence is numbered from the *Hin*dIII site. The sequence between nucleotides 7 and 210 is not shown. The transcription initiation site is indicated. The putative Shine-Dalgarno sequence and ATG start codon are underlined.

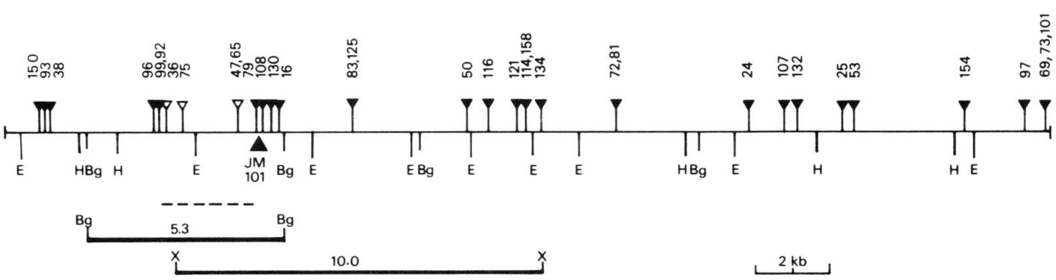

Fig. 4. Location of Tn*5* insertions on the insert of cosmid pMR. Open triangles refer to the Tn*5* insertions that cause a loss of the ability to make WCS374 cells competent for pseudobactin 358 utilization, whereas closed triangles refer to Tn*5* insertions that do not affect pseudobactin 358 utilization by WCS374 cells. The broken line limits the DNA region required for utilization of pseudobactin 358. Positions of the 5.3-kb *Bgl*II- and the 10.0-kb *Xho*I fragments are indicated by the heavy lines. The latter fragment does not make strain WCS374 competent to use pseudobactin 358. The location of the genomic Tn*5* insertion of the siderophore-defective mutant JM101 is indicated by the large triangle. E, *Eco*RI; H, *Hin*dIII; B, *Bam*HI; Bg, *Bgl*II; X, *Xho*I.

gene were unable to transport Fe(III) via pseudobactin 358 (Fig. 6).

Figure 7 (lanes 2 and 9) demonstrates that only the 90- and 92-kilodalton proteins of cell envelopes isolated from iron(III)-limited cells of strain WCS358 reacted in immunoblots with antiserum raised against these two proteins. The antiserum also cross-reacted with a 91-kilodalton protein present in cell envelopes isolated from iron(III)-limited cells of strain WCS374 (Fig. 7, lane 1). A new protein, which comigrated with the WCS358 protein of 90 kilodaltons, was only expressed in cell envelopes of WCS374 cells that harbored cosmid pMR (Fig. 7, lane 3) or its derivatives (Fig. 7, lanes 6 and 7) and that were grown under iron-limitation along with the endogenous 91-kilodalton protein. Under iron-sufficient conditions the 90-kilodalton protein (and the 91-kilodalton protein) was not observed. This indicates that the mechanism of iron regulation in both strains is very much alike or even identical. In cell envelopes of strain WCS374 harboring the mutagenized pMR-derivatives 36 and 47, the 90-kilodalton protein disappeared (Fig. 7, lanes 4 and 5). Together with the fact that these cells were not able to transport Fe(III) via pseu-

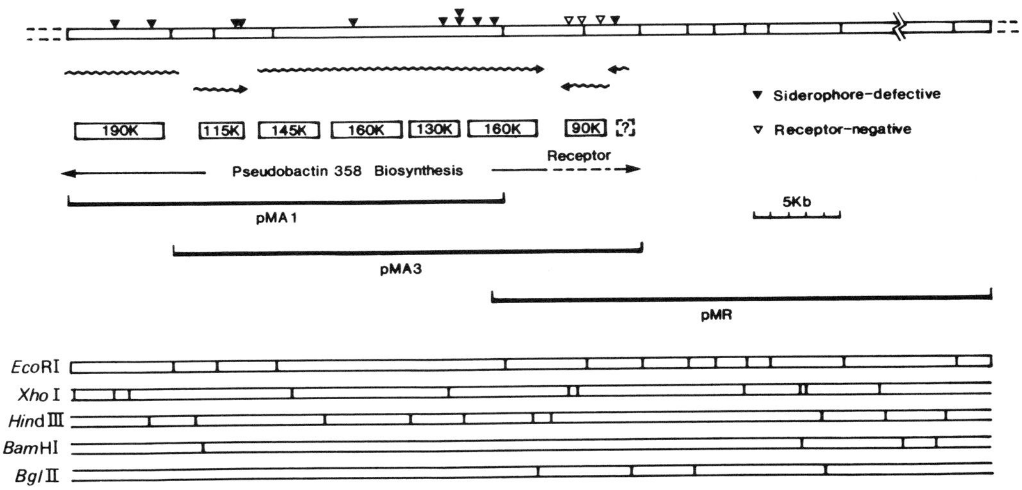

Fig. 5. Genetic organization of the gene cluster A involved in siderophore biosynthesis and transport, present on the genome of *P. putida* WCS358. The orientation and direction of transcriptional units (wavy lines) and polypeptide products (closed boxes) are shown. Open and closed triangles refer to the Tn*5* insertions of siderophore-defective and receptor-negative mutants. The restriction map at the bottom of the figure was combined from the data of the overlapping cosmid clones pMA1, pMA3, and pMR (represented by the heavy lines).

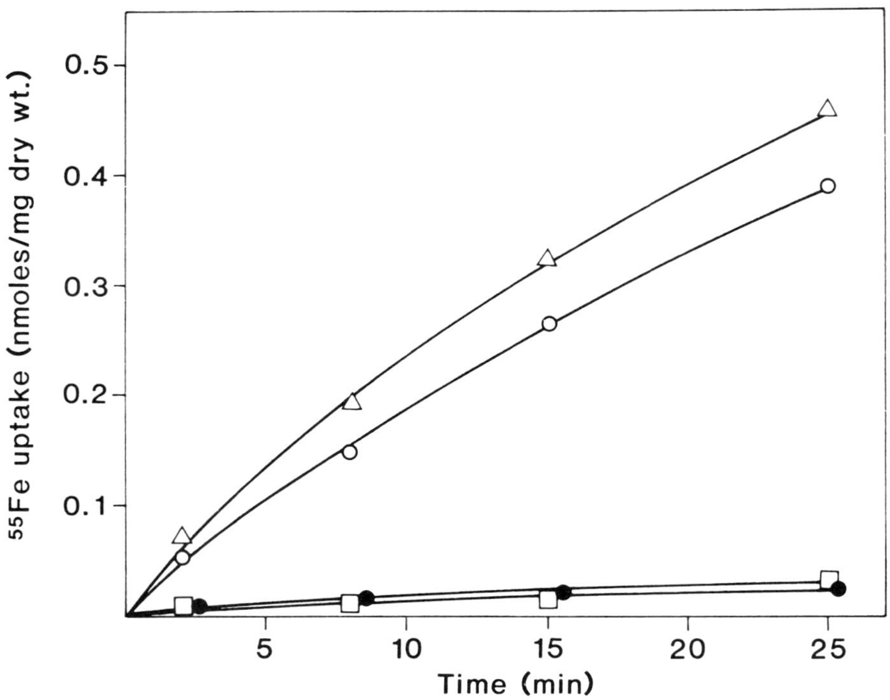

Fig. 6. Pseudobactin 358-mediated ^{55}Fe uptake by cells of WCS374 (-●-●-), WCS374(pMR) (-○-○-), WCS374(pAK22) (-△-△-), and WCS374(pMR-36) (-[]-[]-), grown under iron-limitation.

dobactin 358 any longer (Fig. 6), demonstrates that the 90-kilodalton protein is responsible for pseudobactin 358-specific iron assimilation.

The DNA sequence of this receptor gene has been determined; the mature protein consists of 772 amino acids (86.01 kilodaltons) with a signal sequence of 47 amino acids, which is extremely long for prokaryotes. The receptor protein shares strong homology with four regions of TonB-dependent receptor proteins of *E. coli*, including BtuB (Heller and Kadner, 1985), FecA (Pressler *et al.*, 1988), FepA (Lundrigan and Kadner, 1986), FhuA (Coulton *et al.*, 1986), and IutA (Krone *et al.*, 1985). This suggests the presence of a TonB-like protein in strain WCS358 that is required for iron(III) transport.

Strain WCS358 is able to utilize siderophores produced by many Pseudomonas soil isolates (P A H M Bakker, personal communication). Some of these siderophores were partially characterized and were known to be different from pseudobactin 358. Transposon Tn*5* mutants of strain WCS358, defective in the synthesis of the 90-kilodalton outer membrane receptor protein, are still able to transport Fe(III) delivered by pseudobactin 358, although with a largely decreased efficiency (L A de Weger, personal communication). In contrast no transport of Fe(III) via pseudobactin was observed in *Pseudomonas* B10 Tn*5* mutants lacking the 85-kilodalton outer membrane receptor protein for ferric pseudobactin (Magazin *et al.*, 1986).

The Tn*5* mutants of strain WCS358 are also still able to utilize heterologous siderophores, including pseudobactin 374, pseudobactin BN2, pseudobactin BN7, and pseudobactin BN8, just like the wild-type strain. These results suggest that the 90-kilodalton protein appears to be responsible for specific iron uptake via pseudobactin 358 and that an alternative iron uptake system recognizes ferric pseudobactin 358 and heterologous ferric siderophores. An intriguing question is whether low-affinity iron transport via these other siderophores by strain WCS358 occurs by a single outer membrane receptor protein with low specificity or by several more specific receptors. The iron-regulated 92-kilodalton outer membrane protein may play a promi-

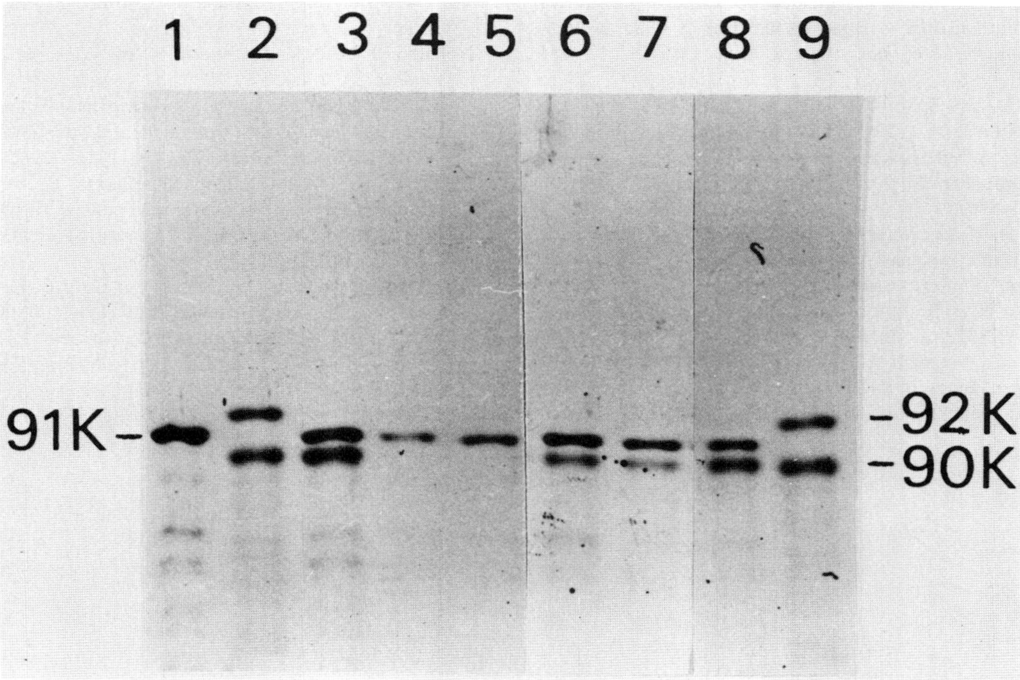

Fig. 7. Immunoblot analysis of cell envelope preparations run on a 9% sodium dodecyl sulfate-polyacrylamide gel using an antiserum raised against a mixture of 90-kilodalton and 92-kilodalton proteins of strain WCS358. Cell envelopes were isolated from cells that were grown under iron-limitation. Lanes: 1, WCS374; 2 and 9, WCS358; 3, WCS374(pMR); 4, WCS374(pMR-36); 5, WCS374(pMR-47); 6, WCS374(pAK21); 7, WCS374(pAK22); 8, WCS374(pMA3). Recombinant plasmid pMA3 also contains the receptor gene.

nent role in this broad specificity uptake system. We are presently identifying gene(s) responsible for this second uptake system.

Acknowledgements

We thank P A H M Bakker and Drs. G A J M van der Hofstad and L A de Weger for communicating results prior to publication. These investigations were supported in part by the Netherlands Technology Foundation and by the European Economic Community grant GBI-4-108 NL from the Bimolecular Engineering Programme.

References

Burr T J and Caesor A J 1984 Beneficial plant bacteria. CRC Crit. Rev. Plant Sci. 2, 1–20.

Coulton J W, Mason P, Cameron D R, Carmel G, Jean R and Rode H N 1986 Protein fusions of β-galactosidase to the ferrichrome-iron receptor of *Escherichia coli* K-12. J. Bacteriol. 165, 181–192.

Gross D C and Cody Y S 1985 Mechanisms of plant pathogenesis by *Pseudomonas* species. Can. J. Microbiol. 31, 403–410.

Heller K and Kadner R J 1985 Nucleotide sequence of the gene for the vitamin B_{12} receptor protein in the outer membrane of *Escherichia coli*. J. Bacteriol. 161, 904–908.

Krone W J A, Stegehuis F, Koningstein G, Doorn C van, Roosendaal B, Graaf F K de and Oudega B 1985 Characterization of the pColV-K30 encoded cloacin DF13/aerobactin outer membrane receptor protein of *Escherichia coli*; isolation and purification of the protein and analysis of its nucleotide sequence and primary structure. FEMS Microbiol. Lett. 26, 153–161.

Leisinger T and Margraff R 1979 Secondary metabolites of the fluorescent pseudomonads. Microbiol. Rev. 43, 422–442.

Leong J 1986 Siderophores: their biochemistry and possible role in the biocontrol of plant pathogens. Annu. Rev. Phytopathol. 24, 187–209.

Lundrigan M D and Kadner R J 1986 Nucleotide sequence of the gene for the ferrienterochelin receptor FepA in *Escherichia coli*. J. Biol. Chem. 261, 10797–10801.

Magazin M, Moores J C and Leong J 1986 Cloning of the

gene coding for the outer membrane receptor protein for ferric pseudobactin, a siderophore from a plant growth-promoting *Pseudomonas* strain. J. Biol. Chem. 261, 795–799.

Marugg J D, van Spanje M, Hoekstra W P M, Schippers B and Weisbeek P J 1985 Isolation and analysis of genes involved in siderophore biosynthesis in plant growth-stimulating *Pseudomonas putida* WCS358. J. Bacteriol. 164, 563–570.

Marugg J D, Nielander H B, Horrevoets A J G, van Megen I, van Genderen I and Weisbeek P J 1988 Genetic organization and transcriptional analysis of a major gene cluster involved in siderophore biosynthesis in *Pseudomonas putida* WCS358. J. Bacteriol. 170, 1812–1819.

Moores J C, Magazin M, Ditta G S and Leong J 1984 Cloning of genes involved in the biosynthesis of pseudobactin, a high-affinity iron transport agent of a plant growth-promoting *Pseudomonas* strain. J. Bacteriol. 157, 53–58.

Neilands J B 1984 Methodology of siderophores. Struct. Bonding 58, 1–24.

Pressler U, Staudenmaier H, Zimmermann L and Braun V 1988 Genetics of the iron dicitrate transport system of *Escherichia coli*. J. Bacteriol. 170, 2716–2724.

Schippers B, Bakker A W and Bakker P A H M 1987 Interactions of deleterious and beneficial rhizosphere microorganisms and the effect of cropping practices. Annu. Rev. Phytopathol. 25, 339–358.

Weger L A de, van Boxtel R, van der Burg B, Gruters R A, Geels F P, Schippers B and Lugtenberg B 1986 Siderophores and outer membrane proteins of antagonistic, plant growth-stimulating, root-colonizing *Pseudomonas* spp. J. Bacteriol. 165, 585–594.

D. L. Keister and P. B. Cregan (Eds.), The rhizosphere and plant growth, 279–287.

PLSO BARC28

Formulation and delivery systems of biocontrol agents with emphasis on fungi

J.A. LEWIS
Biocontrol of Plant Diseases Laboratory, USDA, ARS, Beltsville, MD 20705, USA

Key words: biocontrol bacteria, biocontrol formulations, biocontrol fungi, delivery systems, fermentation media, Gliocladium, Trichoderma

Abstract

For commercial development of microbials for bioncontrol of soilborne pathogens in the spermosphere or rhizosphere, obstacles in growth, formulation, and delivery must be overcome. Most of the research described is with the biocontrol fungi Trichoderma and Gliocladium. Fermentation media for growth of biocontrol fungi or bacteria should contain available, inexpensive agricultural products which result in production of adequate biomass containing effective propagules (*e.g.* chlamydospores). Dry, powdered or moist biomass can be formulated into a variety of dusts, powders, pellets, granules or gels supplemented with inert carriers (*e.g.* Pyrax®) or food bases (*e.g.* bran). Systems are also described in which biocontrol agents are grown on solid fermentation media consisting of various grains or inert carriers amended with food bases. Delivery systems of preparations for seed treatments and soil amendments for the control of various soilborne plant pathogens are discussed.

Introduction

Soilborne diseases, caused by about 50 genera of fungi and a few bacteria and viruses, occur when pathogens infect a seed, stem, or actively growing root. Biological control may be achieved by the inactivation of the pathogen by another microbe as a result of antibiosis, parasitism, or competition for nutrients. The microflora in and around the roots comprise bacteria, fungi, actinomycetes, yeast, and algae. Fungi represent an excellent reservoir for isolating and demonstrating microbials of biocontrol ability. Many diverse physiological groups are found in soil and rhizosphere; the types and amounts dependent largely upon the soil type, nature and age of the plant root, and the methods used for isolation (Rovira, 1965). Biocontrol by augmentation implies application of a microbial preparation to either soil, seed, or phyllosphere to prevent plant infection by a pathogen. Growth, formulation, and delivery of soil and rhizosphere fungi shown to have biocontrol potential represent a feasible approach to disease suppression. They are essential steps to define in developing an augmentation system for disease control.

Potential biocontrol fungi

Some fungal genera routinely isolated from the rhizosphere of various plants which have been implicated in biocontrol are listed in Table 1. Most developmental research to date has been done with isolates of *Trichoderma* spp. and *Gliocladium* spp. effective against various soilborne diseases of many crops in the greenhouse and field (Chet, 1987; Papavizas, 1985). Other fungi with possibilities for development include *Talaromyces flavus*, a rhizosphere colonizer with biocontrol activity against wilt diseases of eggplant, sunflower, and potato (Fravel, 1989) and *Sporidesmium sclerotivorum*, an obligate hyperparasite, which controls diseases caused by *Sclerotina* spp. (Adams and Ayers, 1982). Non-pathogenic isolates of pathogenic fungi are also

Table 1. Some fungi isolated from the rhizosphere of various plants with potential for development in biocontrol

Alternaria	Epicoccum	Paecilomyces
Aspergillus	Fusarium	Penicillium
Cephalosporium	Gliocladium	Rhizopus
Chaetomium	Humicola	sterile mycelia
Cladosporium	Mortierella	Talaromyces
Coniothyrium	Mucor	Trichoderma
Curvularia	mycorrhizal fungi	Verticillium
Cylindrocarpon	Myrothecium	

promising candidates in biocontrol as successful competitors for an ecological niche. Nonpathogenic isolates of Gaeumannomyces and Fusarium effectively suppressed diseases caused by *R. solani*, *G. graminis* and *F. oxysporum*, respectively (Herr, 1988; Lemaire *et al.*, 1986). However, use of these organisms may present additional regulatory problems. Vesicular-arbuscular mycorrhizae have also shown potential for root disease control (Schenk, 1987). Other interesting and little-understood antagonistic fungi currently investigated in the Biocontrol of Plant Diseases Laboratory (BPDL) for development possibilities include Laetisaria, Dendrostilbella, Cladorrhinum and Arachniotus.

Growth of biocontrol agents

The general unavailability of suitable commercial methods for growth, formulation, and delivery of antagonists is a major constraint in the implementation of biocontrol. Problems in these areas have been addressed in recent reviews (Connick *et al.*, 1989; Lumsden and Lewis, 1989). In most cases, adequate and effective biomass formation has been accomplished by modification of liquid (deep tank) or solid (semisolid) fermentations.

Liquid fermentation

The expansion of liquid fermentation technology for the commercial production of organic acids, enzymes, antibiotics, and other drugs by fungi and bacteria has resulted in the production of large scale facilities by industry. Although these advances resulted in the production of bacterial and fungal biomass for use as biocontrol insecticides and herbicides, similar efforts for the production of effective antagonists against plant pathogens are in their infancy (Churchill, 1982). Since deep tank fermentation facilities are already in place, especially in the U.S., liquid fermentation for the production of biomass will be the preferred approach. A suitable medium should consist of inexpensive, readily available agricultural by-products with an appropriate nutrient balance (Zabriskie *et al.*, 1980). Acceptable materials include molasses, brewer's yeast, corn steep liquor, sugar syrups, sulfite waste liquor, whey, and cottonseed and soy flours. In most cases, it is economically unfeasible for a company to use exotic sugars or plant extracts in tanks which hold thousands of liters (Lisanski, 1985).

For successful fermentation, not only must appropriate substrates be used, but sufficient biomass containing adequate amounts of effective propagules of the fungus must be obtained. For example, research in the BPDL with isolates of *Trichoderma* spp. and *G. virens* indicated that preparations containing chlamydospores (the hardy, survival propagules of the fungus) more effectively prevented diseases than preparations containing conidia only (Lewis and Papavizas, 1984; Papavizas *et al.*, 1984). Small scale fermentation of molasses brewer's yeast in the BPDL which resulted in abundant chlamydospore production has been adapted recently to large scale industrial fermentation. Operating conditions (aeration, pH, temperature) as well as media constituents may effect the quality and quantity of the test organisms, especially bacteria (Churchill, 1982). For example, conidia or ascospores of the biocontrol fungus *T. flavus* cannot yet be produced in liquid fermentation, although large amounts of biomass are obtained.

There are several additional factors to consider in liquid fermentation. The rate at which an effective biomass is produced affects cost of production as well as contamination and viability. Satisfactory biomass of isolates of Trichoderma, Gliocladium, and Talaromyces was achieved in 6–7 days, but this time is still long compared to that for obtaining bacteria which normally occurs in less than 2 days (Papavizas *et al.*, 1984). Increased fermentation time also reduced viability of some fungi and increased the risk of contamination (Churchill, 1982).

Solid fermentation

Solid or semisolid fermentation is still used almost exclusively for experimental production of fungal biomass. With the advent of deep tank fermentation, solid state fermentation facilities have become limited in much of the world. In fact, solid state fermentation (except in mushroom production) is not popular in North America because of insufficient consumer demand for the products formed. There are, however, several companies in both the U.S. and Mexico which specialize in solid state fermentations. It appears evident that the process will be used by industry if effective biocontrol preparations can be produced. Substrates for the production of inoculum of biocontrol fungi in the genera Trichoderma, Gliocladium, Coniothyrium, Chaetomium, Laetisaria, and Penicillium include various grain seeds and meals, bagasse, straws, wheat bran, sawdust, and peat individually or in combination (Papavizas, 1985; Papavizas and Lewis, 1981). The system is especially useful for small scale research laboratory, greenhouse, and field tests which require minimal facilities for implementation. Solid fermentations are also suitable for the production of fungi which either do not sporulate in liquid cultures (Lisanski, 1985) or do not survive the liquid fermentation process (Lemaire *et al.*, 1986). The system is also advantageous in countries such as Japan, where fully automated commercial solid fermentation plants are already in existence for other purposes. Also, solid fermentation may be appropriate in underdeveloped countries where agricultural wastes are available, elaborate facilities are limited, and labor is abundant (Carrizales and Jaffe, 1986).

Formulations of biocontrol agents

Preparations from liquid fermentations

After fungal growth, biomass is separated from spent medium by any of various filtration systems (pressure, rotary vacuum drum), centrifugation, or flocculation. The biomass can be incorporated moist into pellets or inert carriers (Celatom, vermiculite) which are subsequently dried (Lewis and Papavizas, 1985; Lewis *et al.*, 1989). Before formulation, the biomass may also be dried by systems utilizing pan, spray or freeze drying, any of which may affect the viability of the propagules in the biomass. Conventional techniques to formulate pesticides may be adapted to biocontrol powders, but care must be taken to avoid drastic treatment during processing (Polon, 1973). As with other formulations containing microbials, commercial preparations to control plant diseases must be stable, possess adequate shelf life, contain high levels of antagonist colony forming units, be easy to prepare and apply, and be cost effective. Experimental dry dust formulations were successfully prepared in the BPDL by diluting powered biomass of any of several isolates of *Trichoderma* spp., *G. virens* or with commercially available pyrophylite clay (Pyrax®) as a carrier (Papavizas *et al.*, 1984). Pyrax was preferred because its neutral pH made it more compatible than alkaline talc for acidophilic fungi.

Formulations of bacterial biomass to control diseases in the spermosphere and rhizosphere include agar cultures and concentrated suspensions of *Agrobacterium radiobacter* (Galltrol®, Norbac®). Another commercial bacterial formulation recently registered by EPA is a peat granule containing at least 10^5 colony-forming units of the bacterium *Pseudomonas fluorescens* per gram (Dagger G®). A commercial product, available in Europe as a wettable powder, contains a streptomycete biomass (Mycostop®). Preparation and delivery of bacterial formulations are discussed elsewhere in this Symposium. Few wettable powder formulations for the control of soilborne pathogens have been reported, but it should not be difficult to develop such products with the proper adjuvants and wetting or suspending agents.

Fungal spores and biomass preparations have also been formulated into various tablets, granules, pastes, and bulk carriers, the exact nature of which is often proprietary information. Fungi formulated in this manner generally are isolates of Trichoderma and Gliocladium. A most intriguing aspect of innovative technology for the formulation of microbials involves the immobilization of biomass within a cross linked matrix of organic polymers such as alginate, polyacrylamide, or carrageenan (Woodward, 1985). These materials have been used to immobilize various plant and animal cells and enzymes. Alginate gel, used most commonly, has been employed successfully by the BPDL to prepare formulations of biocontrol bacteria and fungi (Fravel *et al.*, 1985; Lewis and Papavizas, 1985, 1988; Papavizas *et al.*, 1987) for which U.S. patents were issued (Nos. 4,668,512 and 4,724,147). The angatonists studied included pseudomonads, isolates of *Trichoderma* spp., *G. virens*, *T. flavus*, *Laetisaria arvalis*, and *Dendrostilbella* sp. For all examples, either wet fermentor biomass or the entire fermentation suspension could be used as well as dry, powdered biomass. With fungi, wet biomass was blended before it was added to the polymer. Use of wet biomass or whole suspension was advantageous because a drying and milling step was omitted, and some fungi (*e.g.* Laetisaria, Arachniotus) whose biomass cannot be separated easily from spent liquid because of gum formation were easily processed.

The method consists of mixing fermentor biomass and a carrier (bulking agent) with a sodium alginate solution. The carrier may be inert (*e.g.* Pyrax), a food base (*e.g.* powdered wheat bran), or a combination. The mixture is dripped into a calcium salt gellant solution. Each droplet is transformed into a gel bead by the reaction of the sodium alginate with calcium ions. Gel bead size can be manipulated by adjusting the gauge of the tips in the dispensing apparatus. Drying converts the gel bead into a pellet (granule).

Because of the intrinsic specificity of biological control, it was necessary to adapt pellet ingredients to accommodate the nutritional requirements of the biological agent. For example, since isolates of Trichoderma and Gliocladium need contact with a food base for proliferation and activity, powdered wheat bran rather than an inert clay was incorporated into the pellet as a carrier. Carrageenan gels have also been used for the immobilization of fungi. The gellant used for carrageenan gels is a potassium salt, a material which may be more compatible than calcium salts for certain fungi and bacteria. Carrageenan has the disadvantage that the suspension must be kept warm (40–45°C) while being dripped into the gellant.

Preparations from solid fermentations

In geneal, solid state fermentations do not require sophisticated formulation procedures prior to use. For example, grains or other types of organic matter upon which antagonists are grown are simply dried, ground, and added to the ecosystem. In some instances, these powders are used as seed treatments (Chet *et al.*, 1987; Lemaire *et al.*, 1986). However, there are several inherent problems with solid state fermentation which may make the system inappropriate for commercial product development. The preparations are bulky; they may be subject to a greater risk of contamination; and they may require extensive space for processing, incubation, and storage. In addition, they may require drying and milling with the undesirable formation of dusts containing spores; may require costly shipping and transportation; and special equipment may be necessary for application.

Despite these shortcomings, some solid state fermentation preparations may have potential for commercial development. In the process ued by the BPDL for the biocontrol fungus *Sporidesmium sclerotivorum* (effective against lettuce drop caused by the pathogen *Sclerotinia minor*), vermiculite moistened with a liquid medium is inoculated with the antagonist in a large twin-shell blender, aseptically bagged, and incubated so that *S. sclerotivorum* grows and sporulates (Adams and Ayers, 1982). Recently we have developed a preparation ('germlings') consisting of sterile wheat bran: water (1:1, w/v) and actively growing hyphae of a wide variety of Trichoderma and Gliocladium isolates (Lewis and Papavizas, 1984). In this system, the substrate, inoculated with a conidial suspension of the biocontrol fungus, was allowed to incubate for 3 days. The formulation was immediately mixed into pathogen infested soil whereupon antagonist hyphae in association with the food base (bran) parasitized the pathogen propagules (Lewis and Papavizas, 1987b). In another system from the BPDL, wet fermentor biomass of antagonist isolates was homogenized with dilute acid and mixed with vermiculite (Lewis *et al.*, 1989). After drying, the mixture could be stored for at least 12 weeks at 5 and 25°C. Before addition to soil, the dry vermiculite/biomass was remoistened with dilute acid and incubated for 2 days to induce formation of actively growing hyphae of the antagonist. The major advantage of the system is that aseptic conditions do not have to be maintained for formulation, except during production of the biomass. A formulation with potential commercial applications was devised recently in which conidia of isolates of Trichoderma and Gliocladium were added to a sterile lignite-stillage mixture (Jones *et al.*, 1984). After incubation and drying, the granules were applied to soil.

Delivery systems and efficacy

The problems involved with delivery systems for biocontrol agents are no less important than those concerned with growth of the organism and its subsequent formulation. Delivery must be easy, effective, timely, to the appropriate site of action, and compatible with available agricultural equipment. Formulations can be delivered to soil, wounds, seeds, or as an aerial application. Most of the early, as well as current, research which involves delivery and efficacy of biocontrol agents is done with solid fermentation systems. The biocontrol formulation systems for soil augmentation described up to 1980 were comprehensively reviewed (Papavizas and Lewis, 1981). The application of antagonistic microorganisms to crop seeds, seedpieces, or tubers appears to represent the most suitable method for biocontrol in the spermosphere and rhizosphere. This is especially true if the antagonist can proliferate sufficiently to colonize the developing root system and inactivate the plant pathogen. A review on the introduction of bacteria and fungi to seeds appeared in 1981 (Kommedhal and Windels, 1981).

Bacteria are the most effective organisms in root colonization and a considerable amount of seed treatment research has been done with them, mainly with the genera Pseudomonas and Bacillus. Although seed treatments with fungi have also been investigated, the current practice of applying conidia of fungi, grown on petri plates, to reduce soilborne diseases is impractical and unworkable with regard to commercial production. Seed treatment technology must be developed in which liquid or solid fermentations are used to effectively obtain biomass for these treatments. Generally, seeds of various crops are pelletized by treating the seeds with any of several natural (*e.g.* methyl cellulose) or synthetic (*e.g.* Nu-Film) stickers which contain microbial suspensions (Suslow and Schroth, 1982). Treated seeds are then dried or coated with an inert material before drying. Some formulations are available as dry, hopper box treatments or as wettable powders for commercial slurry application to seeds.

Seed treatment with biocontrol fungi have recently been improved by adding amendments to stimulate antagonists even though conidia were used as the propagules (Nelson *et al.*, 1988). Techniques to improve delivery have also included a solid matrix priming system for seed application in which seeds were hydrated to a controlled level with moistened, finely ground carrier (Harman and Taylor, 1988). Formulating dusts or powders from dried biomass of biocontrol agents represents a rapid and efficient approach in delivery provided the antagonists remain viable. Pyrax dusts containing biomass of *T. viride* and *G. virens* were successfully applied to potato seedpieces for the reduction of disease caused by *R. solani* (Beagle-Ristaino and Papavizas, 1985). Similar preparations containing *T. flavus* were coated on potato seedpieces for the reduction of Verticillium wilt (Davis *et al.*, 1986).

Dust, powder, and granular formulations can also be broadcast or applied in furrow for disease control. Although considerable material may have to be used, soil incorporation and mixing allows for adequate spatial distribution and close contact with pathogen propagules. Formulations containing Pyrax and ground biomass of several isolates of *Trichoderma* spp. and *G. virens* prepared by the BPDL prevented damping-off of cotton seedlings incited by *R. solani* (Lewis and Papavizas, 1985). Biomass of a nonpathogenic *Fusarium oxysporum* mixed with talc for application to soil controlled pathogenic fusaria which attacks tomato and melon (Lemaire *et al.*, 1986).

Other delivery systems have included various sprays and dips containing biomass or spores of biocontrol agents, pellets or tablets, ointments, gels for fluid drill, and bulk materials. Ascospore suspensions of *T. flavus* mixed with potting soil and added to transplant holes in the field significantly reduced eggplant wilt caused by *Verticillium dahliae* (Marois *et al.*, 1982). Five spray and pellet formulations of a Trichoderma based mycofungicide (BINAB-T) have been registered in Sweden, England, and other countries. Some are sprayed on mushroom casing soil in commercial production to control dry bubble disease caused by *Verticillium malthousei*. Others are formulated into pellets for insertion into various trees to prevent and treat foliar diseases (Ricard, 1988). In an interesting delivery system similar to the above, a hypovirulent fungus reported in France to treat chestnut canker caused by *Endothia parasitica* was cultured on malt medium, mixed with corn flour and the concoction placed in aluminum tubes. The paste in the tube was squeezed into holes punched in the tree bark for the control of chestnut canker of an 18,000 ha planting (Dubos, 1987).

Another novel delivery system involves the incorporation of biocontrol agents into fluid drill gels. In one study, vegetable or fruit tree seedlings were dipped into gels incorporated with antagonist so that the root area was surrounded by a thin layer of gel before the seedlings were planted (Conway, 1986). This innovative ap-

Table 2. Examples of formulations of biocontrol fungi effective against soilborne plant pathogens reported since 1980

Formulation (antagonist)	Target pathogen/host	Reference
Pellets		
(Gliocladium)	Pythium, Rhizoctonia/ ornamentals	Lumsden and Locke, 1989
(Gliocladium, Trichoderma)	Rhizoctonia/cotton	Lewis and Papavizas, 1987a
(Gliocladium)	Sclerotium/beans	Papavizas and Lewis, 1989
(Talaromyces)	Verticillium/potato	Fravel *et al.*, 1986
Bulk materials, grains		
(Gliocladium)	Rhizoctonia/peanuts	Jones *et al.*, 1984
(Gliocladium)	Sclerotium/beans	Papavizas and Lewis, 1989
(Rhizoctonia)	Rhizoctonia/ sugar beets	Herr, 1988
(Sporidesmium)	Sclerotinia/lettuce	Adams and Ayers, 1982
(Trichoderma)	Rhizoctonia, Pythium, *etc.*/vegetable, field crops	Chet, 1987
(Trichoderma)	Rhizoctonia/cotton	Lewis and Papavizas, 1987b
Dusts, powders		
(Gliocladium)	Sclerotium/beans	Papavizas and Lewis, 1989
(Gliocladium, Trichoderma)	Rhizoctonia/potato	Beagle-Ristaino and Papavizas, 1985
(Pythium)	Pythium, Rhizoctonia/ carrots, beets, *etc.*	Luchmeah and Cooke, 1985
(Talaromyces)	Verticillium/potato	Davis *et al.*, 1986
(Gaeumannomyces)	Gaeumannomyces/wheat	Lemaire *et al.*, 1986
(Fusarium)	Fusarium/melons, tomato, *etc.*	Lemaire *et al.*, 1986
(Trichoderma)	Verticillium/mushrooms	Ricard, 1988
(Trichoderma)	Pythium/peas	Nelson *et al.*, 1988
Gels, fluid drill		
(Trichoderma)	Rhizoctonia/radish	Mihuta-Grimm and Rowe, 1986
(Laetisaria)	Rhizoctonia/peppers	Conway, 1986
(Trichoderma)	Sclerotium/apple seedlings	Conway, 1986
Spores		
(Talaromyces)	Verticillium/eggplant	Marois *et al.*, 1982

proach, utilizing the benefits derived from fluid drill technology, offers considerable promise for the formulation and application of biocontrol microorganisms. Some examples of biocontrol of soilborne plant pathogens resulting from delivery of various fungal formulations since those described in 1981 (Papavizas and Lewis, 1981) are listed in Table 2.

The efficacy of various preparations for their biocontrol abilities has been evaluated mainly in greenhouse tests, but field data are continuously being reported as research is directed more to this area. In many cases results have been either unrepeatable, inconsistent, or variable from location to location or from year to year. A discussion of the reasons for this does not fall within the scope of this review, but certainly environmental and edaphic factors as well as difficulties in growth, formulation, and delivery of the biocontrol organisms significantly contribute to these problems. Unfortunately, worldwide, there are very few commercially formulated preparations actually available for general agricultural use. This problem may be due to difficulties with government registration, lack of consistent efficacy, or user acceptance. Commercial formulations for the biocontrol of plant pathogens, all of which have EPA registration except BINAB-T and Mycostop, are indicated in Table 3. It is of interest to note that only two of the six formulations contain biocontrol fungi neither of which was developed in the U.S.

Conclusions

Although biological control of soilborne plant pathogens with antagonists has thus far not solved agricultural problems, a turning point in technology to make significant advances has been achieved. The development of a formulation and delivery system of a biocontrol agent, especially one which is effective in the spermosphere or rhizosphere, depends upon close cooperation among formulation chemists, biologists, agricultural engineers, and economists. The adaptations of technology to utilize effective organisms in the biocontrol arsenal needs to be investigated. Development of acceptable, easily prepared, and cost effective formulations for de-

Table 3. Commercial formulations with biocontrol agents effective against plant pathogens

Name	Antagonist	Target pathogen/ host	Source
BINAB-T	*Trichoderma* sp.	*Chondrosterem purpureum*/trees *Endothia parasitica* chestnuts *Verticillium malthousei*/ mushrooms	Bio-Innovation AB, Toreboda, Sweden
Dagger G	*Pseudomonas fluorescens*	*Pythium ultimum*/ cotton *Rhizoctonia solani*/cotton	Ecogen, Inc. Langhorne, PA, USA
Galltrol	*Agrobacterium radiobacter*	*Agrobacterium tumifaciens*/ crucifers	AgBio Chem, Inc. Orinda, CA, USA
Mycostop	*Streptomyces* sp.	*Alternaria brassicicola*/ trees	Kemira Oy, Vass, Finland
Norbac 84-C	*Agrobacterium radiobacter*	*Agrobacterium tumifaciens*/ trees	New BioProducts, Inc. Corvallis, CA, USA
P.g. Suspension	*Phlebia gigantea*	*Heterobasidion annosum*/ trees	Ecological Labs., Ltd. Dover, UK

livery should be a major goal. Increased understanding is essential concerning persistent problems of dependability and reliability of microbials. It is important to demonstrate successful cases of biocontrol to show that this approach is worthy of efforts to register and commercialize products. In achieving these goals, we can demonstrate that microbial biocontrol agents as alternatives to chemicals, are economically feasible, safe, and effective.

References

Adams P B and Ayers W A 1982 Biological control of Sclerotinia lettuce drop in the field by *Sporidesmium sclerotivorum*. Phytopathology 72, 485–488.

Beagle-Ristaino J E and Papavizas G C 1985 Biological control of Rhizoctonia stem canker and black scurf of potato. Phytopathology 75, 560–564.

Carrizales V and Jaffe W 1986 Solid state fermentation: An appropriate biotechnology for developing countries. Interscience 11, 9–15.

Chet I 1987 Trichoderma-application, mode of action, and potential as a biocontrol agent of soil-borne plant pathogenic fungi. *In* Innovative Approaches to Plant Disease Control. Ed. I Chet. pp 137–160. John Wiley and Sons, New York.

Chet I, Sivan A and Elad Y 1987 Novel isolate of Trichoderma, fungicidal compositions containing said isolate and use thereof. U.S. Patent No. 4, 713, 342.

Churchill B W 1982 Mass production of microorganisms for biological control. *In* Biological Control of Weeds with Plant Pathogens. Eds. R Charudattan and H L Walker. pp 157–173. John Wiley and Sons, New York.

Connick W J Jr, Lewis J A and Quimby P C 1990 Formulation of biocontrol agents for use in plant pathology. *In* New Directions in Biological Control. Eds. R Baker and P Dunn. pp 345–372. A. R. Liss, Inc.

Conway K E 1986 Use of fluid-drilling gels to deliver biological control agents to soil. Plant Dis. 70, 835–839.

Davis J R, Fravel D R, Marois J J and Sorensen L H 1986 Effect of soil fumigation and seedpiece treatment with *Talaromyces flavus* on wilt incidence and yield, 1983. Biol. Cult. Tests 1, 18.

Dubos B 1987 Fungal antagonism in aerial agrobiocenoses. *In* Innovative Approaches to Plant Disease Control. Ed. I Chet. pp 107–135. John Wiley and Sons, New York.

Fravel D R 1989 Biocontrol of Verticillium wilt of potato and eggplant. *In* Interaction of Genetic and Environmental Factors in the Development of Vascular Wilt Diseases in Plants. Eds. C. Beckman and E C Tjamos. pp 487–492. Springer-Verlag, Berlin, FRG.

Fravel D R, Davis J R and Sorensen L H 1986 Biological control of Verticillium wilt of potato in the field by *Talaromyces flavus*. Phytopathology 76, 863.

Fravel D R, Marois J J, Lumsden R D and Connick W J Jr 1985 Encapsulation of potential biocontrol agents in alginate. Phytopathology 75, 774–777.

Harman G E and Taylor A G 1988 Improved seedling performance by integration of biological control agents at favorable pH levels with solid matrix priming. Phytopathology 78, 520–525.

Herr L 1988 Biocontrol of crown and root rot of sugar beet by binucleate *Rhizoctonia* spp. and *Laetisaria arvalis*. Ann. Appl. Biol. 113, 107–118.

Jones R W, Pettit R E and Taber R A 1984 Lignite and stillage: Carrier and substrate for application of fungal biocontrol agents to soil. Phytopathology 74, 1167–1170.

Kommedahl T and Windels C E 1981 Introduction of microbial antagonists to specific courts of infection: Seeds, seedlings, and wounds. *In* Biological Control in Crop Production. Ed. G C Papavizas. pp 227–248. Allanheld, Osmun and Co., Totawa, NJ.

Lemaire J M, Alabouvette C, Davet P and Tramier R 1986 Problems posed by the large scale application of microorganisms for biological control of soil-borne plant pathogens. Symbiosis 2, 287–301.

Lewis J A and Papavizas G C 1984 A new approach to stimulate population proliferation of *Trichoderma* spp. and other potential biocontrol fungi introduced into natural soils. Phytopathology 74, 1240–1244.

Lewis J A and Papavizas G C 1985 Characteristics of alginate pellets formulated with *Trichoderma* and *Gliocladium* and their effect on the proliferation of the fungi in soil. Plant Pathol. 34, 571–577.

Lewis J A and Papavizas G C 1987a Application of *Trichoderma* and *Gliocladium* in alginate pellets for control of Rhizoctonia damping off. Plant Pathol. 36, 438–446.

Lewis J A and Papavizas G C 1987b Reduction of inoculum of *Rhizoctonia solani* in soil by germlings of *Trichoderma hamatum*. Soil Biol. Biochem. 19, 195–201.

Lewis J A and Papavizas C G 1988 Biocontrol of *Rhizoctonia solani* (Rs) by some novel soil fungi. Phytopathology 78, 862.

Lewis J A, Papavizas G C and Lumsden R D 1989 A new biocontrol formulation for Trichoderma and Gliocladium. Phytopathology 79, 1160.

Lisanski S G 1985 Production and commercialization of pathogens. *In* Biological Pest Control. Eds. N W Hussey and N Scopes. pp 210–218. Blanford Press, Poole, U.K.

Luchmeah R S and Cooke R C 1985 Pelleting of seed with the antagonist *Pythium oligandrum* for biological control of damping-off. Plant Pathol. 34, 528–531.

Lumsden R D and Lewis J A 1989 Problems and progress in the selection, production, formulation and commercial use of plant disease control fungi. *In* Biotechnology of Fungi for Improved Plant Growth. Eds. J M Whipps and R D Lumsden. pp 171–190. British Mycological Society.

Lumsden R D and Locke J C 1989 Biological control of *Pythium ultimum* and *Rhizoctonia solani* damping off with *Gliocladium virens* in soilless mix. Phytopathology 79, 361–366.

Marois J J, Johnston S A, Dunn M T, and Papavizas G C 1982 Biological control of Verticillium wilt of eggplant in the field. Plant Dis. 66, 1166–1168.

Mihuta-Grimm L and Rowe R C 1986 *Trichoderma* spp. as biocontrol agents of Rhizoctonia damping-off of radish in organic soil and comparison of four delivery systems. Phytopathology 76, 306–312

Nelson E B, Harman G E and Nash G T 1988 Enhancement of Trichoderma-induced biological control of Pythium seed rot and preemergence damping-off of peas. Soil Biol. Biochem. 20, 145–150.

Papavizas G C 1985 *Trichoderma* and *Gliocladium*: Biology, ecology, and potential for biocontrol. Annu. Rev. Phytopathol. 23, 23–54.

Papavizas G C, Dunn M T, Lewis J A and Beagle-Ristaino J 1984 Liquid fermentation technology for experimental production of biocontrol fungi. Phytopathology 74, 1171–1175.

Papavizas G C, Fravel D R and Lewis J A 1987 Proliferation of *Talaromyces flavus* in soil and survival in alginate pellets. Phytopathology 77, 131–136.

Papavizas G C and Lewis J A 1981 Introduction and augmentation of microbial antagonists for the control of soil-borne plant pathogens. *In* Biological Control in Crop Production. Ed. G C Papavizas. pp 305–322. Allanheld Osmun and Co., Totowa, NJ.

Papavizas G C and Lewis J A 1989 Effect of *Gliocladium* and *Trichoderma* on damping-off and blight of snapbean caused by *Sclerotium rolfsii*. Plant Pathol. 38, 277–286.

Polon J A 1973 Formulation of pesticidal dusts, wettable powders and granules. *In* Pesticide Formulations. Ed. W Van Valkenburg. pp 143–234. Marcel Dekker, New York.

Ricard J L 1988 Biocontrol of pathogenic fungi in wood and trees, with particular emphasis on the use of *Trichoderma*. Biocontrol News Inf. 9, 133–142.

Rovira A D 1965 Interactions between plant roots and soil microorganisms. Annu. Rev. Microbiol. 19, 241–266.

Schenk N 1987 Vesicular-arbuscular mycorrhizal fungi and the control of fungal root diseases. *In* Innovative Approaches to Plant Disease Control. Ed. I Chet. pp 179–191. John Wiley and Sons, New York.

Suslow T V and Schroth M N 1982 Rhizobacteria of sugarbeets: Effects of seed application and root colonization on yield. Phytopathology 72, 199–206.

Woodward J 1985 Immobilized Cells and Enzymes: A Practical Approach. IRL Press, Washington, DC. 178 p.

Zabriskie D W, Armiger W B, Phillips D H and Albano P A 1980 Trader's Protein. Memphis, TN. 58 p.

D. L. Keister and P. B. Cregan (Eds.), The rhizosphere and plant growth, 289–295.

PLSO BARC29

Formulation, delivery systems and marketing of biocontrol agents and plant growth promoting rhizobacteria (PGPR)

J.L. McINTYRE[1] and L.S. PRESS[2]
[1]*Ecogen Inc., 2005 Cabot Boulevard West, Langhorne, PA 19047-1810, USA and* [2]*Biotechnology Consultant, 78 Princeton Avenue, Rocky Hill, NJ 08553, USA*

Key words: bacteria, biocontrol, carrier, cotton, Dagger^R G, damping-off, encapsulation, formulation, Pseudomonas, Pythium, Rhizoctonia

Abstract

From formulation through field evaluation and marketing, biological disease control agents and PGPR's provide a special difficulty; they are alive. This fact dictates special consideration for both strain selection and how it will be prepared and packaged for the real world of agriculture. These inoculants must be produced, stored, delivered, and ultimately used by the grower. Therefore, the inoculants must be maintained at proper concentrations on a carrier over significant time periods. Upon delivery the carrier must also enable the inoculant to reproduce itself and colonize the target plant. Upon application, protection from environmental factors may also be required. Additionally, the product must be in a formulation that is compatible with the grower and his practices. These considerations must all be taken into account in order to develop a product that is not only efficacious and cost effective against chemicals with which it may compete, but also a product that will be used by the consumer. Approaches to successfully develop and deliver biopesticides and PGPR's to the agricultural market will be discussed.

Introduction

The use of biological agents for control of arthropod and herbaceous pests celebrates its one hundreth anniversary this year (Morrison, 1989). It was in 1889 that vedalia beetles imported from Australia were released in California to control the important citrus pest, cottonycushion scale. Within a year after their release, the beetles had significantly reduced the scale problem, such that growers were no longer threatened by this formerly serious pest.

In the years that followed, researchers attempted to control other problematic insect pests using this strategy of release of predator insects. And, in so doing, they also began to learn some of the problems inherent in this approach, such as the need to re-release predators to keep their populations high, and thus the need for insect rearing. In attempting to develop alternative control measures, researchers expanded their horizons, looking also at the use of organisms that cause disease in target pests, but positive results were few.

Although the introduction of DDT and related products in the 1940's temporarily deferred enthusiasm away from biocontrol research, the diversion was short-lived as the environmental impact of these noxious chemicals became apparent. The impetus for investigating more biocontrol measures was provided not only by the desire for environmentally safer products, but also by the observation of growing resistance of insects to many chemical pesticides. The Agricultural Research Service (USDA) committed more resources to this area of pest control research with a number of successes, one of the classic ones being the isolation of *Bacillus thuringiensis*

HD1 strain by Howard Dulmage in 1966. Today this organism and related species and strains are sold under a variety of trade names for control of worms, beetles and dipteran pests.

Bacillus popilliae (milky spore disease) for control of Japanese beetles claims fame as the first commercially marketed microbial pesticide. However, this product, although effective, has never realized its commercial potential based on total end user sales of grub control products. The reason that this biocontrol agent does not enjoy a greater market share than it does can be blamed on the economics of production (Weiser *et al.*, 1976). Because *B. popillae* is grown in the host insect rather than using a fermentation process, production costs, and therefore final cost to the consumer, are higher than for competing chemicals. Thus, it is apparent from this example that field efficacy is but one component to the overall formula for success.

Early work with soil organisms for control of plant disease and nematodes was also less than promising. In 1955, after a hiatus of several decades, scientists again began to look seriously at using soil organisms for control of certain plant diseases such as those caused by Rhizoctonia and Pythium. Last year Ecogen Incorporated marketed the first commercial product using a Pseudomonas bacteria to control Rhizoctonia and Pythium diseases in cotton. This product was the first biological fungicide to be registered by the EPA.

Even though a newly discovered biocontrol agent or PGPR may perform well in a laboratory situation, there are many obstacles that need to be addressed for the product to be developed successfully. That it took nearly thirty years of research and development for a product of this type to make it to the market may seem excessive, but scientists were at the beginning of the learning curve. In addition to screening for effective control agents, and characterizing and defining the properties of the most promising ones, researchers had to learn how to apply these unique pesticides and in what form to deliver them to ensure optimal performance. Variable performance is still a concern for both producers and customers.

Although knowledge and experience with microbial control agents has progressed, the development of any novel organism to commercialization is frought with the same litany of technical issues. In addition, there are important commercial issues. Biopesticides are not always cost-competitive with chemicals and are usually pest-group-specific. The latter characteristic makes these pesticides safer to humans and other non-target species including mammals, avian and aquatic organisms, and beneficial microbes and insects. Where there exists only one major pest problem this is an advantage. In some situations, e.g., where disease is caused by a complex of pathogenic fungi, specificity may be disadvantageous for wild-type strains. Genetic improvements including the use of recombinant DNA methods may enable the development of strains that will be able to control an entire pest complex, but still provide the benefit of safety to non-target species.

With all the attendant problems potentially obstructive to the development of biocontrol agents, the incentive for work in this area is even greater today than in the 'Silent Spring' era. The same concerns for environmental safety exist today as they did in the 1950s and 1960s (Newsom *et al.* 1976), and may be even more intense as evidenced by some state and federal regulations that are restricting the use of certain chemical pesticides. Furthermore, costs associated with the registration of chemical pesticides are excessive whereas the registration costs for naturally-derived pesticides are insignificant by comparison, and the process is quicker in the interest of expediting the use of safer products (Betz *et al.*, 1987; Tantillo, 1989). It should also be noted that with the advent of biotechnology, sophisticated techniques are available for selective improvements of active microbial strains, improvements designed to enhance their effectiveness and make their performance competitive with, or better yet, superior to their chemical counterparts.

In economic terms, the worldwide market for biopesticides is wide open (Carlton, 1989). In 1987, all fungicides were a $4.1 billion market, herbicides $8.6 billion and insecticides $6.1 billion. Biofungicides and bioherbicides were each less than $1 million of that market. Although bioinsecticides were a comparatively impressive $45–50 million, that figure represents less than

1% of the total market. Recently, Frost and Sullivan Inc., New York, N.Y., speculated that the world market for genetically engineered microbial and biochemical pesticides may reach $8 billion by the year 2000 (Hoffman, 1989). Thus the market opportunity for an environmentally safe, effective biopesticide is very high.

The condition of 'effectiveness' in assessing market opportunity must be emphasized. Growers have had available to them an arsenal of chemical pesticides, and have to come to expect high levels of pest and disease control. In situations where an effective chemical pesticide has been widely and consistently used by growers, a biological competitor would first of all need to exhibit equivalent or superior performance at a competitive price. To capture market share then, the new biocontrol agent or PGPR would need to offer additional safety, lower cost or other benefits as well as qualities that appeal to consumers *e.g.*, 'pleasant' physical characteristics, ease of application, *etc.*

It is the purpose of this article to illustrate how microbial agents, once identified, characterized and cultured, may be subsequently prepared for commercial agricultural use and introduced to the market. The focus is on the formulation, delivery and marketing of bacterial plant disease control agents and plant growth promoting rhizobacteria. (Fungal biocontrol agents and rhizobia are the subject of other articles in this volume.)

Development of a biocontrol agent or PGPR

Pre-commercialization phase

It is useful here to review the considerations and steps necessary in the development of a biocontrol agent or PGPR. First, a biocontrol agent for the target disease or PGPR for the desired effect (generally yield enhancement) must be identified. This step demands a screening process which often relies on the isolation of organisms from the above-ground surfaces or rhizosphere/rhizoplane of healthy plants located either near diseased ones, or in fields free from the pathogen. Organisms may also be isolated from plants with suppressed disease symptoms. Each isolate must be maintained in pure culture and then bioassayed to determine whether or not activity exists. At times it may be of interest to determine *in vitro* production of metabolites inhibitory to the target pathogen, provided that there is an awareness that *in vitro* and *in planta* effects may not correlate. Those organisms with potential in the laboratory and greenhouse must be field evaluated to see if they are effective under existing environmental conditions.

Once activity is documented, the physiology and biochemistry of the organism is studied so as to understand those factors that may affect viability, growth, storability and field performance. The information gained from these studies not only facilitate the development of production methods, but provide insight for strain improvement (Rowe and Margaritas, 1987). Commercial strains must be stable in culture and grow in high yield.

Fermentation conditions must be developed to enable large-scale production of the organism for demonstration field trials and eventual commercial production. Maintenance of sterility during the culture process is imperative to prevent contamination, which may ultimately affect performance (Churchill, 1982). Media components have also been known to affect field performance through modification of an organism's toxic-potential, and therefore must be selected carefully (Dulmage, 1981). Deep-tank fermentation is the most common approach to mass production (Kenney and Couch, 1981), although some organisms may be more amenable to fermentation on a solid carrier. [Semisolid fermentation methods were used to produce early commercial formulations of *B. thuringiensis* in the U.S. (Dulmage, 1981)]. Where necessary, non-destructive means for harvesting and concentrating organisms from fermentation broth must be developed. Techniques may include centrifugation, filtration, and evaporation. Upon concentration, and dependent upon final formulation, the material may require further processing such as either spray-drying or freeze-drying.

Formulation

Once a reliable fermentation system is available, sufficient quantities of biocontrol agent or PGPR should be available for formulations research. Considerations in developing a formulation for a living biocontrol agent or PGPR may be subdivided into two categories: those that deal with introducing the agent in the field and those that deal with introducing the agent to the dealer's shelf, and keeping it there! Because the second named category is prerequisite to getting the product into the field, it will be discussed first.

For a product to sell, it must be in a form that is acceptable to growers such that they are familiar with the handling properties and can use conventional equipment to apply it, and the rates of application must be consistent with agricultural practices. Second, the material must have a shelf-life of at least one year, but more practically 18 months–2 years, to be successfully integrated into existing agricultural products distribution systems. If the shelf-life is less than two application seasons, manufacturers would be expected to take back product from the distributors, raising their costs and lowering profits. Some products may achieve this shelf-life under refrigeration, but storage at ambient temperatures is preferable not only because it is more economical and convenient, but also because the necessary refrigeration facilities may not be available at the selected distributors or at the growers.

Soil and foliar application

The formulation must also provide for the effective performance of the agent once applied in the field (Bowers, 1982; Connick *et al.*, 1989; Kenney and Couch, 1981; Yap, 1987). A formulation type must be selected based on the crop to be treated, either the disease to be controlled or the regulatory effect desired, and the anticipated application method. Formulation may be either liquid (aqueous or non-aqueous) or dry (wettable powder, dry flowable, granular). Typical additives may include dispersants, humectants, thickeners, surfactants, antimicrobials and stabilizers. Some of these aid in the application and dispersal of the product, others are necessary to protect the biocontrol agent or PGPR from adverse environmental factors such as water stress, unfavorable pH, competitors, metal ions, *etc.* in the soil; and unfavorable pH, dessication, ultraviolet radiation, competitors, *etc.* if applied foliarly. Formulations may also enable the biocontrol agent or PGPR to perform compatibly with chemical pesticides in treatment regimes designed for multiple pest problems.

Dry formulations generally utilize carriers onto which the inoculum is absorbed. The choice of carrier is critical to the performance of the inoculum, as carriers are necessary to ensure adequate dispersal and to protect the organisms from potentially detrimental edaphic or climatic conditions. Examples of carriers are peat and vermiculite. Ecogen's Dagger^R G product is formulated on peat, an acceptable but not optimal carrier. While peat has a high water-retention capacity, water distribution throughout the medium is not consistent. Furthermore, it is not sterile and the shelf-life of the product is limited. Obviously, the development of suitable carriers is not trivial. A comprehensive knowledge of the physiology and biochemistry of the organism to be formulated should help researchers identify carriers with optimal properties.

Liquid formulations (most commonly water or, for spore formers, oil), and slurries (most commonly reconstituted from either dry or moist-carrier based formulations) may also be developed, either as in-furrow applications or as foliar sprays (Paau and McIntyre, 1987). Stickers may need to be incorporated into formulations destined for foliar use. Other innovative formulations such as alginate and acrylamide microcapsules have been evaluated, but their practical potential needs to be demonstrated (Axtell and Guzman, 1987; Marois *et al.*, 1988; Lewis and Papavizas, 1987; Papavizas *et al.*, 1987). Liposome encapsulation may be yet another alternative for delivering viable organisms in protected microenvironments.

Seed treatment

For biocontrol agents intended to protect germinating plants from infection or to provide for growth promotion, the organism must be able to establish itself readily, colonizing the rhizosphere of the plant in sufficient density to suppress the pathogen (Bahme *et al.*, 1988; Cook and Baker,

1983). Application may be made as a seed treatment, in-furrow at time of planting or as a root dip on transplants. Historically, direct introduction of microbial antagonists into the soil for control of a pathogen has been largely unsuccessful due to the inability of the biocontrol agent to establish itself in a new habitat (Papavizas and Lewis, 1981). A very reliable way to have the biocontrol agent or PGPR in close proximity with the germinating seed is to apply it as a seed coating (Kommedahl and Windels, 1981). Factors to consider in selecting a formulation for coating seeds are the inoculum density achievable on the seed, stability of the coating both in organism viability and integrity of the coat, and the economics and feasibility of production. A noteworthy accomplishment is the completion by Agracetus of successful field trails on soybean with their Gold Coat™ product, a vermiculite-based inoculant containing an improved strain of rhizobia (Breitenbeck *et al.*, 1989; Paau, 1989). Agracetus' product is applied to soybean seeds by the grower. A potential drawback using this approach may be variability in performance if the grower applies the seed treatment. Treatment ideally should be done at the seed processor to ensure consistency. Also, seed storage conditions may be harsher than those of packaged goods, perhaps decreasing the inoculant's viability over the storage-life. These issues must be considered, as their importance is relative to the product in question.

Seed encapsulation

A variation on the idea of seed-coating is seed encapsulation, a process whereby a seed is enveloped in a gelatinous pellet which may contain micronutrients, beneficial microbes, pesticides, *etc.* (Paau, 1988). GEL-COAT™, developed by Plant Genetics Incorporated, is a familiar example. While the technique was pioneered for use with tissue culture-generated somatic embryos to make artificial seeds, the potential utility in providing a germinating seedling with a protective environment became apparent to the inventors. GEL-COAT™ is an alginate hydrogel preparation which has also been patented as a delivery system for entomopathogenic nematodes.

Marketing

Once the biological product is packaged and ready to go on the shelf, the manufacturer must be prepared to introduce it to the market. Bringing a new product to market requires careful thought and planning. Knowledge of the existing market and grower concerns, habits and preferences are essential.

The Dagger® G Experience

Dagger® G is a biofungicide having a strain of *Pseudomonas fluorescens* as the active ingredient (Currier *et al.*, 1988). The active strain was isolated from healthy cotton plants growing amidst plants afflicted with damping-off from the fungal pathogens Rhizoctonia and Pythium, and was one of over 200 strains evaluated for this application. Once isolated and cultured, the organism proved its ability to inhibit growth of the pathogens both *in planta* and in the field, making it a good candidate for development and commercialization.

Ecogen completed the development of Dagger® under an R and D limited partnership. Within less than a year, Ecogen finalized the form of the product, arranged for its manufacture, hired a sales force, established a distribution system, performed the required toxicological studies, obtained EPA registration, and made its first sales.

Dagger® is a granular formulation using peat as a carrier, and is applied as an in-furrow treatment at time of planting. Other formulations are in development.

The first step in commercializing Dagger® was to identify the target market. The U.S. cotton market treated with fungicide in-furrow for control of damping-off comprises about 1.0 to 1.3 million acres in the Mississippi delta and mid-South regions. The primary fungicide formulation used is granular, although liquid formulations are gaining in favor in some growing areas. The identified market regions have severe and destructive Pythium/Rhizoctonia problems, and represent a consistent market.

The next step is to position the product in the existing market. In the case of Dagger®, performance, cost and biological derivation were the three most important descriptors, with the first two being of primary importance. Thus our marketing plan was to emphasize performance and cost equivalent to that of competing chemicals, and to promote the characteristics of biologicals as added benefits.

Distributors are key to the marketing of agchem products. Given Ecogen's small size, its participation in the marketing process was critical. In our marketing strategy, distributors would need to participate in the sales of our products, and we would need their facilities and resources to warehouse and deliver our products, and also to collect monies from sales. Ecogen decided to use three distributors, and to use profit as incentive for them to push our product.

Our advertising approach was to mail information directly to an audience of about 6000 growers. These initial mailings were supplemented with advertising in print and on the radio. We also participated in grower meetings, trade shows, and conventions to promote the qualities of Dagger^{R}. As an encouragement to try a new biological product, we offered a rebate to growers who agreed to treat at least 100 acres. This offer stemmed from a concern that the product would be 'tested' by growers on relatively few acres since it was a new product and Ecogen, as a company, was not yet well known.

The first year of sales experience was instructive and rewarding. If anything, we underestimated the marketability of a biological approach to disease control. Dagger^{R} was used by growers, researchers, and extension people in comparison with chemical standards and in evaluation of compatibility with other pesticides like Temik^{R}. Dagger^{R} performed well, and, based on the successes of 1988, we expect a better market in 1989.

Conclusions

It is clear from the foregoing discussion that successful biocontrol agents and PGPR can be developed. It is also apparent that one of the most significant limitations in developing and commercializing these products has been in providing a formulation that will enable biocontrol agents to successfully compete in the marketplace. To capture market share, the new biological products must not only be cost-competitive, but they must also equal or out-perform their chemical counterparts in field efficacy, physical characteristics, handling properties, and shelf-life. Once an organism with either the desired disease control or regulatory activity is isolated, identified, and successfully produced, the key to either achieving or optimizing these attributes is in the formulation. The biocontrol agent must be able to maintain viability for extended periods, and must also be protected from potentially adverse environmental conditions in the field.

Thus, it is possible to have an organism that gives excellent results in the laboratory, greenhouse, and even small field trials, but that fails in practical use because it cannot survive the conditions imposed by existing agricultural products distribution systems, equipment or application techniques of the grower, or prevailing environmental conditions. An increased understanding of the physiology and biochemistry of each potential biocontrol agent should enable researchers to develop formulations that will meet their physiological needs as well as impart the physical characteristics necessary for successful field application. Formulations technology for biological agents is advancing and the time when these products will gain in popularity and sales is at hand.

References

Axtell R C and Guzman D R 1987 Encapsulation of the Mosquito Fungal Pathogen *Lagenidium giganteum* in calcium alginate. J Am Mosquito Control Association 3, 450–459.

Bahme J B, Schroth M N, Van Gundy S D, Weinhold A R, Tolentine D M 1988 Effect of Inocula Delivery Systems on Rhizobacterial Colonization of Underground Organs of Potato. Phytopathology 78, 534–542.

Betz F, Rispin A and Schneider W 1987 Biotechnology Products Related to Agriculture. Overview of Regulatory Decisions at the U.S. Environmental Protection Agency. *In* ACS Symposium Series 334, pp 316–327, American Chemical Society, Washington, DC.

Bowers R C 1982 Commercialization of Microbial Biological Control Agents. *In* Biological Control of Weeds with Plant Pathogens. Eds. R Charudattan and H L Walker. pp 157–173. John Wiley and Sons, New York.

Breitenbeck G A, Boquet D, Griffin J, Hallmark W, Harville B, Hutchinson R and Millhollen E 1987 Comparison of Seed- and Soil-Applied Inoculants for Soybean Production, Louisiana Agriculture, pp 69–70.

Carlton B C, 1989 Economic Considerations in Marketing and Application of Biocontrol Agents, Proceedings; UCLA Symposium, New Directions in Biological Control, Jan. 20–27, Frisco, Col., pp 419–434.

Churchill B W 1982 Mass Production of Microorganisms for Biological Control. *In* Biological Control of Weeds with Plant Pathogens. Eds. R Charudattan and H L Walker. pp 139–156. John Wiley and Sons, New York.

Connick W J, Lewis J A and Quimby P C 1989 Formulation of Biocontrol Agents to Reduce Plant Diseases and Weeds, Proceedings: UCLA Symposium, New Directions in Biological Control, Jan. 20–27, Frisco, Col., pp 345–372.

Cook R J and Baker K F 1983 The Nature and Practice of Biological Control of Plant Pathogens. The American Phytopathological Society, St. Paul, 539 p.

Currier T C, Skwara J E and McIntyre J L 1988 The Development of a *Pseudomonas fluorescens* Product (Dagger™ G) for the Control of *Pythium* and *Rhizoctonia* on Cotton. *In* 1988 Proceedings: Beltwide Cotton Production Research Conferences, pp 18–19. National Cotton Council, Memphis, Tenn.

Dulmage H T 1981 Production of Bacteria for Biological Control of Insects. *In* Biological Control in Crop Production. Ed. G C Papavizas. pp 129–141. BARC Symposium 5. Allanheld and Osmun, Totowa, NJ.

Hoffman J 1989 Biotechnology Spawns Alternative Pesticides. Chemical Marketing Reporter 235 (24) 28.

Kenney D S and Couch T L 1981 Mass Production of Biological Agents for Plant Disease, Weed and Insect Control. *In* Biological Control in Crop Production. Ed. G C Papavizas. pp 143–150. BARC Symposium 5. Allanheld and Osmun, Totowa, NJ.

Kommedahl T and Windels C E 1981 Introduction of Microbial Antagonists to Specific Courts of Infection: Seeds, Seedlings, and Wounds. *In* Biological Control in Crop Production. Ed. G C Papavizas. pp 227–248. BARC Symposium 5. Allanheld and Osmun, Totowa, NJ.

Lewis J A and Papavizas G C 1987 Application of *Trichoderma* and *Gliocladium* in alginate pellets for control of Rhizoctonia damping-off. Plant Pathol. 36, 438–446.

Marois J J, Fravel D R, Connick W J Jr., Walker H C, Quimby P C 1988 Preparation of Pellets Containing Fungi for Control of Soilborne Diseases. U.S. Patent 4, 724, 147.

Morrison J, 1989 Biological Control: Turns 100 This Year. Agricultural Research. U.S.D.A., ARS 37, 4–8.

Newsom L D, Smith R F and Whitcomb W H 1976 Selective Pesticides and Selective Use of Pesticides. *In* Theory and Practice of Biological Control. Eds. C B Huffaker and P S Messenger. pp 565–591. Academic Press, Inc. New York.

Papavizas G C and Lewis J A 1981 Introduction and Augmentation of Microbial Antagonists for Control of Soilborne Plant Pathogens. *In* Biological Control in Crop Production. Ed. G C Papavizas. pp 305–322. BARC Symposium 5. Allanheld and Osmun, Totowa, NJ.

Papavizas G C, Fravel D R, Lewis J A 1987 Proliferation of *Talaromyces flavus* in Soil and Survival in Alginate Pellets. Phytopathology 77, 131–136.

Paau A S 1988 Formulations Useful in Applying Beneficial Microorganisms to Seeds, Trends Biotechnol 6, 276–279.

Paau A S 1989 Improvement of Rhizobium Inoculants, Appl. Environ. Microbiol. 55, 862–865.

Paau A S and McIntyre J L 1987 Development of PGPR Inoculants: Challenges and Priorities, The First International Workshop on Plant Growth-Promoting Rhizobacteria: Proceedings, 44 p.

Rowe G E and Margaritis A 1987 Bioprocess Developments in the Production of Bioinsecticides by *Bacillus thuringiensis*. CRC Crit. Rev. Biotechnol. 6, 87–107.

Tantillo L 1989 The Growing Field of Biopesticides. ChemicalWeek 144, 46–47.

Weiser J, Bucher G E and Poinar G O Jr. 1976 Host Relationships and Utility of Pathogens. *In* Theory and Practice of Biological Control. Eds. C B Huffaker and P S Messenger. pp 169–185. Academic Press, Inc., New York.

Yap H H 1987 Microbial Insecticides in Aquatic Environments: Factors Affecting Efficacy in the Field. Technical Bulletin, Food and Fertilizer Technology Center, Taiwan 104, ii + 8 pp.

Short papers

D. L. Keister and P. B. Cregan (Eds.), The rhizosphere and plant growth, 299.

PLSO BARC63

Effect of a fluorescent-pigment producing Rhizobium on the severity of *Rhizoctonia solani* seed and root rot of *Phaseolus vulgaris*

L.K. BLUM, S.D. FREY and G. SOTO
Department of Environmental Sciences, University of Virginia, Charlottesville, VA 22903, USA

Key words: biological control, rhizobia, root decay, siderophore

Common beans, *Phaseolus vulgaris*, are often inoculated with *Rhizobium leguminosarum* bv. *phaseloi* (RLP) to alleviate the need for extensive nitrogen fertilization. Because rhizobia are competent rhizosphere colonizers and are known to produce bacteriocins and iron chelating compounds (siderophroes), rhizobia have the potential to reduce the severity of seed and root decay diseases. The objective of this study was to compare the ability of RLP strains to reduce the severity of seed decay and damping-off of *P. vulgaris*.

The approach used was to screen rhizobia for their ability to inhibit biomass production in laboratory culture by *Rhizoctonia solani*, a root pathogen of common beans. The fungus was inoculated into media in which the rhizobia had been grown previously and removed. Prior to fungal inoculation, supplemental sugars were added to the spent media to eliminate carbon and energy limitations. Those Rhizobium strains which reduced the fungal biomass by at least 50% were considered to be inhibitory, while those strains that increased growth by greater than 115% were considered to be stimulatory. Of 49 strains tested, six were inhibitory and six were stimulatory.

After the initial screening, several inhibitory and stimulatory rhizobia were examined for the production of fluorescent pigments on King's medium B (KMB) which was used to detect the production of water-soluble, yellow-green pigments that are characteristic of siderophores. Only Rhizobium strains capable of fungal inhibition produced a fluorescent pigment. However, not all (2 of 6) inhibitory strains produced pigment.

Several strains that exhibited the greatest degree of antagonism and fluorescent pigment production and one strain that enhanced fungal biomass production were used to study the effect of rhizobia on disease severity in greenhouse trials with *P. vulgaris*. Five treatments were used in each experiment: (1) no Rhizobium or fungus, (2) fungus plus the fungicide, captan, (3) fungus only, (4) Rhizobium only, and (5) fungus plus Rhizobium. The number of seeds which germinated, number of seedlings surviving to 28 days, disease severity, and foliage dry weight per pot were determined.

Seed inoculation with RLP strain 574-127K17 did not protect the seed from decay, however, for those seeds that germinated, seedling survival was significantly enhanced. For strains 573-127K14 and 576-2535, inoculation with these rhizobia decreased both the incidence of damping-off (increased seedling survival) and the index of disease severity. With strain 4022, a stimulatory strain, disease severity was significantly greater in the presence of the Rhizobium.

The results of these experiments suggest that screening of Rhizobium inoculants for their effect on seed and root disease may be as critical as screening for effectiveness (N_2-Fixation) and competitive ability (nodule occupancy).

D. L. Keister and P. B. Cregan (Eds.), The rhizosphere and plant growth, 300.

PLSO BARC64

Role of ammonia and calcium in lysis of zoospores of *Phytophthora* spp. by *Bacillus cereus* strain UW85

G.S. GILBERT, J. HANDELSMAN and J.L. PARKE
Department of Plant Pathology, University of Wisconsin-Madison, Madison, WI 53706, USA

Key words: ammonia, Bacillus, calcium, zoospores

Cultures and cell-free culture filtrates of *Bacillus cereus* strain UW85 protect alfalfa and soybean seedlings from damping off caused by *Phytophthora megasperma* (Handelsman *et al.*, 1988). Culture filtrate also lyses zoospores of *Phytophthora* spp. *in vitro* (Handelsman *et al.*, 1988). This study was undertaken to determine the role of ionic changes in the growth medium of UW85 in zoospore lysis.

UW85 was grown until it sporulated in casamino acid broth (CB). Spores were then removed by centrifugation and filtration to produce culture filtrate (BcF). CB and BcF were diluted with water or with calcium chloride (0.44 or 2.2 mmol L^{-1}) or ammonium chloride (2.22 or 22.2 mmol L^{-1}) solutions and adjusted to pH 5, 6, 7, 8, or 9. Each solution was then combined with an equal volume of zoospores of *P. megasperma* f. sp. *medicaginis* or *P. cactorum* in each of three replicate microtitre wells. After 1 hr the spores were fixed with glutaraldehyde and classified as zoospores, cysts, germlings, or lysed cells.

Samples of CB, BcF, and filtered zoospore suspensions were analyzed for P, K, Ca, Mg, S, Na, NH_4^+, NO_3^- and Cl. PHREEQE, a computer program for geochemical calculations (Parkhurst *et al.*, 1985) was used to calculate ion speciation in the various treatments.

When BcF was added to Phytophthora zoospores nearly 100% of the zoospores lysed, whereas the same concentration of CB lysed less than 10% of the zoospores. Calcium and ammoniacal-N concentrations and solution pH were correlated with zoospore lysis. BcF contained more than four times more ammoniacal-N and only 3% of the calcium contained in CB. The pH in CB and BcF were 5.9 to 8.8 respectively. PHREEQE modelling indicated that increasing the pH of the solutions led to a decrease in the free calcium ion concentration and an increase in the NH_3 form of ammoniacal-N. Neither calcium nor ammonia concentrations alone were closely correlated with lysis, but zoospores of both *Phytophthora* species lysed only when the ratio of ammonia to calcium was high [$-\log(Ca^{2+}):-\log(NH_3) > 0.8$]. When CB and BcF were adjusted to the same ionic ratio, zoospores responded similarly, suggesting that the ammonia to calcium ratio was the main factor responsible for zoospore lysis by UW85.

These results highlight the potential importance of the ionic component when studying the effects of complex solutions such as bacterial cultures on organisms. We are currently investigating the importance of zoospore lysis in the *B. cereus-P. megasperma* biocontrol system, and whether UW85 can create an ammonia to calcium ratio in the rhizosphere that would lyse zoospores.

References

Handelsman J *et al.* 1988 *In* Molecular Genetics of Plant-Microbe Interactions. Eds. R Palacios and DPS Verma. pp 303–308. APS Press, St. Paul.
Parkhurst D L *et al.* 1985 U.S. Geological Survey Water-Resources Investigation 80–96.

D. L. Keister and P. B. Cregan (Eds.), The rhizosphere and plant growth, 301.

PLSO BARC65

Deleterious rhizobacteria for biocontrol of weed seeds and seedlings in soil

R.J. KREMER[1,2], M.F.T. BEGONIA[2] and L. STANLEY[1]
[1]*Cropping Systems and Water Quality Research, USDA, ARS, and* [2]*Department of Agronomy, University of Missouri, Columbia, MO 65211, USA*

Key words: bacteria, chemotaxis, phytotoxins, rhizosphere

Rhizobacteria isolated from seedling roots of several economically important weeds caused significant decreases in seedling growth when seedlings were inoculated with the rhizobacteria and grown under controlled conditions (Cherrington and Elliott, 1987; Kremer, 1986). For successful development of deleterious rhizobacteria (DRB) as weed biocontrol agents, several key characteristics must be defined including colonizing ability, chemotactic response, and modes of action. The objectives of these studies were to determine the importance of the following characteristics in the effectiveness of DRB on controlling weed seeds and seedlings: (1) phytotoxin production, (2) colonization patterns, and (3) chemotactic ability.

Rhizobacteria were assayed for deleterious effects on several broadleaf weed seedlings in growth pouches and in soil in greenhouse and field trials. Phytotoxicity of cell-free filtrates from selected DRB cultures was bioassayed in water agar using cress and ryegrass indicator species. Velvetleaf seedling roots inoculated with DRB were examined under scanning electron microscopy (SEM) to observe rhizoplane colonization patterns. Selected DRB were also assayed for chemotaxis to weed seed and seedlings exudates using capillary tube and soil plate assays (Begonia and Kremer, 1988).

Various DRB isolates inhibited weed seedling growth and exhibited deleterious effects including stunted, necrotic tap roots, severely inhibited lateral root development, and chlorotic foliage. Soil inoculation with selected DRB resulted in reduced seedling emergence (<55%), decreased seedling vigor, and severely damaged root systems. Cell-free filtrates of DRB significantly inhibited radical growth of both cress and ryegrass (decrease to 30% of control). Subsequent screening on weed seedlings revealed highly inhibitory activity on all tested seedlings. Velvetleaf seedling roots inoculated with DRB examined under SEM had considerably fewer root hairs, confirming inhibitory effects by DRB. Further SEM examination of the rhizoplane revealed characteristic colonization patterns including bacterial cell alignment along cell junctures or formation of distinct microcolonies. Mucigel production in association with DRB was abundant on rhizoplanes. Chemotaxis toward velvetleaf seeds and exudates both *in vitro* and in soil was exhibited by DRB isolates.

These results suggest that selected DRB may produce phytotoxins, which may contribute to deleterious effects on weed seedlings. Selected DRB may colonize weed seedling rhizoplanes via chemotaxis and appear to establish in characteristic patterns.

References

Begonia M F T and Kremer R J 1988 Agron. Abst., p. 210.
Cherrington C A and Elliott L F 1987 Plant and Soil 101, 159–165.
Kremer R J 1986 Am. Nurseryman 164, 162–163.

D. L. Keister and P. B. Cregan (Eds.), The rhizosphere and plant growth, 302–303.

PLSO BARC66

An improved, *in vitro* technique for rapidly assaying rhizosphere bacteria for the production of compounds inhibitory to *Rhizoctonia solani* and *Gaeumannomyces graminis* var. *tritici*

D.A. SCHISLER, M.H. RYDER and A.D. ROVIRA
CSIRO Division of Soils, Glen Osmond, South Australia 5064, Australia

Key words: antagonism, Ggt, Rhizoctonia, technique

With the extraordinary swelling of interest in biological control research in recent years, there has been an increased need for protocols for rapidly assessing large numbers of candidate biological control organisms for antagonistic potential. We have developed an improved, *in vitro* method for efficiently selecting rhizosphere bacteria which are antagonistic to the mycelial growth of fungal pathogens. Assays currently utilized involve dual inoculation of 1–4 putative antagonists and a fungal pathogen onto a nutrient medium. Fungal growth inhibition is assessed up to 10 days later (Fig. 1a). Our assay allows 12 bacterial isolates per Petri plate to be appraised after 4 days (Fig. 1b). Fungal pathogens are grown in pure culture on moistened, pulverized rice hulls (250–500 μm). Bacterial isolates are transferred to a Petri dish containing nutrient medium and the surface of the medium is sprinkled with rice hulls colonized by a fungal pathogen. Zones of inhibited fungal growth around bacterial colonies are distinct after incubating the plates for 4 days at 25°C.

This technique was used to measure the impact of tillage and crop rotation on the proportion of wheat rhizosphere microbes antagonistic to *Rhizoctonia solani* and *Gaeumannomyces graminis* var *tritici*. Tillage (conventional cultivation compared to direct drilling) rarely significantly influenced the percentage of antagonistic "total" aerobic bacteria (TAB), pseudomonads (PS), or heat resistant organisms. A pasture-

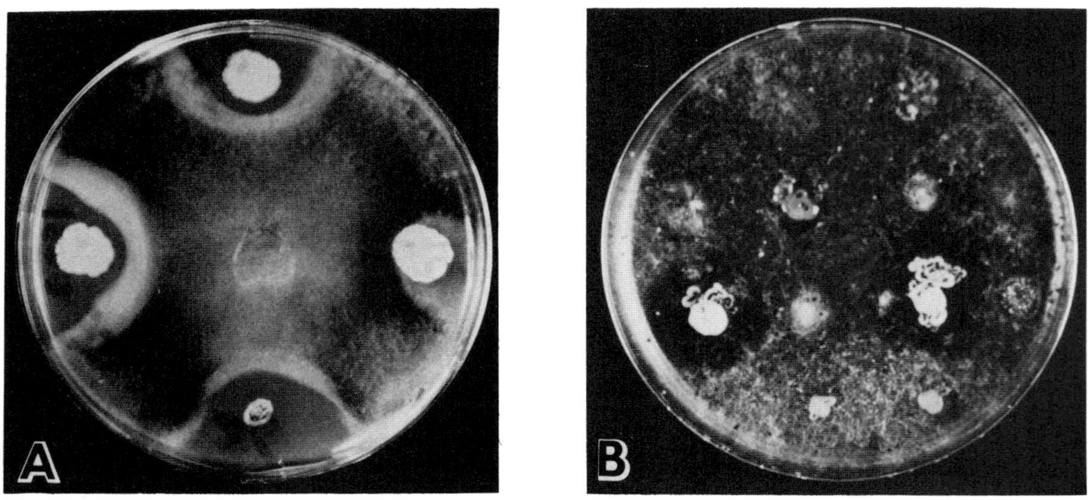

Fig. 1. **A**.) Traditional method and **B**.) Sprinkle technique.

wheat rotation fostered higher percentages of antagonistic TAB and PS than did continuous wheat (8.4% vs 4.4% and 32.0% vs 27.6%, respectively) though the populations of TAB and PS were higher under continuous wheat (8.31 vs 8.18 and 6.74 vs 6.60 {$\log_{10}$ c.f.u./g root}, respectively). The assay medium (PDA vs Tryptic Soy Agar) and test fungus significantly influenced the percentage of isolates deemed antagonistic. In obtaining these results we have demonstrated that the sprinkle technique saves time and labor when assaying for bacterial antagonists. This technique can also be used to assay for the loss of antibiotic production after bacterial mutagenesis.

D. L. Keister and P. B. Cregan (Eds.), The rhizosphere and plant growth, 304.

PLSO BARC67

Improved rhizosphere competence of *Trichoderma harzianum* by protoplast fusion

A. SIVAN and G.E. HARMAN
Department of Horticultural Sciences, Cornell University, Geneva, NY 14456, USA

Key words: protoplast fusion, rhizosphere competence, *Trichoderma harzianum*

Rhizosphere competence enables microorganisms to proliferate and establish in the rhizosphere of the developing plant following seed treatment. Seed treatments with beneficial microorganisms, are an attractive way to apply microbial antagonists, so the ability of such microorganisms to colonize the rhizosphere may be crucial for optimal biological activity. Ideally, a rhizosphere competent strain applied to seeds will colonize the entire root system.

Trichoderma spp. are probably the most studied fungal antagonists however, rhizosphere competence is not wide spread amongst wild type strains. Although rare, rhizosphere competence may be found in wild type strains of *T. harzianum* (Sivan and Chet, 1989). It can also be induced in wild types of mutation (Ahmad and Baker, 1987). Rhizosphere competence was recently demonstrated in protoplast fusion progeny (Harman *et al.*, 1989).

The objectives of this study were to test and quantify the rhizosphere competence of the protoplast fusion progeny 1295-22 which was obtained by fusion protoplasts of *T. harzianum* strains: T95 (a benomyl resistant rhizosphere competent mutant) and T12.

Protoplast fusion between 2 auxotrophic mutants of *Trichoderma harzianum* ($T12his^-$ and $T95lys^-$) resulted in a wide variety of prototrophic progeny (Stasz *et al.*, 1988). One of the strains from this fusion (1295-22) showed also a better biocontrol activity than the parental strains. In order to compare and quantify the rhizosphere competence of this strain with its parental types seeds of cotton and sweet corn were treated with conidia of these strains and planted in a soil with a constant moisture content (−72 mbars). The tested strains could be isolated from the rhizosphere of most portions of corn roots 20–22 cm long. T95 colonized mainly the upper and the lower parts of the roots while most of T12 population densities were found primarily on the upper third of the root. However, 1295-22 colonized the entire root. T95 and T12 along the lower half of corn root were obtained mainly from the rhizosphere soil but not from the rhizoplane. Conversely, strain 1295-22 colonized both the rhizosphere soil and the rhizoplane. Strain 1295-22 colonized more root segments than the parental types.

Population densities of all strains were, in general, lower in cotton rhizosphere than with corn.

References

Ahmad J S and Baker R 1987 Phytopathology 77, 182–189.
Harman G E *et al.* 1989 Plant Disease.
Sivan A and Chet I 1989 Phytopathology 79, 198–203.
Stasz T E *et al.* 1988 Mycologia 80, 141–150.

D. L. Keister and P. B. Cregan (Eds.), The rhizosphere and plant growth, 305.

PLSO BARC68

Sick pathogens make poor pests in biological control of corn diseases

N.G. VAKILI
USDA, ARS, Midwest Area and Department of Plant Pathology, Iowa State University, Ames, IA 50011, USA

Key words: crop management system, mycopathogenic fungi, seed treatment, soilborne plant pathogens

Introduction

Mycopathogenic fungi were isolated from fungi that cause stalk rot of corn (Vakili, 1985). Cultures of mycopathogens with their host fungi resulted in infections, growth retardation, discoloration, and death of the hosts fungi (Vakili, 1985; 1989). Application of mycopathogens to foliage of corn plants, during growing season, resulted in genotype-mycopathogen interaction and yield increase (Vakili and Bailey, 1989). Objectives of this study were 1.) to assess the effect of seed treatment with mycopathogens on biocontrol of seedborne and soilborne fungi, and 2.) to evaluate if field infestation of *M. damnosa* increased the frequency of association of this mycopathogen with *F. moniliforme* in corn pith.

Materials and methods

Corn (*Zea mays* L.) kernels were soaked in aqueous conidial suspension of *Melanospora damnosa*, *Nectria gliocladioides*, or *Sphaeronaemella helvellae*. Untreated kernels served as the standard. To test for seedborne fungi, kernels were germinated in Standard Warm Tests; to test for soilborne fungi, kernels were germinated in Standard Cold Test and in field soil. Seedling vigor was measured as dry weight of root or shoot or both. Fungi colonizing germinated kernels and seminal roots were identified and counted. Random samples of germinated kernels were incubated on acidified potato-dextrose agar and emerging fungi were identified and counted. During the first year, plants in a continuous corn cultivation trial were sampled for the occurrence of *M. damnosa*. The field was then infested by incorporating the *M. damnosa* treated plants into the top soil for 4 successive years. During the fifth year *M. damnosa* was not applied and plants were again sampled to determine the frequency of association of *M. damnosa* with *F. moniliforme* in pith tissues.

Results and discussion

Treatment of corn kernels with mycopathogenic fungi reduced the incidence of seedborne and soilborne fungi, reduced root necrosis, and increased the vigor of the seedlings as compared with untreated kernels. Infestation of soil with *M. damnosa* increased the frequency of assocation of *M. damnosa* with *F. moniliforme* in corn pith samples from 0.5 in year 1 to 80% in year 5.

References

Vakili N G 1985 Phytopathology 75, 1201–1207.
Vakli N D 1989 Mycol. Res. 67–74.
Vakili N G and Bailey T B Jr 1989 Crop Sci. 29, 183–190.

D. L. Keister and P. B. Cregan (Eds.), The rhizosphere and plant growth, 306.

PLSO BARC69

Mutational changes in the O-antigenic side chains of the lipopolysaccharides of *Pseudomonas* spp. affect colonization of but not adhesion to potato roots

L.A. DE WEGER[1], P.A.H.M. BAKKER[2], B. SCHIPPERS[2], M.C.M. VAN LOOSDRECHT[3] and B. LUGTENBERG[1]
[1]*Department of Plant Molecular Biology, Leiden University, Leiden,* [2]*Department of Plant Pathology, Phytopathological Laboratory, Baarn, and* [3]*Department of Microbiology, Wageningen Agricultural University Wageningen, The Netherlands*

Key words: adhesion, cell surface properties, lipopolysaccharides, root colonization

The world wide interest in *Pseudomonas* spp. as biocontrol agents is mainly based on their *in vitro* antagonistic potentials combined with their ability to colonize the roots of many plants efficiently. The properties that confer the colonization abilities on these bacteria are largely unknown. This study focuses on the role of the O-antigenic side chain of the LPS (=lipopolysaccharides) of two plant growth-stimulating Pseudomonas strains, WCS358 and WCS374, in the root colonization process.

After selection with phages specific for either strain WCS358 or strain WCS374, mutants were obtained which lack the O-antigenic side chain, LWP358-43b and LWP374-30b, respectively. Colonization experiments (for details see de Weger *et al.*, 1987) showed that inoculation of the roots with one of the mutant strains resulted in colonization of the deeper root parts in less than 50% of the replicate plants, while the wild type strains colonized the whole root system of more than 90% of the plants.

Further characterization of these LPS mutant strains demonstrated that they were not affected in growth rate, siderophore production and motility, but they were affected in their cell surface properties (for details on the measurements see van Loosdrecht *et al.*, 1987). Compared to the parent strain, the LPS mutant of strain WCS374 showed an increase in hydrophobicity and cell surface charge, while the hydrophobicity of the LPS mutant of strain WCS358 was decreased and the cell surface charge was slightly increased (de Weger *et al.*, 1989). These differences in physicochemical cell surface properties prompted us to look at the interaction of these strains with potato roots. We found that the firm attachment of mutant and parent strain to sterile potato roots did not differ significantly.

The results suggest that the O-antigenic side chain is involved in the efficient colonization of the potato plant root, but that it does not exert this effect at the stage of firm adhesion to the root surface. However, we cannot yet exclude that the mutations that affect the O-antigenic side chains are pleiotrophic and that another molecule than LPS is in fact responsible for the reduced colonization ability of these mutant strains.

References

De Weger L A *et al.* 1987 J. Bacteriol. 169, 2769–2773.
De Weger L A *et al.* 1989 J. Bacteriol. 171, 2756–2761.
Van Loosdrecht M C M *et al.* 1987 Appl. Environ. Microbiol. 53, 1898–1901.

D. L. Keister and P. B. Cregan (Eds.), The rhizosphere and plant growth, 307.

PLSO BARC70

Fungal antagonists of *Pythium ultimum* from a suppressive Sphagnum peat

H. WOLFFHECHEL
The Royal Veterinary and Agricultural University, Department of Plant Pathology, D-1871 Copenhagen, Denmark

Key words: biological control, *Geomyces pannorum*, *Gliocladium virens*, *Trichoderma harzianum*

A commercial Sphagnum peat (limed and fertilized, pH 6.3) has been shown to be suppressive to damping-off and root rot caused by *Pythium ultimum* in an assay employing cucumber seedlings (*Cucumis sativus* cv. "Ideal Nova") as test plants and naturally infested soil, amended with oatmeal as inoculum. The suppressiveness was greatly reduced by heat treatment at 60°C or by addition of benomyl to the peat, whereas several applied antibiotics had no effect on the suppressiveness (Wolffhechel, 1988).

To investigate the possible causal agents of the suppressiveness, several microorganisms were isolated from the peat. Spore suspensions of these (as single strains or mixtures of strains) were added to peat in which the suppressiveness had been destroyed by heat treatment at 96°C (Wolffhechel, 1989).

Of the microorganisms tested, *Trichoderma harzianum*, *Geomyces pannorum*, and a mixture of *Penicillium* spp. and *Aspergillus* spp. re-established the suppressiveness of the peat against Pythium whereas bacteria and actinomycetes had no effect. *Gliocladium virens* inhibited plant growth when added to the heat treated peat.

T. harzianum and *G. virens* were then tested in a Pythium-conducive peat (limed and fertilized, pH 6.1) not subjected to any heat treatment. Both strains induced suppressiveness to Pythium, and no direct effect on plant growth was observed. Apparently, the biological vacuum created by heat treatment of peat had increased the activity of *G. virens* so that it attained population levels inhibitory to plant growth.

The results indicate that fungal antagonists were responsible for the suppression of Pythium disease in the peat. Several antagonists were isolated, but the role of each individual antagonist in the suppression still needs to be determined.

References

Wolffhechel H 1988 Acta Hortic. 221, 217–222.
Wolffhechel H 1989 Växtskyddsnotiser 53, 7–11.

D. L. Keister and P. B. Cregan (Eds.), The rhizosphere and plant growth, 308.

PLSO BARC71

Siderophore-producing bacteria isolated from roots of iron-efficient and inefficient grasses

D.B. ALEXANDER and D.A. ZUBERER
Department of Soil and Crop Sciences, Texas A&M University, Texas Agricultural Experiment Station, College Station, TX 77843, USA

Key words: grasses, iron, siderophores

Siderophores produced by rhizosphere microorganisms may enhance iron acquisition by plants by increasing the availability of iron in the root environment (Crowley *et al.*, 1987). In a greenhouse experiment, we have examined the populations of siderophore-producing bacteria (SPB) colonizing the roots of St. Augustinegrass, a turfgrass which commonly exhibits iron chlorosis, and bermudagrass, a turfgrass which resists iron chlorosis, and have measured the capacity of the bacteria to produce siderophores *in vitro* in an iron-limited medium.

Serial dilutions were prepared with washed roots taken from core samples of bermudagrass and St. Augustinegrass, and inoculated onto Chrome Azurol S (CAS) agar, Tryptic Soy agar (TSA, diluted 1:10), TSA + hexadecyltrimethylammonium bromide (HDTMA), and TSA + 8-hydroxyquinoline (8-OHQ). Bacterial populations were counted by a standard dilution plating method. Production of siderophores was indicated by the development of an orange halo surrounding colonies on CAS agar (Schwyn and Neilands, 1987). The amount of siderophore produced *in vitro* was measured with CAS assay solution after culturing isolates in a mineral salts + glucose medium containing no added iron. After growing the grasses for 4 months in washed sand amended with 50 g $CaCO_3$ kg^{-1} and 2, 3, 4, 5, or 10 mg Fe (ferrihydrite) kg^{-1}, bacterial populations were counted again by inoculating serial dilutions of washed roots onto CAS agar.

Total counts on CAS agar represented only 13–17% of the populations of rhizosphere bacteria which grew on TSA. HDTMA, an ingredient in CAS agar, reduced bacterial growth by 85% when added to TSA. 8-OHQ, a strong iron-chelator, reduced total counts by more than 99% when added to TSA (100 mg mL^{-1}). Isolates from TSA + 8-OHQ all produced halos on CAS agar.

At planting, bermudagrass roots yielded higher total counts on CAS agar than St. Augustinegrass roots, with a lower percentage of SPB. The percentage of colonies which produced visible halos on CAS agar was surprisingly low for both grasses, however colonies which exhibited no halo on the plates used to enumerate SPB often produced small halos when isolated in pure culture on CAS agar.

After 4 months in sand amended with ferrihydrite, bermudagrass roots yielded lower total counts of bacteria than St. Augustinegrass roots, with a higher percentage of SPB. Neither grass exhibited significant differences in total counts or percent SPB at the two levels of iron. Different species of SPB were isolated from the two grasses, though most were members of the genus Pseudomonas. Roots of both grasses were colonized by bacteria which produced substantial concentrations of siderophore *in vitro* in an iron-limited medium.

References

Crowley D E *et al.* 1987 *In* Iron Transport in Microbes, Plants, and Animals. Eds. G Winkelmann *et al.* pp 375–386. VCH Publishers, New York.

Schwyn B and Neilands J B 1987 Anal. Biochem. 160, 47–56.

D. L. Keister and P. B. Cregan (Eds.), The rhizosphere and plant growth, 309.

PLSO BARC72

Production of siderophore-like iron chelators by ericoid and ectomycorrhizal fungi

B.A. CALDWELL[1], R.P. GRIFFITHS[1], R.G. LINDERMAN[2] and J.E. LOPER[2]
[1]*Department of Microbiology, Oregon State University, Corvallis, OR 97331, USA and* [2]*Horticultural Crops Research Laboratory, USDA, ARS, Corvallis, OR 97330, USA*

Key words: mycorrhizal fungi, siderophores

Introduction

Siderophores can play critical roles in both plant nutrition and competition between microorganisms. Production of hydroxamate siderophores by both ericoid and ectomycorrhizal fungi may provide an additional mechanism for the suppression of antagonistic microorganisms. This study was designed to determine if there was a correlation between mycorrhizal fungal antagonism to an actinomycete and siderophore production by these fungi.

Materials and methods

The chrome azural S (CAS) universal siderophore assay of Schwyn and Neilands (1987) was adapted to screen two ericoid and four ectomycorrhizal fungi for production of iron chelators by preparing the CAS-detergent solution with 5 m*M* 3-(dimethyldodecyl-ammonium) propanesulfonate rather than hexadecyltrimethyl-ammonium bromide, as noted in the original paper. Challenge experiments between the mycorrhizal fungi and the actinomycete antagonist, tentatively identified as a *Streptomyces* sp., were performed by preincubating 5 mm fungal plugs for one week to establish growth and then streaking a line of antagonist at a distance of 2 cm.

Results and discussion

Color changes characteristic of removal of Fe(III) from the CAS-detergent complex varied greatly, with the strongest reactions produced by the ericoid mycorrhizal fungi *Oidiodendron griseum* and *Hymenoscyphus ericae*. Decoloration by the four ectomycorrhizal fungi tested was greatest for *Rhizopogon venicolor*, intermediate for *Hebeloma crustuliniforme* and *Cenococcum geophilum*, and least for *Laccaria laccata*. Visual ranking of mycorrhizal inhibition of the actinomycete followed a similar trend with complete inhibition by *O. griseum* and decreasing degrees of inhibition by *H. ericae*, *R. vinicolor*, *H. crustuliniforme*, *C. geophilum* and *L. laccata*. Correspondence between presumptive siderophore production and ability of these mycorrhizal fungi to inhibit growth of a soil actinomycete suggests that iron chelation may be a mechanism of competition within the mycorrhizosphere.

Reference

Schwyn B and Neilands J B 1987 Anal. Biochem. 160, 47–56.

D. L. Keister and P. B. Cregan (Eds.), The rhizosphere and plant growth, 310.

PLSO BARC73

Genetic improvement of siderophore production aimed at enhancing biocontrol in Pseudomonas strains

D.J. O'SULLIVAN and F. O'GARA
Microbiology Department and National Food Biotechnology Centre, University College, Cork, Ireland

Key words: promoter-*lac*Z fusion, siderophore

The exploitation of microorganisms as biological agents to control phytopathogens is a rapidly expanding research area. The secretion of iron binding ligands (siderophores) which decrease the pool of iron available to disease organisms has been proposed as a major factor in the plant growth stimulation by fluorescent *Pseudomonas* spp. (Kloepper *et al.*, 1980). The producing microorganism can utilize the resulting Fe^{3+}-siderophore complex via a specific receptor in its outer membrane (Magazin *et al.*, 1986).

In our laboratory, strains were isolated from the roots of sugarbeet plants and were characterized for: (i) a high affinity siderophore mechanism; (ii) an aggressive colonizing ability; and (iii) the ability to protect sugarbeet roots from disease. Tn*5*-induced mutants in siderophore biosynthesis of strain M114 were complemented by a M114 pLAFR1-based gene bank of the parent strain. This resulted in the isolation of five overlapping cosmid clones. It has been demonstrated that there is a strong repression imposed by iron on both the biosynthesis of the siderophore moiety (Marugg *et al.*, 1985) and uptake of the ferric-siderophore complex in fluorescent *Pseudomonas* spp. (Magazin *et al.*, 1986). This regulation by iron on the expression of the siderophore genes was studied by obtaining iron-regulated promoter-*lac*Z fusions from strain M114. One fusion plasmid (pSP1) was constructed by the shotgun cloning of strain M114 DNA into a promoter probe plasmid pMP190. This iron-regulated promoter/operator sequence was present on a 2.1 kb *Bgl*II fragment and was localized to a 0.4 kb *Pst*I/*Stu*I internal region. Another iron-regulated promoter-*lac*Z fusion plasmid was isolated by inserting a Tn*3-lac* transposon (Tn3HoHo1) into a cloned 21.7 kb DNA fragment containing siderophore biosynthetic genes from strain M114. β-Galactosidase expression from the resulting plasmid (pMSR1) was induced in strain M114 during growth in low iron conditions. This plasmid was exploited to select a Tn*5*-induced Pseudomonas Fe^{3+} uptake regulatory (*fur*) mutant (strain M114FR1) by screening for *lac*Z expression from pMSR1 when iron was present in the growth medium. Subsequent characterization revealed that this mutant could produce both siderophore and ferric-siderophore receptor proteins, irrespective of the presence of iron in the growth medium. By complementation of this mutant, the regulatory (*fur*) gene was cloned. The Fur^- mutant demonstrated inhibition of both bacterial and fungal indicator organisms under conditions where the iron concentration was sufficiently high to render the wild-type strain ineffective.

References

Kloepper J W *et al.* 1980 Nature (London). 286, 885–886.
Magazin M D *et al.* 1986 J. Biol. Chem. 261, 795–799.
Marugg J D *et al.* 1985 J. Bacteriol. 164, 563–570.

D. L. Keister and P. B. Cregan (Eds.), The rhizosphere and plant growth, 311.

PLSO BARC74

Bioinorganic and bio-organic studies of rhizobactin, a structurally novel siderophore related to the opines

M.J. SMITH
Department of Chemistry, Columbia University, New York, NY 10027, USA

Key words: opine, NMR, Rhizobium, siderophore

The recent structure elucidation of rhizobactin, N^2-[2-[D-(1-carboxyethyl)-amino]ethyl]-N^6-L-(3-carboxy-3-)hydroxy-1-oxopropyl)-L-lysine, from *Rhizobium meliloti* DM4 disclosed a third chemical class of microbial siderophores (Gr. *sidero* = iron; *phore* = bearer; Smith *et al.*, 1985); the ethylenediamine ligand also was unprecedented in a natural product. Yet determination of the absolute configuration and proof of the entire structure awaited total synthesis. This has been accomplished (Smith, 1989) and substantiates the conclusion that the siderophore is related in structure and activity to other N^2-substituted amino acids termed opines (Smith and Neilands, 1984a, b). As iminocarboxylic acids, opines, rhizobactin and phytosiderophores possess substantially reduced selectivities for iron(III). Such compounds could enhance the bioavailability of additional essential metal ions in the rhizosphere. To wit, why are these particular ligands elaborated in nature? Molecular recongition phenomena such as these are typically elucidated via studies of structure and reactivity. Even though the crystal structures of several siderophore:metal complexes and one siderophoric receptor protein are under study, a full third of the former defy crystallization, and nonequivalent solution conformations are not uncommon [*e.g.*, bridged *N*-hydroxyethylethylenediaminetriacetato-iron(III) dimers; Lippard *et al.*, 1967]. For these reasons the structure-activity relationship of this double "host-guest" system is being probed with a range of metal ions and diastereomers of rhizobactin, in conjunction with bioassays. The bioactivity, binding affinity, and solution conformation of specific metal complexes are being determined to disclose the basic biological role and chemical mechanism. For example, X-ray and NMR studies have typically found such gallium(III) and iron(III) complexes to be isostructural (Demarco *et al.*, 1978; Garbett *et al.*, 1971). Moreover siderophore:gallium complexes commonly display homogeneous solution conformations (Demarco *et al.*, 1978, and references therein) which are recognized by the cell receptor and transport system (Emery, 1971). Preliminary conformational studies of rhizobactin:gallium(III) reveal that the ^{1}H- and ^{13}C-NMR resonances are nearly equivalent to those of the free ligand, yet better resolved, consistent with a tight chelate which possesses a unique solution conformation. Additional nuclear Overhauser effect (nOe) measurements, such as those between the α-carboxy protons and the adjacent ethylenediamine group, are being compiled to verify this and to deduce the chelate's structure.

References

Emery T 1971 Biochem. 10, 1483–1488.
Demarco A *et al.* 1978 Biopolymers 17, 617–636 and 637–650.
Garbett K *et al.* 1971 J. Chem. Soc. (A) 3433–3436.
Lippard S J *et al.* 1967 Inorg. Chem. 6, 1825–1831.
Smith M J 1989 Tetrahedron Lett. 30, 313–316.
Smith M J *et al.* 1985 J. Am. Chem. Soc. 107, 1739–1743.
Smith M J and Neilands J B 1984a J. Plant Nutri. 7, 449–458.
Smith M J and Neilands J B 1984b *In* Molecular Strategies for Crop Protection, Vol. 48. Eds. C J Arntzen and C Ryan. Alan R. Liss, Inc. New York, pp 157–168.

D. L. Keister and P. B. Cregan (Eds.), The rhizosphere and plant growth, 312.

PLSO BARC75

Uptake of *Bradyrhizobium japonicum* siderophores by soybean cells in culture

MINOCHER REPORTER[1,2], HARRY CALVERT[1] and SARAH MARSH[1]
[1]*Battelle-Kettering Research Laboratory, Yellow Springs, OH 45387, USA and* [2]*NSI Technology Services, EPA Lab Building, Corvallis, OR 97333, USA*

Key words: iron, nodule metabolism

The mechanisms for making iron available for both bacteria and plants during nodule development is not understood. Accumulation of large amounts of iron by bacteroids within plant nodules is required for iron centers of nitrogenase, ferredoxins, cytochromes, and protoheme-XI of the plant leghemoglobin. The soil has organic iron chelators originating with bacteria and fungi. This iron can be reduced and translocated to plants via citrate and microbial siderophores, *e.g.* fusarinins, agrobactins, azotobactins and rhizobactins.

Bradyrhizobium japonicum strains USDA 138, 61A42, and 61A58 excreted yellow-green fluorescent peptides (GYP) into the medium of Valera and Alexander when the iron content of this medium was 2.8 to 3.9 μM. This medium contained $10\,g\,L^{-1}$ gluconate and $7.5\,g\,L^{-1}$ KNO_3 at pH 6.8 (USDA 138) and pH 7.8 (61A42 and 61A58).

GYP was separated using P4 or Sephadex G50; Dowen AG1X2 or QAE-Sephadex followed by Sephadex G50 seperfine, all at pH 8.0.

The peptide tightly bound one mole Fe^{III} ($K_{eq} \geqslant 10^{29}$) but Fe^{II} and other transition metals were loosely held and removable by gel electrophoresis to yield the original peptide fluorescence. The peptide contains ser, thr, gly, glu, asp and Me-asp(?) with a presumed 0-dihydroxyquinoline chromophore.

Peptide-aided iron uptake was shown to be biased in favor of the homologous strain of Bradyrhizobium and an antibody prepared against the $GYP_{USDA138}$ failed to react with peptides from 61A42 or 61A58. The tendency for iron uptake by these strains suggests a mechanism for competitive advantage in the rhizosphere. This may also explain why Bradyrhizobium makes an unusually large siderophore (M_r 2900).

We attempted to determine if cells of leguminous plants could obtain iron from GYP when the GYP was present within plant cells and, by inference, within nodules. Addition of GYP to the medium of suspended plant cells showed linear decrease of its unique fluorescence over 6 hrs. The same amount of plant cells grown for 12 hrs without iron showed a faster (2X) loss of GYP fluorescence and controls, without plant cells, showed none. This was repeated with $^{55}Fe \cdot EDTA$. Plant cell uptake was also confirmed by autoradiography experiments in which ^{55}Fe was found only in the cytoplasm. In conclusion, plant nodule cells may obtain iron with the aid of Rhizobium siderophores.

Part 4

Rhizosphere iteractions and plant growth promotion

D. L. Keister and P. B. Cregan (Eds.), The rhizosphere and plant growth, 315–326.

PLSO BARC30

Plant growth promotion mediated by bacterial rhizosphere colonizers

JOSEPH W. KLOEPPER[1], ROBERT M. ZABLOTOWICZ, ELIZABETH M. TIPPING and RAN LIFSHITZ
[1]*Department of Plant Pathology, Alabama Agricultural Experiment Station, Auburn University, Auburn, AL 36849–5409, USA and Allelix Crop Technologies, 6850 Goreway Drive, Mississauga, Ontario, Canada L4V 1P1*

Key words: bacteria, biological control, growth promotion, PGPR, rhizobacteria

Abstract

Plant growth-promoting rhizobacteria (PGPR) represent a diverse subgroup of rhizosphere-colonizing bacteria. PGPR were first described for root crops in the 1970s when the use of antibiotic resistance made possible the monitoring of introduced bacteria in soil. In recent years, the host list of PGPR has grown to include barley, bean, canola (rapeseed), cotton, maize, peanut, rice, vegetables, wheat, and woody species. In addition to increasing crop yields, different strains of PGPR can exert various effects on plants including biological control of soil-borne pathogens, promotion of legume nodulation by nitrogen-fixing rhizobia, and enhancement of seedling emergence rates. Reported mechanisms of action for PGPR have focussed on the indirect mechanisms of siderophore, antibiotic, or hydrogen cyanide production. Such indirect mechanisms reduce the population densities of deleterious microorganisms and thereby result in increased plant growth. Direct growth promotion by PGPR in the absence of deleterious microorganisms has been recently described.

Background and definition of terms

In the last decade, the term "rhizobacteria" has been accepted to describe rhizosphere bacteria which exhibit root colonization (Schroth and Hancock, 1982). Root colonization is a process whereby the bacteria survive inoculation onto seeds or into soil, multiply in the spermosphere in response to seed exudates rich in carbohydrates and amino acids (Kloepper *et al.*, 1985), attach to the root surface, and colonize the developing root system. Hence, rhizobacteria are efficient microbial competitors in the root zone which displace native root-colonizing microorganisms (Kloepper and Schroth, 1981a,b) and which persist through the mid-stages of the host plant ontogeny at population densities of Log 3 to Log 6 cfu/g root fresh weight (Anderson and Guerra, 1985; Bahme *et al.*, 1988; Howie and Echandi, 1983; Kloepper *et al.*, 1980a,b; Polonenko *et al.*, 1987).

In the past few years, the term "rhizosphere colonization" has been suggested as an alternate term to root colonization. While one may argue that the choice of terms is a matter of semantics, there is also a functional difference. Root colonization applies to those bacteria which are specifically resident in two locations–the outside of root surfaces and the internal areas of roots (endocolonizers). Rhizosphere colonization denotes a larger ecological niche encompassing root colonization and including bacterial populations which are not attached to the root, but which are in close proximity. The term "PGPR" for "plant growth-promoting rhizobacteria" was first coined in 1978 (Kloepper and Schroth, 1978) to describe a subset of rhizobacteria which induce increased plant growth after inoculation onto seeds. The first reports documenting plant growth and yield increases with PGPR involved root crop hosts such as potato (Burr *et al.*, 1978; Geels and Schippers, 1983; Howie and Echandi,

1983; Kloepper *et al.*, 1980a; Vrany and Fiker, 1984), radish (Kloepper and Schroth, 1978), and sugarbeet (Suslow and Schroth, 1982).

Work on mechanisms for growth promotion by these PGPR indicated that the strains increased growth indirectly by changing the microbial balance in the rhizosphere. Iron-chelating siderophores (Kloepper *et al.*, 1980a,b; Schippers, 1988), antibiotics (Gardner *et al.*, 1984; Weller, 1988), and HCN (Ahl *et al.*, 1986; Stutz *et al.*, 1986) are produced by some PGPR and have been implicated in reductions of plant pathogens and deleterious rhizobacteria. HCN has also been implicated in direct growth promotion by the discovery that certain deleterious rhizobacteria produce HCN, which restricts plant growth, and that these deleterious rhizobacteria are inhibited by some PGPR strains (Schippers, 1988).

New directions

In the past 5 years, PGPR research has continued in three additional directions. First, significant work has been done on "nonroot crops" as hosts and has demonstrated that the majority of crop plants are conducive to PGPR-induced growth promotion. A second area of new direction involves the characterization of *specific* effects of PGPR *i.e.* effects other than yield promotion. These studies have resulted in the designation of two new subclasses of PGPR and have shown that PGPR can also be used as biological control agents. The third area, mode of action work, has revealed that some PGPR strains may promote plant growth directly, *i.e.* in the absence of pathogenic or deleterious microorganisms.

Host range for PGPR

Barley
Iswandi *et al.* (1987) investigated the effect of "rhizopseudomonad" strain 7NSK2, isolated from a hydroponic culture of barley, on barley growth in soil. Dry weight of plants following PGPR treatment was increased from 5 to 20% compared to controls without PGPR.

Bean
A *Pseudomonas putida* strain colonized lateral roots and main roots of bean (*Phaseolus vulgaris* L.) in hydroponic culture (Anderson and Guerra, 1985) and resulted in increased lignin content of roots. Plant weight was increased by treatment with *P. putida* after inoculation with *Fusarium solani* f. sp. *phaseoli*.

Canola (Rapeseed)
The potential for obtaining yield increases on canola (*Brassica campestris* L and *B. napus* L.) with PGPR was reported in 1988 (Kloepper *et al.*). Over 4,000 bacterial strains were collected from root zones and were individually evaluated for growth at 4–14°C, metabolism of seed exudates, chemotaxis toward asparagine, and root colonization; 887 of these strains were tested for growth promotion in greenhouse trials with field soils. Thirty-five strains increased leaf area in two of three repeat tests. Thirteen strains increased yields up to 57% compared to controls in one of two test years. Three strains increased yields by 6–13% compared to controls in two-year, all-site averages. PGPR strains identified in this study included *P. putida*, *P. fluorescens*, *Serratia liquefaciens*, *P. putida* biovar B, and *Arthrobacter citreus*

Cotton
Two strains of *P. fluorescens*, which were screened for *in vitro* antagonism to plant pathogenic fungi and bacteria, increased height of 4 week-old cotton in greenhouse trials with field soil by 8 and 40% compared to controls (Sakthivel *et al.*, 1986). *Bacillus subtilis* strain A-13 (Broadbent *et al.*, 1977) has shown promise as a PGPR and a biological control agent on cotton (Backman and Turner, 1989; Greenough and Batson, 1989). The strain is manufactured by Abbott Labs and is licensed to Gustafson Corporation for peanut and cotton. The product is sold under the tradename 'Quantum-4000'.

Maize
The U.S. biotechnology company Agracetus (Middleton, Wisconsin) has conducted field evaluations of pseudomonad PGPR on maize for 5 years (A. Paau, personal communication). The strains were selected as growth promoters in

greenhouse trials under various growth conditions where visual growth promotion and increased dry weight of plants were noted. In field trials, the strains colonize roots at average population densities of Log 3 cfu/cm root and induce yield increases of 3 to 3.5 bu/acre compared to controls in all-site, five-year averages.

Peanut

The A-13 strain of *Bacillus subtilis* (Broadbent *et al.*, 1977) was first found to enhance growth of peanut by field work at Auburn University in 1980 (P.A. Backman, personal communication). Further field trials indicated that the strain induced yield increases averaging 14 to 24% compared to controls in a two-year study in Alabama (Clay, 1986) and 6 to 16% in a three-year study in Texas (Jaks *et al.*, 1985). These results prompted the commercialization of the inoculant, as described above in the cotton section.

Further investigations into the ecology and mechanisms of strain A-13 on peanut indicate that, unlike most strains of *Bacillus* spp., A-13 is a root colonizer, according to the definition above. Turner and Backman (1989) reported root colonization of four peanut cultivars by strain A-13 at average population densities of Log 4.3 to Log 4.7 cfu/g root in field trials 140 days after planting. Root colonization was associated with increased overall plant growth, faster growth of roots, and enhanced plant nutrition. Hence, *B. subtilis* strain A-13 is clearly a PGPR.

Rice

Sakthivel *et al.* (1986) isolated *P. fluorescens* strains from rhizospheres of various crop plants and selected strains which demonstrated broad-spectrum *in vitro* antibiosis toward fungal and bacterial pathogens. When coated onto rice seeds planted in pot trials with field soil, four strains induced increases in plant height ranging from 12 to 24% greater than controls.

Vegetables

The effect of several root-colonizing bacteria on vegetables was reported in a binational collaboration (Elad *et al.*, 1987). Seed treatment with bacteria in pot trials increased dry weight two weeks after planting for tomato, pepper, tobacco, cucumber and melon. Work at Allelix Crop Technologies, a Canadian biotechnology company, has evaluated PGPR, which were initially selected to enhance canola growth, on vegetables (R. Lifshitz, personal communication). Several strains of fluorescent pseudomonads and *Serratia* spp. promoted growth as evidenced by enhanced dry weight of shoots and roots in greenhouse trials with field soil on tomato, cucumber, sweet corn, carrot and celery.

A research team in California investigated the effect of PGPR on celery (M.N. Schroth, personal communication). Thirty root-colonizing bacteria, including some confirmed as PGPR on other crops, were screened directly in the field for growth enhancement of celery in the presence of indigenous *Fusarium oxysporum* f.sp. *apii*. Four strains were selected for use in three follow-up replicated trials. A pronounced genotypic specificity in response to PGPR inoculation was noted. One PGPR strain induced significant increase in early growth (fresh and/or dry weight increases) in all three trials and significant yield increases ranging from 12 to 15% greater than controls in two trials. The same strain had no effect when tested on another cultivar.

Wheat

Several recent reports from four different research groups have confirmed that PGPR can promote growth of wheat. Weller and Cook (1986) found that 17 of 64 bacterial strains isolated from wheat rhizospheres increased growth of wheat in greenhouse trials. In field trials with four strains, significant increases following seed treatment with bacteria were noted in stand, plant height, number of heads and/or grain yield. No PGPR increases occurred in comparison to metalaxyl controls, which suggests that the growth promotion by PGPR resulted from Pythium control.

The importance of Pythium to the observed growth promotion was confirmed in a subsequent study by Becker and Cook (1988). Approximately 120 of 350 strains, which were isolated from wheat rhizospheres exhibited *in vitro* antibiosis toward Pythium. *In vitro* antibiosis was correlated with increased growth of wheat in greenhouse trials with soil naturally infested with Pythium. When 10 of these PGPR were tested in

Pythium-free soil (fumigated), no growth promotion was observed.

In another study with wheat, deFreitas and Germida (1989) developed a consistent growth chamber assay for screening PGPR and found 9 strains which consistently promoted plant growth in field soil. These strains increased plant height, root and shoot biomass, and the number of tillers. No symptoms of Pythium were observed in nontreated controls. The role of the plant in wheat growth promotion by PGPR has also been investigated. Significant genotype variation was observed in response to PGPR inoculation (Ålström and Gerhardson, 1988). In another study, 6 of 7 strains of Bacillus, which were isolated from rhizospheres of cv 'Katepwa', promoted growth on 'Katepwa' but not on the cultivar 'Neepawa' (Chanway *et al.*, 1988). Hence, cultivar specificity must be considered in analysis of PGPR results.

Woody species

Guayule (*Parthenium argentatum*) in a perennial, semi-desert shrub from which latex is extracted. Two pseudomonad strains isolated from roots increased shoot dry weight in greenhouse trials by four-experiment averages of 17 and 55% respectively (Olsen and Misaghi, 1984). After transplanting to the field, no significant differences occurred in shoot dry weight of plants treated with bacteria compared to controls.

Caesar and Burr (1987) tested 226 strains of rhizosphere bacteria isolated from apple and rosaceous wild hosts for growth promotion of apple seedlings six weeks after seed treatment. Thirteen strains induced significant increases in fresh weight ranging from 23 to 47% greater than controls. Four bacterial strains also induced significant weight increases ranging from 33 to 121% when applied to rootstocks in greenhouse trials.

PGPR have also been reported on citrus (Gardner *et al.*, 1984) as part of a study to determine the effect of 43 bacterial strains isolated from rough lemon roots on growth of lemon and orange seedlings. Both stimulatory and inhibitory effects on plant growth resulted 10 mo. after inoculation, with ranges from +116% to −52% compared to controls.

Specific effects of PGPR

Biological control

The distinction between biological control and growth promotion is vague at best. As listed in the above summary by crop, bacteria which are selected as potential biological control agents based on *in vitro* antibiosis toward pathogens may also increase plant growth. The general area of biological control by bacterial agents was recently reviewed (Weller, 1988), as was the use of rhizobacteria in biological control (Schippers, 1988), and therefore, the discussion here focuses on the use of bacteria, initially selected for growth promotion, as biological control agents.

The first reports of PGPR on potato noted that growth promotion was associated with a reduction of total fungal propagules on the rhizoplane (Kloepper and Schroth, 1981b). This suggested that select PGPR could also be used to reduce pathogen populations in the root zone. Some potato PGPR were subsequently shown to reduce populations of the pathogen *Erwinia carotovora* (Kloepper, 1983).

For the past three years, the group at Allelix Crop Technologies Inc has investigated the potential relationship between growth promotion and biological control by evaluating PGPR selected for canola growth promotion in specific disease assays. The disease assays include a sand + vermiculite assay for *Rhizoctonia solani* on canola, a soil assay for Pythium on tomato, cotton, soybean, canola, and whitebean and a sand + vermiculite assay for Pythium on canola and cucumber. In all assays, selected PGPR strains show biological control activity similar to chemical controls, while others have no effect. Some of the PGPR with biocontrol activity show *in vitro* antibiosis toward the pathogen, but others do not.

A representative biocontrol response induced by a PGPR strain in the cucumber/Pythium assay is shown in Figure 1. This strain does not show *in vitro* antibiosis toward Pythium.

Data from a representative canola/*Rhizoctonia* assay are shown in Figure 2. Controls treated with the pathogen, but not with bacteria, increased stand through day 8 and then declined due to post-emergence damping off. Fungicide-

(a)

(b)

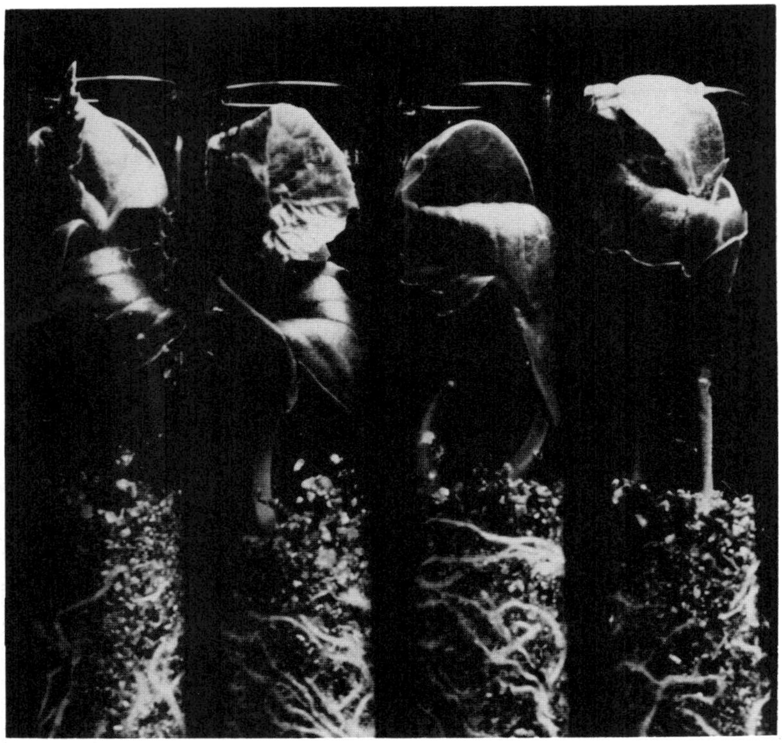

(c)

Fig. 1. Biological control of Pythium with PGPR. Representattive results from a Pythium assay on cucumber are shown. All treatments received Pythium inoculum mixed into the sand + vermiculite planting medium. Treatment **A** is the disease control *i.e.* no other treatment; treatment **B** is the metalaxyl fungicide control; treatment **C** received seed inoculation with a PGPR strain at planting time.

treated controls (Vitavax) also increased through day 8 and then declined slightly. Treatment with different PGPR consistently resulted in increased final stand counts and sometimes resulted in control equal to the fungicide.

Further data on biological control by PGPR are shown in Table 1 for whitebean in raw field soil ammended with Pythium at 100 propagules per gram. Best emergence, final stand, and shoot dry weight resulted from treatment with the metalaxyl control; however, several PGPR strains induced a significant increase in final stand and shoot dry weight compared to the nontreated control.

These examples illustrate that strains selected for growth promotion may also provide protection from plant disease. The work to date suggests that such biocontrol PGPR consist of two groups of strains: those which control disease via antagonism of the pathogen, and those which control the disease via changing the host plant's susceptibility *e.g.* induced resistance. The identification of this second class of strains affords an opportunity to expand the list of known mechanisms for biological control by bacteria beyond the current mechanisms for antibiosis, competition, and exploitation. For example, a PGPR strain, which produces no *in vitro* antagonism toward a fungal pathogen and yet protects against the disease, is more likely to be operating by changing the host's susceptibility than by inhibiting the pathogen.

Nodulation promotion on legumes

Another specific effect which PGPR can have on plants is to increase the symbiotic nodulation of legumes by nitrogen-fixing rhizobia. Enhanced nodulation of *Phaseolus vulgaris* following inoculation with *Pseudomonas putida* strain M 17 was reported by Grimes and Mount (1984). In

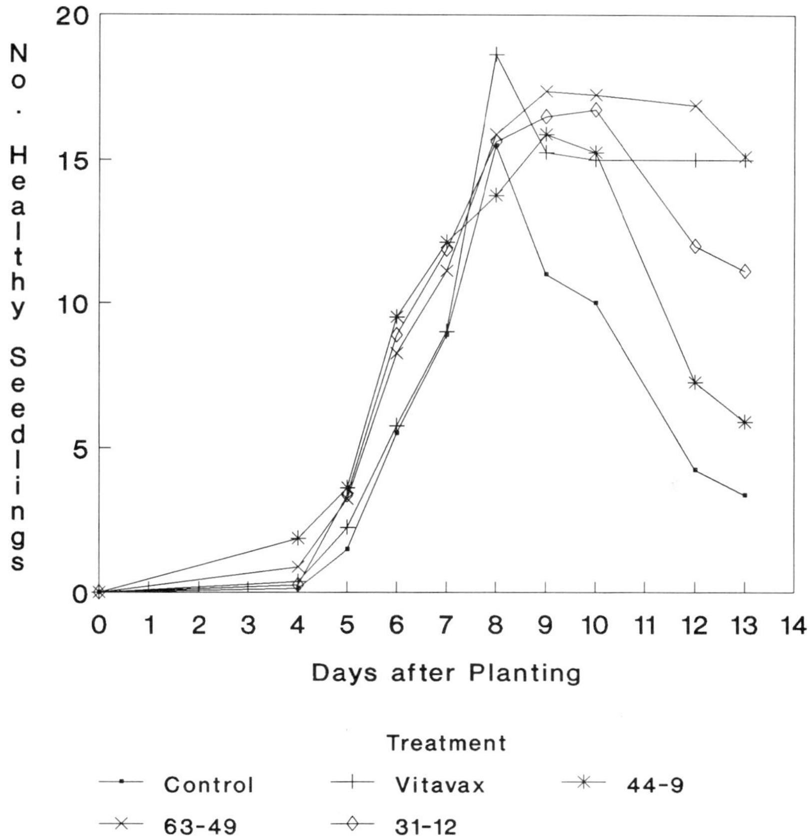

Fig. 2. Biological control of *Rhizoctonia solani* with PGPR. Representative results from a *Rhizoctonia solani* assay on canola are shown. All treatments received Rhizoctonia inoculum mixed into the planting medium. Vitavax is the fungicide control. PGPR strains 63-49, 31-12, and 44-9 all gave early protection from damping off, and strain 63-49 resulted in final stands equal to Vitavax.

greenhouse and field trials, treatment with strain induced increased nodulation by *Rhizobium phaseoli*. M 17 is apparently not a PGPR strain, as growth promotion was, with one exception, not observed in greenhouse or field trials.

In another study with soybean (Polonenko *et al.*, 1987), PGPR strains were found generally not to interfere with nodulation by *Bradyrhizobium japonicum* strain 110. Nine of 18 tested root-colonizers significantly increased the weight of nodules in soil. Subsequent work by this group has confirmed that a subset of PGPR strains act as "Nodulation-Promoting Rhizobacteria (NPR)" in coinoculation studies with rhizobia on legumes. Data showing the effect of three NPR strains on nodulation of soybean by native soil rhizobia are shown in Table 2. Average nodulation from trials at two locations on two cultivars was 19 to 47% greater than controls. Yield was not affected by NPR inoculation in these trials. Mechanisms of nodulation promotion by the NPR strains have not been determined. PGPR strain A-13 (see above sections on cotton and peanut) also shows NPR activity (Turner and Backman, 1989). In field trials with peanut, nodulation was assessed with a 1 to 5 scale for the upper 20 cm of taproot with attached secondary and feeder roots where 1 = no nodules and 5 = 100%. The average nodulation rating for treatment with A-13 + fungicide was 4.3 compared to 3.5 for treatment with the fungicide alone ($LSD_{0.05} = 0.52$). Treatment with A-13 also induced significantly higher levels of nitrogen in leaves as well as yield increases.

Table 1. Screen of plant growth-promoting rhizobacteria strains for biological control of *Pythium ultimum on* whitebean in a soil system[a]

Treatment	After planting (DAP)						Final stand		Dry weight per pot[a]	
	4 DAP		7 DAP		13 DAP					
	No.[b]	%[c]	No.	%	No.	%	No.	%	g	%
61-9A	2.0	82	4.3	48*	4.9	29*	5.1	31*	1.22	59*
GR12-2	1.7	55	3.9	34	4.6	21	4.8	23	1.04	36
63-49	1.8	64	4.6	59*	5.2	37*	5.4	38*	1.04	36
2-114	1.9	73	4.8	66*	5.6	47*	5.4	38*	1.04	36
1-102	3.2	191*	5.1	76*	5.5	45*	5.5	41*	1.23	61*
36-43	1.9	73	3.9	34	4.1	8	4.4	13	0.90	18
31-12	1.3	18	3.9	34	4.6	21	4.2	8	0.91	20
G2-8	1.6	45	3.5	21	4.6	21	4.5	15	0.89	17
G20-18	1.9	73	3.8	31	4.8	26	5.0	28*	1.10	44
44-9	1.0	−11	4.0	38*	4.6	21	4.9	26	1.05	37
34-13	2.0	82	4.4	52*	5.1	34*	5.4	38*	1.29	69*
2-68	2.6	136*	4.8	66*	5.5	45*	5.7	46*	1.29	69*
Methylaxyl[e]	4.6	318*	5.7	97*	5.8	53*	5.9	51*	1.50	97*
Pythium control	1.1	–	2.9	–	3.8	–	3.9	–	0.76	–
Nontreated control	1.5	36	3.6	24	4.2	11	4.4	13	1.00	31
LSD $P = 0.05$	1.1		1.1		1.1		1.1		0.33	

[a] Raw field soil was amended with 100 propagules/g of Pythium using cornmeal/sand inoculum;
[b] Average number of plants per pot based on 10 replications, each with 6 seeds;
[c] Percentage increase over Pythium control;
[d] Dry Weight (g) of shoot;
[e] Apron 70SD seed treatment (350 mg metalazyl and 350 mg thiram ai/kg seed) is the fungicide control.
* Significantly greater than Pythium control at $p = 0.05$.

Emergence promotion by PGPR

Select PGPR strains were found to enhance emergence of soybean and canola in greenhouse studies with field soils maintained at cold temperatures; the bacteria were termed "Emergence-Promoting Rhizobacteria (EPR)" (Kloepper *et al.*, 1986). EPR increased the rate of seedling emergence and the final stand in the absence of disease pressure.

Subsequent work at Allelix Crop Technologies

Table 2. Effect of 3 plant growth-promoting rhizobacteria strains on nodule mass and yield of 2 soybean cultivars at two Ontario field locations[a]

Treatment	Cultivar maple arrow		Cultivar evans		Average Increase Compared To Control
	Georgetown	Milton	Georgetown	Milton	
Nodule mass (mg/plant)					
G11-32	278 a[b]	166 a	146 ab	240 a	+47%
G2-26	266 a	152 ab	63 c	246 a	+27%
86-64	166 b	154 ab	99 b	247 a	+19%
Control	178 b	109 b	118 ab	154 b	–
Yield (Kg/ha)					
G11-32	3604 a	3415 a	nd[c]	3362 a	−2%
G2-26	3695 a	3636 a	nd	3315 a	+1%
86-64	3654 a	3167 a	nd	3443 a	−5%
Control	3654 a	3352 a	nd	3552 a	–

[1] Values are means of eight replications. All plots received granular peat inoculum of *Bradyrhizobium japonicum* strain 122.
[b] Means followed by the same letter do not differ significantly at $P = 0.05$.
[c] nd = not determined.

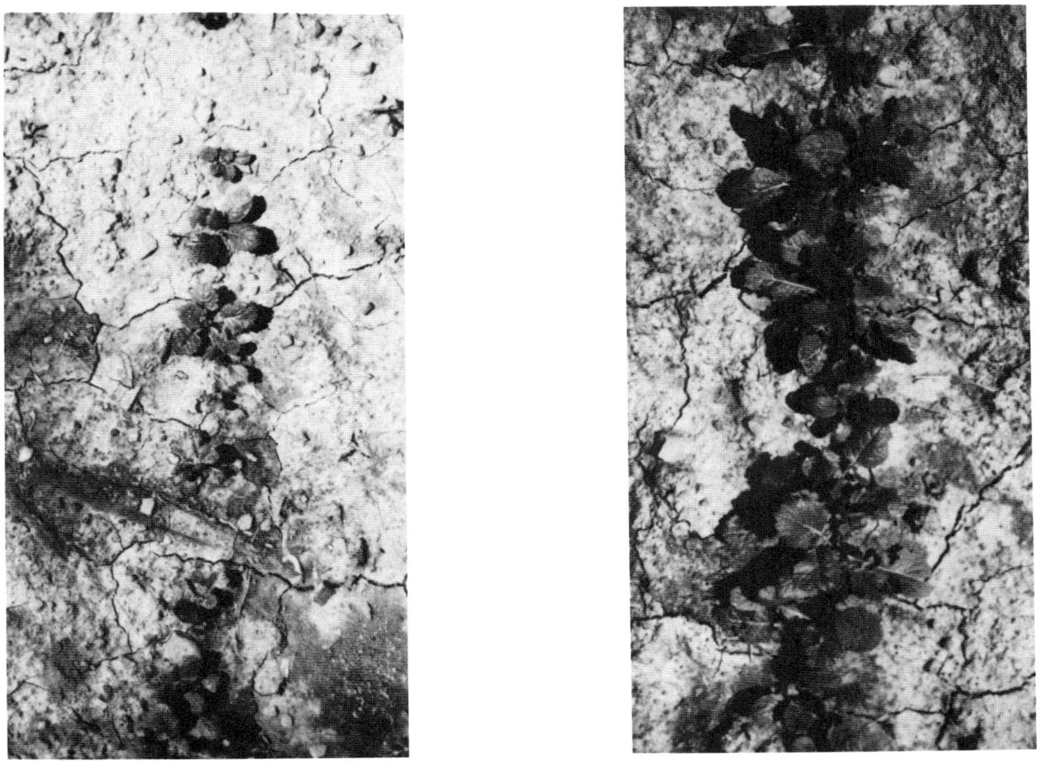

(a) (b)

Fig. 3. Promotion of seedling emergence with PGPR. Figure depicts field results on canola. Treatment **A** is the nonteated control; treatment **B** received seed inoculation of PGPR.

has confirmed that EPR strains may promote emergence under field conditions. Figure 3 shows a typical field response for canola where seedling emergence rate was increased by treatment with EPR. With canola, the emergence enhancement is associated with an observable increase in leaf area. Increased emergence following inoculation with EPR, which were originally selected for growth-promotion activity on canola or soybean, has been detected with tomato, carrot, wheat, maize, whitebean and alfalfa (R. Lifshitz, personal communication). Two other recent studies have confirmed that select PGPR strains can increase seedling emergence. deFreitas and Germida (1989) found that 7 of 9 PGPR strains tested on wheat decreased emergence in a low fertility field soil; however, 2 strains significantly increased emergence in a high-fertility soil. In field trials with PGPR strain A-13 at three planting dates on peanut, Turner and Backman (1989) observed significant emergence enhancement at the earliest planting date but not at the subsequent two dates.

Mode of action work

Studies on mechanisms by which PGPR increase growth have focussed on the indirect mechanisms of siderophore, antibiotic, or hydrogen cyanide production and the associated reductions in population densities of deleterious microorganisms in the rhizosphere. In contrast, direct mechanisms are those whereby the PGPR strain exerts the primary effect on the plant without regard to indigenous rhizosphere microflora.

Direct-acting PGPR were described in 1987 (Lifshitz *et al.*) in a study with canola PGPR. A growth pouch assay was developed which allowed evaluation of PGPR strains for root elongation of seedlings under gnotobiotic conditions.

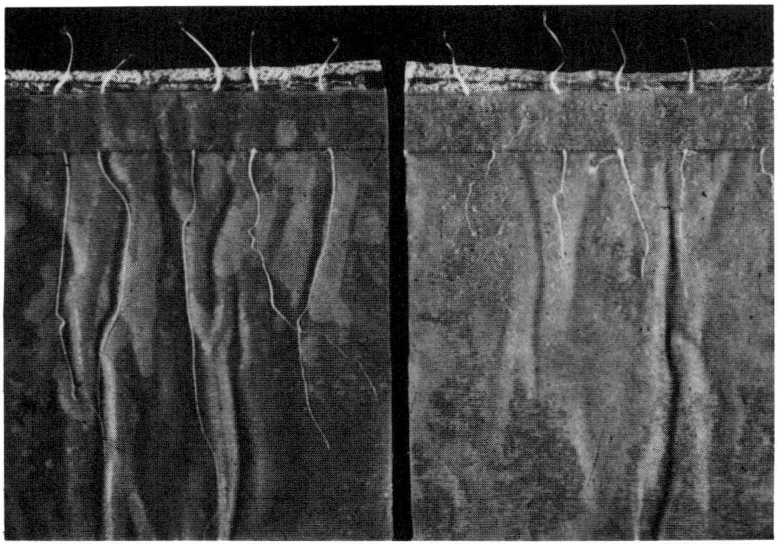

Fig. 4. Direct growth promotion by PGPR. Seeds of canola plants on left were soaked in sterile buffer solution prior to planting. Seeds of plants on right were soaked in PGPR suspension prior to planting. The assay is conducted in sterilized cellophane growth pouches, which are incubated 6 days.

The *P. putida* strain GR12-2 was found to induce consistently a significant elongation of roots compared to controls (Fig. 4). Root elongation was also associated with enhanced shoot height and ^{32}P uptake by roots and shoots. Additional work assessed the direct growth promotion of PGPR from canola in a tube assay using a sterilized mix of sand and vermiculite (Table 3). In this study, EPR strains which increased seedling emergence in field soils retained activity under gnotobiotic conditions. All 6 EPR strains induced a significant ($P = 0.05$) increase in emergence rate and shoot dry weight of seedlings grown at 16/12 C, while no significant increases resulted from treatment with the fungicide Vitavax.

Direct growth promotion by PGPR was reported by two other groups in 1989. Defago *et al.* (1989) demonstrated that PGPR strain CHAO (Stutz *et al.*, 1989) increased root hair formation in a gnotobiotic assay. deFreitas and Germida (1989) observed that under gnotobiotic conditions, some wheat PGPR induce more lateral root hairs and longer roots compared to controls.

Table 3. Emergence enhancement by plant growth-promoting rhizobacterial on canola in sterilized sand + vermiculite at 16/12°C[a]

	5 DAP	6 DAP	7 DAP	8 DAP	9 DAP	10 DAP	11 DAP	12 DAP	Dry weight[b]
Non teated	1	8	15	24	33	35	39	39	111
Vitavax[c]	2	8	12	21	35	39	42	46	89
GR12-2	12	41*	66*	71*	78*	82*	83*	84*	177*
44-9	7	39*	44*	76*	80*	83*	85*	87*	190*
31-12	15	42*	66*	68*	77*	79*	79*	79*	176*
36-43	10	36*	49*	59*	64	65*	66*	69*	160*
61-9A	32*	59*	62*	70*	73*	73*	76*	76*	173*
1-102	20*	40*	47*	49*	64*	68*	69*	74*	156*
L.S.D. ($P = 0.05$)	15	18	17	17	17	18	17	20	42

[a] Average percentage emergence, based on 8 replications, each with 20 seeds;
[b] Dry weight (g) of shoots;
[c] Vitavax RS at 1.26 g ai/kg seed for carbothion and 2.25 g ai/kg seed for thiram.
* Significant increase compared to non-treated control at $P = 0.05$.

Conclusions

Based on the reports summarized above, PGPR research is moving from the descriptive stage to the stage of investigating mechanisms. The descriptive work demonstrated that specific PGPR strains may have diverse beneficial effects on plants, including biological control. Hence, PGPR may be useful components of microbial inoculants for crop improvement and should be included in strategies for increasing crop production.

Early studies on mechanisms suggest that PGPR may produce several different metabolites which stimulate plant growth directly or indirectly. Further elucidation of mechanisms will lay the foundation for understanding and manipulating the genetic control of growth promotion. In addition, details on mechanism will aid efforts to increase consistency of performance by allowing the development of criteria to predict when a plant may benefit from PGPR treatment.

References

Ahl P, Voisard C and Defago G 1986 Iron bound-siderophores, cyanic acid, and antibiotics involved in suppression of *Thielaviopsis basicola* by a *Pseudomonas fluorescens* strain. J. Phytopathol. 116, 121–134.

Åström B and Gerhardson B 1988 Differential reactions of wheat and pea genotypes to root inoculation with growth-affecting rhizosphere bacteria. Plant and Soil 109, 263–269.

Anderson A J and Guerra D 1985 Responses of bean to root colonization with *Pseudomonas putida* in a hydroponic system Phytopathology 75, 992–995.

Backman P A and Turner J T Jr 1989 Plant response and disease control following seed inoculation with *Bacillus subtilis*. Proc. Beltwide Cotton Prod. Res. Conf., Nashville, TN (*In press*).

Bahme J B, Schroth M N, Van Gundy S D, Weinhold A R and Tolentino D M 1988 Effect of inocula delivery systems on rhizobacterial colonization of underground organs of potato. Phytopathology 78, 534–542.

Becker J O and Cook J R 1988 Role of siderophores in suppression of *Pythium* species and production of increased-growth response of wheat by fluorescent pseudomonads. Phytopathology 78, 778–782.

Broadbent P, Baker K F, Franks N and Holland J 1977 Effect of *Bacillus* spp. on increased growth of seedlings in steamed and in nontreated soil. Phytopathology 67, 1027–1034.

Burr T J, Schroth M N and Suslow T V 1978 Increased potato yields by treatment of seedpieces with specific strains of *Pseudomonas fluorescens* and *P. putida*. Phytopathology 68, 1377–1383.

Caesar A J and Burr T J 1987 Growth promotion of apple seedlings and rootstocks by specific strains of bacteria. Phytopathology 77, 1583–1588.

Chanway C P, Nelson L M and Holl F B 1988 Cultivar-specific growth promotion of spring wheat (*Triticum aestivum* L.) by coexistent *Bacillus* species. Can. J. Microbiol. 34, 925–929.

Clay R P 1986 Evaluation of *Bacillus subtilis* as a biological seed treatment for the 'Florunner' peanut plant. M.S. Thesis. Auburn University, 91 p.

Defago G, Berling C H, Burger U, Haas D, Kahr G, Keel C, Voisard C, Wirthner P H and Wothrich B 1989 Suppression of black root rot of tobacco by a *Pseudomonas* strain: potential applications and mechanisms. *In* Biological control of soil-borne plant pathogens. Eds. D Hornby, R J Cook, and Y Henis, CAB International (*In press*).

Elad Y, Chet I and Baker R 1987 Increased growth response of plants induced by rhizobacteria antagonistic to soilborne pathogenic fungi. Plant and Soil 98, 325–330.

de Freitas J R and Germida J J 1989 Plant growth promoting rhizobacteria for winter wheat. Appl. Environ. Microbiol.

Gardner J M Chandler J L and Feldman A E 1984 Growth promotion and inhibition by antibiotic-producing fluorescent pseudomonads on citrus roots. Plant and Soil 77, 103–113.

Geels R P and Shippers B 1983 Reduction of yield depressions in high frequency potato cropping soil after seed tuber treatments with antagonistic fluorescent *Pseudomonas* spp. Phytopathol. Z. 108, 207–238.

Greenough D R and Batson W E 1989 Potential for *Bacillus subtilis* as a bioprotectant of cotton seedlings in soil infested with *Rhizoctonia solani*. (Abstr.) Phytopathology.

Grimes H D and Mount M S 1987 Influence of *Pseudomonas putida* on nodulation of *Phaseolus vulgaris*. Soil Biol. Biochem. 6, 27–30.

Howie W J and Echandi E 1983 Rhizobacteria: influence of cultivar and soil type on plant growth and yield of potato. Soil Biol. Biochem. 15, 127–132.

Iswandi A, Bossier P, Vandenabeele J and Verstraete W 1987 Effect of seed inoculation with the rhizopseudomonad strain 7NSK2 on the root microbiota of maize (*Zea mays*) and barley (*Hordeum vulgare*). Biol. Fertil. Soils 3, 153–158.

Jaks A J, Smith D H, Davis R E and Dolton B D 1985 Effect of *B. subtilis* on seedling emergence and pod yield on spanish market type cultivars and 'Florunner'. Proc. Am. Peanut Res. Educ. Soc. 17, 45 (Abstr).

Kloepper J W 1983 Effect of seed piece inoculation with plant growth-promoting rhizobacteria on populations of *Erwinia carotovora* on potato roots and in daughter tubers. Phytopathology 73, 217–219.

Kloepper J W, Hume D J, Scher F M, Singleton C, Tipping B, Lalibert E M, Frauley K, Kutchaw T, Simonson C, Lifshitz R, Zaleska I and Lee L 1988 Plant growth-promoting rhizobacteria on canola (rapeseed). Plant Disease 72, 42–46.

Kloepper J W, Leong J, Teintze M and Schroth M N 1980a

Enhanced plant growth by siderophores produced by plant growth-promoting rhizobacteria. Nature 286, 885–886.

Kloepper J W, Leong J, Teintze M and Schroth M N 1980b Pseudomonas siderophores: a mechanism explaining disease-suppressive soils. Curr. Microbiol. 4, 317–320.

Kloepper J W, Scher F M, Lalibert E M and Tipping B 1986 Emergence-promoting rhizobacteria: description and implications for agriculture. *In* Iron, Siderophores and Plant Diseases. Ed. T R Swinburne. pp 155–164. Plenum, New York.

Kloepper J W, Scher F M, Lalibert E M and Zaleska I 1985 Measuring the spermosphere colonizing capacity (spermosphere competence) of bacterial inoculants. Can. J. Microbiol. 31, 926–929.

Kloepper J W and Schroth M N 1978 Plant growth promoting rhizobacteria on radishes. Proc. 4th Int. Conf. Plant. Path. Bact. Angers. 879–882

Kloepper J W and Schroth M N 1981a Plant growth-promoting rhizobacteria and plant growth under gnotobiotic conditions. Phytopathology 71, 642–644.

Kloepper J W and Schroth M N 1981b Relationship of *in vitro* antibiosis of plant growth-promoting rhizobacteria to plant growth and the displacement of root microflora.

Kloepper J W, Schroth M N and Miller T D 1980 Effects of rhizosphere colonization by plant growth-promoting rhizobacteria on potato plant development and yield. Phytopathology 70, 1078–1082.

Lifshitz R, Kloepper J W, Kozlowski M, Simonson C, Carlson J, Tipping E M and Zaleska I 1987 Growth Promotion of canola (rapeseed) seedlings by a strain of *Pseudomonas putida* under gnotobiotic conditions. Can. J. Microbiol. 33, 390–395.

Olsen M W and Misaghi I J 1984 Responses or guayule (*Parthenium argentatum*) seedlings to plant growth promoting fluorescent pseudomonads. Plant and Soil 77, 87–101.

Polonenko D R, Scher F M, Kloepper J W, Singleton C A, Lalibert E M and Zaleska I 1987 Effects of root colonizing bacteria on nodulation of soybean roots by *Bradyrhizobium japonicum*. Can. J. Microbiol. 33, 498–503.

Sakthivel N, Sivamani E, Unnamalai N, and Gnanamanickam S S 1986 Plant growth-promoting rhizobacteria in enhancing plant growth and suppressing plant pathogens. Current Science 55, 22–25.

Schippers B 1988 Biological control of pathogens with rhizobacteria. Phil. Trans. R. Soc. Lond. B 318, 283–293.

Schroth M N and Hancock J G 1982 Disease-suppressive soil and root-colonizing bacteria. Science 216, 1376–1381.

Stutz E W, Defago G and Kern H 1986 Naturally occurring fluorescent pseudomonads involved in suppression of black root rot of tobacco. Phytopathology 76, 181–185.

Stutz E W, Kahr G and Defago G 1989 Clays involved in suppression of tobacco black root rot by strain of *Pseudomonas fluorescens*. Soil Biol Biochem.

Suslow T V and Schroth M N 1982 Rhizobacteria of sugar beets: effects of seed application and root colonization on yield. Phytopathology 72, 199–206.

Turner J T and Backman P A 1989 Factors relating to peanut yield increases following *Bacillus subtilis* seed treatment. Plant Disease 73.

Vrany J and Fiker A 1984 Growth and yield of potato plants inodulated with rhizosphere bacteria. Folia Microbiol. 29, 248–253.

Weller D M 1988 Biological control of soilborne plant pathogens in the rhizosphere with bacteria. Annu. Rev. Phytopathol. 26, 379–407.

Weller D M and Cook R J 1986 Increased growth of wheat by seed treatments with fluorescent pseudomonads, and implications of Pythium control. Can. J. Microbiol. 8, 328–334.

D. L. Keister and P. B. Cregan (Eds.), The rhizosphere and plant growth, 327–334.

PLSO BARC31

Microbial production of plant hormones

MUHAMMAD ARSHAD and W.T. FRANKENBERGER Jr*
Department of Soil and Environmental Sciences, University of California, Riverside, CA 92521, USA

Key words: Azotobacter, ectomycorrhizae, plant growth regulating substances, precursor-inoculum interactions

Abstract

Our laboratory has focused on microbial production of plant hormones and precursor-inoculum interactions. Indole-3-acetic acid was detected in soils incubated with L-tryptophan (L-TRP). Inoculation with the ectomycorrhizae, *Pisolithus tinctorius* significantly stimulated the growth of Douglas fir when supplied with low concentrations of L-TRP to soil. Among three *Azotobacter* spp. and two *Pseudomonas* spp., the most prolific producer of cytokinins was *A. chroococcum* and among the precursors tested, adenine (ADE) and isopentyl alcohol (IA) were the most effective. Corn rhizosphere was found to be quite rich with microflora capable of producing ethylene from L-methionine (L-MET). Amino acids, carbohydrates and organic acids typically found in root exudates, were stimulatory to ethylene biosynthesis in soil. Etiolated pea seedlings exhibited the classical 'triple' response when L-MET and *Acremonium falciforme* were applied in combination to sterile soil or when L-MET was added to nonsterile soil.

Introduction

To improve crop yields, inoculation with specific microbial preparations is a common practice in many regions of the world including the Soviet Union, New Zealand, Australia, Belgium, Netherlands, India and U.S.A. In 1895, for the first time, an inoculant 'nitragin' containing rhizobia was marketed as a commercial product. Now rhizobial inoculants are well established as an agricultural product in U.S.A., Europe, Australia and India. In 1937, an Azotobacter preparation, under the trade name 'Azotogen' and later renamed 'Azotobaktrin' was commercialized in the Soviet Union for soil and seed treatment. Increased yields were reported for vegetables and cereals. Later, Soviet microbiologists introduced another product 'phosphobakterin' consisting of *Bacillus megaterium* var. *phosphaticum* which was thought to enhance phosphorus availability. The application of *Azospirillum brasilense* in Brazil (Döbereiner *et al.*, 1976) led to its use as an inoculum for cereals. Today, there are a number of cases where other microbial preparations commercially available under different trade names may benefit crop yield.

It has been assumed that inoculation with Rhizobium, Azotobacter and Azospirillum enhanced plant growth as a result of their ability to fix nitrogen. However, despite extensive research efforts, only rhizobia have been shown to increase yields from dinitrogen fixation. In many cases, nitrogen levels in plant tissues and soils do not increase upon inoculation with these bacteria. Growth promotion may be attributed to other mechanisms such as production of plant growth regulating substances, PGRS (*e.g.* plant hormones) in the rhizosphere. Plants may show a response to inoculation even in nitrogen-rich conditions. Thus, an inoculum should be screened for its ability to synthesize PGRS.

Up to now, an exceedingly large number of studies have indicated the involvement of plant hormones produced by a microsymbiont in altering plant growth and development. Brea and Brown (1974) observed increased growth of several plants including *Paspalum notatum*, without concomitant nitrogen fixation upon inoculation with cultures of *Azotobacter paspali*. Similarly, Hussain *et al.* (1987) reported a yield increase of 16% in maize due to inoculation with *Azotobacter chroococcum* in field studies receiving heavy applications of nitrogen fertilizer. This yield increase was attributed to the bacterial production of plant hormones. Many workers have reported production of plant hormones by Azotobacter (Brea and Brown, 1974; Brown and Burlingham, 1968; Jagnow, 1987; Lee *et al.*, 1970; Nieto and Frankenberger, 1989a), Azospirillum (Hubbell *et al.*, 1979; Tien *et al.*, 1979), and Rhizobium (Phillips and Torrey, 1970; Puppo and Rigaud, 1978) and have considered their physiological action in altering plant growth and development.

Although several studies have shown significant increases in crop yields resulting from the addition of microbial cultures to the rhizosphere, inconsistencies in plant growth are often noted and in most cases lacking in reproducibility. To eliminate this inconsistency and to enhance their reproducibility, precursor-inoculum interactions have proven successful. The production of a specific plant hormone via a microbial transformation of an available precursor can be controlled to which a plant may show a response. In our laboratory, we have successfully established this concept.

Auxins

Indole-3-acetic acid (IAA) is one of the most physiological active auxins. IAA is a common product of L-tryptophan (TRP) metabolism by soil fungi (Gruen, 1959) and bacteria including those which can stimulate plant growth (Lynch, 1985). Frankenberger and Brunner (1983) detected three separate indoles in soil by high performance liquid chromatography (HPLC)-mass spectrometry, including indole-3-acetamide (IAM), indole-3-pyruvic acid (IPyA) and IAA when incubated with L-TRP. Based upon these results, they suggested that IAA is synthesized by two distinct pathways: (i) TRP is converted to IAA by deamination to IPyA followed by decarboxylation to indole-3-acetaldehyde (IAAld) which is further oxidized to IAA and (ii) TRP is decarboxylated to IAM which is hydrolyzed to IAA (Fig. 1). Since IAAld is the immediate substrate of IAA (Libbert and Brunn, 1961), the precursor preceding IAAld could either be tryptamine (TAM) or IPyA. TAM has been reported in some microbial cultures (Perley and Stowe, 1966a,b). IPyA is the most probable intermediate arising from L-TRP by an α-ketoglutarate-dependent transaminase reaction (Frankenberger and Poth, 1988; Gamborg and Wetter, 1963; Kaper and Veldstra, 1958; Sembdner *et al.*, 1980; Truelsen, 1972). Since this reaction is one of the first essential steps in the synthesis of auxins, any parameter which affects the transaminase rate would have an overall influence on the auxins released as secondary metabolites. Frankenberger and Poth (1988) separated and purified a soluble α-ketoglutarate-dependent L-TRP transaminase enzyme from cell-free extracts of a rhizobacterial isolate associated with *Festuca octoflora* Walt. This enzyme catalyzed the conversion of L-TRP to IPyA. Using L-TRP as the substrate in the presence of α-ketoglutarate, the average K_m was 1.22 m*M* and the activation energy 21.0 kJ mol^{-1}. Frankenberger and Poth (1987a) also demonstrated the ability of a rhizobacterium, fluorescent pseudomonad isolated from a grass subfamily Festucoideae, to convert L-TRP into IAA in culture medium by ion suppression reverse-phase HPLC.

Auxins in soil can only influence plant growth if subject to direct uptake by the plant and not metabolized by other soil microorganisms. A precursor-inocuum interaction was recently established by Frankenberger and Poth (1987b) who conducted a greenhouse experiment to test the response of Douglas fir to inoculation of an ectomycorrhizae, *Pisolithus tinctorius* and the IAA precursor, TRP. Prior to initiating this study, they confirmed the ability of this ectomycorrhizae to derive IAA from L-TRP in culture medium by thin-layer chromatography, HPLC, enzyme-linked immunosorbent (monoclonal antibody) assay (ELISA) and unequivocal identification by gas chromatography mass spectrometry. After one growing season, *P. tinctorius* stimulated the growth of potted seedlings of Douglas fir only when supplied with low concentrations of TRP (10^{-8} to 10^{-6} *M*, 0.34 to 34 μg

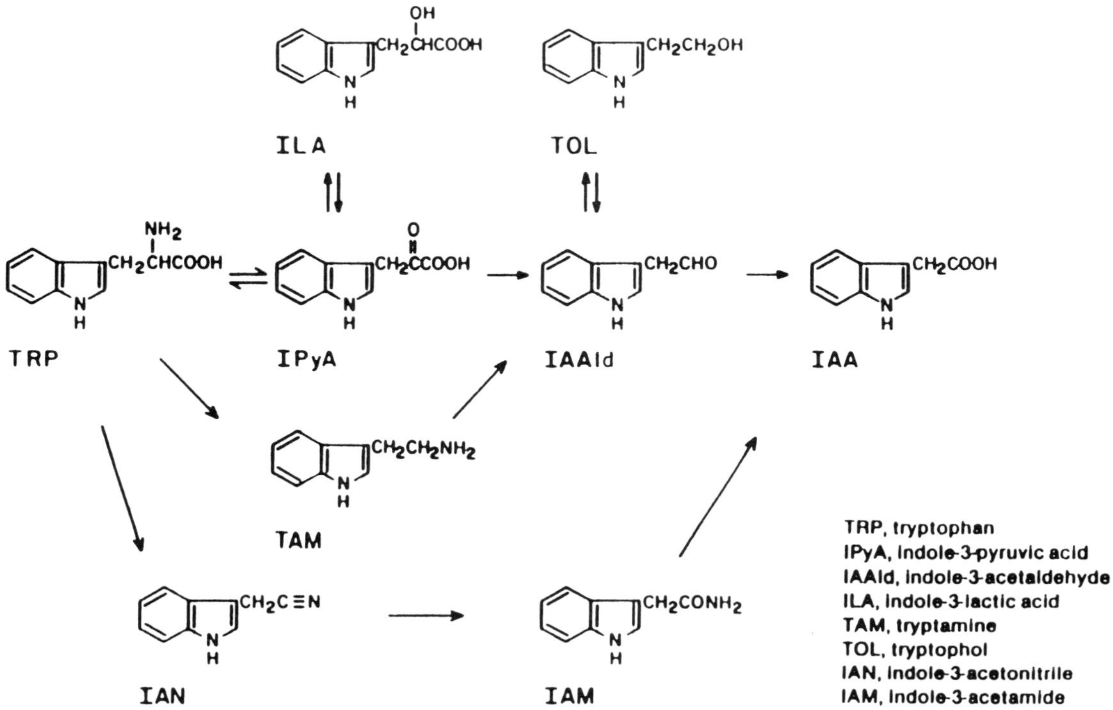

Fig. 1. Proposed pathways for the biosynthesis of indole-3-acetic acid in soil (IAA) (Frankenberger and Brunner, 1983).

kg^{-1} soil). Analyses of variance for seedling height, stem diameter, shoot-root dry weight, and root/shoot ratios showed significant differences among the treatments (Table 1). There was basically no difference in growth between the inoculated and uninoculated treatments; however, the addition of a dilute solution of TRP (10^{-8} to 10^{-6} *M*, 0.34 to 34 μg kg^{-1} soil) with *P. tinctorius* promoted a dramatic increase, indicating a physiological rather than a nutritional effect. In regards to the biomass of shoots and roots, a similar trend was observed with enhanced yield among the 10^{-7} and 10^{-8} *M* TRP treatments.

A higher concentration of TRP applied at 10^{-3} *M* inhibited the dry weights of the seedlings. The most dramatic influence among the interaction of *P. tinctorius* and TRP was the seedling root length, reflecting a phytohormone effect, with elongation being promoted upon 10^{-8} to 10^{-6} *M* TRP applications. Root examination revealed that the mycelial inoculum of *P. tinctorius* was highly effective in establishing colonization but it did not influence plant growth, unless TRP was provided.

Cytokinins

Cytokinins are found in both microbial and plant tissues. Investigations on microbial production of cytokinins have been focused primarily on plant pathogens. The isolation and quantification of cytokinins in nonpathogenic soil bacteria, as well as in soil, have received very little attention with usually cytokinin-like bioactivity via bioassays being reported (Nieto and Frankenberger, 1990). Among the biosynthetic pathways which have been proposed, adenine (ADE) is considered the common precursor for cytokinin production in plant and microbial tissues. Van Andel and Fuchs (1972) had reported the presence of cytokinins in 11 bacteria and 4 fungi. Since Van Andel's and Fuch's report, cytokinins have been detected in the following bacterial cultures, *Agrobacterium tumefaciens*, *Azotobac-*

Table 1. Influence of *Pisolithus tinctorius* (Pt) and tryptophan (TRP) on various growth parameters of Douglas fir (Frankenberger and Poth, 1987b)

Treatments	Shoot height (cm)	Root length (cm)	Shoot weight (g)	Root weight (g)	Root/shoot ratio	Stem diameter (mm)
Control	2.3 a*	42.6 ab	3.6 b	3.5 ab	0.97	0.79 abc
Pt (inoculation)	2.3 a	40.5 ab	3.6 b	3.6 ab	1.00	0.80 abc
Pt + 10^{-3} *M* TRP	2.6 ab	32.2 a	2.8 a	2.9 a	0.98	0.50 a
Pt + 10^{-4} *M* TRP	2.5 ab	38.8 ab	3.6 b	3.8 ab	1.02	0.75 abc
Pt + 10^{-5} *M* TRP	4.5 bc	40.6 ab	3.8 b	4.0 b	1.01	0.82 abc
Pt + 10^{-6} *M* TRP	3.6 abc	53.0 c	4.3 bcd	5.0 c	1.13	1.04 bc
Pt + 10^{-7} *M* TRP	4.1 abc	54.6 c	4.7 cd	5.4 cd	1.12	1.07 bc
Pt + 10^{-8} *M* TRP	5.1 c	54.3 c	5.1 d	5.9 d	1.14	1.12 c

* Treatments followed by the same letter are not significantly different at the 0.05 level according to the Duncan multiple range test.

ter chroococcum, *Bacillus cereus mycoides*, and *Lactobacillus acidophilus*, and the fungi *Taphrina* spp., *Taphrina cerasi* and *Rhizopogon ochraceorubens*, but there is little available evidence on the effect of these exogenously produced cytokinins on plant growth (Lynch, 1985).

Azotobacter chroococcum (ATCC 9043), *A. vinelandii* (ATCC 12981), *A. beijerinckii* (ATCC 12837), and two *Pseudomonads* (*P. fluorescens*, *P. putida*) species, enriched from the rhizosphere of a common grass, *Festuca*, were tested for their ability to produce cytokinins in culture media (Nieto and Frankenberger, 1989a). Among several purine ring constituents and isoprenoid compounds tested for their ability to stimulate bacterial cytokinin production, ADE and isopentyl alcohol (IA) enhanced cytokinin bioactivity to the greatest degree in the culture filtrates of all the bacteria tested. *A. chroococcum* was the most prolific producer of cytokinins. A treatment of 10^{-5} *M* ADE and 10^{-3} *M* IA, 200 mg g^{-1} sucrose, 10^{-5} *M* KNO_3, pH 6.5, 32°C and shaken aerated conditions were found to be optimum. Nieto and Frankenberger (1988) utilized HPLC/UV spectrometry for the identification and quantification of cytokinins produced by *A. chroococcum* (Fig. 2).

The isolation, quantification and bioactivity of cytokinins in soil have received very little attention. One of the major difficulties impeding soil analysis of plant growth regulators is that of sensitive and selective chemical detection. Nieto and Frankenberger (1989b) detected cytokinins in an Arlington soil. Various concentrations of ADE were applied as soil treatments, separately and in combination with 10^{-3} *M* IA and an inoculum of *Azotobacter chroococcum*, over a period of 7 days. Samples were filtered and cytokinins were detected by HPLC/UV spectrometry and the radish cotyledon bioassay. Only two cytokinins, t-zeatin (t-io^{6}Ade) and zeatin riboside (io^{6}Ado), were detected in any appreciable amounts. A treatment of 10^{-6} *M* ADE enhanced cytokinin levels in soil to the greatest degree at 5 days post-treatment, whether applied alone or in combination with IA or an inoculum of *A. chroococcum*. Results indicated that both the indigenous microflora and *A. chroococcum* were capable of producing io^{6}Ado; however, *A. chroococcum* was more effective in its production.

Ethylene

Ethylene (C_2H_4) is well known for its profound effects on plant growth. The biological production of C_2H_4 is not limited to higher plants, since both pathogenic and nonpathogenic microorganisms are capable of synthesizing C_2H_4. Methionine (MET) is the sole precursor for C_2H_4 biosynthesis in plants (Adams and Yang, 1979). In combination with glucose as an energy source, MET serves as an effective substrate for the microbial production of C_2H_4 in a wide diverse group of fungi, yeast and bacteria (Arshad and Frankenberger, 1989; Dasilva *et al.*, 1974; Lynch, 1972; Lynch and Harper, 1974a,b; Primrose, 1976; Primrose and Dilworth, 1976; Swanson *et al.*, 1979; Thomas and Spencer, 1977). Arshad and Frankenberger (1989) ob-

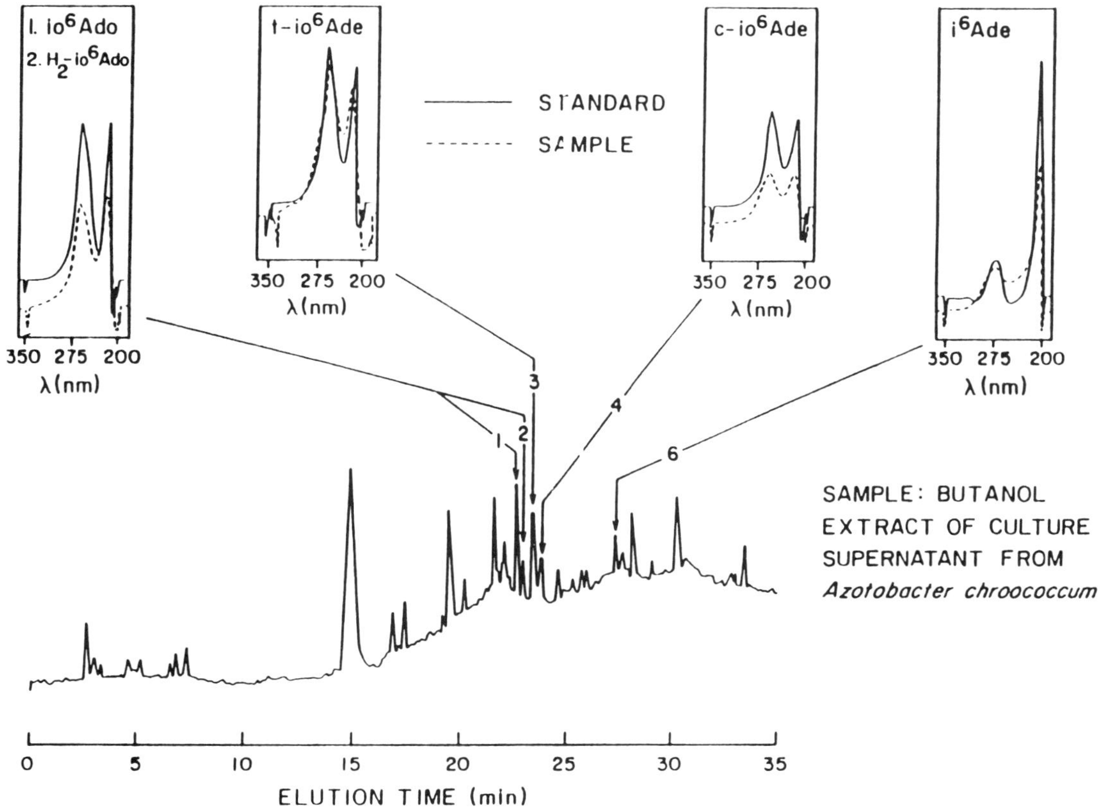

Fig. 2. HPLC chromatogram of a culture supernatant of *Azotobacter chroococcum*. io^{6}Ado, zeatin riboside; H_2-io^{6}Ado, dihydrozeatin riboside; t-io^{6}Ade, t-zeatin; c-io^{6}Ade, c-zeatin; i^{6}Ade, isopentenyl adenine (Nieto and Frankenberger, 1988).

served that corn rhizosphere contained an appreciable number of microflora capable of deriving C_2H_4 from L-MET. Among the soil fungal isolates, *Acremonium falciforme* was an efficient C_2H_4 producer. However, it followed a different pathway of C_2H_4 biosynthesis from that elucidated in higher plants. We found that amino acids, organic acids and carbohydrates typically found in root exudates stimulate C_2H_4 biosynthesis in soil (Arshad and Frankenberger, 1990). This indicates that an excellent site for microbial biosynthesis of C_2H_4 is the rhizosphere, where substrate levels and microflora populations are comparatively very high.

Ethylene has been identified as a common component of the soil atmosphere and under certain conditions has been shown to reach a concentration high enough to influence plant growth and development (Smith, 1976). Ethylene concentrations as low as 10 ppb can evoke plant responses and concentrations of 25 ppb result in decreased fruit and flower development (Primrose, 1979). Jackson and Campbell (1975) were the first to demonstrate rapid movement of radiolabeled [^{14}C]-C_2H_4 applied to the roots of tomato plants producing a specific epinastic response. However, little direct evidence is available in the existing literature on the microbial influence of C_2H_4 production on plant growth. Arshad and Frankenberger (1988) demonstrated the influence of microbial produced L-MET derived C_2H_4 on etiolated pea seedlings by observing the classical 'triple' response which includes reduction in elongation, swelling of the hypocotyl, and a change in the direction of growth (horizontal) (Fig. 3). Etiolated pea seedling responded to C_2H_4 derived from L-MET by the inoculum, *A. falciforme* (Table 2) and the soil indigenous microflora (Table 3). A similar response was observed when seedlings were ex-

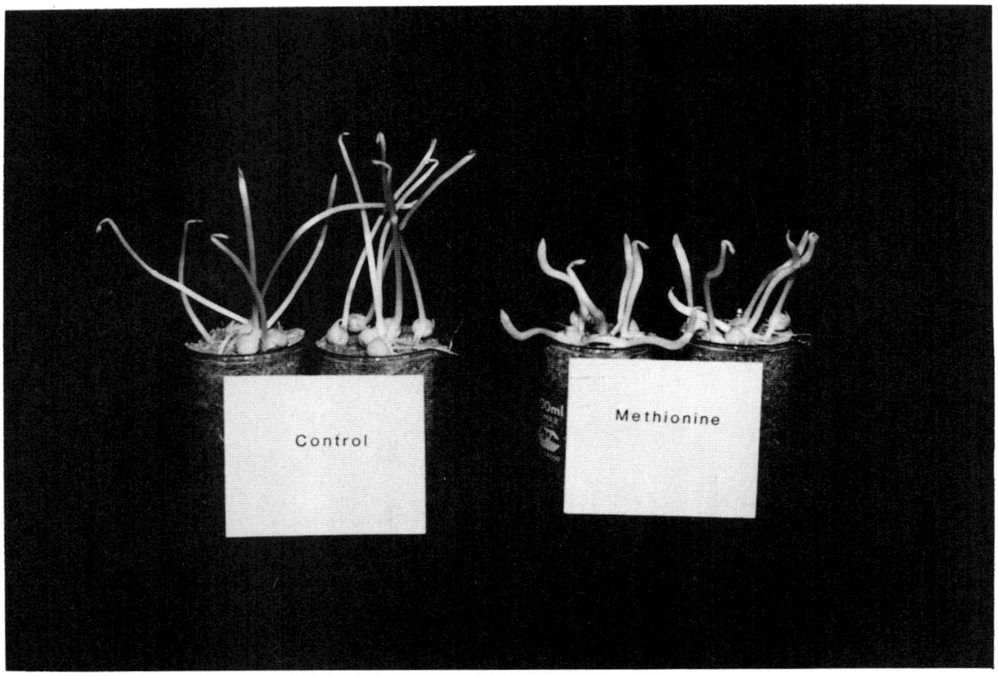

Fig. 3. Response of etiolated pea seedlings to exogenous L-methionine-derived C_2H_4 produced by *A. falciforme* in the absence of $AgNO_3$ (Arshad and Frankenberger, 1988).

posed to C_2H_4 gas. The specific anti-C_2H_4 properties of Ag(I) confirmed that C_2H_4 was the fungal metabolite responsible for the observed effects. Moreover, the culture was incubated outside of the root zone and thus C_2H_4 derived from L-MET in the medium affected the growth of pea seedlings.

Conclusions

The studies reported here reveal the ability of specific inocula as well as the soil indigenous microflora to derive plant hormones from precursors provided in pure culture and in soil. Inoculation with specific organisms may prove

Table 2. Influence of L-methionine-derived C_2H_4 produced by *A. falciforme* on etiolated pea seedlings grown in autoclaved soil (Arshad and Frankenberger, 1988)

Treatment*	Seedling length (cm)	Seedling diameter (mm)
Control	6.23 b***	1.97 ab
$AgNO_3$ (240 mg L^{-1})**	7.57 b	1.87 a
Inoculated (*A. falciforme*)	6.67 b	2.07 b
L-Methionine (10 m*M*)	7.22 b	2.10 b
L-Methionine (10 m*M*) + inoculation	2.58 a	2.44 c
L-Methionine (10 m*M*) + $AgNO_3$ (240 mg L^{-1})	6.11 b	1.93 ab
L-Methionine (10 m*M*) + inoculation + $AgNO_3$ (240 mg L^{-1})	5.77 b	2.07 b

* Samples that did not receive $AgNO_3$ received $NaNO_3$ (240 mg L^{-1})
** $AgNO_3$ is an inhibitor of C_2H_4 action.
*** Values followed by the same letter were not significantly different at the 0.05 level according to the Duncan multiple range test.

Table 3. Influence of L-methionine-derived C_2H_4 produced by soil indigenous microflora on etiolated pea seedlings (Arshad and Frankenberger, 1988)

Treatment*	Seedling length (cm)	Seedling diameter (mm)
Control	6.56 b**	1.87 a
$AgNO_3$ (240 mg L^{-1})	13.50 d	1.93 ab
L-Methionine (5 m*M*)	5.14 ab	2.49 c
L-Methionine (5 m*M*) + $AgNO_3$ (240 mg L^{-1})	11.10 c	2.06 ab
L-Methionine (10 m*M*)	3.90 a	2.75 d
L-Methionine (10 m*M*) + $AgNO_3$ (240 mg L^{-1})	10.10 c	2.11 b

* Samples that did not receive $AgNO_3$ received $NaNO_3$ (240 mg L^{-1})

** Values followed by the same letter were not significantly different at the 0.05 level according to the Duncan multiple range test.

beneficial in affecting plant growth in some cases but results strongly support our hypothesis that inoculation in the presence of a specific physiological precursor is more effective than inoculation alone. Similarly, precursor-inoculum interactions are often much better in creating plant responses than the application of precursors alone, indicating the superiority of particular inocula over the soil indigenous microflora in synthesizing PGRS. Plant responses may be controlled by varying the amount of precursors applied since this response is concentration dependent. Rhizosphere microflora should be extensively screened for their ability to synthesize plant hormones from a given substrate.

References

Adams D O and Yang S F 1979 Ethylene biosynthesis: identification of 1-aminocylopropane-1-carboxylic acid as an intermediate in the conversion of methionine to ethylene. Proc. Natl. Acad. Sci. USA 76, 170–174.

Andel O M van and Fuchs A 1972 Interference with plant growth regulation by microbial metabolites. *In* Phytotoxins in Plant Disease. Eds. R K S Wood, A Ballic and A Graniti. pp 227–249. Academic Press, London.

Arshad M and Frankenberger W T Jr 1988 Influence of ethylene produced by soil microorganisms on etiolated pea seedings. Appl. Env. Microbiol. 54, 2728–2732.

Arshad M and Frankenberger W T Jr 1989 Biosynthesis of ethylene by *Acremonium falciforme*. Soil Biol. Biochem. 21, 633–638.

Arshad M and Frankenberger W T Jr 1990 Ethylene accumulation in soil in response to organic amendments. Soil Sci. Soc. Am. J. (*in press*).

Barea J M and Brown M E 1974 Effects on plant growth produced by *Azotobacter paspali* related to synthesis of plant growth regulating substances. J. Appl. Bacteriol. 37, 583–593.

Brown M E and Burlingham S K 1968 Production of plant growth substances by *Azotobacter chroococcum*. J. Gen. Microbiol. 53, 135–144.

Dasilva E J, Henriksson E and Henriksson L A 1974 Ethylene production by fungi. Plant Sci. Lett. 2, 63–66.

Döbereiner J, Marriel I E and Nery M 1976 Ecological distribution of *Spirillum lipoferum*. Beyerinck. Can. J. Microbiol. 22, 1464–1473.

Frankenberger W T Jr and Brunner W 1983 Methods of detection of auxin-indole-3-acetic acid in soils by high performance liquid chromatography. Soil Sci. Soc. Am. J. 47, 237–241.

Frankenberger W T Jr and Poth M 1987a Determination of substituted indole derivatives by ion suppression-reverse-phase high-performance liquid chromatography. Anal. Biochem. 165, 300–308.

Frankenberger W T Jr and Poth M 1987b Biosynthesis of indole-3-acetic acid by the pine ectomycorrhizal fungus *Pisolithus tinctorius*. Appl. Environ. Microbiol. 53, 2908–2913.

Frankenberger W T Jr and Poth M 1988 L-tryptophan transaminase of a bacterium isolated from the rhizosphere of *Fustuca octoflora* (Graminae). Soil Biol. Biochem. 20, 299–304.

Gamborg O I and Wetter L R 1963 An aromatic amino acid transaminase from mung bean. Can. J. Biochem. Physiol. 41, 1733–1740.

Gruen H E 1959 Auxins and fungi. Annu. Rev. Plant Physiol. 10, 405–440.

Hubbell D H, Tien T M, Gaskin M H and Lee J 1979 Physiological interaction in the *Azospirillum* – grass root association. *In* CRC Associative Symbiosis. Eds. P B Vose and A P Ruschel. 1, 1–6.

Hussain A, Arshad M, Hussain A and Hussain F 1987 Response of maize (*Zea mays*) to *Azotobacter* inoculation under fertilized and unfertilized conditions. Biol. Fertil. Soils 4, 73–77.

Jackson M B and Campbell D J 1975 Movement of ethylene from roots to shoots, a factor in the responses of tomato plants to waterlogged conditions. New Phytol. 74, 397–406.

Jagnow G 1987 Inoculation of cereal crops and forage grasses with nitrogen-fixing rhizosphere bacteria: Possible causes of success and failure with regard to yield response – A review. Z. Pflanzenernaehr. Bodenkd. 150, 361–368.

Kaper J M and Veldstra H 1958 Metaboism of tryptophan by *Agrobacterium tumefaciens*. Biochim. Biophys. Acta 30, 401–420.

Lee M, Breckenridge C and Knowles R 1970 Effect of some culture conditions on the production of indole-3-acetic acid and a gibberellin-like substance by *Azotobacter vinelandii*. Can. J. Microbiol. 16, 1325–1330.

Libbert E and Brunn K 1961 Nachweis von indol-3-brenztrauben saure und indol-3-athanol (tryptophol) bei der enzymatischen auxin bildung aus tryptophan *in vitro*. Naturwissenschaft 48, 741.

Lynch J M 1972 Identification of substrates and isolation of microorganisms responsible for ethylene production in the soil. Nature 240, 45–46. London.

Lynch J M 1985 Origin, nature and biological activity of aliphatic substances and growth hormones found in soil. *In* Soil Organic Matter and Biological Activity. Eds. D Vaughan and R E Malcolm. pp 151–174. Martinus Nijhoff/ Dr W Junk Publishers. Dordrecht/Boston/Lancaster.

Lynch J M and Harper S H T 1974a Formation of ethylene by a soil fungus. J. General Microbiol. 80, 187–195.

Lynch J M and Harper S H T 1974b Fungal growth rate and the formation of ethylene in soil. J. Gen. Microbiol. 85, 91–96.

Nieto K F and Frankenberger W T Jr 1988 Determination of cytokinins by ion suppression-reverse phase high performance liquid chromatography. J. Liq. Chrom. 11, 2907–2925.

Nieto K F and Frankenberger W T Jr 1989a Biosynthesis of cytokinins produced by *Azotobacter chroococcum*. Soil Biol. Biochem. 21, 967–972.

Nieto K F and Frankenberger W T Jr 1989b Biosynthesis of cytokinins in soil. Soil Sci. Soc. Am. J. 53, 735–740.

Nieto K F and Frankenberger W T Jr 1990 Microbial production of cytokinins. *In* Soil Biochemistry. Vol. 6. pp 191–248. Eds. J M Bollag and G Stotzky. Marcel Dekker, New York.

Perley J E and Stowe B B 1966a The production of tryptamine from tryptophan by *Bacillus cereus* (KVT). Biochem. J. 100, 169–174.

Perley J E and Stowe B B 1966b On the ability of *Taphrina deformans* to produce indole acetic acid from tryptophan by way of triptomine. Plant Physiol. 41, 234–237.

Phillips D A and Torrey J G 1970 Cytokinin production by *Rhizobium japonicum*. Physiol. Plant. 23, 1057–1063.

Primrose S B 1976 Formation of ethylene by *Escherichia coli*. J. Gen. Microbiol. 95, 159–165.

Primrose S B 1979 A review, ethylene and agriculture: the role of the microbes. J. Appl. Bacteriol. 46, 1–25.

Primrose S B and Dilworth M J 1976 Ethylene production by bacteria. J. Gen. Microbiol. 93, 177–181.

Puppo A and Rigaud J 1978 Cytokinins and morphological aspects of French-bean roots in the presence of *Rhizobium*. Physiol. Plant. 42, 205–206.

Sembdner G, Gross D, Liebisch H W and Schneider G 1980 Biosynthesis and metabolism of plant hormones. *In* Hormonal Regulation of Development. I. Molecular Aspects of Plant Hormones. Ed. J MacMillan. pp 281–444. Springer, New York.

Smith A M 1976 Ethylene in soil biology. Annu. Rev. Phytopathol. 14, 53–73.

Swanson B T, Wilkins H F and Kennedy B 1979 Factors affecting ethylene production by some plant pathogenic bacteria. Plant and Soil 51, 19–26.

Thomas K C and Spencer M 1977 L-methionine as an ethylene precursor in *Saccharomyces cerevisiae*. Can. J. Microbiol. 23, 1669–1674.

Tien T M, Gaskins M H and Hubbell D H 1979 Plant growth substances produced by *Azospirillum bransilense* and their effect on the growth of pearl millet (*Pennisetum americanum* L.). Appl. Env. Microbiol. 37, 1016–1024.

Truelsen T A 1972 Indole-3-pyruvic acid as an intermediate in the conversion of tryptophan to indole-3-acetic acid. I. Some characteristics of tryptophan transaminase from mung bean seedling. Physiol. Plant. 26, 289–295.

D. L. Keister and P. B. Cregan (Eds.), The rhizosphere and plant growth, 335–342.

Characterization and use of vesicular-arbuscular mycorrhizae in agricultural production systems

P.D. MILLNER
Soil Microbial Systems Laboratory, USDA-ARS-BARC, Beltsville, MD 20705, USA

Key words: host and soil factors, identification, inoculation, RFLP, serotaxonomy, VAM structure and function, water relationships

Abstract

Vesicular-arbuscular mycorrhizal (VAM) fungi are the most common fungal endophytic symbionts associated with plant roots. They form associations with a wide range of hosts in diverse soil types and climatic regions of the world. This review describes VAM occurrence in agricultural production systems and the effects that cultural practices have on them. The role of VAM fungi in improving the growth of plants by increasing the absorption of phosphorus and micronutrients and enhancing tolerance to drought is reviewed. The potential for commercial use of VAM is described along with the most recent advances in inoculum production technology. New approaches to identification of VAM fungi, including ELISA with polyclonal and monoclonal antibodies, isozyme analysis, and restriction fragment length polymorphism are discussed in the context of a model for research and development of VAM-fungus inoculation.

Introduction

Vesicular-arbuscular mycorrhizae (VAM) are symbiotic associations between fungi and plant roots; they can be considered the normal state of the roots of most agronomically important crops except Cruciferae, Chenopodiaceae, Cyperaceae, and Caryophyllaceae. Paddy rice is often non-mycorrhizal, probably because the hyphae cannot meet their oxygen demands. Many trees have VAM associations in addition to ectomycorrhizae. Numerous reports have substantiated the positive influence of VAM fungi on plant growth. About 126 species of VAM fungi have been described (Morton, 1988). Ecophysiological and effectiveness data are available for less than 15 percent of them and 43 percent of *Glomus* species have never been shown to form typical vesicles and arbuscules and thus are unverified as VAM fungi. Several excellent reviews provide detailed accounts of the structure, culture, ecology, (Powell and Bagyaraj, 1984), ecophysiology (Safir, 1987) and taxonomy (Morton, 1988; Schenck and Perez, 1987) of the VAM fungi.

An examination of the fundamental effects of VAM on plant growth frequently leads to the question: How can VAM be used in agricultural production systems? One approach to their use involves management of indigenous VAM, as described in the Chapter by Abbott and Robson (this symposium). Another approach is to inoculate with superior isolates under appropriate conditions. Both approaches require detailed knowledge of the characteristics of both the host plant and endophyte as well as of the biotic and abiotic components of the soil. However, the characterization of an individual VAM fungus as a biological entity is limited because of its obligate dependence on a host. The ecology of VAM fungi, effects on the host, and factors affecting spore germination have dominated the characterization of their functional capacity, while morphology has dominated their taxonomic characterization.

No correlations are currently available between the functional and taxonomic characterizations of VAM fungi.

In this paper, only the major aspects of structure-function relationships of VAM and the ecophysiological factors affecting them will be reviewed. Research areas which will provide information or methods critical to the ultimate objective of management or inoculation with VAM fungi are identified. The relevance of recent advances in inoculum production technologies and methods of species identification is examined.

Characterization of VAM

VAM structure-function relationships

The VAM fungal hyphae proliferate both outside and inside the root, with the internal hyphae differentiating into arbuscules and vesicles. The arbuscules are formed only on internal hyphae in response to contact with host cortical cells; they are the sites of nutrient exchange between the plant and the fungus. Vesicles, which contain high concentrations of triglycerides at maturity, are thought to serve as storage and survival organs although evidence of hyphal germination from them is lacking. Thick-walled chlamydospores, the common propagules, are formed on external and, in some species, on internal hyphae. Morphology of the chlamydospores (hereafter referred to as spores) is the current basis for identification of VAM fungi.

Benefits from VAM symbiosis occur because VAM hyphae extend beyond the root hair zone and increase the absorptive surface area of the root. VAM can improve uptake of certain relatively immobile soil elements, *e.g.*, P, Zn, and Cu when availability of these nutrients limits plant growth. The external VAM hyphae reach beyond the P depletion zone around the root hairs, absorb soil P and translocate it, perhaps in the form of polyphosphate granules, to the arbuscules where P is transferred to the plant cell in exchange for carbon. Mycorrhizal and nonmycorrhizal plants absorb the same ionic form of P (Gianinazzi-Pearson *et al.*, 1981). Specific mechanisms for absorption, transport, and exchange of Zn, Cu and other microelements have not been clearly established for VAM fungi but probably also involve translocation.

In contrast to soil P, soil N availability is not diffusion limited and hence the role of VAM fungi in enhancement of plant uptake is not clearly discernible. However, recent studies using ^{15}N strongly implicate VAM hyphal connections between roots of companion plants as conduits for the transfer of N from N-fixing plants to nonfixing plants (Van Kessel *et al.*, 1985). The practical impact of this phenomenon on intercropping systems has yet to be determined. The effect of the nitrogen source on VAM fungi is different in different hosts. Mycorrhization of onions and maize is favored by NO_3–N (Hayman, 1983; Thompson, 1986), whereas NH_4–N reduces colonization, spore production, pH, and plant dry matter. In contrast, $NH_4 - N$ enhances germination and hyphal growth of *G. mosseae* at pH 7.0 and 8.5 in the rhizosphere of soybean and on agar (Wong and Marschner, 1988).

Water relations and soil aggregation

Mycorrhizal plants recover from brief periods of water stress more quickly than do nonmycorrhizal plants (Nelsen, 1987). However, long periods of water stress affect the larger mycorrhizal plants more severely than the smaller nonmycorrhizal plants. Theoretical explanations for these observations include effects on stomatal regulation with accompanying increases in transpiration and photosynthesis, direct hyphal transport of water, and improved P status of plants (Nelsen, 1987). The latter is an often cited explanation, but it still is considered inconclusive. The mycorrhizal effect on water stress is important by itself even though the mechanism by which it occurs has not been resolved. The water stress amelioration trait may well be related to the capacity of an isolate to produce copious external hyphae. In this case, the uptake of soil water would be facilitated by isolates that enhance the formation of soil aggregates (Bethlenfalvay *et al.*, 1988) since soil aggregation improves water-holding capacity. At present, comparisons are needed among VAM fungi of their effect on water stress under a variety of soil conditions with a variety of hosts. Reduction in water stress may be a significant beneficial trait for VAM fungal isolates that colonize plants grown in soils containing relatively high levels of available P in which the VAM effect on plant uptake of P is not demonstrable. The identification of isolates especially valuable for water stress reduction could be used ultimately for inoculation of transplants or for targeting management practices to encourage associations with particular species of VAM fungi.

Infection measurement

The symbiotic association is usually quantified by measurements of the percent infection defined as percentage of root segments (often 1 cm lengths) infected or root length infected, as determined by a grid-line intersect method (Giovanetti and Mosse, 1980). However, the degree of infection by different species of VAM fungi is not always predictive of the increase in plant growth (McGonigle, 1988), especially when comparisons between different hosts in different soils are made. The weak predictive relationship may be associated with differences in the surface area of external hyphae produced by the fungi or in inherent differences in metabolism and rates of nutrient translocation. The dehydrogenase reduction assay (Sylvia, 1988) might be used to measure physiologically active lengths of external VAM hyphae if adequate methods for separation of hyphae from test soils can be achieved. Measurement of the ratio between external and internal hyphal formation during the early phases of plant development when P demand is high might be useful in preliminary assessments of isolate effectiveness and response to edaphic factors.

Host factors

Aside from the plant families that do not form mycorrhizae (see introduction), there is no evidence of host: fungus species specificity among VAM fungi. However, coarse rooted plants (ultimate roots >1.0 mm diam) generally benefit more from VAM than do fine rooted ones (<0.5 mm ultimate root diam, see Baylis, 1970). In addition, C-4 plants, which are more photosynthetically efficient than C-3 plants, are more mycotrophic (more attractive to VAM fungi) perhaps because more carbohydrates are available to VAM fungi in the roots (Hayman, 1983). Differences in cultivar susceptibility to VAM have been reported for alfalfa (Lambert *et al.*, 1979; O'Bannon *et al.*, 1980), citrus (Menge *et al.*, 1978), soybean (Heckman and Angle, 1987), and wheat (Azcon and Ocampo, 1981). The comparative mycotrophy of various cultivars of cotton, maize, sorghum, other grains, alternative crops, and horticultural rootstocks needs to be studied. Also, the lack of suitable host plants (Ocampo and Hayman, 1980) and/or long-term fallowing (Thompson, 1987) can severely reduce the amount of indigenous VAM fungi present in the soil for subsequent crops. These host factors may be advantageous in cropping situations in which it is desirable to introduce inoculum or they may be deleterious if no inoculum introduction is planned.

Soil factors

Fertility, pH, temperature, and pesticide/fungicide history of a soil are all important in mycorrhization (Daniels Hetrick, 1984; Nemec, 1987). High levels of soil P and N can reduce VAM fungal spore germination and plant colonization, although the P status of the root itself (Menge *et al.*, 1978) or the sugar and organic acid concentration in the root cortex (Siqueira *et al.*, 1982) influences root colonization more than does the soil P level. The root P status is decreased even in P sufficient soils during drought (Nelsen and Safir, 1981; Sieverding, 1981) because low soil moisture results in reduced diffusion of soil P. The optimal pH for spore germination often coincides with that for root colonization (Porter *et al.*, 1987).

Model for VAM inoculation research and development

A decision diagram for research on field inoculation of VAM fungi in the tropics was proposed by Sieverding (1986). In his approach, fungus, crop, and edpahic conditions are matched after an isolate has been determined to be potentially effective. In contrast, Figure 1 depicts sequential steps in a research and development program for VAM-fungal inoculation that is directed at specific cropping systems based on pretest decisions regarding hosts, climatic zones, crop rotations, *etc.* The model is applicable to 1) inoculation of containerized plants prior to transplanting into soils with high or low levels of indigenous VAM fungi or into VAM-fungi depleted soils, or 2) direct inoculation into VAM-fungi depleted soils. Transplanating of inoculated plants into soils with high indigenous VAM fungus populations will require more extensive testing at stages II, III and V than would be needed for VAM-fungus depleted soils.

Current strategies and practices rely on an isolate-by-isolate evaluation simply because no information exists on generic or specific features of VAM fungi which are predictive of fungal effectiveness. It is ironic that we are seeking functional information about a group of fungi for which the species concept is based solely on morphology because so little evaluation of phenotype has been done. The bulk of the evidence suggests that no single isolate or species of VAM fungi will be universally effective; a combination of complementary isolates will have a greater chance of meeting the needs of the variety of conditions typically encountered in a cropping situation.

STAGE PRETEST DECISION: CROP(S), CLIMATIC ZONE, CROP SYSTEM

I. ISOLATION OF VAM FUNGUS SPORES
INOCULUM POT CULTURE TESTS

II. EFFECTIVENESS EVALUATION
1. NONCOMPETITIVE
2. COMPETITIVE

III. EDAPHOTYPE EVALUATION
FUNGUS-EDAPHIC RANGE MATCHING

IV. SCALE-UP INOCULUM PRODUCTION
INOCULUM STORAGE AND DELIVERY

V. FIELD STUDIES
1. EFFECTIVENESS
2. SURVIVAL
3. DISPERSAL

VI. APPROPRIATE VAM TECHNOLOGY

Fig. 1. Sequential processes for research and development of VAM fungus inoculation.

It is impractical to test the effectiveness of very many species with all their possible eco- and edaphotypic adaptions for application in all cropping systems. An alternative strategy might be to target certain cropping systems that would benefit from VAM fungus inoculation because of low indigenous populations, low soil fertility, or dependence of the host on mycorrhizal fungi, *e.g.*, citrus (Menge, 1982), grapes, peaches, and horticultural transplants like asparagus. VAM fungi, isolated from the most vigorous plants growing in edaphically and climatically similar cropping situations would be established in single-species pot cultures (Stage 1) on asparagus, onions, or sudan grass; the potting soil should be consistent with the source from which the VAM fungus was isolated particularly with regard to pH and fertility. The possibility of cross-contamination of VAM fungi in pot culture necessitates a periodic microscopic examination of pot inocula to verify single species integrity.

Since there are no wholly satisfactory methods for long-term preservation of VAM propagules the maintainence of pot cultures remains the backbone of this technology. Thus, at Stage I, research and development on reliable storage and cryopreservation methods for VAM fungal spores/inocula, axenic culture methods, and identification of the fungal species are needed.

At Stage II, VAM fungi should be screened first in sterilized soils, compatible with the source of the isolate, and then in nonsterilized soils in order to assess the competitive ability of the fungus. All hosts that are to be used in the ultimate cropping system should be used in the tests. Measurements of plant dry matter accumulation, P uptake, and fungal colonization of the rhizosphere and root will provide data for a primary evaluation of the potential effectiveness of the isolate. Sieverding (1986) found that nearly 60 percent of isolates from cassava were ineffective (i.e., did not enhance plant growth or P uptake). Verification of infection by the inoculated species in competitive tests is the most difficult aspect of work at this stage, since little distinction can be made between arbuscules and internal hyphae among different VAM fungi. However, verification is fundamental to completion of Koch's postulates.

Identification of VAM fungi

Genera of VAM fungi are morphologically distinct and their identification is generally not a problem However, identification of species, especially in *Glomus*, has become increasingly complex because very subtle and in some cases ephemeral differences in size, ornamentation, and wall microstructure of spores have been used to delimit new species. Since hyphae, not spores, are the major structural components important in functional symbioses, methods for distinguishing species by their internal and external hyphae are needed. This will also aid in comparative evaluations of the other traits thought to be important, *e.g.*, spatial density of the external hyphae.

Enzyme-linked immunosorbent assay (ELISA) using polyclonal sera from rabbits immunized with external hyphae collected from pot cultures was used to detect hyphal antigens of *G. clarum*, *G. mosseae*, *Sclerocystis dussii*, *Acaulospora laevis*, *Gigaspora calospora*, and *Gi. margarita* (Aldwell *et al.*, 1985) Cross-reactions between genera and species were common. Kough *et al.* (1983) used spores of *G. epigeum* to produce rabbit polyclonal antibodies which were then used for immunofluorescent staining. Cross-reactions were observed with spores and hyphae of *G. mosseae* and *G. deserticola* but not with Gigaspora or Acaulospora. Wright *et al.*, 1987 have produced monoclonal antibodies (Mab) to *Glomus occultum* spores and hyphae that do not cross react with other VAM or soil fungi. In modified immunoblots of infected roots, this Mab recognized *G. occultum* spores and hyphae external, but not internal, to the root (Wright *et al.*, 1989). This is most likely a problem of inaccessibility of the antibody to the fungal antigen caused by the impermeability of root cell walls to the antibody. Enzymatic digestion of roots prior to immunoblotting or cryosectioning might improve accessibility of the antibodies to the fungal antigents.

We have used *G. mosseae* (INVAM 156) spores/sporocarps and external hyphae of *G. mosseae* and *G. etunicatum* (Beltsville No. 1) to produce murine Mabs using methods similar to those of Wright *et al.* (1987). We have one Mab to *G. mosseae* spores, which is an exceedingly sticky IgM that is very difficult to use in ELISA, but it is clearly species specific to *G. mosseae* spores but not hyphae in immunofluorescent staining tests. We have several Mabs which secrete antibodies that recognize Glomus hyphae. Several of these are also IgMs and recognize Pythium hyphae also. However, at least one is specific for Glomus external hyphae and will be further characterized for ability to recognize internal hyphae and arbuscules.

Isozymes have been used to investigate relationships among *G. clarum*, *G. manihotis*, *G. mosseae*, *G. monosporum*, and *G. caledonium* (Hepper *et al.*, 1988). This technique is very useful along with serotaxonomic and possibly genetic techniques for confirming suspected taxonomic relationships based on morphologic traits. However, practical application would be limited to research, since gel electrophoresis and enzyme staining cannot provide information on proportional infections by various fungi. Also, reliance on enzyme patterns from asynchronously produced spores can lead to false negatives and, thus, to misidentifications if too few of variously matured entitites are included in the samples to be analyzed.

A serotaxonomic identification system also may be limited by the availability of specific immunogenic antigens. With no *a priori* knowledge about these antigens, immunizations will continue with biomass of whole organisms (hyphae/spores/combinations) but certain cellular components may be so immunogenic that they dominate the immune response to the practical exclusion of others. A relatively simple type of immunoassay for identification of field collections is ultimately needed to work for spore, hyphal, and root crushs or extracts. However, the validity of the identifications and serotaxonomy must reflect a phylogenetically based species concept. Fortunately, advances in molecular genetic technology provide some useful tools for an alternate approach to conventional resolution of species concepts in VAM fungi. Restriction fragment length polymorphism analysis (RFLP) (Michelmore and Hulbert, 1987) would provide data on genetic relatedness of these fungi even if their phenotypic traits are unavailable or incompletely understood.

Some preliminary work along this line has been done in our laboratory. As with most molecular genetic involving fungi, methods for extraction and purification of DNA must be developed and adapted to the particular organisms being studied since their properties differ enough to reduce purity or yields when standardized techniques are used. In the case of VAM fungi, there is an additional problem, *i.e.*, the need to remove soil and microbial contaminants from the biomass before the actual extraction and purification. We have used the closely related fungus *Endogone pisiformis* as a model for the extraction and purification steps, since this fungus can be cultured in a flask and reasonable quantities of biomass are available. Recently, we have applied these extraction and purification procedures successfully to *G. etunicatum*, *G. mosseae*, and *Scutellospora coralloidea* and we have obtained DNA that is sufficiently pure to be digested by restriction endonucleases. The next step will be the hybridization of the blotted electrophoresed digests of the restriction fragments with gene probes directed at the highly conserved rDNA.

Edaphotype effectiveness

In Stage III (Fig. 1), the need to assess effectiveness of isolates in widely divergent circumstances is reduced because the range of hosts, rotations, and pest management practices are limited by the pretest decisions. The range of soil pH, density, P fertility, temperature, and water availability in which the VAM fungus is effective would need to be tested. Special consideration of the tolerance to high levels of trace elements would need to be included for inocula intended for use in reclamation sites. Results will provide data for matching isolates to cropping situations and for selection of those most suited for certain field tests.

Inoculum Production (Stage IV)

VAM fungus inoculum is commonly produced in pots of pasteurized soil on alternate hosts in greenhouses. Newer methods are based on sand culture with nutrient solution (Ojala and Jarrell, 1980), aeroponic (Sylvia and Hubbell, 1986) and nutrient film culture (Mosse and Thompson, 1984), and culture with calcined clays (Dehne and Backhaus, 1986). All these methods have eliminated soil which reduces much of the variability in production. Also, these methods provide relatively cleaner biomass useful in immunologic, physiologic, and genetic studies. Spores from sand and calcined clay particle production can be wet-seived and collected in large numbers for use or storage. Hyphae and spores can be easily obtained from aeroponic or mistigated cultures by wet-seiving and gentle sonication of roots. For field trials, in which separation of fungal biomass from the roots and the culture production medium is unnecessary, mixtures of calcined clay particles and lightweight inert porous horticultural media would have several advantages; these include low density, reduced variability, high porosity for colonization, and ease of delivery. With all these media and the production systems, quality control and vigilance concerning cross-contamination is still required. Standardization of inoculum density will be needed and procedures such as the "Most Probable Number" method (Porter, 1979) might be used.

Field tests (Stage V)

Isolates that are effective in noncompetitive and/or competitive tests and have been matched with edaphic and cropping conditions for which they are suited must be proven effective in field tests. Previously, production of large amounts of inoculum was a bulky operation, but many of the above described methods can be used to reduce the bulk of the end product. The effects of timing and method of inoculation will need to be developed particularly for cases of field applied inocula. Inoculum pellets containing seeds and a mixture of VAM fungi such as Hall and Kelson (1981) used to establish VAM on white clover might be particularly suited to small seeded plants. Field studies (Stage V) should include at least those measurements used in primary evaluations in addition to any that might be pertinent to the special cropping systems being evaluated, *e.g.*, water stress reduction at transplanting. Assessments of inoculum spread and survival would help determine the need for reinoculation. Multiple site tests that include the range of tolerance to pH, P fertility *etc*. Will be needed to confirm initial efficacy of isolates as determined in the edaphotype evaluation, Stage III.

References

Aldwell F E B, Hall I R and Smith J M 1985 Enzyme-linked immunosorbent assay as an aid to taxonomy of the endogonaceae. Trans. Br. Mycol. Soc. 84, 399–402.

Azcon R and Ocampo J A 1981 Factors affecting the vesicular-arbuscular infection and mycorrhizal dependency of thirteen wheat cultivars. New Phytol. 87, 677–685.

Baylis G T S 1970 Root hairs and phycomycetous mycorrhizas in phosphorus-deficient soil. Plant and Soil 33, 713–716.

Bethlenfalvay G J, Thomas R S, Dakessian S, Brown M S and Ames R N 1988 Mycorrhizae in stressed environments: effects on plant growth, endophyte development, soil stability and soil water. *In* Arid Lands Today and Tomorrow. Eds. E E Whitehead, C F Hutchinson, B N Timmerman and R G Varady. pp. 1015–1029. Westview Press, Boulder, CO.

Daniels Hetrick B 1984 Ecology of VA mycorrhizal fungi. *In* VA mycorrhizae. Eds. C L Powell and D J Bagyaraj. pp 35–55. CRC Press, Boca Raton, FL.

Dehne H W and Backhaus G F 1986 The use of vesicular-arbuscular mycorrhizal fungi in plant production. I. Inoculum production. J. Pl. Dis. Prot. 93, 415–424.

Gianinazzi-Pearson V, Fardeau J C, Asimi S and Gianinazzi S 1981 Source of additional phosphorus adsorbed from soil by vesicular-arbuscular mycorrhizal soybeans. Physiol. Veg. 19, 33–43.

Giovannetti M and Mosse B 1980 An evaluation of techniques for measuring vesicular-arbuscular mycorrhizal infection in roots. New Phytol. 84, 489–500.

Hall I R and Kelson A 1981 An improved technique for the production of endomycorrhizae infested soil pellets. N.Z. J. Agric. Res. 24, 221–222.

Hayman D S 1983 The physiology of vesicular-arbuscular endomycorrhizal symbiosis. Can. J. Bot. 61, 944–963.

Heckman J R and Angle J S 1987 Variation between soybean cultivars in vesicular-arbuscular mycorrhiza fungi colonization. Agron. J. 79, 428–430.

Hepper C M, Sen R. Azcon-A C and Grace C 1988 Variation in certain isozymes amongst different geographical isolates of the vesicular-arbuscular mycorrhizal fungi *Glomus clarum*, *Glomus monosporum* and *Glomus mosseae*. Soil Biol. Biochem. 20, 51–59.

Kough J, Malajczuk N and Linderman R G 1983 Use of the indirect immunofluorescent technique to study the vesicular-arbuscular fungus *Glomus epigaeum* and other *Glomus* species. New Phytol. 94, 57–62.

Lambert D H, Baker D E and Cole H Jr 1979 The role of mycorrhizae in the interactions of phosphorus with zinc, copper, and other elements. Soil Sci. Soc. Am. J. 43, 976–980.

McGonigle R P 1988 A numerical analysis of published field trials with vesicular-arbuscular mycorrhizal fungi. Funct. Ecol. 2, 473–478.

Menge J A 1983 Utilization of vesicular-arbuscular mycorrhizal fungi in agriculture. Can. J. Bot. 61, 1015–1024.

Menge J A, Johnson E L V and Platt R G 1978 Mycorrhizal dependency of several citrus cultivars under three nutrient regimes. New Phytol. 81, 553–559.

Michelmore R W and Hulbert S H 1987 Molecular makers for genetic analysis of phytopathogenic fungi. Annu. Rev. Phytopathol. 25, 383–404.

Morton J B 1988 Taxonomy of VA mycorrhizal fungi: classification nomenclature and identification. Mycotaxon 32, 267–324.

Mosse B, and Thompson J P 1984 Vesicular-arbuscular endomycorrhizal inoculum production. I. Exploratory experiments with beans (*Phaseolus vulgaris*) in nutrient flow culture Can. J. Bot. 62, 1523–1530.

Nelsen C E 1987 The water relations of vesicular-arbuscular mycorrhizal systems. *In* Ecophysiology of VA Mycorrhizal Plants. Ed. G R Safir. pp 71–91. CRC Press, Boca Raton, FL.

Nelsen C E and Safir G R 1981 Increased drought resistance of mycorrhizal onion plants. Phytopathology 71, 896–897.

Nemec S 1987 VA mycorrhizae in horticulture systems. *In* Ecophysiology of VA Mycorrhizal Plants. Ed. G R Safir. pp 171–211. CRC Press, Boca Raton, FL.

O'Bannon J H, Evans D W and Peaden R N 1980 Alfalfa varietal response to seven isolates of vesicular-arbuscular mycorrhizal fungi. Can. J. Plant Sci. 60, 859–863.

Ocampo J A and Hayman D S 1981 Influence of plant interactions on vesicular-arbuscular mycorrhizal infections. II. Crop rotations and residual effects of non-host plants. New Phytol. 87, 333–343.

Ojala J C and Jarrell W M 1980 Hydroponic sand culture system for mycorrhizal research. Plant and Soil 57, 297–303.

Porter W M 1979 The 'most probable number' method for enumerating infective propagules of vesicular-arbuscular mycorrhizal fungi in soil. Aust. J. Soil. Res. 17, 515–519.

Porter W M, Robson A D and Abbott L K 1987 Factors controlling the distribution of vesicular-arbuscular mycorrhizal fungi in relation to soil pH. J. App. Ecol. 24, 663–672.

Powell C L and Bagyaraj D J 1984 VA Mycorrhiza. CRC Press, Inc., Boca Raton, FL.

Safir G R 1987 Ecophysiology of VA Mycorrhizal Plants. CRC Press, Inc., Boca Raton, FL.

Schenck N C and Perez Y 1987 Manual for the Identification of VA Mycorrhizal Fungi. INVAM, Univ. Florida Gainesville, FL.

Sieverding E 1981 Influence of soii water regimes on VA mycorrhiza. J. Agronomy and Crop Science 150, 400–411.

Sieverding E 1986 Research model towards practical application of VA mycorrhizal fungi in tropical agriculture. *In* Physiological and Genetical Aspects of Mycorrhizae. Eds. V. Gianinazzi-Pearson and S Gianinazzi. pp 475–478, INRA, Paris.

Siqueira J O, Hubbell D H and Mahmud A W 1984 Effect of liming on spore germination, germ tube growth and root colonization by vesicular-arbuscular mycorrhizal fungi. Plant and Soil. 76, 115–124.

Sylvia D M 1988 Activity of external hyphae of vesicular-arbuscular mycorrhizal fungi. Soil Biol. Biochem. 20, 38–43.

Sylvia D M and Hubbell D H 1986 Growth and sporulation of vesicular-arbuscular mycorrhizal fungi in aeroponic and membrane systems. Symbiosis 1, 259–267.

Thompson J P 1986 Soilless culture of vesicular-arbuscular mycorrhizae of cereals: effects of nutrient concentration and nitrogen source. Can. J. Bot. 64, 2282–2294.

Thompson J P 1987 Decline of vesicular-arbuscular mycorrhizae in long fallow disorder of field crops and its expression in phosphorus deficiency of sunflower. Aust. J. Agric. Res. 38, 847–867.

Van Kessel C, Singleton P W and Hoben H J 1985 Enhanced N-transfer from a soybean to maize by vesicular-arbuscular mycorrhizal (VAM) fungi. Plant Physiol. 79, 562–563.

Wright S F and Morton J B 1989 Detection of vesicular-arbuscular mycorrhizal fungus colonization of roots by using a dot-immunoblot assay. Appl. Environ. Microbiol. 55, 761–763.

Wright S F, Morton J B and Sworobuk J E 1987 Identification of a vesicular-arbuscular mycorrhizae fungus by using monoclonal antibodies in an enzyme-linked immunosorbent assay. Appl. Environ. Microbiol. 53, 2222–2225.

Wong J-Y and Marschner H 1988 Effect of nitrogen supply on germination and hyphal growth of *Glomus mosseae in vitro* and in the rhizosphere in soybean. Angew. Bot. 62, 169–182.

D. L. Keister and P. B. Cregan (Eds.), The rhizosphere and plant growth, 343–348.

PLSO BARC33

Mycorrhizal interactions in the rhizosphere

R.G. LINDERMAN
USDA-ARS, Horticultural Crops Research Laboratory, Corvallis, OR 97330, USA

Key words: biocontrol, microbial combinations, mycorrhizae, mycorrhizosphere, nitrogen fixation, plant growth enhancement, phosphate solubilization

Abstract

Mycorrhizal fungi form a symbiotic association with the roots of most plants, and significantly alter the morphology and physiology of the plant. Among other physiological changes, root exudation is qualitatively and quantitatively altered, and thus the microbial composition of the rhizosphere changes. The kinds of microbial interactions that occur in the 'mycorrhizosphere' are often reflected in plant responses as a result of the combination of mycorrhizae and their microbial associates. Mycorrhiza-microbial interactions may be direct between organisms, or indirect as mediated by the host plant. For example, plant growth and health may be influenced when mycorrhizal fungi interact with the general microflora, or with specific microbes such symbiotic and free-living nitrogen-fixing bacteria, nutrient solubilizing bacteria, fungal and bacterial biocontrol agents, and other fungi. These associations and interactions occur naturally but to varying degrees in most soils. Those with positive effects on crop plants can be managed by inoculation with microbial combinations selected for their compatibility and combined effects.

Introduction

Mycorrhizal fungi form a symbiotic relationship with roots of most plants, the exceptions being members of some plant families, very young plants, and plants grown in mycorrhizal fungus-free (fumigated) soils or soilless potting mixes. Of the two major types of mycorrhizae – ectomycorrhizae and endomycorrhizae – ectomycorrhizae form a visible mantle of fungal hyphae around root tips that may then become greatly altered morphologically. Endomycorrhizae, and most commonly the vesicular-arbuscular (VA) mycorrhizae, occur on most agricultural crop plants, and cause little morphological change to roots. Both types of mycorrhizae also have an external hyphal phase that extends out into the soil, and both types have profound effects on the physiology of their host plants. Mycorrhizal plants are known to have altered nutritional status, increased photosynthetic rates, altered levels of growth-regulating substances, and altered patterns of root exudation due to changes in membrane permeability. These physiological changes plus the physical and chemical presence of external hyphae of the mycorrhizal fungi significantly alter the chemical, physical, and microbiological composition of the rhizosphere soil (Katznelson *et al.*, 1962; Meyer and Linderman, 1986b; Schisler and Linderman, 1989), appropriately referred to as the 'mycorrhizosphere' (Linderman, 1988; Rambelli, 1973).

The microbial shifts that occur in the mycorrhizosphere reflect a new equilibrium compared to the rhizosphere of non-mycorrhizal plants, and these organisms interact directly with each other and with the extramatrical hyphae of the mycorrhizal fungi. In addition, all these organisms may respond indirectly to the presence of others in the mycorrhizosphere via changes induced in the host plant. Selective population shifts may then occur that can affect plant growth and development in concert with the mycorrhizal fungi. These microbial interactions are the subject of this report.

Mycorrhiza formation may be influenced by the general microflora of soils, with some soils inhibiting and others stimulating the process. In addition, some specific microbial types interact positively with mycorrhizal fungi, as indicated by observations and analyses of naturally occurring mycorrhizospheres, or by data derived from experimental co-inoculations. Selected examples of these interactions will be discussed to illustrate their nature and to provide some insight into the mechanisms involved.

General microflora effects on mycorrhiza formation

Certain soils contain microorganisms that inhibit formation of mycorrhizae (Wilson *et al.*, 1988). The mechanisms involved in this suppression are not known, but early colonization dynamics, beginning with spore or other propagule germination, are most likely to be involved. For example, it has been shown that spores of one VA mycorrhizal fungus *Glomus versiforme* are colonized by bacteria that can either stimulate or inhibit spore germination (Mayo *et al.*, 1986). Similarly, certain bacteria have been shown to stimulate or inhibit hyphal growth of VA mycorrhizal fungi (Azcon, 1987; Azcon and Barea, 1985; Azcon-Aguilar *et al.*, 1986; Mugnier and Mosse, 1987) and ectomycorrhizal fungi (Garbaye and Bowen, 1987; McAfee and Fortin, 1988), thus influencing the total potential of the mycorrhizal association to influence plant growth (Graham *et al.*, 1982). Some reports indicate that stimulatory bacteria produce volatile compounds that affect spore germination (Mugnier and Mosse, 1987) and germ tube orientation relative to plant roots (Gemma and Koske, 1988; St. John *et al.*, 1983). Therefore, ecological or cultural factors that influence the microbial balance of a soil may favor mycorrhiza stimulators or inhibitors. To enhance the potential for successful mycorrhiza formation, stimulatory bacteria could be selected and co-inoculated with mycorrhizal fungi.

Interaction with nitrogen-fixing bacteria

Most leguminous plants grown in natural soils are symbiotic with both Rhizobium and VA mycorrhizal fungi, and plants with both symbionts generally grow better than plants with either alone. The literature strongly indicates that nitrogen fixation by Rhizobium nodules is greatly enhanced by the presence of compatible strains of VA mycorrhizal fungi (Barea and Azcon-Aguilar, 1985). These fungi stimulate formation of more and larger nodules, while imcompatible strains may have little or no effect on nodulation (Bayne and Bethlenfalvay, 1987; Bethlenfalvay *et al.*, 1985; Ianson and Linderman, 1990). Thus, maximum nodulation depends on having both effective strains of Rhizobium and compatible strains of VA mycorrhizal fungi. If the wrong strains of mycorrhizal fungi are present, the net result may be reduced nitrogen fixation in response to inoculation with Rhizobium.

Some have hypothesized that the enhanced nitrogen fixation response is due to improved uptake of phosphorus and/or minor elements by mycorrhizae (Pacovsky, 1986), which affects photosynthesis and the allocation of more photosynthate to nodules (Harris *et al.*, 1985; Kucey and Paul, 1982). The relationship of increased phosphorus uptake to enhanced nitrogen fixation has been challenged by several studies, including that by Ames and Bethlenfalvay (1987) who used a split-root system to show that the enhancement of nitrogen fixation was localized in the portion of the root system with mycorrhizae. The presence of mycorrhizae could have increased the strength of the photosynthate demand, or localized mycorrhizosphere effects might have stimulated the activity of nodule-promoting bacteria (Grimes and Mount, 1984; Meyer and Linderman, 1986a). These questions await further investigation.

Multiple symbioses also occur in mycorrhizal plants that also form Frankia nodules. Nitrogen fixation and plant growth are increased when both nodules and VA mycorrhizae occur together in the same root system (Gardner *et al.*, 1984; Rose and Youngberg, 1981). This relationship has not been studied in detail, so one could only speculate that the mechanisms of interaction are similar to those involved with the Rhizobium-VA mycorrhiza system on legumes.

Free-living or associative nitrogen-fixing bacteria such as Azospirillum and Azotobacteria have positive combined effects on plant growth when co-inoculated with VA mycorrhizal fungi on the same plant (Bagyaraj and Menge, 1978; Barea *et al.*, 1983; Subba-Rao *et al.*, 1985). There is some uncertainty, however, whether the enhanced plant growth is due to increased nitrogen fixation, or to the production of growth-promoting substances produced by the bacteria in close proximity to the mycorrhizal fungus hyphae (Azcon and Barea, 1985; Strzelczyk and Pokojska-Burdziej, 1984; Tien *et al.*, 1979). Such growth-promoting substances could readily be absorbed by the hyphae and translocated to the host plant tissues. Azospirillum has been found within the mantle of ectomycorrhizae (Li and Castellano, 1987; Li and Hung, 1987), and nitrogen fixation has been demonstrated to occur by this microbial combination *in situ*. In this case, fixed nitrogen could readily be transported to the host tissues via mycorrhizal hyphae.

Interaction with mineral nutrient solubilizing bacteria

A number of soil and rhizosphere bacteria have been shown to produce phosphatase enzymes that solubilize phosphorus from mineral sources (Barea *et al.*, 1975; Raj *et al.*, 1981). When these bacteria have been co-inoculated with VA mycorrhizal fungi, plant growth enhancement has resulted. VA mycorrhizal fungi apparently do not produce phosphatase enzymes, but enzyme levels are elevated in their presence (Dodd *et al.*, 1987), indicating that their hyphae may support bacteria that produce phosphatases. Other studies indicate that the growth enhancement may be due to bacterial production of growth substances such as phytohormones or vitamins (Barea *et al.*, 1976; Baya *et al.*, 1981).

Interactions with biocontrol agents

Most soils contain bacteria and fungi that can antagonize root pathogens (Linderman *et al.*, 1983) by competition, antibiosis, or hyperparasitism. Undoubtedly, these microbes interact in some way with mycorrhizal fungi in the mycorrhizosphere, some positively and some negatively (Krishna *et al.*, 1982). To effectively suppress root pathogens, however, the numbers and activity of biocontrol agents must be sufficiently high to block ingress and development of the pathogen. Their numbers, however, may be greatly affected by the presence of mycorrhizal fungi. Numbers of potential biocontrol agents have been shown to increase in pot cultures of VA mycorrhizae (Secilia and Bagyaraj, 1987), and Meyer and Linderman (1986b) showed increased suppressiveness of mycorrhizosphere soil to *Phytophthora cinnamomi* as indicated by reduced zoospore production, compared to soil around non-mycorrhizal roots. Various workers have shown an increase in pathogen antagonists associated with ectomycorrhizae (Katznelson *et al.*, 1962; Neal *et al.*, 1968). Malajczuk (1988) has shown bacteria embedded in the mantle of ectomycorrhizae of eucalyptus that antagonized Phytophthora zoospores. Thus, formation of mycorrhizae can impose a selective pressure on the soil microflora that favours populations of potential biocontrol agents. This microbial combination may be responsible for disease suppression in many situations. This is undoubtedly a disease control strategy worthy of further exploration.

Studies are being conducted in which mycorrhizal fungi are combined with biocontrol agents to determine compatibility between the two types of organisms. Some have predicted that biocontrol agents that produce potent antibiotics or siderophores which effectively inhibit fungal pathogens would also inhibit mycorrhizal fungi. Our studies have indicated little or no such response. Combinations of fluorescent pseudomonads targeted to control Pythium and Rhizoctonia pathogens had no significant effect on any aspect of VA mycorrhiza formation (Paulitz and Linderman, 1989). Similarly, seed treatment with the biocontrol agent *Trichoderma harzianum* had no significant effect on formation and growth response of VA mycorrhizal pepper and onion plants (Fig. 1) (Linderman *et al.*, 1989). The lack of inhibition of mycorrhizae by biocontrol agents has led to the hypothesis that

Fig. 1. Response of onion to inoculation with the VA mycorrhizal fungus *Glomus intraradices* (left and middle) with (left) or without (middle) the biocontrol agent *Trichoderma harzianum*, compared to the non-inoculated control (right). Note that the mycorrhizal growth response under phosphorus-limiting conditions was not influenced by the fungal biocontrol agent.

mycorrhizal fungi have evolved a tolerance to anti-fungal substances that sets them apart from other fungi, including fungal pathogens.

Potential for microbial inoculum combinations

It is now well documented that mycorrhizal fungi naturally interact with certain other rhizosphere microbes to positively affect plant growth and health. It only makes good sense to use this knowledge to enhance commercial crop production systems. The only way to guarantee that microbes have a chance to benefit plant growth is to inoculate plants with them. Legumes have been successfully inoculated with Rhizobium, and biocontrol agents are being inoculated to control some root diseases. The application technology and commercial inoculum of mycorrhizal fungi are now available. Both could be combined in the same inoculum; *i.e.* Rhizobium and VA mycorrhizal spores could be drilled beneath legume seed at planting. Biocontrol agents could be seed-coated or drilled beneath the seed along with mycorrhizal fungi. As long as these associates are checked for compatibility, there is no reason why they shouldn't be combined to give multiple benefits to plants. To further increase efficacy, it has been suggested that microbial combinations with compatible fungicides could be used. Fungicides such as metalaxyl and fosetyl-Al have been shown to enhance the development of VA mycorrhizae (Jabaji-Hare and Kendrick, 1987), and could give additional disease control of Pythium and Phytophthora pathogens when coupled with biocontrol agents.

Conclusions

Formation of mycorrhizae is the rule, not the exception. When the symbiosis is established, very significant morphological and physiological changes take place in the plant, and significant changes occur in the mycorrhizosphere soil. The microbial composition in that soil reaches a new equilibrium as a result of the selective pressure induced by mycorrhizae. For this reason, mycorrhizae should be considered a cornerstone foundation component in rhizosphere studies. Mycorrhizal fungi are present in most soils, but their establishment may be delayed due to cultural practices or to distribution of sparce inoculum in the soil. Thus, we should develop the technology required to inoculate crop plants with mycorrhizal fungi early in the production cycle. At the same time, other beneficial microbes could be added to the inoculum to increase the over-all enhancement of growth and health of plants due to increased nitrogen fixation, biological control of disease, nutrient acquisition potential, and drought tolerance. Managing the mycorrhizosphere is a target strategy whose time has come.

References

Ames R N and Bethlenfalvay G J 1987 Localized increase in nodule activity but no competitive interaction of cowpea rhizobia due to pre-establishment of vesicular-arbuscular mycorrhiza. New Phytol. 206, 207–215.

Azcon R 1987 Germination and hyphal growth of *Glomus mosseae in vitro*: effects of rhizosphere and cell-free culture media. Soil Biol. Biochem. 19, 417–419.

Azcon R and Barea J M 1975 Synthesis of auxins, gibberellins and cytokinins by *Azotobacter vinelandii* and *Azotobacter beijerinckii* related to effects produced on tomato plants. Plant and Soil 43, 609–619.

Azcon-Aguilar C and Barea J M 1985 Effect of soil micro-organisms on formation of vesicular-arbuscular mycorrhizas. Trans. Brit. Mycol. Soc. 84, 536–537.

Azcon-Aguilar C, Diaz-Rodriquez R M and Barea J M 1986 Effect of soil micro-organisms on spore germination and growth of the vesicular-arbuscular mycorrhizal fungus *Glomus mosseae*. Trans. Brit. Mycol. Soc. 86, 337–340.

Bagyaraj D J and Menge J A 1978 Interactions between a VA mycorrhiza and *Azotobacter* and their effects on rhizosphere microflora and plant growth. New Phytol. 80, 567–573.

Barea J M and Azcon-Aguilar C 1985 Mycorrhizas and their significance in nodulating nitrogen-fixing plants. Adv. Agron. 36, 1–54.

Barea J M, Azcon R and Hayman D 1975 Possible synergistic interactions between *Endogone* and phosphate-solubilizing bacteria in low-phosphate soil. *In* Endomycorrhizas. Eds. F E Sanders, B Mosse and P B Tinker. pp 409–417. Acad. Press, London.

Barea J M, Navarro E and Montoya E 1976 Production of plant growth regulators by rhizosphere phosphate-solubilizing bacteria. J. Appl. Bacteriol. 40, 129–134.

Barea J M, Bonis A F and Olivares J 1983 Interactions between *Azospirillum* and VA mycorrhiza and their effects on growth and nutrition of maize and ryegrass. Soil Biol. Biochem. 15, 705–709.

Baya A M, Boethling R S and Ramos-Cormenzana A 1981 Vitamin production in relation to phosphate solubilization by soil bacteria. Soil Biol. Biochem. 13, 527–531.

Bayne H G and Bethlenfalvay G J 1987 The *Glycine-Glomus-Rhizobium* symbiosis. IV. Interactions between the mycorrhizal and nitrogen-fixing endophytes. Plant, Cell Environ. 10, 607–612.

Bethlenfalvay G J, Brown M S and Stafford A E 1985 *Glycine-Glomus-Rhizobium* symbiosis. II. Antagonistic effects between mycorrhizal colonization and nodulation. Plant Physiol. 79, 1054–1058.

Bowen G D and Theodorou C 1979 Interaction between bacteria and ectomycorrhizal fungi. Soil Biol. Biochem. 11, 119–126.

Dodd J C, Burton C C, Burns R G and Jeffries P 1987 Phosphatase activity associated with the roots and the rhizosphere of plants infected with vesicular-arbuscular mycorrhizal fungi. New Phytol. 107, 163–172.

Garbaye J and Bowen G D 1987 Effect of different microflora on the success of ectomycorrhizal inoculation of *Pinus radiata*. Can. J. For. Res. 17, 941–943.

Gardner I C, Clelland D M and Scott A 1984 Mycorrhizal improvement in non-leguminous nitrogen fixing associations with particular reference to *Hippophae rhamnoides*. Plant and Soil 78, 189–199.

Gemma J N and Koske R E 1988 Pre-infection interactions between roots and the mycorrhizal fungus *Gigaspora gigantea*: Chemotropism of germ-tubes and root growth response. Trans. Brit. Mycol. Soc. 91, 123–132.

Graham J H, Linderman R G and Menge J A 1982 Development of external hyphae by different isolates of mycorrhizal *Glomus* spp. in relation to root colonization and growth of Troyer citrange. New Phytol. 91, 183–189.

Grimes H D and Mount M S 1984 Influence of *Pseudomonas putida* on nodulation of *Phaseolus vulgaris*. Soil Biol. Biochem. 16, 27–30.

Harris D, Pacovsky R S and Paul E A 1985 Carbon economy of soybean-*Rhizobium-Glomus* associations. New Phytol. 101, 427–440.

Ianson D C and Linderman R G 1990 Variation in VA mycorrhizal strain interactions with *Rhizobium* on pigeon pea. *In* The Rhizosphere and Plant Growth. Eds. D L Keister and P B Cregan. p 371. Kluwer Academic Publishers, Dordrecht, The Netherlands.

Jabaji-Hare S H and Kendrick W B 1987 Response of an endomycorrhizal fungus in *Allium porrum* L. to different concentrations of the systemic fungicides metalaxyl (Ridomil) and fosetyl-Al (Aliette). Soil Biol. Biochem. 19, 95–99.

Katznelson H, Rouatt J W and Peterson E A 1962 The rhizosphere effect of mycorrhizal and nonmycorrhizal roots of yellow birch seedlings. Can. J. Bot. 40, 377–382.

Krishna K R, Balakrishna A N and Bagyaraj D J 1982 Interactions between a vesicular-arbuscular mycorrhizal fungus and *Streptomyces cinnamomeous* and their effects on finger millet. New Phytol. 92, 401–405.

Kucey R M N and Paul E A 1982 Carbon flow, photosynthesis, and N_2 fixation in mycorrhizal and nodulated faba beans (*Vicia faba* L.). Soil Biol. Biochem. 14, 407–412.

Li C Y and Castellano M A 1987 *Azospirillum* isolated from within sporocarps of the mycorrhizal fungi *Hebeloma crustuliniforme*, *Laccaria laccata* and *Rhizopogon vinicolor*. Trans. Brit. Mycol. Soc. 88, 563–566.

Li C Y and Hung L L 1987 Nitrogen-fixing (acetylene-reducing) bacteria associated with ectomycorrhizae of Douglas-fir. Plant and Soil 98, 425–428.

Linderman R G 1988 Mycorrhizal interactions with the rhizosphere microflora: The mycorrhizosphere effect. Phytopathology 78, 366–371.

Linderman R G, Moore L W, Baker K F and Cooksey D A 1983 Strategies for detecting and characterizing systems for biological control of soilborne plant pathogens. Pl. Disease 67, 1058–1064.

Linderman R G, Paulitz T C, Mosier N J, Griffiths R P, Loper J E, Caldwell B A and Henkels M E 1990 Evaluation of the effects of biocontrol agents on mycorrhizal fungi. *In* The Rhizosphere and Plant Growth. Eds. D L Keister and P B Cregan. p 379. Kluwer Academic Publishers, Dordrecht, The Netherlands.

Malajczuk N 1988 Interaction between *Phytophthora cinnamomi* zoospores and micro-organisms on non-mycorrhizal and ectomycorrhizal roots of *Eucalyptus marginata*. Trans. Brit. Mycol. Soc. 90, 375–382.

Mayo K, Davis R E and Motta J 1986 Stimulation of

germination of spores of *Glomus versiforme* by spore-associated bacteria. Mycologia 78, 426–431.

McAfee B J and Fortin J A 1988 Comparative effects of the soil microflora on ectomycorrhizal inoculation of conifer seedlings. New Phytol. 108, 443–449.

Meyer J R and Linderman, R G 1986a. Response of subterranean clover to dual inoculation with vesicular-arbuscular mycorrhizal fungi and a plant growth-promoting bacterium, *Pseudomonas putida*. Soil Biol. Biochem. 18, 185–190.

Meyer J R and Linderman, R G 1986b. Selective influence on populations of rhizosphere or rhizoplane bacteria and actinomycetes by mycorrhizas formed by *Glomus fasciculatum*. Soil Biol. Biochem. 18, 191–196.

Mugnier J and Mosse, B 1987. Spore germination and viability of a vasicular arbuscular mycorrhizal fungus, *Glomus mosseae*. Trans. Brit. Mycol. Soc. 88, 411–413.

Neal J L, Lu K C, Bollen W B and Trappe J M 1968. A comparison of rhizosphere microfloras associated with mycorrhizae of red alder and Douglas-fir. pp 57–71 *In* Biolgy of Alder. Eds. J M Trappe, J F Franklin, R F Tarrant and G M Hansen, USDA For. Serv, Pac. Northwest For. Range Exp. Sta. 292 p.

Pacovsky R S 1986. Micronutrient uptake and distribution in mycorrhizal and phosphorus-fertilized soybeans. Plant and Soil 95, 379–388.

Paulitz T C and Linderman R G 1989. Interactions between fluorescent pseudomonads and VA mycorrhizal fungi. New Phytol. 113, 37–45.

Raj J, Bagyaraj, D J and Manjunath A 1981. Influence of soil inoculation with vesicular-arbuscular mycorrhiza and a phosphate-dissolving bacterium on plant growth and ^{32}P-uptake. Soil Biol. Biochem. 13, 105–108.

Rambelli A 1973. The rhizosphere of mycorrhizae. *In* Ectomycorrhizae. Eds. G L Marks and T T Kozlowski. pp 299–343. Academic Press, New York.

Rose S L and Youngberg C T 1981 Tripartite associations in snowbrush (*Ceanothus velutinus*): effect of vesicular-arbuscular mycorrhizae on growth, nodulation, and nitrogen fixation. Can. J. Bot. 59, 34–39.

Schisler D A and Linderman R G 1989 The influence of volatiles purged from soil around Douglas-fir ectomycorrhizae on soil microbial populations. Soil Biol. Biochem. 21, 389–396.

Secilia J and Bagyaraj D J 1987 Bacteria and actinomycetes associated with pot cultures of vesicular-arbuscular mycorrhizas. Can. J. Microbiol. 33, 1069–1073.

St. John T V, Hays R I and Reid C P P 1983 Influence of a volatile compound on formation of vesicular-arbuscular mycorrhizas. Trans. Brit. Mycol. Soc. 81, 153–154.

Strzelczyk E and Pokojska-Burdziej A 1984 Production of auxins and gibberellin-like substances by mycorrhizal fungi, bacteria and actinomycetes isolated from soil and the mycorrhizosphere of pine (*Pinus sylvestris* L.). Plant and Soil 81, 185–194.

Subba Rao N S, Tilak K V B R and Singh C S 1985 Synergistic effect of vesicular-arbuscular mycorrhizas and *Azospirillum brasilense* on the growth of barley in pots. Soil Biol. Biochem. 17, 119–121.

Tien T M, Gaskins M H and Hubbell D H 1979 Plant growth substances produced by *Azospirillum brasilense* and their effect on the growth of pearl millet (*Pennesetum americanum* L.). Appl. Environ. Microbiol. 37, 1016–1024.

Wilson G W T, Hetrick B A D and Kitt D G 1988 Suppression of mycorrhizal growth response by big bluestem by non-sterile soil. Mycologia 80, 338–343.

D. L. Keister and P. B. Cregan (Eds.), The rhizosphere and plant growth, 349–354.

PLSO BARC34

Agro-ecological aspects of the mycorrhizal, nitrogen-fixing legume symbiosis

G.J. BETHLENFALVAY and W.E. NEWTON
U.S. Department of Agriculture, Agricultural Research Service, Western Regional Research Center, Albany, CA 94710, USA

Key words: legumes, mycorrhizal fungi, plant nutrition, plant-soil interactions, rhizobium, root nodules, soil nutrition

Abstract

Vesicular-arbuscular mycorrhizal (VAM) fungi have fundamental effects on the ecophysiology of nodulated legumes, on the biota of the surrounding soil and on associated non-legumes. The effects are seen to be mediated through the direct transfer of nutrients from plant to plant by live hyphae of the soil mycelium, and through improving the competitiveness of legumes in nutrient uptake. As envisioned, the main role of legumes, in the open source-sink system between associated plants and the soil biota, is that of a nitrogen source. The export of nitrogen may be balanced by flows of carbon and phosphorus from plant to plant along gradients of sink demand. Soil microbes are beneficiaries of inter-plant nutrient transport, utilizing exudates as well as the soil mycelium as substrates, while the products of microbial metabolism serve to aggregate the abiotic component of the soil. Thus, the tripartite legume association is seen to function not only as a self-contained system, but as one involved in nutrient fluxes between adjacent plants and the soil that surrounds its roots.

Introduction

Following a lag phase in research on the tripartite association between legumes, rhizobia and vesicular-arbuscular mycorrhizal (VAM) fungi (Asai, 1944; Jones, 1924; Stahl, 1900), evidence for a synergistic relationship between the endophytes (see Mosse, 1977) has led to a mushrooming in interest. The findings on N_2-fixing VAM legumes were presented in detail by Barea and Azcón-Aguilar (1983), and excellent reviews on the ecology (Allen and Allen, 1989), physiology (Smith and Gianinazzi-Pearson, 1988) and ecophysiology (Safir, 1987) of the VAM symbiosis and on progress in N_2-fixation research (Bothe *et al.*, 1988) in general are also available. We will therefore not summarize information already available, but will focus on the changes in perception of the role and rationale for the tripartite VAM association in the agro-ecosystem.

Changing concepts

Since first described a century ago (Frank, 1885), the mycorrhiza (fungus-root) was considered to be a plant symbiosis, whose obligately biotrophic VAM-fungal component formed an integral part of the plant (Gianinazzi *et al.*, 1982). Focus on the plant was expressed even by the choice of terms describing fungal organs: the mycelia are called intra- or extraradical, and the vast majority of the literature describes VAM effects on plants only. While this aspect will no doubt remain important, changing agricultural priorities (Paul *et al.*, 1985) are re-focussing the role of VAM fungi as key components of the rhizosphere ecosystem (Linderman, 1988a). This process has gained additional impetus during the past decade as a result of findings which demonstrated the influence of VAM fungi both on the microbiota of the bulk soil (see Linderman, 1988b) and on adjacent plants due to inter-plant linkages formed by the live VAM-fungal hyphae of the soil mycelium (Newman, 1988). Thus, we are coming to view the VAM fungus as a soil symbiont as well as a plant symbiont, but more importantly, also as an interface between plant and soil communities.

Mycorrhizae: agents of plant cooperation

The symplastic mycorrhizal connections between plants have special significance for the tripartite legume association because of the latter's role as a source of N (Barea *et al.*, 1987) to the associated soil ecosystem. Traditionally, the transfer of N from a legume to its environment was thought to occur only by the 'soil-pool pathway' (*sensu* Newman and Ritz, 1986) through the eventual mineralization of N-rich legume biomass. By contrast, the 'direct transfer pathway' of nutrients from plant to plant by VAM-fungi provides for the immediate export of N (van Kessel *et al.*, 1985) and C (Francis and Reed, 1984) to adjacent plants. The resulting open source-sink system between plants of varying N- and C-reduction capabilities and requirements would conceptually extend the influence of the tripartite legume association to associated non-legumes, and also to the organisms of the surrounding soil ecosystem which are in trophic interchange with the VAM soil mycelium (Bagyaraj, 1984). While the evidence on mycorrhizal linkages and their possible roles is still tentative (Newman, 1988), the implications for plant to plant interactions are profound (Fitter, 1985).

When grown in nutrient-deficient substrates, the fungal, rhizobial and plant components of the tripartite association may be regarded as the sources of P, N, and C, respectively, and as sinks for the products of their co-symbionts (Bayne *et al.*, 1984). In the restricted, intra-plant system, the acquisition and incorporation of nutrients is determined largely by source and sink limitations. In the expanded, VAM-linked, inter-plant system, such source-sink relationships are likely to be influenced by differential P-uptake and CO_2-fixation capabilities and utilization requirements, as influenced by differences in the mycotrophic status (Trappe, 1987) of the associated plants. Transfer of P (Newman and Ritz, 1986) has been demonstrated and the potential effects of inter-plant fluxes of P and C on N_2 fixation may be inferred from the well-known effects within the intra-plant symbiosis (Brown and Bethlenfalvay, 1987).

Our appreciation of the impact of VAM-mediated nutrient transfer on agriculture is limited at present, since the literature of multiple cropping systems (Francis, 1986) and that of legume crops (Wilcox, 1987) has taken little note, if any, of the existence of mycorrhizae. Gains to non-legumes, as well as legumes, in intercrop systems are thought to be due to different competitive situations (Haynes, 1980) of which the ecology of weeds and their mycorrhizal status (Trappe, 1987) is perhaps the least understood. Conflicting data on benefits to non-legumes from N supplied by an intercropped legume (see van Kessel and Roskoski, 1988; Abaido and van Kessel, 1989) may be resolved as inter-plant VAM effects on factors affecting N_2 fixation, such as soil N availability (Morris and Weaver, 1987), P nutrition (Barea *et al.*, 1987) and CO_2 fixation (Brown *et al.*, 1988; Sheehy, 1987) become better known.

Mycorrhizae: agents of plant competition

The effectiveness of VAM fungi in enhancing plant growth is dependent on fungal, plant and soil factors (Bethlenfalvay *et al.*, 1985b; Sainz and Arines, 1988). Plant species and cultivars within species differ in their dependence on the fungi (Reeves, 1987). While symbiotic plant response may vary with fungal species and even with edaphotypes of the same morphospecies (Bethlenfalvay *et al.*, 1989), plants most dependent on mycorrhizae are those with coarse roots and few root hairs (Howeler *et al.*, 1987; Mosse, 1981). Plants best able to extract limiting nutrients in marginal soils have a competitive advantage (Mosse, 1977). Thus, legumes with less extensive root systems are less efficient nutrient absorbers than grasses (Haynes, 1980), and may show large responses to VAM colonization (see Abbott and Robson, 1984a). Legumes can be highly dependent on colonization by native VAM fungi (Sainz and Arines, 1988), but enhanced growth responses to more effective introduced fungi have also been observed (see Hall, 1987). The co-evolution of plants with their VAM-fungal endophytes over geological time (see Trappe, 1987) would have favored the establishment of best-adapted host-endophyte combinations in natural habitats. Recent cultural disturbances of soils are often detrimental to VAM fungi (tillage, Evans and Miller, 1988; chemicals, Trappe *et al.*, 1984) and the introduction of exotic crop plants make the indigenous VAM fungi poorly suited for the new conditions (Hall, 1984) and agricultural requirements (Hayman, 1987).

The data on VAM influence on inter-plant competition (Allen and Allen, 1989; Safir, 1987) raise the question whether the cooperative and competitive VAM effects on plant-to-plant interactions may be counteractive. The answer is likely to be particularly complex for legumes, because of the complexity of the tripartite association (Daft *et al.*, 1985) both as a strong sink for P and a source of N. Thus, when the N source (legume) is out-competed for P, the resulting inhibition of N_2 fixation will be felt by the successful competitor (non-legume) as N stress, to the detriment of both. This simple scenario is in reality complicated by differences in plant-fungus preferences (Louis and Lim, 1988), inter-fungal competition (Abbott and Robson, 1984b), and antagonistic (Bayne and Bethlenfalvay, 1987; Bethlenfalvay *et al.*, 1985a) or stimulatory (Dagoberto, 1986) effects between VAM fungi and rhizobia in the legume. Soil effects (Hayman, 1982) and differences in the production of reduced C, and its allocation within and among associated plants and its loss to (gain by) the soil (Oades, 1984; Francis and Read, 1984; Paul and Clark, 1988) add a further little-known dimension to the role of VAM fungi in plant competition.

Recent reports of adverse effects of disturbance of VAM mycelial systems (Read and Birch, 1988) on plant growth and nutrition add further credence to the role of fungal hyphae in nutrient transfer. Should future work establish the cooperative model as a significant mechanism to supplement the competitive model in plant-to-plant interactions, it will probably also provide another reason for the common occurrence of multiple VAM colonization: the many desirable properties of the one 'ideal endophyte' (Daft, 1983), which are probably distributed among the several actual (less than ideal) ones, would have to include differential, bi-directional nutrient-transfer capabilities.

Mycorrhizae: agents of soil nutrition

The effectiveness of VAM fungi and rhizobia as legume symbionts has been measured in terms of host-plant development (Barea and Azcón-Aguilar, 1983) characteristic of production-oriented agriculture. With the shift in priorities in the 1980s to productivity and resource conservation (Gibbs and Carlson, 1985; Healey *et al.*, 1986), the soil, and the dynamics of its biotic component (Curl and Truelove, 1986) may become additional focal points for research on the legume association. Since the soil mycelium of VAM fungi represents a significant portion of the soil microbial biomass (Hayman, 1978), and since the cell walls of its hyphae are composed mainly of the amino-sugar chitin (Weijman and Meuzelaar, 1978), the soil hyphae of legume mycorrhizae may be one of the most important vehicles for C and N input into the soil. We therefore expect soil organic matter content and, as a consequence, aggregate stability (Lynch and Bragg, 1984; Oades, 1984) to become new measures of VAM effectiveness, complementing that of plant response.

The literature of VAM effects on soil aggregation is brief and had been recently reviewed by Thomas *et al.* (1986), who found a highly significant (60%) increase in water-stable aggregate formation in VAM soil (onions) and disaggregation in nonVAM controls over a five-month growth period. In VAM legumes (Phaseolus), Sutton and Sheppard (1976) showed that the binding of sand to extensive VAM-fungal mycelia was the mechanism of linking sand grains in aggregates, and isolated an adhesive polysaccharide as the binding agent (Clough and Sutton, 1978). Such adhesion of soil particles to fungal hyphae may have important consequenes for soil stabilization (Lynch and Bragg, 1984). It provides sites where microorganisms flourish (Linderman, 1988b) and where their organic products contribute to the formation of aggregates at different levels of size and organization (Gupta and Geminda, 1988; Tisdall and Oades, 1982). While each organism may have a necessary function in the aggregation process, fungi and filamentous actimomycetes (streptomyces) were shown to be most effective and rhizobia were intermediate in binding soil particles into crumbs (Harris *et al.*, 1966). Of these, the largest (>2 mm) are considered to be fungal products, while the actinomycete-produced crumbs are slightly smaller (Focht and Martin, 1979).

The interest in actinomycetes, as important 'outsiders' of the legume association, is based not only on their role in soil aggregation. As chitin decomposers, they are associated with VAM fungi (Meyer and Linderman, 1986), producing both antagonistic (Krishna and Bagyaraj, 1982) and positive (Mugnier and Mosse, 1987) effects. Recently, the closeness of the actinomycete-VAM fungus association had been emphasized by Ames *et al.* (1989) who showed that 82% of 190 spores of a VAM fungus were colonized by actinomycetes. Using selected strains from his isolates, Ames (1989) demonstrated that dual inoculation with VAM fungi and actinomycetes could significantly increase the development of both the root and soil mycelia of the fungus, without adversely affecting plant (onion) growth. Reports on interactions of actinomycetes with rhizobia are few (see Barrion and Habte, 1988), and indicate that inhibitory effects by actinomycetes on rhizobia vary with the isolate-strain combinations investigated. In view of the ability of actinomycetes to control microbial populations (Trinick *et al.*, 1983), it is conceivable that proliferation of *Streptomyces* spp. on VAM hyphae may explain antagonistic effects observed between VAM colonization and nodulation (Bethlenfalvay *et al.*, 1985).

Summary

Seventy-five years ago, Jones (1924), working with pea plants, remarked that 'the root-inhabiting fungus is so abundant in plants that is it surprising to find that its presence is not common knowledge among botanists'. Today, the 'presence' has become known among plant scientists, but while we realize that a plant does not end at its root tips, an integrated view of the role of mycorrhizae in the tripartite legume association as a component of sustainable agro-ecosystems (Mosse, 1986; Sprent, 1986) is still lacking.

References

Abaido R C and van Kessel C 1989 ^{15}N-uptake, N_2 fixation and rhizobial inter-strain competition in soybean and bean, intercropped with maize. Soil Biol. Biochem. 21, 155–159.

Abbott L K and Robson A D 1984a The effect of VA mycorrhizae on plant growth. *In* VA Mycorrhiza. Eds. L C Powell and D J Bagyaraj. pp 113–130. CRC Press, Boca Raton, FL.

Abbott L K and Robson A D 1984b Colonization of the root system of subterranean clover by three species of vesicular-arbuscular mycorrhizal fungi. New Phytol. 96, 275–281.

Allen E B and Allen M F 1989 The mediation of competition by mycorrhizae in successional and patchy environments. *In* Perspectives in Plant Competition. Eds. J B Grace and G D Tilman. Academic Press, London.

Ames R N, Mihara K L and Bayne H G 1989 Chitin-decomposing actinomycetes associated with a vesicular-arbuscular mycorrhizal fungus from a calcareous soil. New Phytol. 111, 67–71.

Ames R N 1989 Mycorrhiza development in onion in response to inoculation with chitin-decomposing actinomycetes. New Phytol. 112, 423–427.

Asai T 1944 Über die Mykorrhizenbildung der leguminösen Pflanzen. Jpn. J. Bot. 13, 463–485.

Bagyaraj D J 1984 Biological interactions with mycorrhizal fungi. *In* VA Mycorrhiza. Ed. G R Safir. pp 131–152. CRC Press, Boca Raton, FL.

Barea J M and Azcón-Aguilar C 1983 Mycorrhizas and their significance in nodulating nitrogen-fixing plants. Adv. Agron. 36, 1–54.

Barea J M, Azcón-Aguilar C and Azcón R 1987 Vesicular-arbuscular mycorrhiza improve both symbiotic N_2 fixation and N uptake from soil as assessed with a ^{15}N technique under field conditions. New Phytol. 106, 717–725.

Barrion M and Habte M 1988 Interaction of *Bradyrhizobium* sp. with an antagonistic actinomycete in culture medium and in soil. Biol. Fertil. Soils 6, 306–310.

Bayne H G and Bethlenfalvay G J 1987 The *Glycine-Glomus-Rhizobium* symbiosis. IV. Interactions between the mycorrhizal and nitrogen-fixing endophytes. Plant Cell Environ. 10, 607–612.

Bayne H G, Brown M S and Bethlenfalvay G J 1984 Defoliation effects on mycorrhizal colonization, nitrogen fixation and photosynthesis in the *Glycine-Glomus-Rhizobium* symbiosis. Physiol Plant. 62, 576–580.

Bethlenfalvay G J, Brown M S and Stafford A E 1985a The *Glycine-Glomus-Rhizobium* symbiosis. II. Antagonistic effects between mycorrhizal colonization and nodulation. Plant Physiol. 79, 1054–1058.

Bethlenfalvay G J, Franson R L, Brown M S and Mihara K L 1989 The *Glycine-Glomus-Bradyrhizobium* symbiosis. IX. Nutritional, morphological and physiological responses of nodulated soybean to geographic isolates of the mycorrhizal fungus *Glomus mosseae*. Physiol. Plant. 76, 226–232.

Bethlenfalvay G J, Ulrich J M and Brown M S 1985b Plant response to mycorrhizal fungi: Host, endophyte and soil effects. Soil Sci. Soc. Am. J. 49, 1164–1168.

Brown M S and Bethlenfalvay G J 1987 The *Glycine-Glomus-Rhizobium* symbiosis. VI. Photosynthesis in nodulated, mycorrhizal, or N- and P-fertilized soybean plants. Plant Physiol. 85, 120–123.

Brown M S, Thamsurakul S and Bethlenfalvay G J 1988 The *Glycine-Glomus-Bradyrhizobium* symbiosis. VIII. Phosphorus-use efficiency of CO_2 and N_2 fixation in mycorrhizal soybean. Physiol. Plant. 74, 159–163.

Bothe H, de Bruijn F J and W E Newton (Eds.) 1988 Nitrogen Fixation: Hundred Years Later. Gustav Fischer, Stuttgart, 878 p.

Clough K S and Sutton J C 1978 Direct observation of fungal aggregates in sand dune soil. Can. J. Bot. 24, 333–335.

Curl E A and Truelove B 1986 The Rhizosphere. Springer-Verlag, Heidelberg, 288 p.

Daft M J 1983 The influence of mixed inocula on endomycorrhizal development. Plant and Soil 71, 331–337.

Daft M J, Clelland D M and Gardner I C 1985 Symbiosis with endomycorrhizas and nitrogen-fixing organisms. Proc. Roy. Soc. Edinburgh 89B, 283–298.

Dagoberto A, Bojórquez A, Ferrera-Cerrato R, Trinidad-Santos A and Volke-Haller V 1986 Fertilización e inoculatión con *Rhizobium* y endomycorrhizas (V-A) en garbanzo blanco (*Cicer arietinum* L.) en suelos del noroeste de Mexico. Agrociencia 65, 141–160.

Evans D G and Miller M H 1988 VA-mycorrhiza and the soil disturbance induced reduction of nutrient absorption in maize. New Phytol. 110, 67–75.

Fitter A H 1985 Functioning of vesicular-arbuscular mycorrhizas under field conditions. New Phytol. 99, 257–265.

Focht D D and Martin J P 1979 Microbiological and biochemical aspects of semi-arid agricultural soils. *In* Agriculture in Semi-Arid Environments. Eds. A E Hall, G H Cannell and Lawton H W. pp 118–147. Springer-Verlag, Berlin.

Francis C A (Ed.) 1986 Multiple Cropping Systems. Macmillan Publishing Company, New York, 383 p.

Francis R and Read D J 1984 Direct transfer of carbon between plants connected by vesicular-arbuscular mycorrhizal mycelium. Nature 307, 53–56.

Frank A B 1885 Über die auf Wurzelsymbiose beruhende Ernährung gewisser Bäume durch unterirdische Pilze. Ber. dtsch. Bot. Ges. 3, 128–145.

Gibbs M and Carlson C 1985 (Eds.) 1985. Crop Productivity – Research Imperatives Revisited, an international conference held at Boyne Highlands Inn, October 13–18.

Gianinazzi S, Gianinazzi-Pearson V and Trouvelot A (Eds.) 1982 Les Mycorrhizes, Partie Intégrante de la Plante: Biologie et Perspectives d'Utilisation. Les Colloques d'INRA, No. 13, Paris, 397 p.

Gupta V V S R and Germida J J 1988 Distribution of microbial biomass and its activity in different soil aggregate size classes as affected by cultivation. Soil Biol. Biochem. 20, 777-786.

Hall I R 1987 A review of VA mycorrhizal growth responses in pastures. Angew. Bot. 61, 127–134.

Hall I R 1984 Effect of inoculant mycorrhizal fungi on white clover growth in soil cores. J. Agric. Sci. 102, 719–723.

Harris R F, Chesters G and Allen O N 1966 Dynamics of soil aggregation. Adv. Agron. 18, 107–169.

Haynes R J 1980 Competitive aspects of the grass-legume association. Adv. Agron. 33, 227–261.

Hayman D S 1987 VA mycorrhizas in field crop systems. *In* Ecophysiology of VA Mycorrhizal Plants. Ed. G R Safir. pp 171–192. CRC Press, Boca Raton, FL.

Hayman D S 1982 Influence of soils and fertility on activity and survival of vesicular-arbuscular mycorrhizal fungi. Phytopathology 72, 1119–1123.

Hayman D S 1978 Endomycorrhizae. *In* Interactions between Non-Pathogenic Soil Microorganisms and Plants. Eds. Y R Dommergues and S V Krupa. pp 401–442. Elsevier Scientific Publishing Co., Amsterdam.

Healy R G, Waddell T E and Cook K A 1986 Agriculture and Environment in a Changing World Economy. The Conservation Foundation, Washington, DC, 66 p.

Howeler R H, Sieverding E and Saif S 1987 Practical aspects of mycorrhizal technology in some tropical crops and pastures. Plant and Soil 100, 249–283.

Jones F R 1924 A mycorrhizal fungus in the roots of legumes and other plants. J. Agric. Res. 29, 459–470.

Krishna K R and Bagyaraj D J 1982 Interaction between a vesicular-arbuscular mycorrhizal fungus and *Streptomyces cinnamomeus* and their effects on finger millet. New Phytol. 93, 401–405.

Linderman R G 1988a VA (vesicular-arbuscular) mycorrhizal symbiosis. ISI Atlas of Science: Plants and Animals, Vol. 1, 1988, 183–188.

Linderman R G 1988b Mycorrhizal interactions with the rhizosphere microflora: the mycorrhizosphere effect. Phytopathology 78, 366–371.

Louis I and Lim G 1988 Differential growth and mycorrhizal colonization of soybean to inoculation with two isolates of *Glomus clarum* in soils of different P availability. Plant and Soil 112, 37–43.

Lynch J M and Bragg E 1985 Microorganisms and soil aggregate stability. Adv. Soil Sci. 2, 133–171.

Meyer J R and Linderman R G 1986 Selective influence on populations of rhizosphere or rhizoplane bacteria and actinomycetes by mycorrhizas formed by *Glomus fasciculatum*. Soil Biol. Biochem. 18, 191–196.

Morris D R and Weaver R W 1988 Competition for Nitrogen-15 depleted ammonium nitrate and nitrogen fixation in arrowleaf clover-gulf ryegrass mixtures. Soil Sci. Soc. Am. J. 51, 115–119.

Mosse B 1986 Mycorrhiza in a sustainable agriculture. Biol. Agric. Hortic. 3, 191–209.

Mosse B 1981 Vesicular-arbuscular mycorrhiza research for tropical agriculture. Res. Bul. 194. Hawaii Inst. of Trop. Agric. and Human Resources. Univ. of Hawaii, 82 p.

Mosse B 1977 The role of mycorrhiza in legume nutrition on marginal soils. *In* Exploiting the Legume-Rhizobium Symbiosis in Tropical Agriculture. Eds. J M Vincent, A S Whitney and J Bose. pp 275–292. Univ. College of Trop. Agric. Misc. Publ. 145, University of Hawaii, Honolulu.

Mugnier J and Mosse B 1987 Spore germination and viability of a vesicular-arbuscular mycorrhizal fungus, *Glomus mosseae*. Trans. Brit. Mycol. Soc. 88, 411–413.

Newman E I 1988 Mycorrhizal links between plants: their

functioning and ecological significance. Adv. Ecol. Res. 18, 243–270.

Newman E I and Ritz K 1986 Evidence on the pathways of phosphorus transfer between vesicular-arbuscular mycorrhizal plants. New Phytol. 104, 77–87.

Oades J M 1984 Soil organic matter and structural stability: mechanisms and implications for management. *In* Biological Processes and Soil Fertility. Eds. J Tinsley and J F Darbyshire. pp 319–337. Martinus Nijhoff/Dr W. Junk Publishers, The Hague.

Paul E A and Clark F E 1988 Soil Microbiology and Biochemistry. pp 198–221. Academic Press, San Diego, CA.

Paul E A, Tinker P B and Bauer W D 1985 Rhizosphere dynamics. *In* Crop Productivity – Research Imperatives Revisited. Eds. M Gibbs and C Carlson. pp 152–176. An International Conference held at Boyne Highlands Inn, October 13–18, 1985.

Read D J and Birch C P D 1988 The effects and implications of disturbance of mycorrhizal mycelial systems. Proc. Roy. Soc. Edinburgh 94B, 13–24.

Reeves F B 1987 Mycorrhizal responsivenesss and ecology of VAM in arid and semiarid ecosystems. *In* Mycorrhizae in the Next Decade. Eds. D M Sylvia, L L Hung and J H Graham. pp 136–138. Proc 7th North Am. Conf. on Mycorrhizae, Univ. Florida, Gainesville.

Safir G R, ed. 1987 Ecophysiology of VA Mycorrhizal Plants. CRC Press, Boca Raton, 224 p.

Sainz M J and Arines J 1988 Effect of indigenous and introduced vesicular-arbuscular mycorrhizal fungi on growth and phosphorus uptake of *Trifolium pratense* and on inorganic phosphorus fractions in a cambisol. Biol. Fert. Soils 6, 55–60.

Sheehy J E 1987 Photosynthesis and nitrogen fixation in legume plants. CRC Crit. Rev. Plant Sci. 5, 121–159.

Smith S E and Gianinazzi-Pearson V 1988 Physiological interactions between symbionts in vesicular-arbuscular mycorrhizal plants. Annu. Rev. Plant Physiol. 39, 221–244.

Sprent J 1986 Nitrogen fixation in sustainable agriculture. Biol. Agric. Hortic. 3, 153–165.

Stahl E 1900 Der Sinn der Mycorrhizenbildung. Jahrb. Wiss. Bot. 34, 540–668.

Sutton J C and Sheppard B R 1976 Aggregation of sand-dune soil by endomycorrhizal fungi. Can. J. Bot. 54, 326–333.

Thomas R S, Dakessian S, Ames R N, Brown M S and Bethlenfalvay G J 1986 Aggregation of a silty clay loam soil by mycorrhizal onion roots. Soil Sci. Soc. Am. J. 50, 1494–1499.

Tisdall J M and Oades J M 1982 Organic matter and water-stable aggregates in soils. J. Soil Sci. 33, 141–163.

Trappe J M 1987 Phylogenetic and ecologic aspects of mycotrophy in the Angiosperms from an evolutionary standpoint. *In* Ecophysiology of VA Mycorrhizal Plants. Ed. G R Safir. pp 5–25. CRC Press, Boca Raton, FL.

Trappe J M, Molina R and Castellano M 1984. Reactions of mycorrhizal fungi and mycorrhiza formation to pesticides. Ann. Rev. Phytopathol. 22, 331–359.

Trinick M J, Parker C A and Palmer M J 1983 Interaction of the microflora from nodulation problem and non-problem soils toward *Rhizobium* spp. on agar culture. Soil Biol. Biochem. 15, 295–301.

Van Kessel C and Roskoski J P 1988 Row spacing effects on N_2-fixation, N-yield and soil N uptake of intercropped cowpea and maize. Plant and Soil 111, 17–23.

Van Kessel C, Singleton P W and Hoben H J 1985 Enhanced N-transfer from a soybean to maize by vesicular-arbuscular mycorrhizal (VAM) fungi. Plant Physiol. 79, 562–563.

Weijman A C and Meuzelaar H L 1978 Biochemical contributions to the taxonomic status of the Endogonaceae. Can. J. Bot. 57, 284–291.

Wilcox J R (Ed.) 1987 Soybeans: Improvement, Production, and Uses, 2nd Edition. Am. Soc. Agron. Monograph No. 16, 888 p.

D. L. Keister and P. B. Cregan (Eds.), The rhizosphere and plant growth, 355–362.

PLSO BARC35

Field management of VA mycorrhizal fungi

L.K. ABBOTT and A.D. ROBSON
Soil Science and Plant Nutrition, School of Agriculture, The University of Western Australia, Nedlands, W.A. 6009, Australia

Key words: agricultural practices, inoculation, mediterranean climate, phosphate, predicting benefits, vesicular arbuscular mycorrhizas

Abstract

The potential for management of VA mycorrhizal fungi at a particular site depends on the factors that limit plant growth and the extent and timing of mycorrhiza formation. We are attempting to develop a procedure for predicting those sites where the benefits from the mycorrhizas are less than optimal. This involves gaining an understanding of the relationships between plant growth and soil and fungus parameters, predicting mycorrhiza formation using a bioassay, and subsequently quantifying the benefits that could result from management of the fungi. With existing knowledge it is not possible to (i) identify sites used for broadscale agriculture where the symbiosis is operating suboptimally, (ii) quantify the benefits that would result from increasing the rate and extent of colonization of roots by effective fungi, and (iii) identify the most cost-effective method for managing the fungi either by choice of agricultural practice or by inoculation with selected fungi. Our attempt to overcome these limitations has led to the identification of specific areas where further research is necessary. These include studies of the factors that affect the formation and functioning of mycorrhizas in field soils, the relationship between mycorrhiza development in the field and plant growth for soils that differ in their phosphate adsorbing capacity, and the ecology of the fungi in field soils.

Introduction

Vesicular arbuscular (VA) mycorrhizal fungi occur in almost every soil, but species abundance and diversity differ among soils (Abbott and Robson, 1989). The symbiotic association formed between these fungi and host roots can lead to enhanced plant growth via the alleviation of phosphorus deficiency (Stribley, 1987). Other benefits may be derived from the symbiosis (*e.g.* drought tolerance, resistance to pathogens) but these may or may not be related to phosphate nutrition (Bagyaraj, 1984; McGonigle and Fitter, 1988; Nelson, 1987). Some detrimental effects have been reported (see review by Cooper 1984) but are poorly understood.

Research on VA mycorrhizal fungi and their effects on plant growth has been principally conducted under glasshouse conditions. Benefits of the symbiosis in field soils are less well studied (Hayman, 1987; Powell, 1984), although they have been clearly demonstrated where low levels of mycorrhizas were formed after soil erosion (Hall, 1980), fallowing (Black and Tinker, 1979) and fumigation (Haas *et al.*, 1987). The benefits contributed by VA mycorrhizas formed by indigenous fungi to growth of field-grown plants are not easy to quantify because elimination of the fungi (*e.g.* by fumigation) may change other soil properties that affect plant growth. One approach to overcoming this experimental limitation was to grow non-mycorrhizal plants concurrently to assess the effects of fumigation on factors which may affect plant growth other than by elimination the VA mycorrhizal fungi (Yost and Fox, 1979). More recently, benomyl has been used to eliminate these fungi in an attempt to assess the mycorrhizal benefit (Fitter, 1986).

Fitter (1985) emphasised the need to estimate the benefits of VA mycorrhizas in the field. He suggested that benefits in the field may be less than benefits observed in the glasshouse because of ineffective VA mycorrhizal fungi in field soils, a greater importance of hyphal connections between plants in the field, and grazing of hyphae by soil animals in field soils. While there is little data to evaluate these suggestions, there is a clear need to increase our understanding of the formation and functioning of VA mycorrhizas under field conditions.

In assessing the potential for managing VA mycorrhizal fungi in agricultural soils, it is first necessary to identify the current limitations to plant growth. If plant growth is limited by factors that would prevent the alleviation of phosphorus deficiency, management of the VA mycorrhizal fungi would be premature. For agricultural soils in our mediterranean climate, the most important criterion for considering managing the fungi is that phosphate is the major limitation to plant growth. A deficiency in phosphorus for a particular host cannot usually be accurately predicted from a soil test alone. In our environment, phosphorus fertilizer recommendations for plant growth are made by first predicting a Mitscherlich response curve to applied phosphorus using data for rainfall, a soil test of phosphorus and phosphate adsorption capacity of the soil (Bowden and Bennett, 1974). However, the extent of mycorrhiza formation also contributes to the relationship between extractable soil phosphorus and plant growth (Stribley *et al.*, 1980). Thus there is potential for including a mycorrhiza component in the model which could be used to assess the effectiveness of the indigenous VA mycorrhizal fungi and the potential for increasing their contribution to plant growth.

In a soil where the indigenous VA mycorrhizal fungi infect rapidly and extensively, there is little chance of benefits from managing the fungi. One reason for this is that there appears to be a close relationship between mycorrhiza formation and

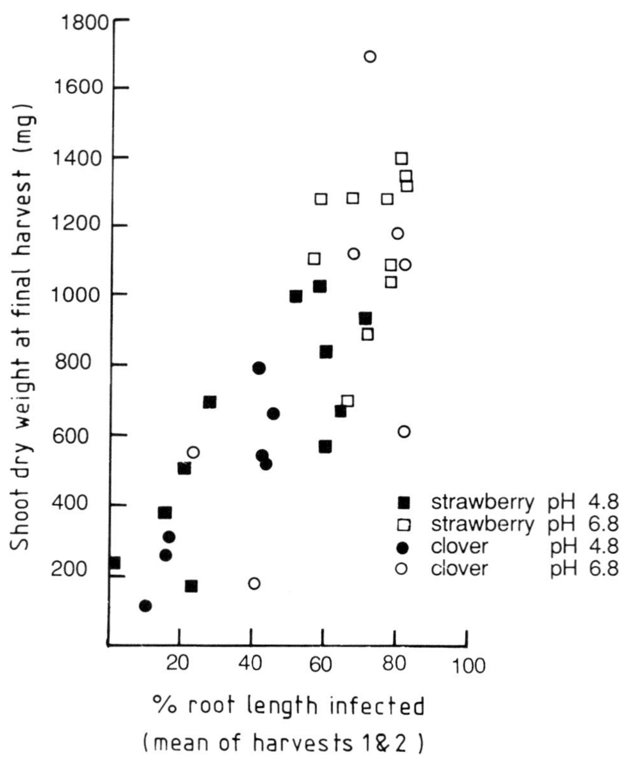

Fig. 1. Relationship between plant growth and VA mycorrhizal infection for strawberry and clover plants grown under glasshouse conditions and inoculated with various combinations of fungal species at two levels of soil pH (after Koomen *et al.*, 1987).

plant growth for a range of fungi (Koomen *et al.*, 1987, Fig. 1). Therefore, if mycorrhizas form rapidly and extensively, it is likely that maximum benefits are already being gained. On the other hand, there have been claims that species may differ in their effectiveness at increasing plant growth in ways that are unrelated to infectivity (see reviews Smith, 1985; Gianinazzi-Pearson, 1985). If this is so, direct substitution of more effective fungi would be a logical approach but this is likely to be difficult to achieve. Extensive mycorrhiza formation by fungi introduced into soils already containing highly infective populations of these fungi does not usually occur (Abbott and Robson, 1984a; Hepper *et al.*, 1988; Lopez-Aguillon and Mosse, 1987). In such circumstances, inoculation may be successful if the indigenous fungi are first managed to reduce their infectivity.

The potential for managing VA mycorrhizal fungi depends on an understanding of (i) factors limiting for plant growth, (ii) the extent of mycorrhiza formation likely in the host to be grown, and (iii) how agricultural practices alfter the formation and benefits from the symbiosis. Additional information on the selection of inoculant fungi is required where such an option is appropriate. With existing knowledge and approaches, it is not possible to (i) identify sites used for broadscale agriculture where the symbiosis is operating sub-optimally, (ii) quantify the benefits that would result from increasing the rate and extent of colonization of roots by effective VA mycorrhizal fungi, and (iii) identify the most cost-effective method for managing the fungi. In this review, we present our approach to overcoming these limitations to assessing the potential for managing VA mycorrhizal fungi in pasture soils. This approach has led to the identification of specific areas where further research is necessary.

Modelling response to phosphorus and VA mycorrhizas

The response of both mycorrhizal and non-mycorrhizal plants to phosphorus application is frequently well described by the Mitscherlich equation (see Abbott and Robson 1984b, 1987; Fig. 2). In the simplest case, where A and B have the same value for mycorrhizal and non-mycorrhizal plants, at any yield level, the relative effectiveness of phosphorus for mycorrhizal compared to non-mycorrhizal plants is c_m/c_{nm} (Fig. 2).

The values of c_m/c_{nm} represent the dependence of the host on mycorrhizas in that soil and have been calculated to range from 1.1 (no mycorrhiza benefit) to 5.8 (a large benefit from mycorrhizas) (Abbott and Robson, 1987). High values of c_m/c_{nm} will be obtained either when highly mycorrhiza-dependent hosts are grown or when the source of phosphorus is extremely unavailable in phosphate-deficient soils.

We are attempting to identify relationships between mycorrhiza formation and parameters of the Mitscherlich equation so that growth can be predicted for plants which differ in their degree of root colonization (Fig. 3). In the case where A and B are the same for mycorrhizal and non-mycorrhizal plants, the parameter that would be affected by the degree of mycorrhiza formation is c. This parameter (c) is related to the adsorption capacity of the soil (Fig. 4) and mycorrhizas would be expected to change the shape of the curve as indicated. The ratio c_m/c_{nm} is dependent upon the degree of mycorrhiza formation. An understanding of how this ratio varies with the extent of mycorrhiza development is a key relationship within the model that has to be determined.

In order to use this approach to calculate the current and potential benefits from VA mycorrhizal fungi for a particular host in a particular soil, the field data required is: (i) rainfall (for A), (ii) P adsorption capacity (for c), (iii) soil P test (for B), and (iv) a predicted value for mycorrhiza development (for c). The relationships between these properties and A, B and c must also be available (see Bowden and Bennett, 1974).

It should be possible to predict response curves to applied phosphorus for plants which have (i) no mycorrhizas, (ii) the maximum possible level of mycorrhizas, and (iii) the expected level of mycorrhiza formation at a particular site for the plant species that is to be grown. On the basis of these curves, the likely value from increasing the level of mycorrhiza development by

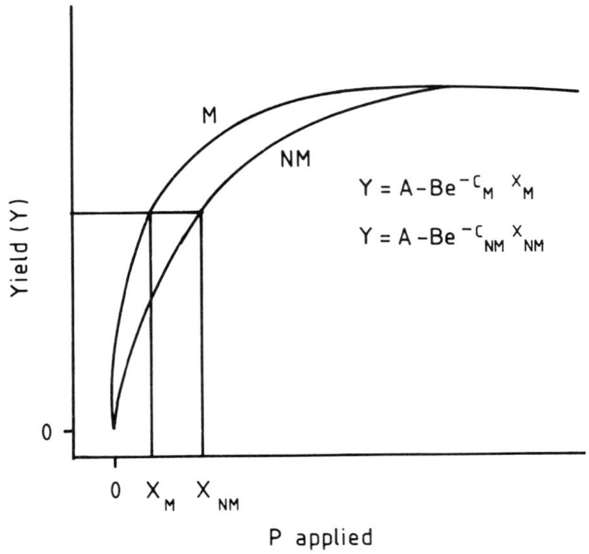

Fig. 2. Mitscherlich response of plant growth to phosphate application with (M) and without (NM) VA mycorrhizal fungi where A is yield for non-limiting phosphate, B is a constant reflecting the adequacy of phosphate for plant growth, c is the curvature or Mitscherlich constant and x is any level of phosphate application. When A and B are the same for mycorrhizal and non-mycorrhizal plants, $c_m/c_{nm} = x_{nm}/x_m$.

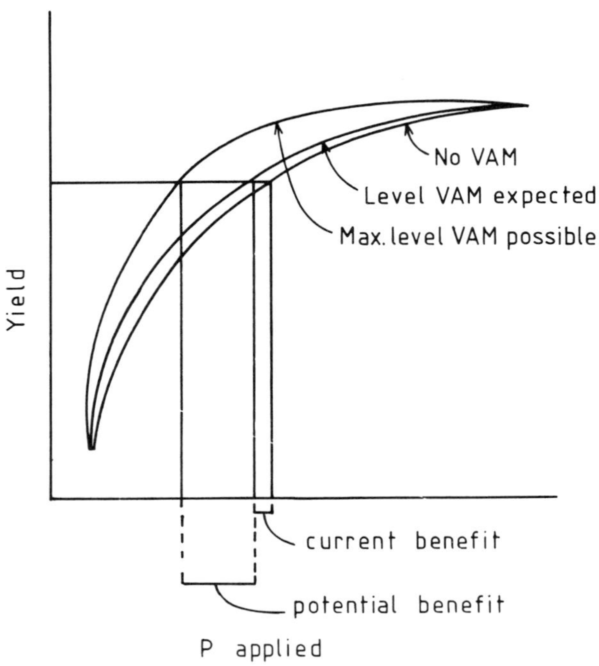

Fig. 3. Predicting benefits of mycorrhizal infection from the Mitscherlich equations (from Abbott and Robson, 1987).

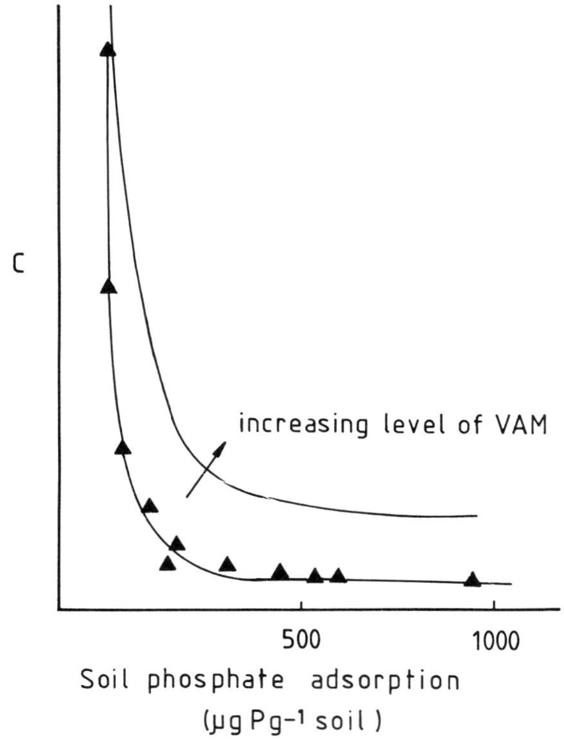

Fig. 4. Relationship between soil phosphate adsorption capacity and the Mitscherlich coefficient (c) for subterranean clover (after Barrow, 1975) and a prediction for the effect of increasing the level of colonization of roots by VA mycorrhizal fungi.

management practices can be calculated (Fig. 3). Alternatively, a recommendation for phosphorus fertilizer can be made which takes account of the potential of the VA mycorrhizal fungi in the soil to form mycorrhizas and contribute to plant growth.

Predicting mycorrhiza development

The best parameter useful for prediction of mycorrhizal development is not known. From glasshouse studies, a measure of infectivity (incorporating the potential rate and extent of infection) would be appropriate. However, relationships between the rate and extent of infection and plant growth have not been well established for field-grown plants.

We have attempted to determine this relationship for plants grown in pots of steamed soil placed under field conditions. Plant growth was not correlated with the rate and extent of mycorrhiza formation for three species of fungi inoculated at two levels of phosphate addition. However, plant growth appeared to be related to the extent to which the inoculant fungus spread to unexplored soil below a depth of 10 cm. *Acaulospora laevis* colonized more roots at that depth than the other species of fungi used and also increased plant growth to the greatest extent.

Once the appropriate measure of mycorrhiza development is identified, a procedure for predicting it is required. It is not possible to predict mycorrhiza formation in the field from either spore counts (Abbott and Robson, 1982a; Hayman and Stovold, 1979) or estimates of the number of infective propagules (Scheltema *et al.*, 1987). We are attempting to develop a glasshouse bioassay in which relationships are being identified between mycorrhiza formation in bait plants grown in soil samples from field sites and actual mycorrhiza formation in the field.

We have varied conditions during the bioassay (undisturbed soil compared with mixed soil, soil maintained at different soil temperatures, soil maintained at either field capacity or allowed to fluctuate between wilting point and field capacity). Infectivity as assessed by the bioassay was largely unaffected by these changes in conditions (Abbott *et al.*, unpublished data). Additionally infectivity as assesed by the bioassay did not change markedly in the period between cessation of plant growth in one season and the commencement of plant growth in the next season (Abbott *et al.*, unpublished). For a useful bioassay, it would be necessary to collect soil and conduct the bioassay prior to the start of the growing season.

At this stage in the development of a bioassay to be used to predict mycorrhizal formation in pasture plants in our mediterranean environment, the best relationship so far has been found between percentage root length infected in three week old clover plants in the glasshouse bioassay and percentage root length in field grown plants towards the end of the growing season (Fig. 5). The two main areas that require further clarification before a useful bioassay is available are (i) identification of the critical time for mycorrhiza formation and its functional state in relation to the phosphate requirement of the plant growing in the field, and (ii) identification of the reasons for a poor relationship between infection in the glasshouse bioassay and that in the field at some sites. Variation in species composition of the pasture is one factor that may influence the relationship and this requires further study. Clearly, if the factors that limit the development of a broadly applicable bioassay can be identified, much progress will have been made towards understanding the ecology of VA mycorrhizal fungi, an essential prerequisite to their management.

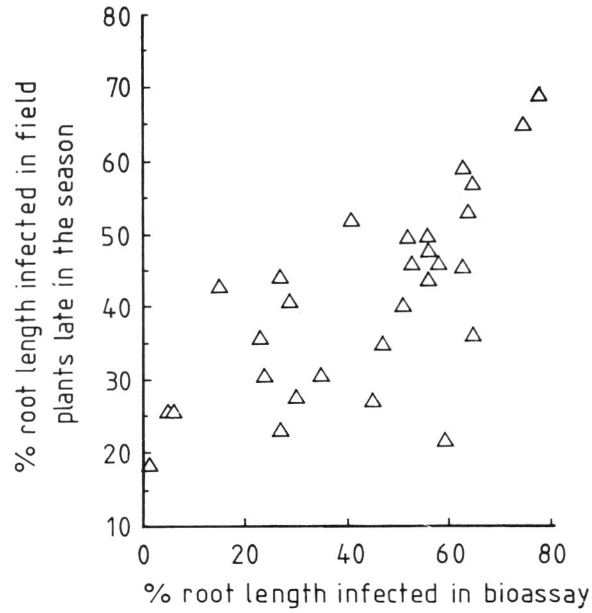

Fig. 5. Relationship between the extent of mycorrhizal infection formed in subterranean clover grown for three weeks in a glasshouse bioassay for soils collected prior to plant growth in the field and that actually formed in pasture plants grown in the same field sites towards the end of the growing season.

Mycorrhiza formation can be changed by altering agricultural practice (Abbott and Robson, 1989). Thus it may be necessary to develop relationships between infectivity assessed in a glasshouse bioassay and mycorrhiza formation in plants grown with particular agricultural practices. In our case, the bioassay is developed to predict mycorrhiza formation in a mixed pasture, using subterranean clover as the bait plant (Fig. 5). From a knowledge of the effect of changes in agricultural practice on mycorrhiza formation in other hosts, it may be possible to temper the prediction from a standard bioassay for a range of hosts or agricultural practices.

Although much research is required for selecting and predicting the best parameter of infection for inclusion in the model, another important requirement is to predict the maximum value possible for this parameter. That is, can the current level of mycorrhizas be increased and to what extent? This may be answered using an empirically defined relationship between c_m/c_{nm} for the highest level of mycorrhizas that can be achieved by inoculation in soils ranging in phosphate adsorption capacities.

Management of indigenous VA mycorrhizal fungi

The two options for managing VA mycorrhizal fungi if the expected level of mycorrhizas is below the maximum possible for a soil/host combination are (i) to alter the agricultural practice with the aim of increasing the contribution from the population of fungi already in the soil and (ii) to inoculate with selected fungi.

Changing agricultural practices

Agricultural practices change soil physical, chemical and biological properties that could alter the ability of VA mycorrhizal fungi to colonize roots. From an understanding of the effects of these changes on (i) the number and types of propagules and (ii) the species diversity, agricultural practices could be selected which maximize mycorrhiza formation (Table 1). The practicality of changing agricultural practice merely to gain maximum benefits from these fungi is limited, because mycorrhizas are only one of many factors associated with soil fertility (Robson and Abbott, 1988).

Selection of inoculant VAM fungi

VA mycorrhizal fungi need to be selected for particular soil and environmental conditions (Abbott and Robson, 1982b). Among species, there is evidence for variation in their ability to infect rapidly. This in turn is related to differences in the ability of fungi to (i) form propagules in high numbers, (ii) form extensive mycelium in soil, and (iii) grow rapidly from germinated spores or root fragments. Evidence that species differ in their ability to absorb phosphate from soil may or may not be simply related to their ability to form extensive mycorrhizas (see Smith and Gianinazzi-Pearson, 1987). There could be differences among fungi in the length and distribution of external hyphae formed and/or the absorption and transport of phosphate from the soil solution to the plant root. Further research is necessary to clarify this, however, in selecting inoculant fungi, there would be little value in a fungus with an exceptional ability to enhance phosphate uptake if it were unable to infect extensively as well.

Although at least 120 species of VA mycorrhizal fungi have been described (Morton, 1988), limited comparative information on the ecology of these fungi in field soils is available. However, there are sufficient data available to show that considerable genetic variation occurs among species and strains to enable the selection of inoculant fungi suited to particular soil conditions. Genetic engineering to provide further variation is unwarranted until existing variation is assessed. Ecological studies of indigenous VA mycorrhizal fungi in soils are basic to the selec-

Table 1. Agricultural practices that may affect the species diversity and abundance of VA mycorrhizal fungi (see review by Abbott and Robson, 1989)

Practices that lead to decreased infectivity	Practices that lead to increased infectivity	Practices that alter species dominance
(i) Erosion (ii) Fallowing (iii) Fumigation (iv) Growth of non-mycorrhizal plants (v) Phosphate application	(i) Growth of plant species with a high length of infected root	(i) Lime application (ii) Phosphate application (iii) Changed plant species

tion of a range of potential inoculant fungi and to predict their likely competitive ability when inoculated into field soils.

There may also be differences among VA mycorrhizal fungi which affect their suitability for large-scale inoculum production. Some of these differences may confer advantages in colonization of soil (for example formation of a large number of propagules) but others may not (for example maintenance of infectivity when dried). There is a need to select fungi for their suitability for large scale inoculum production.

Conclusion

Considerable scope exists to increase the efficiency of use of nutrients from soils and fertilizers by manipulating the symbiosis between VA mycorrhizal fungi and plants. Areas of high priority for future research to achieve this aim are (i) studies on the formation and functioning of VA mycorrhizas in the field to quantify the relationship between benefit and the rate and extent of mycorrhiza formation for different plant species for soils with different capacities to adsorb phosphate, (ii) development of an understanding of the life cycle of many species of VA mycorrhizal fungi so that the production, survival and infectivity of different propagules can be predicted, and (iii) the development of mechanistic models of the effect of agricultural practices on VA mycorrhizal fungi within roots and in soil.

Acknowledgements

We thank Martine Scheltema and Chris Gazey for conducting the experiments, the Australian Wool Research Trust Fund for financial support and Risø National Laboratory, Denmark for support of study leave for LKA.

References

Abbott L K and Robson A D 1982a Infectivity of vesicular arbuscular mycorrhizal fungi in agricultural soils. Aust. J. Agric. Res. 33, 1049–1059.

Abbott L K and Robson A D 1982b The role of vesicular-arbuscular mycorrhizal fungi in agriculture and the selection of fungi for inoculation. Aust. J. Agric. Res. 33, 389–408.

Abbott L K and Robson A D 1984a Colonization of the root system of subterranean clover by three species of vesicular-arbuscular mycorrhizas. New Phytol. 96, 275–281.

Abbott L K and Robson A D 1984b The effect of mycorrhizae on plant growth. *In* VA Mycorrhiza. Eds. C L Powell and D J Bagyaraj. pp 113–130. CRC Press, Boca Raton, FL.

Abbott L K and Robson A D 1987 Managing VA mycorrhizas in pasture. *In* Proceedings of the 7th North American Conference on Mycorrhizae. Eds. D M Sylvia, L L Hung and J H Graham. pp 10–12. University of Florida, Gainesville.

Abbott L K and Robson A D 1989 Factors affecting the occurrence of vesicular-arbuscular mycorrhizas. Agric. Ecosystems Environ. (*In press*.)

Bagyaraj D J 1984 Biological interactions with VA mycorrhizal fungi. *In* VA Mycorrhiza. Eds. C L Powell and D J Bagyaraj. pp 131–153. CRC Press, Boca Raton, FL.

Barrow N J 1975 The response to phosphate of two annual pasture species. I. The effect of the soil's ability to adsorb phosphate on comparative phosphate requirement. Aust. J. Agric. Res. 26, 137–159.

Black R and Tinker P B 1979 The development of endomycorrhizal root systems. II. Effect of agronomic factors and soil conditions on the development of vesicular-arbuscular mycorrhizal infection in barley and on the endophyte spore density. New Phytol. 83, 401–413.

Bowden J W and Bennett D 1974 The 'Decide' model for predicting superphosphate requirements. *In* Proceedings of Phosphorus in Agriculture Symposium. Australian Institute of Agricultural Science, Melbourne.

Cooper K M 1984 Physiology of VA mycorrhizal associations. *In* VA Mycorrhiza. Eds. C L Powell and D J Bagyaraj. pp 155–186. CRC Press, Boca Raton, FL.

Fitter A H (1985) Functioning of vesicular-arbuscular mycorrhizas under field conditions. New Phytol. 99, 257–265.

Fitter A H 1986 Effect of benomyl on leaf phosphorus concentrations in alpine grasslands: A test of mhycorrhizal benefit. New Phytol. 103, 767–776.

Gianinazzi-Pearson V (1985) Mycorrhizal effectiveness in phosphate nutrition: How, when and where? *In* Proceedings 6th North American Conference on Mycorrhizae. Ed. R Molina. pp 150–154. Forest Research Laboratory, Corvallis, OR.

Haas J H, Bar-Yosef B, Krikun J, Barak R, Markovitz T and Kramer S 1987 Vesicular-arbuscular mycorrhizal infestation and phosphorus fertilization to overcome pepper stunting after methyl bromide treatment. Agron. J. 79, 905–910.

Hall I R 1980 Growth of *Lotus pedunculatus* Car in an eroded soil containing soil pellets infested with endomycorrhizal fungi. N.Z. J. Agric. Res. 23, 103–105.

Hayman D S 1987 VA mycorrhizas in field crop systems. *In* Ecophysiology of VA mycorrhizal plants. Ed. G R Safir. pp 171–192. CRC Press, Boca Raton, FL.

Hayman D S and Stovold G S 1979 Spore populations and infectivity of vesicular-arbuscular mycorrhizal fungi in New South Wales. Aust. J. Bot. 27, 227–233.

Hepper C M, Azcon-Aguilar C, Rosendahl S and Sen R 1988 Competition between three species of *Glomus* used as spatially separated introduced and indigenous mycorrhizal inocula for leek (*Allium porrum* L.). New Phytol. 110, 207–215.

Koomen I, Grace C and Hayman D S 1987 Effectiveness of single and multiple inocula on growth of clover and strawberry plants at two soil pHs. Soil Biol. Biochem. 19, 539–544.

Lopez-Aguillon R and Mosse B 1987 Experiments on competitiveness of three endomycorrhizal fungi. Plant and Soil 97, 155–170.

McGonigle T P and Fitter A H 1988 Water relations of red clover (*Trifolium pratense* L) as affected by VA mycorrhizal infection and phosphorus supply before and after drought. J. Exp. Bot. 39, 595–603.

Morton J B 1988 Taxonomy of VA mycorrhizal fungi: classification, nomenclature and identification. Mycotaxon 32, 267–324.

Nelson C E 1987 The water relations of vesicular-arbuscular mycorrhizal systems. *In* Ecophysiology of VA mycorrhizal plants. Ed. G R Safir. pp 71–91. CRC Press, Boca Raton, FL.

Powell C L 1984 Field inoculation with VA mycorrhizal fungi. *In* VA Mycorrhiza. Eds. C L Powell and D J Bagyaraj. pp 205–222. CRC Press, Boca Raton, FL.

Robson A D and Abbott L K 1988 Improving the biological and chemical fertility of temperate area soils. *In* National Soils Conference 1988 – Review Papers. Ed. J Loveday. pp 73–86. Australian Society of Soil Science, Nedlands.

Scheltema M A, Abbott L K and Robson A D 1987 Seasonal variation in the infectivity of VA mycorrhizal fungi in annual pastures in a mediterranean environment. Aust. J. Agric. Res. 38, 707–715.

Smith S E 1985 The concept of effectiveness in symbiotic relationships. *In* Proceedings of the 6th North American Conference on Mycorrhizas. Ed. R Molina. pp 146–149. Forest Research Laboratory, Corvallis.

Smith S E and Gianinazzi-Pearson V 1987 Physiological interactions between symbionts in vesicular-arbuscular mycorrhizal plants. Ann. Rev. Plant Physiol. and Plant Mol. Biol. 39, 221–244.

Stribley D 1987 Mineral nutrition. *In* Ecophysiology of VA mycorrhizal plants. Ed. G R Safir. pp 59–69. CRC Press, Boca Raton, FL.

Stribley D P, Tinker P B and Snellgrove R C 1980 Effect of vesicular-arbuscular mycorrhizal fungi on the relations of plant growth, internal phosphorus concentration and soil phosphate analysis. J. Soil Sci. 31, 655–672.

Yost R S and Fox R L 1979 Contribution of mycorrhizae to P nutrition of crops growing on an oxisol. Agron. J. 71, 903–909.

Short papers

D. L. Keister and P. B. Cregan (Eds.), The rhizosphere and plant growth, 365.

PLSO BARC76

Characterization of cultivar specific growth promotion of spring wheat by *Bacillus* sp.

C.P. CHANWAY[1] and L.M. NELSON[2]
[1]*Department of Forest Sciences, University of British Columbia, Vancouver, B.C., Canada V6T 1W5, and* [2]*Plant Biotechnology Institute, National Research Council of Canada, Saskatoon, Saskatchewan, Canada S7N 0W9*

Key words: Bacillus, plant growth promotion, *Triticum aestivum*

The potential of isolating beneficial plant growth promoting organisms should be increased by selecting isolates from the rhizosphere of a crop plant that has been successfully grown in a soil for a long period of time. Seven strains of *Bacillus* sp. were isolated from the rhizosphere soil of spring wheat (*Triticum aestivum* L.) cv. Katepwa grown in a field that has been continuously cropped to spring wheat for 27 years; to cv. Ketepwa for the last 5 years and to cv. Neepawa for the previous 12 years. Six of the seven isolates promoted root growth of cv. Katepwa but none promoted that of cv. Neepawa in sterile glass tubes. In this paper we present more data on the characterization of the cultivar-specific growth promotion and results from an initial field trial.

In the laboratory, plants were grown in sterile glass tubes, sterile Leonard jars on non-sterile styrofoam pots and inoculated with 10^7 cells of Bacillus strains 5A1 or 3A in nutrient broth or $MgSO_4$. In some experiments autoclaved cells or filter-sterilized culture supernatant of 5A1 replaced live cells. Plants were harvested 2 to 4 weeks after planting for determination of root and shoot dry weights. In the field, seeds of cv. Katepwa and Neepawa were inoculated with 10^7 cells of strains 3A or 5A1. Emergence, shoot dry weight and tiller number were determined.

Root growth of cv. Katepwa but not Neepawa was promoted in the presence of live cells or supernatant of strain 5A1 but no effect was seen with heat-killed cells. IAA was detected in the culture supernatant of 5A1 using GCMS and concentrations appear to be in the ng mL^{-1} range. Addition of IAA to glass tubes or pots only partially duplicated the effect of adding cells or supernatant of strain 5A1. Cultivar-specific growth promotion of roots by Bacillus strains 5A1 and 3A was observed in non-sterile pots as well as sterile tubes. Although the cells were suspended in nutrient broth in most experiments, the same growth promotion effects were also seen when $MgSO_4$ was the suspending medium. Strain 5A1 also promoted root growth of cv. Katepwa in Leonard jars.

Strain 3A promoted emergence of cv. Katepwa in the field. Strain 5A1 inhibited shoot growth of both cultivars 19 days after sowing but by 70 days 5A1-inoculated plants of cv. Katepwa had 38% greater shoot dry weight ($P = 0.10$) and 79% more tillers ($P = 0.05$) than control.

Cultivar-specific growth promotion of spring wheat by Bacillus isolates was observed in sterile and non-sterile *in vitro* systems as well as in the field. A heat labile factor may be involved in the mode of action with strain 5A1 and IAA has been identified in the supernatant. Strains such as these show potential as PGPR but more extensive field testing and research on mode of action are required.

D. L. Keister and P. B. Cregan (Eds.), The rhizosphere and plant growth, 366.

PLSO BARC77

Effect of Bacillus strains on growth of pine (*Pinus contorta* Dougl.), spruce (*Picea glauca* Voss.), and Douglas-fir (*Pseudotsuga menziesii* (Mirb. Franco)

C.P. CHANWAY[1], R.A. RADLEY[2], F.B. HOLL[3] and P.E. AXELROOD[2]
[1]*Department of Forest Sciences,* [3]*Department of Plant Science, University of British Columbia, Vancouver, B.C., Canada V6T 1W5, and* [2]*Forest Biotechnology Centre, B.C. Research Corp., 3650 Westbrook Mall, Vancouver, B.C., Canada*

Key words: Bacillus, conifers, growth promotion

Bacteria that are applied to seeds, roots, or tubers and subsequently colonize roots and enhance plant growth have been termed plant growth promoting rhizobacteria (PGPR) (Kloepper *et al.*, 1980). PGPR have been studied extensively in agricultural systems and have been shown to promote growth of various deciduous species. (Caesar and Burr, 1987). However, little work has been done describing PGPR effects on coniferous forest species. We have tested the effects of two Bacillus strains previously demonstrated to promote growth of forage species (Holl *et al.* 1988; Chanway *et al.*, 1988). Bacterial strains were grown in nutrient or tryptic soy broth, harvested by centrifugation, and resuspended in phosphate buffer. A small volume of the suspension (ca. 10^8 cells) was added to seed or seedlings in containers of a peat/venmiculite mixture. Seedlings were grown under artificial light in growth chambers. Strain L5 significantly increased root surface area of both lodgepole pine (28%) and white spruce (55%) after 12 weeks growth. Root and shoot dry weights were not affected. Douglas-fir was unaffected by inoculation with strain L5. Strain L6 caused significant increases in lodgepole pine root (18%) and shoot (21%) dry weight, root surface area (36%), and root collar diameter (15%) after 8 weeks growth, but these effects were not detectable after 16 weeks. When 1-0 containerized lodgepole pine seedlings were tested in a 12-week growth assay, shoot weight increased (17%) but root growth was unaffected. White spruce was unaffected by inoculation with strain L6. Douglas-fir did not show a growth response to strain L6 after 8 weeks, but root surface area was significantly higher (22%) 12 weeks after sowing. Our results suggest that the effect of bacterial inoculation was primarily on root surface area and to a lesser degree on root biomass. The specific plant/microbe combination was an important determinant of the growth response therefore effective combinations must be identified through screening programs. The decline of effects with time may be related to inoculum decline in the rhizosphere and re-inoculation experiments and colonization studies are underway to explore this possibility. We feel these results hold promise for incorporation of bacterial inoculants into current forestry practices because production of seedlings with greater root surface area or biomass may significantly enhance survival and performance of outplanted stock.

References

Caesar A J and Burr T J 1987 Phytopathology 77, 1583–1588.
Holl F B *et al.* 1988 Soil Biol. Biochem. 20, 19–24.
Kloepper J W *et al.* 1980 Phytopathology 70, 1078–1082.
Chanway C P 1988 Plant and Soil 106, 281–284.

D. L. Keister and P. B. Cregan (Eds.), The rhizosphere and plant growth, 367.

PLSO BARC78

Sulfur oxidizing microorganisms for growth promotion of canola

S.J. GRAYSTON and J.J. GERMIDA
Department of Soil Science, University of Saskatchewan, Saskatoon, Saskatchewan, Canada S7N 0W0

Key words: canola, growth promotion, sulfur

Canola has a high sulfur requirement during vegetative growth and exhibits symptoms of sulfur deficiency when grown on Saskatchewan soils low in plant available sulfur. Elemental sulfur is frequently used as a fertilizer to alleviate sulfur deficiency in soils. The oxidation of elemental sulfur to plant available sulfate is primarily a heterotrophic microbial process in Saskatchewan soils. Inoculation of canola with rhizobacteria has resulted in increased yields of canola. The aim of this study was to determine the potential of sulfur oxidizing microorganisms isolated from the canola rhizosphere to enhance canola growth through increased sulfur uptake.

Sulfur oxidizing bacteria and fungi isolated from the rhizosphere and rhizoplane of canola, grown in four Saskatchewan soils under growth chamber conditions, were assessed for their sulfur oxidizing ability *in vitro*. The eighteen best *in vitro* sulfur oxidizing bacteria were tested as seed inoculants for canola under growth chamber conditions. In addition the isolates were screened for their ability to inhibit the growth of canola pathogens.

Of 273 bacteria isolated from the canola rhizosphere 89.7% oxidized sulfur to thiosulfate or tetrathionate and 48.7% oxidized sulfur to sulfate *in vitro*. All 70 fungi isolated oxidized sulfur to sulfate. Thirteen of the eighteen isolates assessed as PGPR increased canola leaf size at the bud stage of growth. Seven isolates increased root and pod dry weight at maturity. Analysis of plant tissue from this experiment has shown increases in the sulfur and iron content of shoots and pods from canola inoculated with sulfur oxidizing isolates. Three isolates also inhibited the growth of the canola pathogens *Rhizoctonia solani* AG2-1, *R. solani* AG4 and *Leptosphaeria maculans* 'Leroy'. One isolate was antagonistic towards both *R. solani* strains and one isolate inhibited the growth of *R. solani* AG2-1 and *L. maculans* 'Leroy'.

Some of the sulfur oxidizing isolates appear to be able to stimulate canola growth through the enhancement of sulfur, iron and possibly other nutrient uptake. Some isolates also inhibited the growth of canola pathogens which suggests antibiosis may be another mechanism of growth stimulation.

D. L. Keister and P. B. Cregan (Eds.), The rhizosphere and plant growth, 368.

PLSO BARC79

Microbe enhanced P uptake by corn under no till and conventional till

H. M. KUNISHI[1] and V.A. BANDEL[2]
[1]*Environmental Chemistry Laboratory, USDA, ARS, Beltsville, MD 20705, USA and*
[2]*Department of Agronomy, University of Maryland, College Park, MD 20742, USA*

Key words: fumigation, P fertility, grain yield, Plant growth, vesicular-arbuscular-mycorrhiza

Microbial enhancement of P uptake by plants could have significant relevance to low-input sustainable agriculture. Our 1986–87 study, conducted with soil that tested deficient in P and positive for the presence of vesicular-arbuscular-mycorrhizae (VAM), examined plant performance in no-till (NT) and conventional-till (CT) plots fertilized with P at 0, 19, and 78 kg ha^{-1}. Microbial effects were determined by comparison of plant data obtained from check plots and test plots fumigated to lower or eliminate VAM activity.

Test and check plots were planted to corn, *Zea mays* L., about 3 weeks after date of fumigation. Above-ground plant samples were taken at selected intervals, and the following indices of plant performance were measured: P uptake rate, plant-top growth, and grain yield.

The unfertilized plots were clearly deficient in P. The combined NT and CT grain-yield data for these plots and corresponding P-fertilized plots have been recorded from 1984 to 1988 as part of a larger study; and they show that each year, the yields from the unfertilized plots were lower than the combined yields from the unfertilized and fertilized plots by an average of 14% (range = 6% to 17%).

In the NT unfertilized plots, fumigation lowered P uptake rate by 70%, plant-top growth by 75%, and grain yield by 90%. In the corresponding CT plots, fumigation lowered these respective indices by 55%, 53%, and 77%. Fumigation decreased P uptake and inhibited plant growth early in the season, and the stunted plants were unable to overcome this growth deficit even though P uptake efficiency increased later.

In the NT plots with 19 kg P ha^{-1}, fumigation did not reduce plant performance. But in the corresponding CT plots, fumigation lowered P uptake rate by 53%, plant-top growth by 50%, and grain yield by 73%. These differences in effect of fumigation on the NT and CT plots likely indicate a higher P fertility status of the NT plots.

In the plots receiving 78 kg P ha^{-1}, fumigation did not reduce plant performance in either NT or CT plots. Nor would a reduction be expected, since a higher enough amount of P was added to prevent restricted P uptake and plant growth even in the absence of microbial activity.

Fumigation effects were the most pronounced at the lowest level of P fertility, a level that may be approached in soils under low-input management. Thus, in low-input sustainable agriculture, care should be taken to avoid practices that inhibit or destroy microbes that enhance P uptake. In corn production, a modest addition of P fertilizer coupled with NT management would appear to provide some protection against chemicals that destroy such microbes.

D. L. Keister and P. B. Cregan (Eds.), The rhizosphere and plant growth, 369.

PLSO BARC80

Exposure of *Phleum pratense* to SO_2 and its subsequent effect on VAM root infectivity

M.J. CLAPPERTON and D.M. REID
Plant Physiology Research Group, Department of Biological Sciences, University of Calgary, Calgary, Alberta, Canada T2N 1N4

Key words: Phleum, SO_2, VAM infection

Two experiments were conducted to test the indirect effects of SO_2 (sulphur dioxide) fumigation on the ability of VAM to infect and proliferate in the roots of *Phleum pratense* which had been continuously exposed to 0.05–0.07 ppm SO_2. *Phleum pratense* seedlings (6 day-old) were planted into 65 ml syringe barrels which had been filled with field soil. The field soil contained known sources of VAM inoculum. The seedlings were then placed into either an SO_2 exposure chamber or control chamber. The chambers were designed in such a way that only the shoots of plants were exposed to SO_2. In the first experiment, seedlings were continuously exposed to SO_2 for 6 weeks. In the second experiment, seedlings were allowed to grow under control (non-SO_2) conditions for 3 weeks before being continuously exposed to SO_2 fumigation for the remaining 3 weeks. The results showed that (1) SO_2 fumigation of *Phleum pratense* shoots inhibited the ability of VAM to infect and proliferate within host roots; and (2) the greater the length of root infected with VAM the greater the carbohydrate drain on the plant, possibly amplifying the stress caused to host photosynthesis by SO_2 exposure.

D. L. Keister and P. B. Cregan (Eds.), The rhizosphere and plant growth, 370.

PLSO BARC81

Promotion of maize growth by legume soil factors

A. FYSON and A. OAKS
Department of Biology, McMaster University, Hamilton, Ontario, Canada L8S 4K1. Present addresses: Department of Botany, University of Guelph, Ontario, Canada N1G 2W1

Key words: mycorrhizae, plant growth promotion

In southern Ontario, continuous maize production is widely practiced but may lead to deterioration of soil structure and reduction in yields (Bolton, 1976; Ketcheson, 1980). Rotation with legumes such as alfalfa can reverse this trend.

Greenhouse experiments have established that inoculation with a variety of legume soils (4.1% by volume) but not the maize soils tested can enhance the growth of maize by 100–300% after 5–8 weeks growth in a standard test soil. The growth promotion was completely destroyed by gamma irradiation (3.6 Mrad) and therefore likely involves microbial factors. A bacteriocide, streptomycin (500 mg L^{-1}) and two fungicides, etridiazol (0.18 ml L^{-1}) and metalaxyl (0.18 ml L^{-1}) which are selective against Oomycetes (*e.g.* Pythium and Phytophthora) did not reduce the growth response. On the other hand, the less selective fungicides benomyl (36 mg L^{-1}) and PCNB (500 mg L^{-1}) eliminated the growth response. The latter results and the observation that these treatments also eliminated vesicular arbuscular mycorrhizal (VAM) infection suggest a role for mycorrhizae in the growth response. Exposure of alfalfa soil to low level gamma irradiation (100 krad) has been shown (Jacobsen, 1984) to eliminate the infective capacity of VAM fungi. In our experiments such treatment of alfalfa soil eliminated both maize growth promotion and VAM infection after inoculation with this soil. This result adds more weight to the suggestion that VAM infection is an important component in our growth response. Experiments are being carried out to establish whether a growth response can be attained from inocula of VAM fungi isolated from our alfalfa soil. An experiment carried out in 1988 at the Agriculture Canada Experimental Farm, Woodslee, Ontario established that inoculation with our alfalfa soil can enhance early growth (3 fold after 7 weeks growth) and increased VAM infection of maize (cv W64A × W182E) on an unfertilized soil planted to maize for 42 years.

References

Bolton E F *et al.* 1976 Can. J. Soil Sci. 56, 21–25.
Jacobsen I 1984 Soil Biol. Biochem. 16, 281–282.
Ketcheson J W 1980 Can. J. Soil. Sci. 60, 403–4120.

D. L. Keister and P. B. Cregan (Eds.), The rhizosphere and plant growth, 371–372.

PLSO BARC82

Variation in VA mycorrhizal strain interactions with Rhizobium on pigeon pea

D.C. IANSON[1] and R.G. LINDERMAN[2]
[1]*Department of Botany and Plant Pathology, Oregon State University, Corvallis, OR 97330, USA and*
[2]*USDA-ARS, Horticultural Crops Research Laboratory, Corvallis, OR 97330, USA*

Key words: drought stress, nitrogen fixation, nodulation, VA mycorrhizae

Legumes generally form symbioses with both VA mycorrhizal (VAM) fungi and nodulating, nitrogen-fixing Rhizobium. The presence of VAM generally enhances plant growth by reducing response to nutrient deficiency or other stresses, and by enhancing the nitrogen fixation process. The microbial and/or physiological interactions in these tripartite associations, however, are complex and poorly understood, and little is known on how different VAM fungi may affect these interactions. The objective of our study was to evaluate the variation in response of pigeon pea to dual inoculation with Rhizobium and 7 different VAM fungi, in terms of plant growth, the nitrogen fixation processes, and soil drought.

Pigeon pea (*Cajanus cajan* (L.) Millsp.) germlings were inoculated with a cowpea strain of Rhizobium, and planted into inoculum of the VAM fungi *Glomus mosseae* (Gm), *G. deserticola*-Utah (GdU), *G. etunicatum* (Ge), *G. intraradices* (Gi), *G. deserticola*-Calif. (GdC), *G. Aggregatum/microcarpum* mix (Gmix), or *Gigaspora margarita* (Gma). Plants were fertilized with low-nitrogen LANS with either 20, 40, or 60 ppm P. After 12 weeks, plant height, stem caliper, foliar P levels, and drought stress measurements were made. Plants were grouped ac-

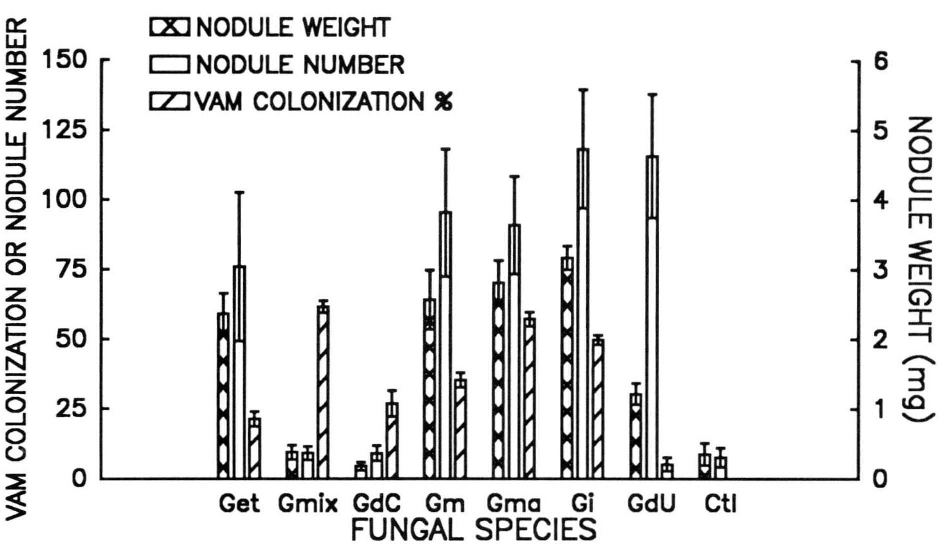

Fig. 1. Nodule weight, number and VAM colonization as per fungal species in pigeon pea.

cording to size to compare photosynthetic rates and soil water potential (measured with soil psychrometers).

Pigeon pea growth enhancement and drought stress responses varied with VAM strain used, and were independent of P nutrition. Growth increases were highly correlated with increased nitrogen fixation resulting from a dramatic increase in number and size of Rhizobium nodules induced by some VAM strains and not others (Fig. 1). Plant response to soil drought also varied with VAM strain used, and was not correlated with nodule stimulation.

These results demonstrate that inter-endophyte compatibility between Rhizobium and VAM fungi is critical to optimize nitrogen fixation in legumes, and that VAM strains should be selected for that reason and for their capacity to reduce drought stress.

D. L. Keister and P. B. Cregan (Eds.), The rhizosphere and plant growth, 373.

PLSO BARC83

Use of histochemical techniques to quantify mass/activity relations of mycorrhizal roots and soil

B.N. JOHNSON and W.B. McGILL
Department of Soil Science, University of Alberta, Edmonton, Canada T6G 2E3

Key words: biotic activity, ergosterol, mycelia mass, ornithine decarboxylase

In mycorrhizal research, morphology-based methods are used to estimate the proportion of ectomycorrhizae on tree seedlings. Extramatrical mycelial mass and activity of the mycorrhiza-soil system are not commonly quantified. We used the quantity of ergosterol and the activity of ornithine decarboxylase (ODC) in rhizosphere soil, non-rhizosphere soil and roots of *Pinus contorta* seedlings ectomycorrhizal with *Hebeloma crustuliniforme* as analogues of mycelial mass and biotic activity, respectively. The three components were sampled at 3 week intervals for 12 weeks to describe ontogeny of mycorrhizae.

Ergosterol content of pure-cultured *H. crustuliniforme* mycelia was used as a conversion factor to estimate mycelial mass. Percent recovery of ergosterol and ornithine decarboxylase activity from soil were also established in preliminary trials.

The total mass of rhizosphere soil increased by 25% from 6 to 12 weeks but, rhizosphere soil mass mg^{-1} dry root decreased 2.4 fold. Root mycelia content (mg mycelia mg^{-1} root) decreased by 10% while that of the rhizosphere was constant over time. Ornithine decarboxylase activities of roots and rhizosphere soils declined over time and were correlated ($r = 0.65$; significant at $P < 0.01$). Rhizosphere mycelia content and enzyme activity were also correlated ($r = 0.60$; $P < 0.01$). The rhizosphere contained up to 6 times more mycelia and 5 times greater ODC activity than nonrhizosphere soil. On an equal mass basis, roots harbored up to 70 times more mycelia than soils and enzyme activity was 45 times greater on average.

The assays were capable of distinguishing fungal mass and biotic activity both spatially and temporally within a mycorrhizal root-soil system. Elevated values of rhizosphere enzyme activity, compared to non-rhizosphere soil reflect the extent to which roots stimulated soil microflora. The contribution of fungi to biotic activity was greater in the rhizosphere than in non-rhizosphere soil and thus, community composition of the two soil zones differed. The data support the gradual dilution of intramatrical hyphae into larger masses of structural tissues as roots age.

We have demonstrated the utility of these assays to quantify fungi in various components of a mycorrhizal root-soil system. The impact of mycorrhizal fungi on soil and root biotic activity was also described. The assays may have future applications to studies of plant and microflora responses to soil chemical and biological amendments.

D. L. Keister and P. B. Cregan (Eds.), The rhizosphere and plant growth, 374.

PLSO BARC84

Genetic transformation of protoplasts from an ectomycorrhizal fungus

P.A. LEMKE[1], V. BARRETT[1] and R.K. DIXON[2]
[1]*Molecular Genetics Program, Department of Botany and Microbiology, Auburn University, AL 36849, USA, and* [2]*School of Forestry, Auburn University, AL 36849, USA*

Key words: ectomycorrhizae, mycorrhizae, protoplasts, transformation

While the majority of ectomycorrhizal fungi grow vegetatively in culture, they do not sporulate or complete a sexual cycle in the laboratory. These limitations are principal reasons why these fungi have not been studied genetically. We have evaluated conditions for the generation of protoplasts from several species in an effort to identify and develop systems with potential for transformation (Barrett *et al.*, 1989).

Strains of the basidiomycete *Laccaria laccata* yield up to 1.8×10^8 protoplasts per gram wet mycelia following treatment with Novozyme 234 ($5\ \mathrm{mg\ ml^{-1}}$) at 32°C. Protoplasts were transformed at frequencies of 5-50 transformants per μg of DNA. Transformation involved plasmid pAN7-1 (Punt *et al.*, 1987) and was based on positive selection for resistance to hygromycin B (HmB) using the *Escherichia coli* hygromycin phosphotransferase (*hph*) gene bracketed by an *Aspergillus nidulans* glyceraldehyde-3-phosphate dehydrogenase (*gpd*) promoter and the transcription terminator region of the *A. nidulans* tryptophan synthetase (*trpC*) gene. Southern blot hybridization revealed that HmB-resistant transformants of *L. laccata* have integrated vector sequences involving one or more insertional events. DNA from the wildtype or untransformed *L laccata*, probed with ^{32}P-labeled pAN7-1, failed to hybridize to the plasmid and thus presumably lacks regions homologous to the *A. nidulans* control sequences, the only pAN7-1 sequences of fungal origin.

These results provide the first evidence for genetic transformation of a mycorrhizal fungus and further demonstrate the ability of promoter and termination signals of ascomycetous origin to function in a basidiomycete. The potential to improve ectomycorrhizal symbiosis through transgenic manipulation of the fungal component can now be realized, provided specific genes beneficial to this symbiosis can be identified.

Successful transformation of *L. laccata* also portends the ability to transform other species of basidiomycetes, especially edible mushroom species such as *Agaricus bisporus* and *Lentinus edodes*, wherein increased nutritional value might be attained by incorporating genes for specific proteins from plant or animal sources.

References

Barrett V *et al.* 1989 Appl. Microbiol. Biotechnol. 30, 381–387.

Punt P J *et al.* Gene 56, 117–124.

This contribution is an extended abstract for a paper presented at the Symposium. The extended paper on this subject has been published in *Applied Microbiology and Biotechnology* 33 (1990), 313–316.

D. L. Keister and P. B. Cregan (Eds.), The rhizosphere and plant growth, 375.

PLSO BARC85

Response of two *Acacia* species to drought and inoculation with an ectomycorrhizal fungus

O. OSONUBI[1] and K. MULONGOY[2]
[1]*University of Ibadan, and* [2]*International Institute of Tropical Agriculture (IITA), Ibadan, Nigeria*

Key words: *Acacia auriculiformis*, *Acacia mangium*, *Boletus suillus*, drought, P uptake

Introduction

Inclusion of trees in cropping systems is an efficient way to recycle nutrients and replenish soil organic matter. In the tropics, P deficiency is commonplace, and rains are erratic in many areas. Mycorrhizal fungi are known to enhance P uptake of the host-plant in low-P soils, and to improve its water relations under drought conditions.

Materials and methods

A pot experiment was carried out to assess P uptake, growth and drought tolerance of *Acacia auriculiformis* and *A. mangium* in response to inoculation with the ectomycorrhizal fungus *Boletus suillus* in a sterile soil having 5.5 mg P kg^{-1}, 0.14% N, and 0.50% C. Both *Acacia* species have potential in agroforestry systems. We used a factorial combination of two mycorrhiza treatments (uninoculated, and inoculated with *B. suillus* at 5 weeks after planting) and two watering regimes (daily-watering, and at weekly intervals) arranged in a randomized block design with 5 replications. Observations were made at 16 weeks after inoculation.

Results and discussion

Inoculation with *B. suillus* increased dry matter production and % P of *A. auriculiformis* for both daily- and weekly-watered plants by 85 and 112% respectively. High biomass was significantly correlated with plant P ($r = 0.539$). Inoculation also increased P and plant dry matter of *A. mangium* but only for daily-watered plants. Correlations between shoot P and plant or root dry weights of *A. mangium* were poor. For weekly-watered *A. auriculiformis*, inoculated plants had higher leaf area values than non-mycorrhizal plants, and there was no difference in stomatal conductance and leaf area between mycorrhizal plants watered weekly and non-mycorrhizal plants watered daily. Inoculation with *B. suillus* likely increased drought tolerance of *A. auriculiformis*. For *A mangium*, daily-watered plants and more so after inoculation with *B. suillus* had higher leaf area and stomatal conductance values than weekly-watered plants. But inoculation did not increase leaf area and stomatal conductance of weekly-watered plants of *A. mangium*, and thus did not improve their drought tolerance.

D. L. Keister and P. B. Cregan (Eds.), The rhizosphere and plant growth, 376.

PLSO BARC86

Specific interactions of Bradyrhizobium and four VA-mycorrhizal isolates in soybean

H. von ALTEN, A. TANNEBERG and F. SCHÖNBECK
Institut für Pflanzenkrankheiten und Pflanzenschutz, Universität Hannover, Herrenhäuser Str. 2, D-3000 Hannover, FRG

Key words: *Bradyrhizobium*, nodulation, plant growth, soybean cultivars, VA mycorrhiza,

Introduction

Mycorrhized legumes may be of great economic importance. The selection of the right plant cultivars and microsymbiont isolates is a very important aspect for a purposeful use of the tripartite symbiosis. It is necessary to test the specificity of interactions between host cultivars and microsymbiont isolates and to screen for compatible microorganisms which are more efficient and competitive than the autochthonic population in the fields.

Materials and methods

Plant culture: Soybean cultivars 'Sito', 'Corsoy', 'Williams', 'Union', 'Davis', 'Hobbit' and 'SRF 250' were used. Plants were grown in the greenhouse in plastic pots with 20% sieved commercial compost and 80% sand without fertiliza- and were harvested after 5 weeks. VA-mycorrhiza: Four isolates from different locations (no 'one spore' or 'one species' cultures) were used. Inoculum was produced on expanded clay ('LECADAN' 2-4 mm; LECA Co., FRG) according to Dehne and Backhaus (1986). Inoculation was with 5% (v/v). Mycorrhization was checked with the microscope and expressed as % (100% = all checked roots were mycorrhized with highest intensity). Bacterial symbionts: Seeds were dipped into a thick suspension of *Bradyrhizobium japonicum* (NITRAGIN Co., USA).

Results and discussion

Seven soybean cultivars were inoculated simultaneously with Bradyrhizobium and four VAM isolates for testing specific interactions and the symbiotic efficiency of the microsymbionts. Plants were intensely mycorrhized (40–70%) with slight differences between three cultivars. With one VAM isolate mycorrhization was lower in all cultivars.

In six of seven cultivars, mycorrhization caused a significant increase in the number of nodules. The increase was highest in cultivar 'Union' (+100%). The extent of the effect also depended on the VAM isolate. One VAM isolate which showed the lowest mycorrhization, only increased nodulation of 'Union'. Simultaneous inoculation with Bradyrhizobium and VAM promoted shoot growth significantly in four of seven cultivars ('Sito', 'Davis', 'Hobbit', 'SRF250'; max. +40%). The increase depended on the combination of plant cultivar and microsymbiont isolates. There was no correlation of growth promotion with increased nodulation. Bradyrhizobium inoculated 'Union', which had distinctly increased nodulation in tripartiate symbiosis, showed no growth response and cultivar 'SRF250' reacted vice versa. The VAM isolate showing the lowest mycorrhization also had no significant effect on plant growth.

Reference

Dehne H W and Backhaus G F 1986 Z.PflKrankh.-PflSchutz. 93, 415–424.

D. L. Keister and P. B. Cregan (Eds.), The rhizosphere and plant growth, 377.

PLSO BARC87

Ozone-induced alteration of biomass allocation and nitrogen derived from air in the leguminous plant-Rhizobium-VAM system

S.R. SHAFER[1] and M.M. SCHOENEBERGER[2]
[1]*USDA-ARS and Department of Plant Pathology, North Carolina State University, Raleigh, NC 27695-7616, and* [2]*Southeastern Forest Experiment Station, USDA-FS, Research Triangle Park, NC 27709, USA*

Key words: *Gigaspora margarita*, ^{15}N isotope dilution, *Trifolium subterraneum*, VA mycorrhizae

Ozone (O_3), a photochemical air pollutant, suppresses C availability to roots and may alter interactions among leguminous plants, rhizobia, and vesicular-arbuscular mycorrhizal (VAM) fungi. Objectives of this experiment were to quantify O_3-induced changes in plant N status and determine the influence of a VAM fungus on the response of the plant-Rhizobium system to O_3.

Germinating seeds of subterranean clover (*Trifolium subterraneum*) were inoculated with *Rhizobium leguminosarum* ("*R*"), a VAM fungus (*Gigaspora margarita*, "*G*"), both ("*R* + *G*") or neither ("*N*") and transplanted into pots containing a low-P, sandy loam soil amended with ^{15}N-urea. Plants were grown in a greenhouse for 16 d and exposed during the next 12 wk (Mon–Fri, 6 hr/d) to O_3 (0, 0.05, 0.10, or 0.15 $\mu L.L^{-1}$ of air) in chambers.

Inoculation with *R* or *R* + *G* stimulated shoot dry weight (SDW) and root dry weight (RDW) compared to inoculation with *G* or *N*. Ozone did not affect SDW but suppressed RDW, and RDW of plants inoculated with *R* or *R* + *G* was suppressed more than that of plants that did not receive *R* (*i.e.*, *N* or *G*). Thus, shoot:root ratios increased with increasing O_3. The ratio of nitrogen derived from air (Ndfa) to nitrogen derived from soil (Ndfs) in shoots, but not roots, of plants inoculated with either *R* or *R* + *G* significantly decreased ($P < 0.05$) with increasing O_3 and indicated that, with increasing O_3 stress, an increasing proportion of shoot N was derived from the soil pool (Fig. 1).

The influence of O_3 on C allocation was evident in the suppression of root growth without changes in shoot growth. The O_3-related decrease in the ratio Ndfa:Ndfs may reflect changes in sink strengths or altered relationships among the three symbionts.

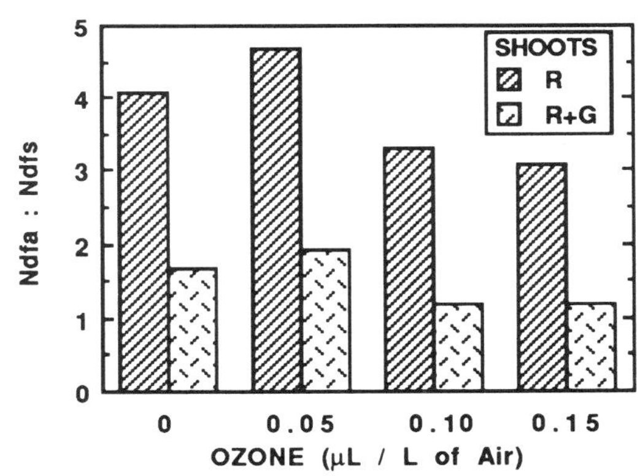

Fig. 1. Influence of O_3 and *Gigaspora margarita* on the ratio of N derived from air (Ndfa) to N derived from soil (Ndfs) in shoots of subterranean clover. R = plants inoculated with *Rhizobium leguminosarum*. R + G = plants inoculated with *R. leguminosarum* and *G. margarita*.

D. L. Keister and P. B. Cregan (Eds.), The rhizosphere and plant growth, 378.

PLSO BARC88

Strategy for inoculating nursery-grown sea oats with vesicular-arbuscular mycorrhizal fungi

D.M. SYLVIA and A.G. JARSTFER
Soil Science Department, University of Florida, Gainesville, FL 32611-0151, USA

Key words: aeroponics, inoculation, *Uniola paniculata*, VA mycorrhiza

Coastal sand dunes of the southeastern U.S. are stabilized naturally by perennial grasses such as sea oats (Woodhouse, 1982). Roots of these pioneer grasses are colonized by vesicular-arbuscular mycorrhizal (VAM) fungi (Koske, 1987; Sylvia, 1986). In established dunes, sea oats are colonized extensively by VAM fungi, but in many Florida coastal areas building construction has destroyed this source of inoculum for replants. Sylvia and Will (1988) reported that when replenishment sand is devoid of VAM fungi, beach grasses transplanted into it are colonized only slowly. In Florida, sea oats are produced in container nurseries which should afford an opportunity to inoculate a large number of plants (Johnson, 1987). Our objective is to demonstrate the feasibility of commercially producing VAM sea oats. Two studies were conducted under standard protocol at a commercial nursery. In the first, inoculation with a VAM fungus resulted in 27% colonization after 8 weeks, even with routine applications of soluble fertilizer and Banrot and Daconil fungicides. The noninoculated plants remained uncolonized. A second study showed a near-linear decrease in root colonization with dilution of the inoculum to a dilution rate of 1:256. Plants from the first nursery inoculation study were outplanted at Miami Beach. The low-level of colonization established in the nursery resulted in well-colonized plants in the replenishment sand. Twenty months after outplanting, sea oats colonized with VAM fungi in the nursery had 219% greater biomass compared to sea oats outplanted without VAM fungi.

Commercial inoculation strategy

1) Aeroponic culture of an effective VAM fungus in a commercial nursery.
2) Determine the minimum inoculum density required to establish colonization which can be maintained after outplanting.
3) Develop root-inoculum processing and application technology.

References

Johnson C R 1987 *In* Mycorrhizae in the Next Decade: Practical Applications and Research Priorities. Proc. of the 7th N. Amer. Conf. on Mycorrhizae. Eds. D M Sylvia, L L Hung and J H Graham pp 275–277. Univ. of Fla., Gainesville, FL.

Koske R E 1987 Mycologia 79, 55–68.

Sylvia D M 1986 Mycologia 78, 728–734.

Sylvia D M and Will M E 1988 Appl. Environ. Microbiol. 54, 348–352.

Woodhouse W W 1982 *In* Creation and Restoration of Coastal Plant Communities. Ed. R R Lewis. pp 1–44. CRC Press, Boca Raton, FL.

D. L. Keister and P. B. Cregan (Eds.), The rhizosphere and plant growth, 379.

PLSO BARC89

Evaluation of the effects of biocontrol agents on mycorrhizal fungi

R.G. LINDERMAN[1], T.C. PAULITZ[1], N.J. MOSIER[2], R.P. GRIFFITHS[3], J.E. LOPER[1], B.A. CALDWELL[3] and M.E. HENKELS[1]
[1]*USDA-ARS, Horticultural Crops Research Laboratory,* [2]*Department of Botany and Plant Pathology and* [3]*Department of Microbiology, Oregon State University, Corvallis, OR 97330, USA*

Key words: biocontrol, ectomycorrhizae, microbial interactions, VA mycorrhizae

Bacterial and fungal biocontrol agents are currently being developed by researchers and commercial companies for the control of soilborne pathogens, but little is known about the impact of those microbes on non-target beneficial microbes such as symbiotic mycorrhizal fungi which may significantly enhance plant growth and health. The objective of our research is to determine whether introduced biocontrol agents may adversely affect ectomycorrhizae, ericoid mycorrhizae, or VA mycorrhizae.

Effects of bicontrol agents on mycorrhizae were evaluated in terms of (1) spore germination, (2) root colonization dynamics, (3) development of external hyphae or rhizomorphs, (4) plant growth enhancement under nutrient deficiency stress, (5) changes in root respiration, and (6) *in vitro* inhibition. Biocontrol agents were introduced into the soil with the mycorrhizal fungi or were applied as seed treatments. Fungal biocontrol agents tested were species of Talaromyces, Trichoderma, and Gliocladium; bacterial agents tested were species of Pseudomonas, Bacillus, Agrobacterium, Enterobacter, Alcaligenes, Streptomyces, and Serratia.

In studies with VA mycorrhizae on onion or cucumber, some biocontrol agents caused a slight delay in spore germination, but largely had no effect on root colonization and plant growth enhancement.

Ectomycorrhiza formation by *Rhizopogon vinicolor* was significantly reduced by all the fungal biocontrol agents, but they had no effect on mycorrhiza formation by *Hebeloma crustuliniforme*. Similar tests are currently being conducted with bacterial biocontrol agents.

Our studies have resulted in the development of protocols to detect adverse effects of biocontrol agents on mycorrhizae, protocols which can be used in the development of regulatory guidelines. These protocols focus on mycorrhiza developmental stages such as spore germination, colonization, and plant growth enhancement, but future assays being investigated could consider physiological responses such as respiration or enzyme production.

Results to date indicate that most biocontrol agents are quite compatible with mycorrhizal fungi, which appear not to be sensitive to the antagonistic metabolites that inhibit root pathogens. This finding suggests that combinations of biocontrol agents and mycorrhizal fungi could be applied to plants as dual inoculum to accomplish the dual purpose of increased plant growth and health.

D. L. Keister and P. B. Cregan (Eds.), The rhizosphere and plant growth, 380.

PLSO BARC90

A misting apparatus for studying plant-microbe interactions and nutrient utilization

C.F. TESTER, P.D. MILLNER and D. GERSCHEFSKE KITT
Soil Microbial Systems Laboratory, NRI, USDA, ARS, Beltsville, MD 20705, USA

Key words: hyphae, misting, spores, VAM

Among the most widespread associations between microorganisms and higher plants are mycorrhizas. Endomycorrhizal associations are known to play a significant role in the growth of plants in soils worldwide. The major benefit of the fungus-plant symbiosis is that the mycorrhizal fungi assimilate and transport P, Zn, and Cu to the host plant roots. Determining the effects of field and cultural practices and of competitive microbial interactions on the populations of and infections by different VAM fungi has not been possible because methods for their study have not been available. Development of appropriate and cost effective inoculation technology would expand the use of VAM fungi to a far wider scale of agricultural production systems. Progress in basic and applied studies on VAM fungi in agriculture requires a reliable source of germplasm uncontaminated with other microorganisms. This is especially true when selecting and testing potential inoculants free of plant pathogens.

Thus, to enhance our biochemical and genetic studies of VAM fungi, a misting (fogging) apparatus was designed and fabricated for culturing VAM infected plants. The apparatus consists of: a microprocessor which controls timing of solenoid operation; nutrient containers; 0.2 μ filter; fogging or misting nozzles (Bete PJ8 Fog or Hago B37 Humidifier) and containers with supports for plants. The entire construction can be made from stainless steel or plastic for micronutrient studies. The system is an intermittent (as little as 0.4 sec 4 min^{-1}) and non-recycling design with the hollow cone nozzles placed at the top of the plant supports. The basic nutrient solution consisted of 2 mM $Ca(NO_3)_2$, 2 mM KNO_3, 1 mM $MgSO_4$, 1 μM KH_2PO_4, 32 μM NaFeEDTA, 10 μM H_3BO_3, 2 μM $MnCl_2$, 1 μM $ZnSO_4$, 0.5 μM $CuSO_4$, and 0.2 μM each of the following Na_2MoO_4, $CoCl_2$, and $NiSO_4$.

In this first report of culturing VAM infected plants with misting of non-recycling nutrients, we obtained exceptionally clean VAM hyphae and spores free of plant and soil debris in a 2 to 3 month period. Continuous harvesting was possible yielding a reliable source of biomass for immunological, biochemical, and molecular genetic studies. Leek, corn, and other desirable plants adapted to the nutrient fogging technique with prolific root production. Numerous fogging and misting nozzles were evaluated for specific applications (*e.g.* root hair were produced on both Fe-sufficient and Fe-deficient tomatoes for root membrane studies, VAM biomass, and plant micronutrient utilization studies). It was extremely important to have the 0.2 μ filter inline between the nutrient container and misting nozzles to prevent blockage of the small nozzle ports.

D. L. Keister and P. B. Cregan (Eds.), The rhizosphere and plant growth, 381.

PLSO BARC91

Increasing VA-mycorrhization with applications of rhizosphere bacteria

H. von ALTEN, A. LINDEMANN and F. SCHÖNBECK
Institut für Pflanzenkrankheiten und Pflanzenschutz, University of Hannover, Herrenhäuser Str. 2, D-3000 Hannover, FRG

Key words: *Bacillus cereus*, bacteria, inoculation, rhizosphere, VA mycorrhiza.

Introduction

In mycorrhized root systems, rhizosphere flora has a special quality (Meyer and Linderman, 1986). It appeared useful to search for microorganisms, able to stimulate mycorrhizal infection. Especially in the early phase of symbiosis an enhancement of VAM might increase its known positive effects. When such VAM-increasing bacterial effects can be verified in native soils, perhaps it will be possible to combine mycorrhizal and bacterial inocula. Thus, the inoculated mycorrhizal fungus might be more competitive in the soil.

Materials and methods

Plants were cultured in the greenhouse in plastic pots with sand at 24°C. Three VA-mycorrhizae isolates (Nr. 6, 13, 32b) (no 'one spore' cultures) were used. The inoculum was produced on expanded clay ('LECADAN' 2–4 mm; LECA Co., FRG) (Dehne and Backhaus, 1986). Inoculation was with 5% (v/v). Mycorrhization was checked with the microscope. The bacterial culture for screening for VAM-stimulating effects was grown in complex medium. *Bacillus cereus* for the following tests was grown in synthetic medium (30°C). Application of bacteria by dipping of root systems (screening) or VAM inoculum (tests with *B. cereus*) in baterial culture (five min.).

Results and discussion

Several bacterial isolates improved the frequency but not the intensity of VAM in Linum (isolate 6, max +50%). An isolate of *B. cereus* showed the best results. A method was developed to inoculate the bacteria together with VAM using the same clay particles. When plants are inoculated in this way, increases of mycorrhization can be due to the direct effects of the bacteria on the mycorrhizal propagules or to an influence on root physiology. The duration of the VAM-stimulation depended on the plant. In Linum, the effects disappeared after 3 weeks. In barley ('Gimpel'), a plant with slow mycorrhization, they were more durable. All plants were harvested after 4 weeks. The further influence of the bacteria is assumed to be non-significant. The enhancement of the mycorrhization appeared in all checked plant species (tagetes, tobacco, flax, cucumber, tomato, bean, barley, wheat) and with 3 VAM isolates (6, 13, 32b). As expected the soil quality has a limiting influence; with increasing amounts of commercial compost in the sand, the effects vanished.

References

Dehne H W and Backhaus G F 1986 Z. PflKrankh.PflSchutz. 93, 415–424.
Meyer J R and Linderman R G 1986 Soil Biol. Biochem. 18, 183–190.
Meyer J R and Linderman R G 1986 Soil Biol. Biochem. 18, 191–196.

D. L. Keister and P. B. Cregan (Eds.), The rhizosphere and plant growth, 382.

PLSO BARC92

Mycorrhizal fungi increase yields of winter wheat

D.H. YOCOM and M.G. BOOSALIS
Department of Biology, Millersville University, Millersville, PA 17551, USA, and Department of Plant Pathology, University of Nebraska, Lincoln, NE 68583-0722, USA

Key words: grain yield, mycorrhizae, mycorrhizal fungi, VAM, wheat.

In western Nebraska, declines in soil populations of vesicular-arbuscular mycorrhizal (VAM) fungi with continued wheat production in wheat-fallow rotations were observed. The significance of these declines was difficult to evaluate since the effect of VAM on grain yields under field conditions was unknown. This project was initiated to determine the effect of VAM fungi on the yield of winter wheat.

Winter wheat, Centurk, was drilled into plots (2.4 m × 3.6 m) that had been fumigated with methyl bromide or not treated (nonfumigated). The fumigation treatments were arranged in a split-plot design with five or six inoculation treatments applied to plots arranged within fumigation sub-blocks. Each treatment combination was replicated five times. The inoculation treatments were species of VAM fungi isolated from the study site (Sidney, Nebraska) and a sterile control. The inocula were produced in pot cultures, which were ground into flowable mixtures and were banded under the seed. For the three years of the study, wheat was planted around September 15th and was harvested in mid-July.

Fumigation destroyed most of the viable propagules of VAM fungi as determined by the absence of mycorrhizal colonization of month-old wheat roots. VAM colonization of wheat roots in nonfumigated plots was significantly greater than colonization of roots in fumigated plots. Introduced inocula did not significantly affect the colonization levels of month-old roots.

Fumigation had a significant effect on grain yield. In 1984, nonfumigated plots produced a mean of 2919 kg/ha of wheat grain; compared with 2544 kg ha^{-1} produced in fumigated plots. The increased yield was due to the larger size of wheat seeds (kernels). The mean weight of 1000 kernels from nonfumigated plots was 23.3 g; compared with 18.0 g for 1000 kernels from fumigated plots. Inoculation treatments did not have a significant effect on grain yield or the weight of 1000 kernels. Results observed in 1986 were very similar to those observed in 1984.

In 1985, inoculation treatments had a significant effect on grain yield. For example, fumigated plots inoculated with *Glomus fasciculatum* produced a mean of 2311 kg ha^{-1} of wheat; compared with 2116 kg ha^{-1} produced in fumigated plots inoculated with the sterile control. In 1985, fumigation did not have a significant effect on grain yield.

Under field conditions in western Nebraska, wheat colonized by indigenous VAM fungi in nonfumigated plots or by reintroduced VAM fungi produced more grain than nonmycorrhizal wheat plants.

INDEX